The Kubernetes Bible

Second Edition

The definitive guide to deploying and managing Kubernetes across cloud
and on-prem environments

Gineesh Madapparambath

Russ McKendrick

The Kubernetes Bible
Second Edition

Senior Publishing Product Manager: Rahul Nair

Acquisition Editor – Peer Reviews: Gaurav Gavas, Jane D'Souza

Project Editor: Amisha Vathare

Content Development Editor: Shikha Parashar

Copy Editor: Safis Editing

Technical Editor: Simanta Rajbangshi

Proofreader: Safis Editing

Indexer: Pratik Shirodkar

Presentation Designer: Ganesh Bhadwalkar

Developer Relations Marketing Executive: Maran Fernandes

First published: January 2022
Second edition: November 2024

Production reference: 1271124

Published by Packt Publishing Ltd.
Grosvenor House
11 St Paul's Square
Birmingham
B3 1RB, UK.

ISBN 978-1-83546-471-7

www.packt.com

Contributors

About the authors

Gineesh Madapparambath has over 15 years of experience in IT service management and consultancy, specializing in Linux, automation, and containerization. He has worked extensively in planning, deploying, and supporting automation solutions with Ansible and the Ansible Automation Platform, across private clouds, public clouds, bare metal, and network environments. His experience spans globally, with roles including Systems Engineer, Automation Specialist, and Infrastructure Designer. Gineesh is also the author of *Ansible for Real Life Automation*.

To my wife, Deepthy, for supporting and motivating me as always. To my son, Abhay, and my daughter, Anu, for allowing me to take time away from playing with them to write the book. To my parents and my ever-supportive friends, for their motivation and help.

- Gineesh Madapparambath

Russ McKendrick is an experienced DevOps practitioner and system administrator with a passion for automation and containers. He has been working in IT and related industries for the better part of 30 years. During his career, he has had responsibilities in many different sectors, including first-line, second-line, and senior support in client-facing and internal teams for small and large organizations.

He works primarily with Linux, using open-source systems and tools across dedicated hardware and virtual machines hosted in both public and private clouds at Node4, where he is the practice manager (SRE and DevOps). He also buys way too many records!

About the reviewers

Rom Adams (né Romuald Vandepoel) is an open source and C-Suite advisor with 20 years of experience in the IT industry. He is a cloud-native expert who helps organizations to modernize and transform with open-source solutions. He advises companies and lawmakers on their open- and inner-source strategies. He has previously worked as a Principal Architect at Ondat, a cloud-native storage company acquired by Akamai, where he designed products and hybrid cloud solutions. He has also held roles at Tyco, NetApp, and Red Hat, becoming a Subject Matter Expert in hybrid cloud.

Adams has been a moderator and speaker for several events, sharing his insights into culture, process, and technology adoption, as well as his passion for open innovation.

To my grandmother, for her kindness; my grandfather, for his wisdom; and my partner and best friend, Mercedes Adams, for her love, patience, and continuous support.

– Rom Adams

Shane Boulden is a Solution Architect at Red Hat, supporting organisations to deploy, manage, and secure open-source platforms and technologies.

He is a contributor to several open-source security projects, including the "Compliance as Code" project, building SCAP profiles, and Ansible playbooks aligned with Australian security guidelines. He also supported the Keycloak project to achieve Australian Consumer Data Right (CDR) certification through the OpenID foundation and has contributed a number of new policies to the StackRox Kubernetes-native security project.

Shane has a keen interest in artificial intelligence and machine learning, publishing peer-reviewed papers on genetic algorithms and their applications and running sessions at conferences helping others get started with open-source Generative AI frameworks.

Shane regularly publishes articles on platform operations and security on his blog, stb.id.au.

Foreword

You can bundle and run your applications by using containers. You get to package your dependencies, libraries, and frameworks into a single unit. But then, you'll need to manage those containers and, well, orchestrate them. For example, you'll need to schedule containers across a cluster of nodes, automatically restart your failed containers, and replace the failed nodes. That was the need that Google recognized, and Kubernetes was born out of that need.

For me, I first got into Kubernetes right after it was created (late 2015). I was helping run customer feedback programs for developers who were building solutions on Microsoft Azure. We needed our services to work with container orchestration systems, and that especially included Kubernetes. By the time **Azure Kubernetes Services (AKS)** launched in 2018, I was publishing architectural content on the Azure Architecture Center, as well as whitepapers, e-books, and blog posts. Naturally, I worked with Microsoft's top solution architects to help publish AKS architecture design guidance and reference architectures. For example, in 2022, I was heavily involved in a series of content called "AKS for Amazon EKS professionals." I later helped publish content about **Google Kubernetes Engine (GKE)** on the Google Cloud Architecture Center.

> *"What makes Kubernetes so incredible is its implementation of Google's own experience with Borg. Nothing beats the scale of Google. Borg launches more than 2-billion containers per week, an average of 3,300 per second... Kubernetes was born in a cauldron of fire, battle-tested and ready for massive workloads."*
>
> *– Swapnil Bhartiya, OpenSource.com*

Swapnil Bhartiya gets to the heart of the success of Kubernetes... it was born out of need and refined to become incredibly effective. But words alone won't save your application. Let's see what Scott Adams has his characters say about cloud-native application design, containerization, and, yes, Kubernetes.

"I need to know why moving our app to the cloud didn't automatically solve all our problems."

– Pointy-Haired Boss (PHB)

"You wouldn't let me re-architect the app to be cloud native."

– Dilbert

"Just put it in containers."

– PHB

"You can't solve a problem just by saying techy things."

– Dilbert

"Kubernetes."

– PHB

And that's where *The Kubernetes Bible* comes in. As Scott Adams explained, you can't just say the terms; you have to do the work. That means you need to know how to get up and running on Kubernetes, how to design and deploy large clusters, and, in general, how to deploy, debug, and recover containerized applications. In this book, Gineesh Madapparambath and Russ McKendrick take you through everything from installing your first Kubernetes cluster to using and configuring pods to deploying stateless and stateful applications. You'll learn how to launch your Kubernetes clusters on Google Cloud (GKS), Amazon Web Services (EKS), and Microsoft Azure (AKS). You'll also explore further techniques, including using Helm charts, security, advanced Pod techniques, traffic management, and much more!

Be sure to study each chapter of the Kubernetes Bible. Soak in the wisdom of Gineesh and Russ like a sponge. The more you engage with the content of this book, the more you'll remember and apply it. And keep this book close. You're going to need it as you continue to manage and orchestrate your containers.

Ed Price

Technical Writer, Cloud Architecture Content Manager

Co-author of 8 books, including The Azure Cloud Native Architecture Mapbook from Packt

Join our community on Discord

Join our community's Discord space for discussions with the authors and other readers:

`https://packt.link/cloudanddevops`

Table of Contents

Chapter 3: Installing Your First Kubernetes Cluster   69

Chapter 5: Using Multi-Container Pods and Design Patterns   129

Chapter 6: Namespaces, Quotas, and Limits for Multi-Tenancy in Kubernetes   163

Chapter 13: DaemonSet – Maintaining Pod Singletons on Nodes 389

Chapter 14: Working with Helm Charts and Operators 405

Chapter 17: Kubernetes Clusters on Microsoft Azure with Azure Kubernetes Service

Preface

Containers have allowed a real leap forward since their massive adoption in the world of virtualization because they have allowed greater flexibility, especially these days, when buzzwords such as **cloud**, **agile**, and **DevOps** are on everyone's lips.

Today, almost no one questions the use of containers—they're basically everywhere, especially after the success of Docker and the rise of Kubernetes as the leading platform for container orchestration.

Containers have brought tremendous flexibility to organizations, but they have remained questionable for a very long time when organizations face the challenge of deploying them in production. For years, companies have been using containers for proof-of-concept projects, local development, and similar purposes, but the idea of using containers for real production workloads was inconceivable for many organizations.

Container orchestrators were the game-changer, with Kubernetes in the lead. Originally built by Google, today, Kubernetes is the leading container orchestrator that provides you with all the features you need in order to deploy containers in production at scale. Kubernetes is popular, but it is also complex. This tool is so versatile that getting started with it and progressing to advanced usage is not an easy task: it is not an easy tool to learn and operate.

As an orchestrator, Kubernetes has its own concepts independent of those of a container engine. But when both container engines and orchestrators are used together, you get a very strong platform ready to deploy your cloud-native applications in production. As engineers working with Kubernetes daily, we were convinced, like many, that it was a technology to master, and we decided to share our knowledge in order to make Kubernetes accessible by covering most of this orchestrator.

This book is entirely dedicated to Kubernetes and is the result of our work. It provides a broad view of Kubernetes and covers a lot of aspects of the orchestrator, from pure container Pod creation to deploying the orchestrator on the public cloud. We didn't want this book to be a *Getting Started* guide.

We hope this book will teach you everything you want to learn about Kubernetes!

Who this book is for

This book is for people who intend to use Kubernetes with container runtimes. Although Kubernetes supports various container engines through the **Container Runtime Interface (CRI)** and is not tied to any specific one, the combination of Kubernetes with containerd remains one of the most common use cases.

This book is highly technical, with a primary focus on Kubernetes and container runtimes from an engineering perspective. It is intended for engineers, whether they come from a development or system administration background and is not aimed at project managers. It is a Kubernetes bible for people who are going to use Kubernetes daily, or for people who wish to discover this tool. You shouldn't be afraid of typing some commands on a terminal.

Being a total beginner to Kubernetes or having an intermediate level is not a problem for following this book. While we cover some container fundamentals within the chapters, it's helpful to have basic technical familiarity with containers. This book can also serve as a guide if you are in the process of migrating an existing application to Kubernetes.

The book incorporates content that will allow readers to deploy Kubernetes on public cloud offerings such as Amazon EKS or Google GKE. Cloud users who wish to add Kubernetes to their stack on the cloud will appreciate this book.

What this book covers

Chapter 1, Kubernetes Fundamentals, is an introduction to Kubernetes. We're going to explain what Kubernetes is, why it was created, who created it, who keeps this project alive, and when and why you should use it as part of your stack.

Chapter 2, Kubernetes Architecture – from Container Images to Running Pods, covers how Kubernetes is built as a distributed software and is technically not a single monolith binary but built as a set of microservices interacting with each other. We're going to explain this architecture and how Kubernetes proceeds to translate your instructions into running containers in this chapter.

Chapter 3, Installing Your First Kubernetes Cluster, explains that Kubernetes is really difficult to install due to its distributed nature, so as to make the learning process easier, it is possible to install Kubernetes clusters by using one of its distributions. Kind and minikube are two options we're going to discover in this chapter to have a Kubernetes cluster working on your machine.

Chapter 4, Running Your Containers in Kubernetes, is an introduction to the concept of Pods.

Chapter 5, Using Multi-Container Pods and Design Patterns, introduces multi-container Pods and design patterns such as a proxy, adapter, or sidecar that you can build when running several containers as part of the same Pod.

Chapter 6, Namespaces, Quotas, and Limits for Multi-Tenancy in Kubernetes, explains how using namespaces is a key aspect of cluster management and, inevitably, you'll have to deal with namespaces during your journey with Kubernetes. Though it's a simple notion, it is a key one, and you'll have to master namespaces perfectly in order to be successful with Kubernetes. We will also learn how to implement multi-tenancy in Kubernetes using Namespaces, Quotas and Limits.

Chapter 7, Configuring Your Pods Using ConfigMaps and Secrets, explains how, in Kubernetes, we separate Kubernetes applications from their configurations. Both applications and configurations have their own life cycle, thanks to the ConfigMap and Secret resources. This chapter will be dedicated to these two objects and how to mount data in a ConfigMap or Secret as environment variables or volumes mounted on your Pod.

Chapter 8, *Exposing Your Pods with Services*, teaches you about the notion of services in Kubernetes. Each Pod in Kubernetes gets assigned its own IP address dynamically. Services are extremely useful if you want to provide a consistent one to expose Pods within your cluster to other Pods or to the outside world, with a single static DNS name. You'll learn here that there are three main service types, called ClusterIp, NodePort, and LoadBalancer, which are all dedicated to a single use case in terms of Pod exposition.

Chapter 9, *Persistent Storage in Kubernetes*, covers how, by default, Pods are not persistent. As they're just managing raw containers, destroying them will result in the loss of your data. The solution to that is the usage of persistent storage thanks to the PersistentVolume and PersistentVolumeClaim resources. This chapter is dedicated to these two objects and the StorageClass object: it will teach you that Kubernetes is extremely versatile in terms of storage and that your Pods can be interfaced with a lot of different storage technologies.

Chapter 10, *Running Production-Grade Kubernetes Workloads*, takes a deep dive into high availability and fault tolerance in Kubernetes using ReplicationController and ReplicaSet.

Chapter 11, *Using Kubernetes Deployments for Stateless Workloads*, is a continuation of the previous chapter and explains how to manage multiple versions of ReplicaSets using the Deployment object. This is the basic building block for stateless applications running on Kubernetes.

Chapter 12, *StatefulSet – Deploying Stateful Applications*, takes a look at the next important Kubernetes object: StatefulSet. This object is the backbone of running stateful applications on Kubernetes. We'll explain the most important differences between running stateless and stateful applications using Kubernetes.

Chapter 13, *DaemonSet – Maintaining Pod Singletons on Nodes*, covers DaemonSet, which are a special Kubernetes object that can be used for running operational or supporting workloads on Kubernetes clusters. Whenever you need to run precisely one container Pod on a single Kubernetes node, DaemonSet is what you need.

Chapter 14, *Working with Helm Charts and Operators*, covers Helm Charts, which is a dedicated packaging and redistribution tool for Kubernetes applications. Armed with knowledge from this chapter, you will be able to quickly set up your Kubernetes development environment or even plan for the redistribution of your Kubernetes application as a dedicated Helm Chart. In this chapter, we will also introduce the Kubernetes operators and how they will help you to deploy application stacks.

Chapter 15, *Kubernetes Clusters on Google Kubernetes Engine*, looks at how we can move our Kubernetes workload to Google Cloud using both the native command-line client and the Google Cloud console.

Chapter 16, *Launching a Kubernetes Cluster on Amazon Web Services with Amazon Elastic Kubernetes Service*, looks at moving the workload we launched in the previous chapter to Amazon's Kubernetes offering.

Chapter 17, *Kubernetes Clusters on Microsoft Azure with Azure Kubernetes Service*, looks at launching a cluster in Microsoft Azure.

Chapter 18, *Security in Kubernetes*, covers authorization using built-in role-based access control and authorization schemes together with user management. This chapter also teaches you about admission controllers, TLS certificates based communication, and security context implementations.

Chapter 19, Advanced Techniques for Scheduling Pods, takes a deeper look at Node affinity, Node taints and tolerations, and advanced scheduling policies in general.

Chapter 20, Autoscaling Kubernetes Pods and Nodes, introduces the principles behind autoscaling in Kubernetes and explains how to use Vertical Pod Autoscaler, Horizontal Pod Autoscaler, and Cluster Autoscaler.

Chapter 21, Advanced Kubernetes: Traffic Management, Multi-Cluster Strategies, and More, covers Ingress objects and IngressController in Kubernetes. We explain how to use nginx as an implementation of IngressController and how you can use Azure Application Gateway as a native IngressController in Azure environments. We will also explain advanced Kubernetes topics including Cluster Day 2 tasks, best practices, and troubleshooting.

To get the most out of this book

While we cover container fundamentals in the chapters, this book is focused on Kubernetes. Although Kubernetes supports multiple container engines, the content primarily discusses using Kubernetes with containerd as the runtime. You don't need to be an expert, but having a basic understanding of launching and managing applications with containers will be helpful before diving into this book.

While it is possible to run Windows containers with Kubernetes, most of the topics covered in this book will be Linux-based. Having a good knowledge of Linux will be helpful, but is not required. Again, you don't have to be an expert: knowing how to use a terminal session and basic Bash scripting should be enough.

Lastly, having some general knowledge of software architecture such as REST APIs will be beneficial.

Software/hardware covered in the book	OS Requirements
Kubernetes >= 1.31	Windows, macOS, Linux
kubectl >= 1.31	Windows, macOS, Linux

We strongly advise you to not attempt to install Kubernetes or kubectl on your machine for now. Kubernetes is not a single binary but is distributed software composed of several components and, as such, it is really complex to install a complete Kubernetes cluster from scratch. Instead, we recommend that you follow the third chapter of this book, which is dedicated to the setup of Kubernetes.

If you are using the digital version of this book, we advise you to type the code yourself or access the code via the GitHub repository (link available in the next section). Doing so will help you avoid any potential errors related to the copying and pasting of code.

Please note that kubectl, helm, etc. are the tools we're going to use most frequently in this book, but there is a huge ecosystem around Kubernetes and we might install additional software not mentioned in this section. This book is also about using Kubernetes in the cloud, and we're going to discover how to provision Kubernetes clusters on public cloud platforms such as Amazon Web Services and Google Cloud Platform. As part of this setup, we might install additional software dedicated to these platforms that are not strictly bound to Kubernetes, but also to other services provided by these platforms.

Download the example code files

You can download the example code files for this book from GitHub at `https://github.com/PacktPublishing/The-Kubernetes-Bible-Second-Edition`. In case there's an update to the code, it will be updated on the existing GitHub repository.

We also have other code bundles from our rich catalogue of books and videos available at `https://github.com/PacktPublishing/`. Check them out!

Download the color images

We also provide a PDF file that has color images of the screenshots/diagrams used in this book. You can download it here: `https://static.packt-cdn.com/downloads/9781835464717_ColorImages.pdf`.

Conventions used

There are a number of text conventions used throughout this book.

`Code in text`: Indicates code words in the text, database table names, folder names, filenames, file extensions, pathnames, dummy URLs, user input, and Twitter handles. Here is an example: "Now, we need to create a `kubeconfig` file for our local `kubectl` CLI."

A block of code is set as follows:

```
apiVersion: v1
kind: Pod
metadata:
  name: nginx-Pod
```

When we wish to draw your attention to a particular part of a code block, the relevant lines or items are set in bold:

```
apiVersion: v1
kind: ReplicationController
metadata:
  name: nginx-replicationcontroller-example
```

Any command-line input or output is written as follows:

```
$ kubectl get nodes
```

Bold: Indicates a new term, an important word, or words that you see on screen. For example, words in menus or dialog boxes appear in the text like this. Here is an example: "On this screen, you should see an **Enable Billing** button."

> **IMPORTANT NOTES**
>
> appear like this

> **Tips**
>
> appear like this.

Get in touch

Feedback from our readers is always welcome.

General feedback: If you have questions about any aspect of this book, mention the book title in the subject of your message and email us at customercare@packtpub.com.

Errata: Although we have taken every care to ensure the accuracy of our content, mistakes do happen. If you have found a mistake in this book, we would be grateful if you would report this to us. Please visit www.packtpub.com/support/errata, selecting your book, clicking on the Errata Submission Form link, and entering the details.

Piracy: If you come across any illegal copies of our works in any form on the internet, we would be grateful if you would provide us with the location address or website name. Please contact us at copyright@packt.com with a link to the material.

If you are interested in becoming an author: If there is a topic that you have expertise in and you are interested in either writing or contributing to a book, please visit authors.packtpub.com.

Leave a Review!

Thank you for purchasing this book from Packt Publishing—we hope you enjoy it! Your feedback is invaluable and helps us improve and grow. Once you've completed reading it, please take a moment to leave an Amazon review; it will only take a minute, but it makes a big difference for readers like you.

https://packt.link/r/1835464718

Scan the QR code below to receive a free ebook of your choice.

https://packt.link/NzOWQ

Download a free PDF copy of this book

Thanks for purchasing this book!

Do you like to read on the go but are unable to carry your print books everywhere?

Is your eBook purchase not compatible with the device of your choice?

Don't worry, now with every Packt book you get a DRM-free PDF version of that book at no cost.

Read anywhere, any place, on any device. Search, copy, and paste code from your favorite technical books directly into your application.

The perks don't stop there, you can get exclusive access to discounts, newsletters, and great free content in your inbox daily.

Follow these simple steps to get the benefits:

1. Scan the QR code or visit the link below:

https://packt.link/free-ebook/9781835464717

2. Submit your proof of purchase.
3. That's it! We'll send your free PDF and other benefits to your email directly.

1

Kubernetes Fundamentals

Welcome to *The Kubernetes Bible*, and we are happy to accompany you on your journey with Kubernetes. If you are working in the software development industry, you have probably heard about Kubernetes. This is normal because the popularity of Kubernetes has grown a lot in recent years.

Built by Google, Kubernetes is the leading container orchestrator solution in terms of popularity and adoption: it's the tool you need if you are looking for a solution to manage containerized applications in production at scale, whether it's on-premises or on a public cloud. Be focused on the word. Deploying and managing containers at scale is extremely difficult because, by default, container engines such as Docker do not provide any way on their own to maintain the availability and scalability of containers at scale.

Kubernetes first emerged as a Google project, and Google has put a lot of effort into building a solution to deploy a huge number of containers on their massively distributed infrastructure. By adopting Kubernetes as part of your stack, you'll get an open source platform that was built by one of the biggest companies on the internet, with the most critical needs in terms of stability.

Although Kubernetes can be used with a lot of different container runtimes, this book is going to focus on the Kubernetes and containers (Docker and Podman) combination.

Perhaps you are already using Docker on a daily basis, but the world of container orchestration might be completely unknown to you. It is even possible that you do not even see the benefits of using such technology because everything looks fine to you with just raw Docker. That's why, in this first chapter, we're not going to look at Kubernetes in detail. Instead, we will focus on explaining what Kubernetes is and how it can help you manage your application containers in production. It will be easier for you to learn a new technology if you understand why it was built.

In this chapter, we're going to cover the following main topics:

- Understanding monoliths and microservices
- Understanding containers
- How can Kubernetes help you to manage containers?
- Understanding the history of Kubernetes

- Exploring the problems that Kubernetes solves

You can download the latest code samples for this chapter from the official GitHub repository at `https://github.com/PacktPublishing/The-Kubernetes-Bible-Second-Edition/tree/main/Chapter01`

Understanding monoliths and microservices

Let's put Kubernetes and Docker to one side for the moment, and instead, let's talk a little bit about how the internet and software development evolved together over the past 20 years. This will help you to gain a better understanding of where Kubernetes sits and the problems it solves.

Understanding the growth of the internet since the late 1990s

Since the late 1990s, the popularity of the internet has grown rapidly. Back in the 1990s, and even in the early 2000s, the internet was only used by a few hundred thousand people in the world. Today, almost 2 billion people are using the internet for email, web browsing, video games, and more.

There are now a lot of people on the internet, and we're using it for tons of different needs, and these needs are addressed by dozens of applications deployed on dozens of devices.

Additionally, the number of connected devices has increased, as each person can now have several devices of a different nature connected to the internet: laptops, computers, smartphones, TVs, tablets, and more.

Today, we can use the internet to shop, to work, to entertain, to read, or to do whatever. It has entered almost every part of our society and has led to a profound paradigm shift in the last 20 years. All of this has given the utmost importance to software development.

Understanding the need for more frequent software releases

To cope with this ever-increasing number of users who are always demanding more in terms of features, the software development industry had to evolve in order to make new software releases faster and more frequent.

Indeed, back in the 1990s, you could build an application, deploy it to production, and simply update it once or twice a year. Today, companies must be able to update their software in production, sometimes several times a day, whether to deploy a new feature, integrate with a social media platform, support the resolution of the latest fashionable smartphone, or even release a patch to a security breach identified the day before. Everything is far more complex today, and you must go faster than before.

We constantly need to update our software, and in the end, the survival of many companies directly depends on how often they can offer releases to their users. But how do we accelerate software development life cycles so that we can deliver new versions of our software to our users more frequently?

IT departments of companies had to evolve, both in an organizational sense and a technical sense. Organizationally, they changed the way they managed projects and teams in order to shift to agile methodologies, and technically, technologies such as cloud computing platforms, containers, and virtualization were adopted widely and helped a lot to align technical agility with organizational agility. All of this is to ensure more frequent software releases! So, let's focus on this evolution next.

Understanding the organizational shift to agile methodologies

From a purely organizational point of view, agile methodologies such as Scrum, Kanban, and DevOps became the standard way to organize IT teams.

Typical IT departments that do not apply agile methodologies are often made of three different teams, each of them having a single responsibility in the development and release process life cycle.

> Rest assured, even though we are currently discussing agile methodologies and the history of the internet, this book is really about Kubernetes! We just need to explain some of the problems that we have faced before introducing Kubernetes for real!

Before the adoption of agile methodologies, development and operations often worked in separate silos. This could lead to inefficiency and communication gaps. Agile methodologies helped bridge these gaps and foster collaboration. The three isolated teams are shown below.

- **The business team:** They're like the voice of the customer. Their job is to explain what features are needed in the app to meet user needs. They translate business goals into clear instructions for the developers.
- **The development team:** These are the engineers who bring the app to life. They translate the business team's feature requests into code, building the functionalities and features users will interact with. Clear communication from the business team is crucial. If the instructions aren't well defined, it can be like a game of telephone – misunderstandings lead to delays and rework.
- **The operation team:** They're the keepers of the servers. Their main focus is keeping the app running smoothly. New features can be disruptive because they require updates, which can be risky. In the past, they weren't always aware of what new features were coming because they weren't involved in the planning.

These are what we call silos, as illustrated in *Figure 1.1*:

Figure 1.1: Isolated teams in a typical IT department

The roles are clearly defined, people from the different teams do not work together that much, and when something goes wrong, everyone loses time finding the right information from the right person.

This kind of siloed organization has led to major issues:

- A significantly longer development time
- Greater risk in the deployment of a release that might not work at all in production

And that's essentially what agile methodologies and DevOps fixed. The change agile methodologies made was to make people work together by creating multidisciplinary teams.

> DevOps is a collaborative culture and set of practices that aims to bridge the gap between development (Dev) and operations (Ops) teams. DevOps promotes collaboration and automation throughout the software lifecycle, from development and testing to deployment and maintenance.

An agile team consists of a product owner describing concrete features by writing them as user stories that are readable by the developers who are working in the same team as them. Developers should have visibility of the production environment and the ability to deploy on top of it, preferably using a **continuous integration and continuous deployment (CI/CD)** approach. Testers should also be part of agile teams in order to write tests.

With the collaborative approach, the teams will get better and clearer visibility of the full picture, as illustrated in the following diagram.

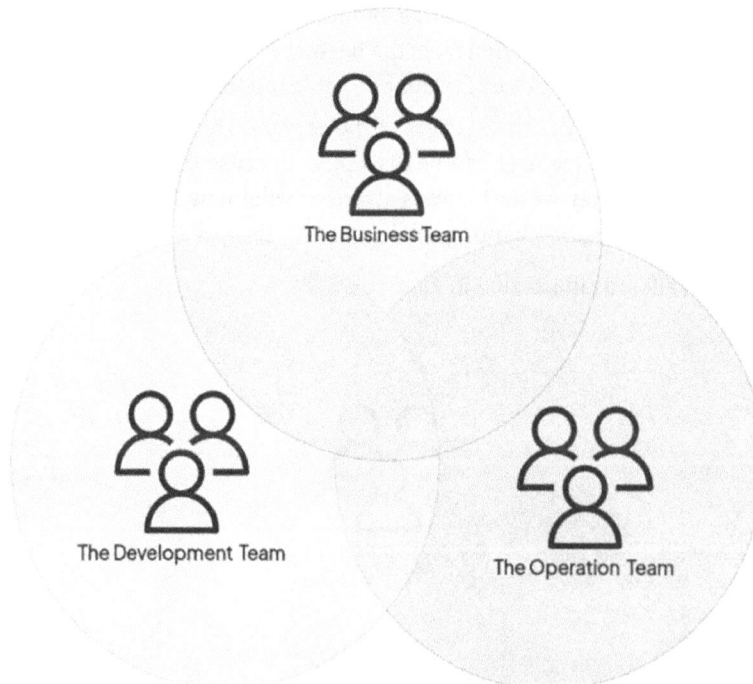

Figure 1.2: Team collaboration breaks silos

Simply understand that, by adopting agile methodologies and DevOps, these silos were broken and multidisciplinary teams capable of formalizing a need, implementing it, testing it, releasing it, and maintaining it in the production environment were created. *Table 1.1* presents a shift from traditional development to agile and DevOps methodology.

Feature	Traditional Development	Agile & DevOps
Team Structure	Siloed departments (Development, Operations)	Cross-functional, multi-disciplinary teams
Work Style	Isolated workflows, limited communication	Collaborative, iterative development cycles
Ownership	Development hands off to Operations for deployment and maintenance	"You Build It, You Run It" - Teams own the entire lifecycle
Focus	Features and functionality	Business value, continuous improvement
Release Cycle	Long release cycles, infrequent deployments	Short sprints, frequent releases with feedback loops
Testing	Separate testing phase after development	Integrated testing throughout the development cycle
Infrastructure	Static, manually managed infrastructure	Automated infrastructure provisioning and management (DevOps)

Table 1.1: DevOps vs traditional development – a shift in collaboration

So, we've covered the organizational transition brought about by the adoption of agile methodologies. Now, let's discuss the technical evolution that we've gone through over the past several years.

Understanding the shift from on-premises to the cloud

Having agile teams is very nice, but agility must also be applied to how software is built and hosted.

With the aim to always achieve faster and more recurrent releases, agile software development teams had to revise two important aspects of software development and release:

* Hosting
* Software architecture

Today, apps are not just for a few hundred users but potentially for millions of users concurrently. Having more users on the internet also means having more computing power capable of handling them. And, indeed, hosting an application became a very big challenge.

In the early days of web hosting, businesses primarily relied on two main approaches to housing their applications: one of these approaches is on-premises hosting. This method involved physically owning and managing the servers that ran their applications. There are two main ways to achieve on-premises hosting:

1. **Dedicated Servers**: Renting physical servers from established data center providers: This involved leasing dedicated server hardware from a hosting company. The hosting provider would manage the physical infrastructure (power, cooling, security) but the responsibility for server configuration, software installation, and ongoing maintenance fell to the business. This offered greater control and customization compared to shared hosting, but still required significant in-house technical expertise.

2. **Building Your Own Data Center**: Constructing and maintaining a private data center: This option involved a massive investment by the company to build and maintain its own physical data center facility. This included purchasing server hardware, networking equipment, and storage solutions, and implementing robust power, cooling, and security measures. While offering the highest level of control and security, this approach was very expensive and resource-intensive and was typically only undertaken by large corporations with significant IT resources.

Also note that on-premises hosting also encompasses managing the operating system, security patches, backups, and disaster recovery plans for the servers. Companies often needed a dedicated IT staff to manage and maintain their on-premises infrastructure, adding to the overall cost.

When your user base grows, you need to get more powerful machines to handle the load. The solution is to purchase a more powerful server and install your app on it from the start or to order and rack new hardware if you manage your data center. This is not very flexible. Today, a lot of companies are still using an on-premises solution, and often, it's not very flexible.

The game-changer was the adoption of the other approach, which is the public cloud, which is the opposite of on-premises. The idea behind cloud computing is that big companies such as Amazon, Google, and Microsoft, which own a lot of datacenters, decided to build virtualization on top of their massive infrastructure to ensure the creation and management of virtual machines was accessible by APIs. In other words, you can get virtual machines with just a few clicks or just a few commands.

The following table provides high-level information about why cloud computing is good for organizations.

Feature	On-Premises	Cloud
Scalability	Limited – requires purchasing new hardware when scaling up	Highly scalable – easy to add or remove resources on demand
Flexibility	Inflexible – changes require physical hardware adjustments	Highly flexible – resources can be provisioned and de-provisioned quickly
Cost	High upfront cost for hardware, software licenses, and IT staff	Low upfront cost – pay-as-you-go model for resources used

Maintenance	Requires dedicated IT staff for maintenance and updates	Minimal maintenance required – cloud provider manages infrastructure
Security	High level of control over security, but requires significant expertise	Robust security measures implemented by cloud providers
Downtime	Recovery from hardware failures can be time-consuming	Cloud providers offer high availability and disaster recovery features
Location	Limited to the physical location of datacenter	Access from anywhere with an internet connection

Table 1.2: Importance of cloud computing for organizations

We will learn how cloud computing technology has helped organizations scale their IT infrastructure in the next section.

Understanding why the cloud is well suited for scalability

Today, virtually anyone can get hundreds or thousands of servers, in just a few clicks, in the form of virtual machines or instances created on physical infrastructure maintained by cloud providers such as **Amazon Web Services**, **Google Cloud Platform**, and **Microsoft Azure**. A lot of companies decided to migrate their workloads from on-premises to a cloud provider, and their adoption has been massive over the last few years.

Thanks to that, now, computing power is one of the simplest things you can get.

Cloud computing providers are now typical hosting solutions that agile teams possess in their arsenal. The main reason for this is that the cloud is extremely well suited to modern development.

Virtual machine configurations, CPUs, OSes, network rules, and more are publicly displayed and fully configurable, so there are no secrets for your team in terms of what the production environment is made of. Because of the programmable nature of cloud providers, it is very easy to replicate a production environment in a development or testing environment, providing more flexibility to teams, and helping them face their challenges when developing software. That's a useful advantage for an agile development team built around the DevOps philosophy that needs to manage the development, release, and maintenance of applications in production.

Cloud providers have provided many benefits, as follows:

- Elasticity and scalability
- Helping to break up silos and enforcing agile methodologies
- Fitting well with agile methodologies and DevOps
- Low costs and flexible billing models
- Ensuring there is no need to manage physical servers
- Allowing virtual machines to be destroyed and recreated at will
- More flexible compared to renting a bare-metal machine monthly

Due to these benefits, the cloud is a wonderful asset in the arsenal of an agile development team. Essentially, you can build and replicate a production environment over and over again without the hassle of managing the physical machine by yourself. The cloud enables you to scale your app based on the number of users using it or the computing resources they are consuming. You'll make your app highly available and fault tolerant. The result is a better experience for your end users.

> **IMPORTANT NOTE**
>
> Please note that Kubernetes can run both on the cloud and on-premises. Kubernetes is very versatile, and you can even run it on a Raspberry Pi. Kubernetes and the public cloud are a good match, but you are not required or forced to run it on the cloud.

Now that we have explained the changes the cloud produced, let's move on to software architecture because, over the years, a few things have also changed there.

Exploring the monolithic architecture

In the past, applications were mostly composed of monoliths. A typical monolith application consists of a simple process, a single binary, or a single package, as shown in *Figure 1.3*.

This unique component is responsible for the entire implementation of the business logic, to which the software must respond. Monoliths are a good choice if you want to develop simple applications that might not necessarily be updated frequently in production. Why? Well, because monoliths have one major drawback. If your monolith becomes unstable or crashes for some reason, your entire application will become unavailable:

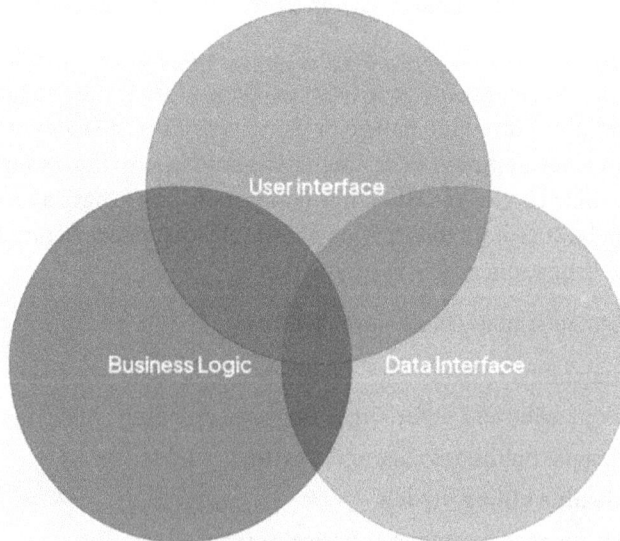

Figure 1.3: A monolith application consists of one big component that contains all your software

The monolithic architecture can allow you to gain a lot of time during your development and that's perhaps the only benefit you'll find by choosing this architecture. However, it also has many disadvantages. Here are a few of them:

- A failed deployment to production can break your whole application.
- Scaling activities become difficult to achieve; if you fail to scale, all your applications might become unavailable.
- A failure of any kind on a monolith can lead to a complete outage of your app.

In the 2010s, these drawbacks started to cause real problems. With the increase in the frequency of deployments, it became necessary to think of a new architecture that would be capable of supporting frequent deployments and shorter update cycles, while reducing the risk or general unavailability of the application. This is why the microservices architecture was designed.

Exploring the microservices architecture

The microservices architecture consists of developing your software application as a suite of independent micro-applications. Each of these applications, which is called a **microservice**, has its own versioning, life cycle, environment, and dependencies. Additionally, it can have its own deployment life cycle. Each of your microservices must only be responsible for a limited number of business rules, and all your microservices, when used together, make up the application. Think of a microservice as real full-featured software on its own, with its own life cycle and versioning process.

Since microservices are only supposed to hold a subset of all the features that the entire application has, they must be accessible in order to expose their functions. You must get data from a microservice, but you might also want to push data into it. You can make your microservice accessible through widely supported protocols such as HTTP or AMQP, and they need to be able to communicate with each other.

That's why microservices are generally built as web services that expose their functionality through well-defined APIs. While HTTP (or HTTPS) REST APIs are a popular choice due to their simplicity and widespread adoption, other protocols, such as GraphQL, AMQP, and gRPC, are gaining traction and are used commonly.

The key requirement is that a microservice provides a well-documented and discoverable API endpoint, regardless of the chosen protocol. This allows other microservices to seamlessly interact and exchange data.

This is something that greatly differs from the monolithic architecture:

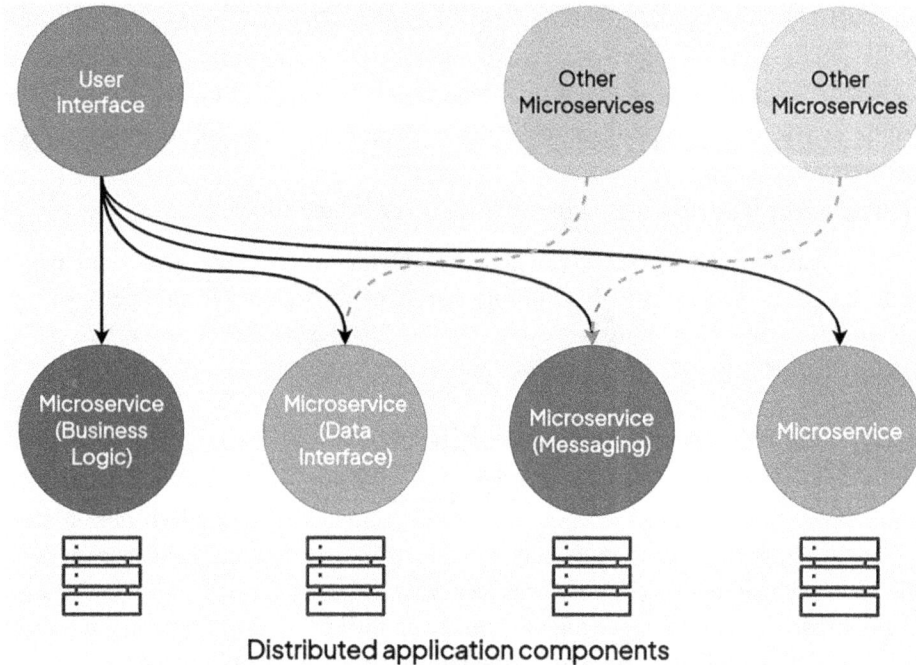

Distributed application components

Figure 1.4: A microservice architecture where different microservices communicate via the HTTP protocol

Another key aspect of the microservice architecture is that microservices need to be decoupled: if a microservice becomes unavailable or unstable, it must not affect the other microservices or the entire application's stability. You must be able to provision, scale, start, update, or stop each microservice independently without affecting anything else. If your microservices need to work with a database engine, bear in mind that even the database must be decoupled. Each microservice should have its own database and so on. So, if the database of **microservice A** crashes, it won't affect **microservice B**:

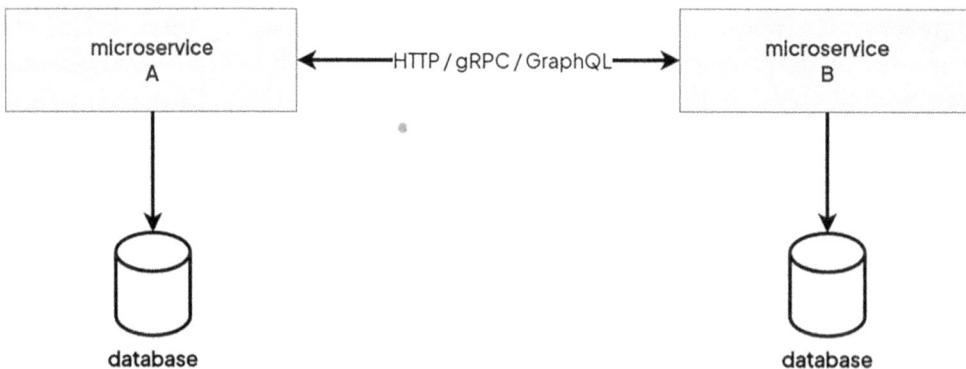

Figure 1.5: A microservice architecture where different microservices communicate with each other and with a dedicated database server; this way, the microservices are isolated and have no common dependencies

The key rule is to decouple as much as possible so that your microservices are fully independent. Because they are meant to be independent, microservices can also have completely different technical environments and be implemented in different languages. You can have one microservice implemented in Go, another one in Java, and another one in PHP, and all together they form one application. In the context of a microservice architecture, this is not a problem. Because HTTP is a standard, they will be able to communicate with each other even if their underlying technologies are different.

Microservices must be decoupled from other microservices, but they must also be decoupled from the operating system running them. Microservices should not operate at the host system level but at the upper level. You should be able to provision them, at will, on different machines without needing to rely on a strong dependency on the host system; that's why microservice architectures and containers are a good combination.

If you need to release a new feature in production, you simply deploy the microservices that are impacted by the new feature version. The others can remain the same.

As you can imagine, the microservice architecture has tremendous advantages in the context of modern application development:

- It is easier to enforce recurring production deliveries with minimal impact on the stability of the whole application.
- You can only upgrade to a specific microservice each time, not the whole application.
- Scaling activities are smoother since you might only need to scale specific services.

However, on the other hand, the microservice architecture has a couple of disadvantages too:

- The architecture requires more planning and is hard to develop.
- There are problems in managing each microservice's dependencies.

Microservice applications are considered hard to develop. This approach might be hard to understand, especially for junior developers. Dependency management can also become complex since all microservices can potentially have different dependencies.

Choosing between monolithic and microservices architectures

Building a successful software application requires careful planning, and one of the key decisions you'll face is which architecture to use. Two main approaches dominate the scene: monoliths and microservices:

- **Monoliths:** Imagine a compact, all-in-one system. That's the essence of a monolith. Everything exists in a single codebase, making development and initial deployment simple for small projects or teams with limited resources. Additionally, updates tend to be quick for monoliths because there's only one system to manage.
- **Microservices:** Think of a complex application broken down into independent, modular components. Each service can be built, scaled, and deployed separately. This approach shines with large, feature-rich projects and teams with diverse skillsets. Microservices provide flexibility and potentially fast development cycles. However, they also introduce additional complexity in troubleshooting and security management.

Ultimately, the choice between a monolith and microservices hinges on your specific needs. Consider your project's size, team structure, and desired level of flexibility. Don't be swayed by trends – pick the architecture that empowers your team to develop and manage your application efficiently.

Kubernetes provides flexibility. It caters to both fast-moving monoliths and microservices, allowing you to choose the architecture that best suits your project's needs.

In the next section, we will learn about containers and how they help microservice software architectures.

Understanding containers

Following this comparison between monolithic and microservice architectures, you should have understood that the architecture that best combines agility and DevOps is the microservice architecture. It is this architecture that we will discuss throughout the book because this is the architecture that Kubernetes manages well.

Now, we will move on to discuss how Docker, which is a container engine for Linux, is a good option for managing microservices. If you already know a lot about Docker, you can skip this section. Otherwise, I suggest that you read through it carefully.

Understanding why containers are good for microservices

Recall the two important aspects of the microservice architecture:

1. Each microservice can have its own technical environment and dependencies.
2. At the same time, it must be decoupled from the operating system it's running on.

Let's put the latter point aside for the moment and discuss the first one: two microservices of the same app can be developed in two different languages or be written in the same language but as two different versions. Now, let's say that you want to deploy these two microservices on the same Linux machine. That would be a nightmare.

The reason for this is that you'll have to install all the versions of the different runtimes, as well as the dependencies, and there might also be different versions or overlaps between the two microservices. Additionally, all of this will be on the same host operating system. Now, let's imagine you want to remove one of these two microservices from the machine to deploy it on another server and clean the former machine of all the dependencies used by that microservice. Of course, if you are a talented Linux engineer, you'll succeed in doing this. However, for most people, the risk of conflicts between the dependencies is huge, and in the end, you might just make your app unavailable while running such a nightmarish infrastructure.

There is a solution to this: you could build a machine image for each microservice and then put each microservice on a dedicated virtual machine. In other words, you refrain from deploying multiple microservices on the same machine. However, in this example, you will need as many machines as you have microservices. Of course, with the help of AWS or GCP, it's going to be easy to bootstrap tons of servers, each of them tasked with running one and only one microservice, but it would be a huge waste of money to not mutualize the computing power provided by the host.

You have similar solutions in the container world, but not with the default container runtimes because they don't guarantee complete isolation between microservices. This is exactly how the **Kata runtime** and the **Confidential Container** projects come into play. These technologies provide enhanced security and isolation for containerized applications. We'll delve deeper into these container isolation concepts later in this book.

We will learn about how containers help with isolation in the next section.

Understanding the benefits of container isolation

Container engines such as Docker and Podman play a crucial role in managing microservices. Unlike **virtual machines (VMs)** that require a full guest operating system, containers are lightweight units that share the host machine's Linux kernel. This makes them much faster to start and stop than VMs.

Container engines provide a user-friendly API to build, deploy, and manage containers. Container engines don't introduce an additional layer of virtualization. Instead, they use the built-in capabilities of the Linux kernel for process isolation, security, and resource allocation. This efficient approach makes containerization a compelling solution for deploying microservices.

The following diagram shows how containers are different from virtual machines:

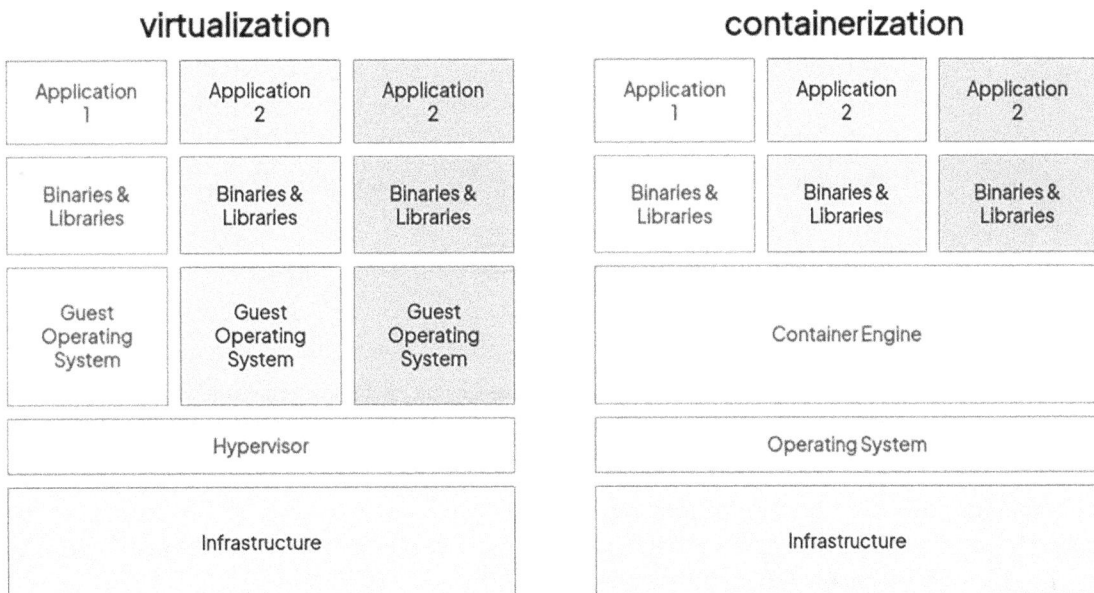

Figure 1.6: The difference between virtual machines and containers

Your microservices are going to be launched on top of this layer, not directly on the host system whose sole role will be to run your containers.

Since containers are isolated, you can run as many containers as you want and have them run applications written in different languages without any conflicts. Microservice relocation becomes as easy as stopping a running container and launching another one from the same image on another machine.

The usage of containers with microservices provides three main benefits:

- It reduces the footprint on the host system.
- It mutualizes the host system without conflicts between different microservices.
- It removes the coupling between the microservice and the host system.

Once a microservice has been containerized, you can eliminate its coupling with the host operating system. The microservice will only depend on the container in which it will operate. Since a container is much lighter than a real full-featured Linux operating system, it will be easy to share and deploy on many different machines. Therefore, the container and your microservice will work on any machine that is running a container engine.

The following diagram shows a microservice architecture where each microservice is wrapped by a container:

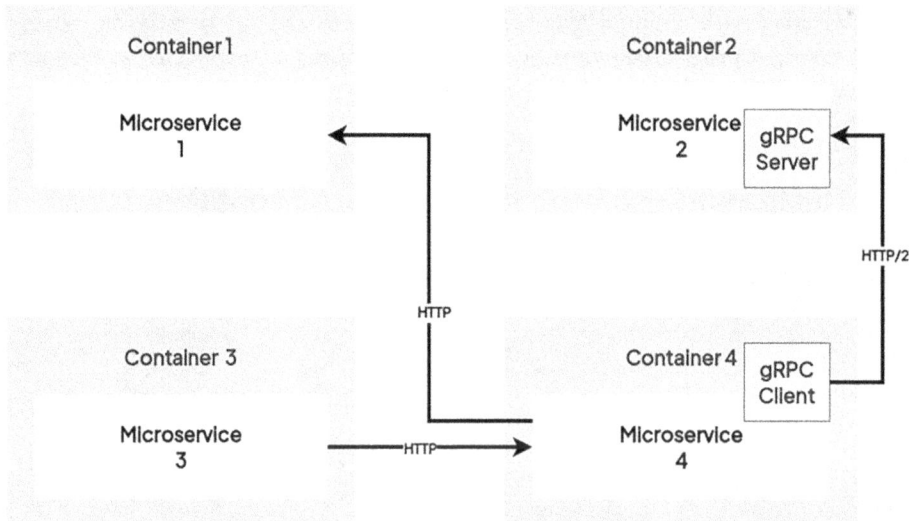

Figure 1.7: A microservice application where all microservices are wrapped by a container; the life cycle of the app becomes tied to the container, and it is easy to deploy it on any machine that is running a container engine

Containers fit well with the DevOps methodology too. By developing locally in a container, which would later be built and deployed in production, you ensure you develop in the same environment as the one that will eventually run the application.

Container engines are not only capable of managing the life cycle of a container but also an entire ecosystem around containers. They can manage networks, and the intercommunication between different containers, and all these features respond particularly well to the properties of the microservice architecture that we mentioned earlier.

By using the cloud and containers together, you can build a very strong infrastructure to host your microservice. The cloud will give you as many machines as you want. You simply need to install a container engine on each of them, and you'll be able to deploy multiple containerized microservices on each of these machines.

Container engines such as Docker or Podman are very nice tools on their own. However, you'll discover that it's hard to run them in production alone, just as they are.

Container engines excel in development environments because of their:

- **Simplicity**: Container engines are easy to install and use, allowing developers to quickly build, test, and run containerized applications.
- **Flexibility**: Developers can use container engines to experiment with different container configurations and explore the world of containerization.
- **Isolation**: Container engines ensure isolation between applications, preventing conflicts and simplifying debugging.

However, production environments have strict requirements. Container engines alone cannot address all of these needs:

- **Scaling**: Container engines (such as Docker or Podman) don't provide built-in auto-scaling features to dynamically adapt container deployments based on resource utilization.
- **Disaster Recovery**: Container engines don't provide comprehensive disaster recovery capabilities to ensure service availability in case of outages.
- **Security**: While container engines provide basic isolation, managing security policies for large-scale containerized deployments across multiple machines can be challenging.
- **Standardization:** Container engines require custom scripting or integrations for interacting with external systems, such as CI/CD pipelines or monitoring tools.

While container engines excel in development environments, production deployments demand a more robust approach. Kubernetes, a powerful container orchestration platform, tackles this challenge by providing a comprehensive suite of functionalities. It manages the entire container lifecycle, from scheduling them to run on available resources to scaling deployments up or down based on demand and distributing traffic for optimal performance (load balancing). Unlike custom scripting with container engines, Kubernetes provides a well-defined API for interacting with containerized applications, simplifying integration with other tools used in production environments. Beyond basic isolation, Kubernetes provides advanced security features such as role-based access control and network policies. This allows the efficient management of containerized workloads from multiple teams or projects on the same infrastructure, optimizing resource utilization and simplifying complex deployments.

Before we dive into the Kubernetes topics, let's discuss the basics of containers and container engines in the next section.

Container engines

A **container engine** acts as the interface for end-users and REST clients, managing user inputs, downloading container images from container registries, extracting downloaded images onto the disk, transforming user or REST client data for interaction with container engines, preparing container mount points, and facilitating communication with container engines. In essence, container engines serve as the user-facing layer, streamlining image and container management, while the underlying container runtimes handle the intricate low-level details of container and image management.

Docker stands out as one of the most widely adopted container engines, but it's important to note that various alternatives exist in the containerization landscape. Some notable ones are **LXD, Rkt, CRI-O**, and **Podman**.

At its core, Docker relies on the `containerd` container runtime, which oversees critical aspects of container management, including the container life cycle, image transfer and storage, execution, and supervision, as well as storage and network attachments. `containerd`, in turn, relies on components such as `runc` and `hcsshim`. Runc is a command-line tool that facilitates creating and running containers in Linux, while `hcsshim` plays a crucial role in the creation and management of Windows containers.

It's worth noting that `containerd` is typically not meant for direct end-user interaction. Instead, container engines, such as Docker, interact with the container runtime to facilitate the creation and management of containers. The essential role of `runc` is evident, serving not only `containerd` but also being used by Podman, CRI-O, and indirectly by Docker itself.

The basics of containers

As we learned in the previous section, Docker is a well-known and widely used container engine. Let's learn the basic terminology related to containers in general.

Container image

A container image is a kind of template used by container engines to launch containers. A container image is a self-contained, executable package that encapsulates an application and its dependencies. It includes everything needed to run the software, such as code, runtime, libraries, and system tools. Container images are created from a `Dockerfile` or `Containerfile`, which specify the build steps. Container images are stored in image repositories and shared through container registries such as Docker Hub, making them a fundamental component of containerization.

Container

A container can be considered a running instance of a container image. Containers are like modular shipping containers for applications. They bundle an application's code, dependencies, and runtime environment into a single, lightweight package. Containers run consistently across different environments because they include everything needed. Each container runs independently, preventing conflicts with other applications on the same system. Containers share the host operating system's kernel, making them faster to start and stop than virtual machines.

Container registry

A container registry is a centralized repository for storing and sharing container images. It acts as a distribution mechanism, allowing users to push and pull images to and from the registry. Popular public registries include Docker Hub, Red Hat Quayi, Amazon's **Elastic Container Registry** (**ECR**), Azure Container Registry, Google Container Registry, and GitHub Container Registry. Organizations often use private registries to securely store and share custom images. Registries play a crucial role in the Docker ecosystem, facilitating collaboration and efficient management of containerized applications.

Dockerfile or Containerfile

A Dockerfile or Containerfile is a text document that contains a set of instructions for building a container image. It defines the base image, sets up the environment, copies the application code, installs the dependencies, and configures the runtime settings. Dockerfiles or Containerfiles provide a reproducible and automated way to create consistent images, enabling developers to version and share their application configurations.

A sample Dockerfile can be seen in the following code snippet:

```
# syntax=docker/dockerfile:1

FROM node:18-alpine
WORKDIR /app
COPY . .
RUN yarn install --production
CMD ["node", "src/index.js"]
EXPOSE 3000
```

And, here's a line-by-line explanation of the provided Dockerfile:

1. `# syntax=docker/dockerfile:1`: This line defines the Dockerfile syntax version used to build the image. In this case, it specifies version 1 of the standard Dockerfile syntax.

2. `FROM node:18-alpine`: This line defines the base image for your container. It instructs the container engine to use the official Node.js 18 image with the Alpine Linux base. This provides a lightweight and efficient foundation for your application.

3. `WORKDIR /app`: This line sets the working directory within the container. Here, it specifies /app as the working directory. This is where subsequent commands in the Dockerfile will be executed relative to.

4. `COPY . .`: This line copies all files and directories from the current context (the directory where you have your Dockerfile) into the working directory (/app) defined in the previous step. This essentially copies your entire application codebase into the container.

5. `RUN yarn install --production`: This line instructs the container engine to execute a command within the container. In this case, it runs `yarn install --production`. This command uses the `yarn` package manager to install all production dependencies listed in your `package.json` file. The `--production` flag ensures that only production dependencies are installed, excluding development dependencies.

6. CMD ["node", "src/index.js"]: This line defines the default command to be executed when the container starts. Here, it specifies an array with two elements: "node" and "src/index.js". This tells Docker to run the Node.js interpreter (node) and execute the application's entry point script (src/index.js) when the container starts up.

7. EXPOSE 3000: This line exposes a port on the container. Here, it exposes port 3000 within the container. This doesn't map the port to the host machine by default, but it allows you to do so later when running the container with the -p flag (e.g., docker run -p 3000:3000 my-image). Exposing port 3000 suggests your application might be listening on this port for incoming connections.

> **IMPORTANT NOTE**
>
> To build the container image, you can use a supported container engine (such as Docker or Podman) or a container build tool, such as Buildah or kaniko.

Docker Compose or Podman Compose

Docker Compose is a tool for defining and running multi-container applications. It uses a YAML file to configure the services, networks, and volumes required for an application, allowing developers to define the entire application stack in a single file. Docker Compose or Podman Compose simplifies the orchestration of complex applications, making it easy to manage multiple containers as a single application stack.

The following compose.yaml file will spin up two containers for a WordPress application stack using a single docker compose or podman compose command:

```yaml
# compose.yaml
services:
  db:
    image: docker.io/library/mariadb
    command: '--default-authentication-plugin=mysql_native_password'
    volumes:
      - db_data:/var/lib/mysql
    restart: always
    environment:
      - MYSQL_ROOT_PASSWORD=somewordpress
      - MYSQL_DATABASE=wordpress
      - MYSQL_USER=wordpress
      - MYSQL_PASSWORD=wordpress
    expose:
      - 3306
      - 33060
    networks:
      - wordpress
  wordpress:
```

```
      image: wordpress:latest
      ports:
        - 8081:80
      restart: always
      environment:
        - WORDPRESS_DB_HOST=db
        - WORDPRESS_DB_USER=wordpress
        - WORDPRESS_DB_PASSWORD=wordpress
        - WORDPRESS_DB_NAME=wordpress
      networks:
        - wordpress
  volumes:
    db_data:
  networks:
    wordpress: {}
```

In the next section, we will learn how Kubernetes can efficiently orchestrate all these container operations.

How can Kubernetes help you to manage your containers?

In this section, we will focus on Kubernetes, which is the purpose of this book.

Kubernetes — designed to run workloads in production

If you open the official Kubernetes website (at https://kubernetes.io), the title you will see is **Production-Grade Container Orchestration**:

Figure 1.8: The Kubernetes home page showing the header and introducing Kubernetes as a production container orchestration platform

Those four words perfectly sum up what Kubernetes is: it is a container orchestration platform for production. Kubernetes does not aim to replace Docker or any of the features of Docker or other container engines; rather, it aims to manage the clusters of machines running container runtimes. When working with Kubernetes, you use both Kubernetes and the full-featured standard installations of container runtimes.

The title mentions **production**. Indeed, the concept of production is central to Kubernetes: it was conceived and designed to answer modern production needs. Managing production workloads is different today compared to what it was in the 2000s. Back in the 2000s, your production workload would consist of just a few bare-metal servers, if not even one on-premises. These servers mostly ran monoliths directly installed on the host Linux system. However, today, thanks to public cloud platforms such as **Amazon Web Services (AWS)** or **Google Cloud Platform (GCP)**, anyone can now get hundreds or even thousands of machines in the form of instances or virtual machines with just a few clicks. Even better, we no longer deploy our applications on the host system but as containerized microservices on top of Docker Engine instead, thereby reducing the footprint of the host system.

A problem will arise when you must manage Docker installations on each of these virtual machines on the cloud. Let's imagine that you have 10 (or 100 or 1,000) machines launched on your preferred cloud and you want to achieve a very simple task: deploy a containerized Docker app on each of these machines.

You could do this by running the docker run command on each of your machines. It would work, but of course, there is a better way to do it. And that's by using a **container orchestrator** such as **Kubernetes**. To give you an extremely simplified vision of Kubernetes, it is a **REST API** that keeps a registry of your machines executing a Docker daemon.

Again, this is an extremely simplified definition of Kubernetes. In fact, it's not made of a single centralized REST API, because as you might have gathered, Kubernetes was built as a suite of microservices.

Also note that while Kubernetes excels at managing containerized workloads, it doesn't replace virtual machines (**VMs**) entirely. VMs can still be valuable for specific use cases, such as running legacy applications or software with complex dependencies that are difficult to containerize. However, Kubernetes is evolving to bridge the gap between containers and VMs.

KubeVirt — a bridge between containers and VMs

KubeVirt is a project that extends Kubernetes' ability to manage virtual machines using the familiar Kubernetes API. This allows users to leverage the power and flexibility of Kubernetes for VM deployments alongside containerized applications. KubeVirt embraces **Infrastructure as Code (IaC)** principles, enabling users to define and manage VMs declaratively within their Kubernetes manifests. This simplifies VM management and integrates it seamlessly into existing Kubernetes workflows.

By incorporating VMs under the Kubernetes umbrella, KubeVirt provides a compelling approach for organizations that require a hybrid environment with both containers and VMs. It demonstrates the ongoing evolution of Kubernetes as a platform for managing diverse workloads, potentially leading to a more unified approach to application deployment and management.

We have learned about containers and the complications of managing and orchestrating containers at a large scale. In the next section, we will learn about the history and evolution of Kubernetes.

Understanding the history of Kubernetes

Now, let's discuss the history of the Kubernetes project. It will be useful for you to understand the context in which the Kubernetes project started and the people who are keeping this project alive.

Understanding how and where Kubernetes started

Since its founding in 1998, Google has gained huge experience in managing high-demanding workloads at scale, especially container-based workloads. Since the mid-2000s, Google has been at the forefront of developing its applications as Linux containers. Well before Docker simplified container usage for the general public, Google recognized the advantages of containerization, giving rise to an internal project known as Borg. To enhance the architecture of Borg, making it more extensible and robust, Google initiated another container orchestrator project called Omega. Subsequently, several improvements introduced by Omega found their way into the Borg project.

Kubernetes was born as an internal project at Google, and the first commit of Kubernetes was in 2014 by Brendan Burns, Joe Beda, and Craig McLendon, among others. However, Google didn't open source Kubernetes on its own. It was the efforts of individuals like Clayton Coleman, who was working at Red Hat at the time, and who played a crucial role in championing the idea of open-sourcing Kubernetes and ensuring its success as a community-driven project. Kelsey Hightower, an early Kubernetes champion at CoreOS, became a prominent voice advocating for the technology. Through his work as a speaker, writer, and co-founder of KubeCon, he significantly boosted Kubernetes' adoption and community growth.

Today, in addition to Google, Red Hat, Amazon, Microsoft, and other companies are also contributing to the Kubernetes project actively.

> **IMPORTANT NOTE**
>
> Borg is not the ancestor of Kubernetes because the project is not dead and is still in use at Google. It would be more appropriate to say that a lot of ideas from Borg were reused to make Kubernetes. Bear in mind that Kubernetes is not Borg or Omega. Borg was built in C++ and Kubernetes in Go. In fact, they are two entirely different projects, but one is heavily inspired by the other. This is important to understand: Borg and Omega are two internal Google projects. They were not built for the public.

Kubernetes was developed with the experience gained by Google to manage containers in production. Most importantly, it inherited Borg's and Omega's ideas, concepts, and architectures. Here is a brief list of ideas and concepts taken from Borg and Omega, which have now been implemented in Kubernetes:

- The concept of Pods to manage your containers: Kubernetes uses a logical object, called a pod, to create, update, and delete your containers.
- Each pod has its own IP address in the cluster.
- There are distributed components that all watch the central Kubernetes API to retrieve the cluster state.

- There is internal load balancing between Pods and Services.
- Labels and selectors are metadata that are used together to manage and orchestrate resources in Kubernetes.

That's why Kubernetes is so powerful when it comes to managing containers in production at scale. In fact, the concepts you'll learn from Kubernetes are older than Kubernetes itself. Although Kubernetes is a young project, it was built on solid foundations.

Who manages Kubernetes today?

Kubernetes is no longer maintained by Google because Google handed over operational control of the Kubernetes project to the **Cloud Native Computing Foundation** (CNCF) on August 29, 2018. CNCF is a non-profit organization that aims to foster and sustain an open ecosystem of cloud-native technologies.

Google is a founding member of CNCF, along with companies such as Cisco, Red Hat, and Intel. The Kubernetes source code is hosted on GitHub and is an extremely active project on the platform. The Kubernetes code is under Apache License version 2.0, which is a permissive open source license. You won't have to pay to use Kubernetes, and if you are good at coding with Go, you can even contribute to the code.

Where is Kubernetes today?

In the realm of container orchestration, Kubernetes faces competition from various alternatives, including both open-source solutions and platform-specific offerings. Some notable contenders include:

- Apache Mesos
- HashiCorp Nomad
- Docker Swarm
- Amazon ECS

While each of these orchestrators comes with its own set of advantages and drawbacks, Kubernetes stands out as the most widely adopted and popular choice in the field.

Kubernetes has won the fight for popularity and adoption and has become the standard way of deploying container-based workloads in production. As its immense growth has made it one of the hottest topics in the IT industry, it has become crucial for cloud providers to come up with a Kubernetes offering as part of their services. Therefore, Kubernetes is supported almost everywhere now.

The following Kubernetes-based services can help you get a Kubernetes cluster up and running with just a few clicks:

- Google Kubernetes Engine (GKE) on Google Cloud Platform
- Elastic Kubernetes Service (Amazon EKS)
- Azure Kubernetes Service on Microsoft Azure
- Alibaba Cloud Container Service for Kubernetes (ACK)

It's not just about the cloud offerings. It's also about the Platform-as-a-Service market. **Red Hat** started incorporating Kubernetes into its OpenShift container platform with the release of OpenShift version 3 in 2015. This marked a significant shift in OpenShift's architecture, moving from its original design to a Kubernetes-based container orchestration system, providing users with enhanced container management capabilities and offering a complete set of enterprise tools to build, deploy, and manage containers entirely on top of Kubernetes. In addition to this, other projects, such as Rancher, were built as **Kubernetes distributions** to offer a complete set of tools around the Kubernetes orchestrator, whereas projects such as **Knative** manage serverless workloads with the Kubernetes orchestrator.

> **IMPORTANT NOTE**
>
> AWS is an exception because it has two container orchestrator services. The first one is Amazon ECS, which is entirely made by AWS. The second one is Amazon EKS, which was released later than ECS and is a complete Kubernetes offering on AWS. These services are not the same, so do not be misguided by their similar names.

Where is Kubernetes going?

Kubernetes isn't stopping at containers! It's evolving to manage a wider range of workloads. KubeVirt extends its reach to virtual machines, while integration with AI/ML frameworks such as TensorFlow could allow Kubernetes to orchestrate even machine learning tasks. The future of Kubernetes is one of flexibility, potentially becoming a one-stop platform for managing diverse applications across containers, VMs, and even AI/ML workflows.

Learning Kubernetes today is one of the smartest decisions you can take if you are into managing cloud-native applications in production. Kubernetes is evolving rapidly, and there is no reason to wonder why its growth would stop.

By mastering this wonderful tool, you'll get one of the hottest skills being searched for in the IT industry today. We hope you are now convinced!

In the next section, we will learn how Kubernetes can simplify operations.

Exploring the problems that Kubernetes solves

Now, why is Kubernetes such a good fit for DevOps teams? Here's the connection: Kubernetes shines as a container orchestration platform, managing the deployment, scaling, and networking of containerized applications. Containers are lightweight packages that bundle an application with its dependencies, allowing faster and more reliable deployments across different environments. Users leverage Kubernetes for several reasons:

- **Automation:** Kubernetes automates many manual tasks associated with deploying and managing containerized applications, freeing up time for developers to focus on innovation.
- **Scalability:** Kubernetes facilitates easy scaling of applications up or down based on demand, ensuring optimal resource utilization.

- **Consistency**: Kubernetes ensures consistent deployments across different environments, from development to production, minimizing configuration errors and streamlining the delivery process.
- **Flexibility**: Kubernetes is compatible with various tools and technologies commonly used by DevOps teams, simplifying integration into existing workflows.

You can imagine that launching containers on your local machine or a development environment is not going to require the same level of planning as launching these same containers on remote machines, which could face millions of users. Problems specific to production will arise, and Kubernetes is a great way to address these problems when using containers in production:

- Ensuring high availability
- Handling release management and container deployments
- Autoscaling containers
- Network isolation
- Role-Based Access Control (RBAC)
- Stateful workloads
- Resource management

Ensuring high availability

High availability is the central principle of production. This means that your application should always remain accessible and should never be down. Of course, it's utopian. Even the biggest companies experience service outages. However, you should always bear in mind that this is your goal. Kubernetes includes a whole battery of functionality to make your containers highly available by replicating them on several host machines and monitoring their health on a regular and frequent basis.

When you deploy containers, the accessibility of your application will directly depend on the health of your containers. Let's imagine that for some reason, a container containing one of your microservices becomes inaccessible; with Docker alone, you cannot automatically guarantee that the container is terminated and recreated to ensure the service restoration. With Kubernetes, it becomes possible as Kubernetes will help you design applications that can automatically repair themselves by performing automated tasks such as health checking and container replacement.

If one machine in your cluster were to fail, all the containers running on it would disappear. Kubernetes would immediately notice that and reschedule all the containers on another machine. In this way, your applications will become highly available and fault tolerant as well.

Release management and container deployment

Deployment management is another of these production-specific problems that Kubernetes solves. The process of deployment consists of updating your application in production to replace an old version of a given microservice with a new version.

Deployments in production are always complex because you have to update the containers that are responding to requests from end users. If you miss them, the consequences could be severe for your application because it could become unstable or inaccessible, which is why you should always be able to quickly revert to the previous version of your application by running a rollback. The challenge of deployment is that it needs to be performed in the least visible way to the end user, with as little friction as possible.

Whenever you release a new version of the application, there are multiple processes involved, as follows:

1. Update the `Dockerfile` or `Containerfile` with the latest application info (if any).
2. Build a new Docker container image with the latest version of the application.
3. Push the new container image to the container registry.
4. Pull the new container image from the container registry to the staging/UAT/production system (Docker host).
5. Stop and delete the existing (old version) of the application container running on the system.
6. Launch the new container image with the new version of the application container image in the staging/UAT/production system.

Refer to the following image to understand the high-level flow in a typical scenario (please note that this is an ideal scenario because, in an actual environment, you might be using different and isolated container registries for development, staging, and production environments).

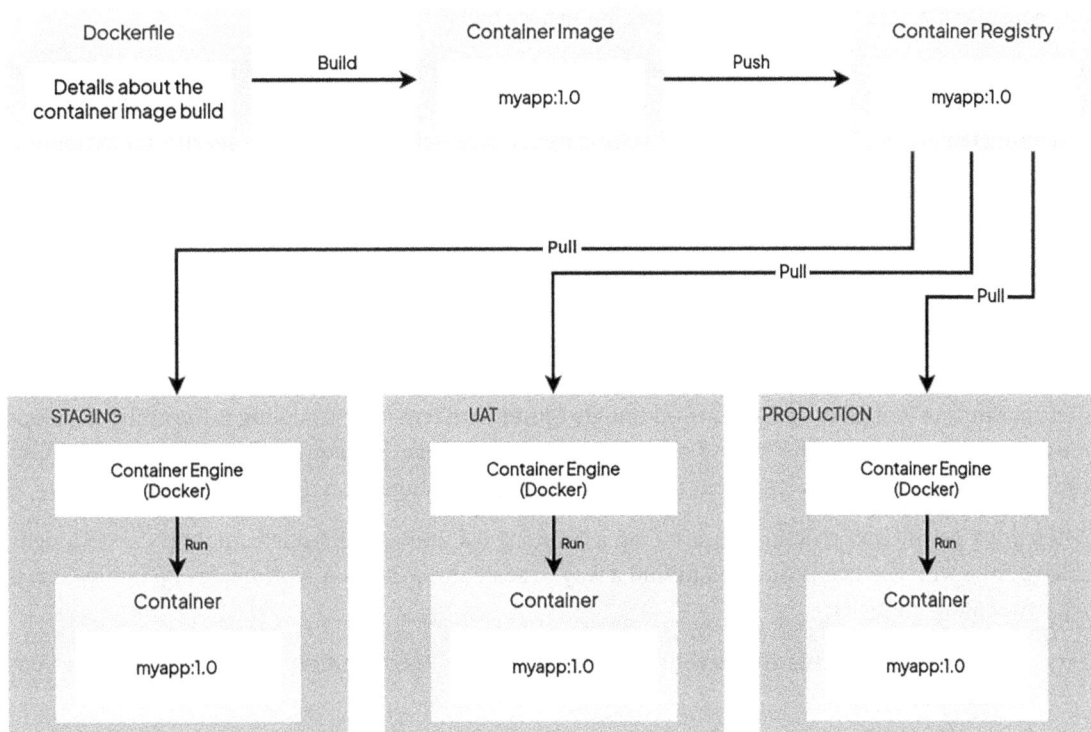

Figure 1.9: High-level workflow of container management

> **IMPORTANT NOTE**
>
> The container build process has absolutely nothing to do with Kubernetes: it's purely a container image management part. Kubernetes will come into play later when you have to deploy new containers based on a newly built image.

Without Kubernetes, you'll have to run all these operations including `docker pull`, `docker stop`, `docker delete`, and `docker run` on the machine where you want to deploy a new version of the container. Then, you will have to repeat this operation on each server that runs a copy of the container. It should work, but it is extremely tedious since it is not automated. And guess what? Kubernetes can automate this for you.

Kubernetes has features that allow it to manage deployments and rollbacks of Docker containers, and this will make your life a lot easier when responding to this problem. With a single command, you can ask Kubernetes to update your containers on all of your machines as follows:

```
$ kubectl set image deploy/myapp myapp_container=myapp:1.0.0
```

On a real Kubernetes cluster, this command will update the container called `myapp_container`, which is running as part of the application deployment called `myapp`, on every single machine where `myapp_container` runs to the `1.0.0` tag.

Whether it must update one container running on one machine or millions over multiple datacenters, this command works the same. Even better, it ensures high availability.

Remember that the goal is always to meet the requirement of high availability; a deployment should not cause your application to crash or cause a service disruption. Kubernetes is natively capable of managing deployment strategies such as rolling updates, which aim to prevent service interruptions.

Additionally, Kubernetes keeps in memory all the revisions of a specific deployment and allows you to revert to a previous version with just one command. It's an incredibly powerful tool that allows you to update a cluster of Docker containers with just one command.

Autoscaling containers

Scaling is another production-specific problem that has been widely democratized using public clouds such as **Amazon Web Services** (**AWS**) and **Google Cloud Platform** (**GCP**). Scaling is the ability to adapt your computing power to the load you are facing, again to meet the requirement of high availability and load balancing. Never forget that the goal is to prevent outages and downtime.

When your production machines are facing a traffic spike and one of your containers is no longer able to cope with the load, you need to find a way to scale the container workloads efficiently. There are two scaling methods:

- **Vertical scaling**: This allows your container to use more computing power offered by the host machine.

- **Horizontal scaling**: You can duplicate your container in the same or another machine, and you can load-balance the traffic between the multiple containers.

Docker is not able to respond to this problem alone; however, when you manage Docker with Kubernetes, it becomes possible.

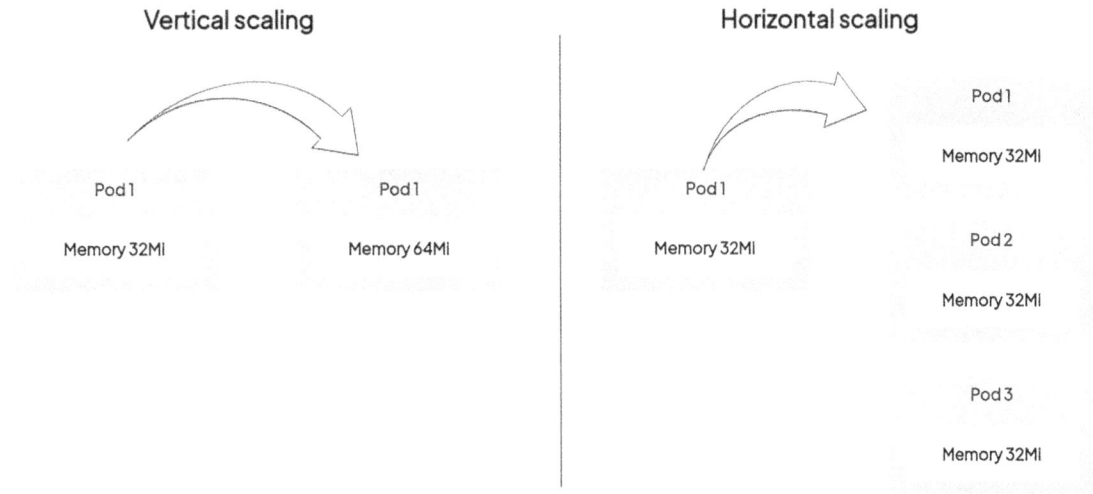

Figure 1.10: Vertical scaling versus horizontal scaling for pods

Kubernetes can manage both vertical and horizontal scaling automatically. It does this by letting your containers consume more computing power from the host or by creating additional containers that can be deployed on the same or another node in the cluster. And if your Kubernetes cluster is not capable of handling more containers because all your nodes are full, Kubernetes will even be able to launch new virtual machines by interfacing with your cloud provider in a fully automated and transparent manner by using a component called a **cluster autoscaler**.

IMPORTANT NOTE

The cluster autoscaler only works if the Kubernetes cluster is deployed on a supported cloud provider (a private or public cloud).

These goals cannot be achieved without using a container orchestrator. The reason for this is simple. You can't afford to do these tasks; you need to think about DevOps' culture and agility and seek to automate these tasks so that your applications can repair themselves, be fault-tolerant, and be highly available.

Contrary to scaling out your containers or cluster, you must also be able to decrease the number of containers if the load starts to decrease to adapt your resources to the load, whether it is rising or falling. Again, Kubernetes can do this, too.

Network isolation

In a world of millions of users, ensuring secure communication between containers is paramount. Traditional approaches can involve complex manual configuration. This is where Kubernetes shines:

- **Pod networking:** Kubernetes creates a virtual network overlay for your pods. By default, containers within the same Pod can communicate directly, while containers in different Pods are isolated by default. This prevents unintended communication between containers and enhances security.

- **Network policies:** Kubernetes allows you to define granular network policies that further restrict how pods can communicate. You can specify allowed ingress (incoming traffic) and egress (outgoing traffic) for pods, ensuring they only access the resources they need. This approach simplifies network configuration and strengthens security in production environments.

Role-Based Access Control (RBAC)

Managing access to container resources in a production environment with multiple users is crucial. Here's how Kubernetes empowers secure access control:

- **User roles:** Kubernetes defines user roles that specify permissions for accessing and managing container resources. These roles can be assigned to individual users or groups, allowing granular control over who can perform specific actions (such as viewing pod logs and deploying new containers).

- **Service accounts:** Kubernetes utilizes service accounts to provide identities for pods running within the cluster. These service accounts can be assigned roles, ensuring pods only have the access they require to function correctly.

This multi-layered approach of using user roles and service accounts strengthens security and governance in production deployments.

Stateful workloads

While containers are typically stateless (their data doesn't persist after they stop), some applications require persistent storage. Kubernetes provides solutions to manage stateful workloads: **Persistent Volumes (PVs)** and **Persistent Volume Claims (PVCs)**. Kubernetes introduces the concept of PVs, which are persistent storage resources provisioned by the administrator (e.g., host directory, cloud storage). Applications can then request storage using PVCs. This abstraction decouples storage management from the application, allowing containers to leverage persistent storage without worrying about the underlying details.

Resource management

Efficiently allocating resources to containers becomes critical in production to optimize performance and avoid resource bottlenecks. Kubernetes provides functionalities for managing resources:

- **Resource quotas:** Kubernetes allows you to set resource quotas (limits and requests) for CPU, memory, and other resources for namespaces or pods. This ensures fair resource allocation and prevents individual pods from consuming excessive resources that could starve other applications.

- **Resource limits and requests:** When defining deployments, you can specify resource requests (minimum guaranteed resources) and resource limits (maximum allowed resources) for containers. These ensure your application has the resources it needs to function properly while preventing uncontrolled resource usage.

We will learn about all of these features in the upcoming chapters.

Should we use Kubernetes everywhere? Let's discuss that in the next section.

When and where is Kubernetes not the solution?

Kubernetes has undeniable benefits; however, it is not always advisable to use it as a solution. Here, we have listed several cases where another solution might be more appropriate:

- **Container-less architecture:** If you do not use a container at all, Kubernetes won't be of any use to you.
- **A very small number of microservices or applications:** Kubernetes stands out when it must manage many containers. If your app consists of two to three microservices, a simpler orchestrator might be a better fit.

Summary

This first chapter gave us room for a big introduction. We covered a lot of subjects, such as monoliths, microservices, Docker containers, cloud computing, and Kubernetes. We also discussed how this project came to life. You should now have a global vision of how Kubernetes can be used to manage your containers in production. You have also learned why Kubernetes was introduced and how it became a well-known container orchestration tool.

In the next chapter, we will discuss the process Kubernetes follows to launch a Docker container. You will discover that you can issue commands to Kubernetes, and these commands will be interpreted by Kubernetes as instructions to run containers. We will list and explain each component of Kubernetes and its role in the whole cluster. There are a lot of components that make up a Kubernetes cluster, and we will discover all of them. We will explain how Kubernetes was built with a focus on the distinction between master nodes, worker nodes, and control plane components.

Further reading

- Kubernetes documentation: `https://kubernetes.io/docs/home/`
- Podman documentation: `https://docs.podman.io/en/latest/`
- Docker docs: `https://docs.docker.com/`
- Kata containers: `https://katacontainers.io/`
- kaniko: `https://github.com/GoogleContainerTools/kaniko`
- Buildah: `https://buildah.io`
- KubeVirt: `https://kubevirt.io`
- Knative: `https://knative.dev/docs/`

- Kubernetes: The Documentary [PART 1]: `https://www.youtube.com/watch?v=BE77h7dmoQU`
- Kubernetes: The Documentary [PART 2]: `https://www.youtube.com/watch?v=318elIq37PE`
- Technically Speaking: Clayton Coleman on the History of Kubernetes: `https://www.youtube.com/watch?v=zUJTGqWZtq0`

Join our community on Discord

Join our community's Discord space for discussions with the authors and other readers:

`https://packt.link/cloudanddevops`

2

Kubernetes Architecture — from Container Images to Running Pods

In the previous chapter, we laid the groundwork regarding what Kubernetes is from a functional point of view. You should now have a better idea of how Kubernetes can help you manage clusters of machines running containerized microservices. Now, let's go a little deeper into the technical details. In this chapter, we will examine how Kubernetes enables you to manage containers that are distributed on different machines. Following this chapter, you should have a better understanding of the anatomy of a Kubernetes cluster. In particular, you will have a better understanding of Kubernetes components and know the responsibility of each of them in the execution of your containers.

Kubernetes is made up of several distributed components, each of which plays a specific role in the execution of containers. To understand the role of each Kubernetes component, we will follow the life cycle of a container as it is created and managed by Kubernetes: that is, from the moment you execute the command to create the container to the point when it is actually executed on a machine that is part of your Kubernetes cluster.

In this chapter, we're going to cover the following main topics:

- The name – Kubernetes
- Understanding the difference between the control plane nodes and compute nodes
- Kubernetes components
- The control plane components
- The compute node components
- Exploring the `kubectl` command-line tool and YAML syntax
- How to make Kubernetes highly available

Technical requirements

The following are the technical requirements to proceed with this chapter:

- A basic understanding of the Linux OS and how to handle basic operations in Linux
- One or more Linux machines

The code and snippets used in the chapter are tested on the Fedora workstation. All the code, commands, and other snippets for this chapter can be found in the GitHub repository at `https://github.com/PacktPublishing/The-Kubernetes-Bible-Second-Edition/tree/main/Chapter02`.

The name – Kubernetes

Kubernetes derives its name from Greek origins, specifically from the word "**kubernētēs**," which translates to helmsman or pilot. This nautical term signifies someone skilled in steering and navigating a ship. The choice of this name resonates with the platform's fundamental role in guiding and orchestrating the deployment and management of containerized applications, much like a helmsman steering a ship through the complexities of the digital landscape.

In addition to its formal name, Kubernetes is commonly referred to as "K8s" within the community. This nickname cleverly arises from the technique of abbreviating the word by counting the eight letters between the "K" and the "s." This shorthand not only streamlines communication but also adds a touch of informality to discussions within the Kubernetes ecosystem.

Understanding the difference between the control plane nodes and compute nodes

To run Kubernetes, you will require Linux machines, which are called nodes in Kubernetes. A node could be a physical machine or a virtual machine on a cloud provider, such as an EC2 instance. There are two types of node in Kubernetes:

- Control plane nodes (also known as master nodes)
- Compute nodes (also known as worker nodes)

The master and worker nodes

In various contexts, you might encounter the terms "master nodes" and "worker nodes," which were previously used to describe the conventional hierarchical distribution of roles in a distributed system. In this setup, the "master" node oversaw and assigned tasks to the "worker" nodes. However, these terms may carry historical and cultural connotations that could be perceived as insensitive or inappropriate. In response to this concern, the Kubernetes community has chosen to replace these terms with "control plane nodes" (or controller nodes), denoting the collection of components responsible for managing the overall state of the cluster. Likewise, the term "node" or "compute node" is now used in lieu of "worker" to identify the individual machines in the cluster executing the requested tasks or running the application workloads. The control plane is responsible for maintaining the state of the Kubernetes cluster, whereas compute nodes are responsible for running containers with your applications.

Linux and Windows containers

You have the flexibility to leverage Windows-based nodes to launch containers tailored for Windows within your Kubernetes cluster. It's worth noting that your cluster can harmoniously accommodate both Linux and Windows machines; however, attempting to initiate a Windows container on a Linux worker node, and vice versa, is not feasible. Striking the right balance between Linux and Windows machines in your cluster ensures optimal performance.

In the next sections of this chapter, we will learn about different Kubernetes components and their responsibilities.

Kubernetes components

Kubernetes, by its inherent design, functions as a distributed application. When we refer to Kubernetes, it's not a standalone, large-scale application released in a single build for installation on a dedicated machine. Instead, Kubernetes embodies a compilation of small projects, each crafted in Go (language), collectively constituting the overarching Kubernetes project.

To establish a fully operational Kubernetes cluster, it's necessary to individually install and configure each of these components, ensuring seamless communication among them. Once these prerequisites are fulfilled, you can commence running your containers using the Kubernetes orchestrator.

For development or local testing, it is fine to install all of the Kubernetes components on the same machine. However, in production, to meet requirements like high availability, load balancing, distributed computing, scaling, and so on, these components should be spread across different hosts. By spreading the different components across multiple machines, you gain two benefits:

- You make your cluster highly available and fault-tolerant.
- You make your cluster a lot more scalable. Components have their own life cycle; they can be scaled without impacting others.

In this way, having one of your servers down will not break the entire cluster but just a small part of it, and adding more machines to your servers becomes easy.

Each Kubernetes component has its own clearly defined responsibility. It is important for you to understand each component's responsibility and how it articulates with the other components to understand the overall working of Kubernetes.

Depending on its role, a component will have to be deployed on a control plane node or a compute node. While some components are responsible for maintaining the state of a whole cluster and operating the cluster itself, others are responsible for running our application containers by interacting with the container runtime directly (e.g., `containerd` or Docker daemons). Therefore, the components of Kubernetes can be grouped into two families: control plane components and compute node components.

You are not supposed to launch your containers by yourself, and therefore, you do not interact directly with the compute nodes. Instead, you send your instructions to the control plane. Then, it will delegate the actual container creation and maintenance to the compute node on your behalf.

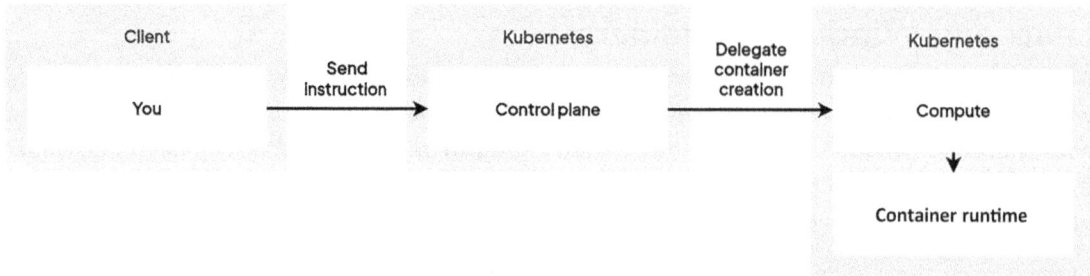

Figure 2.1: A typical Kubernetes workflow

Due to the distributed nature of Kubernetes, the control plane components can be spread across multiple machines. There are two ways to set up the control plane components:

- You can run all the control planes on the same machine or on different machines. To achieve maximum fault tolerance, it's a good idea to spread the control plane components across different machines. The idea is that Kubernetes components must be able to communicate with each other, and this still can be achieved by installing them on different hosts.

- Things are simpler when it comes to compute nodes (or worker nodes). In these, you start from a standard machine running a supported container runtime, and you install the compute node components next to the container runtime. These components will interface with the local container engine that is installed on said machine and execute containers based on the instructions you send to the control plane components. Adding more computing power to your cluster is easy; you just need to add more worker nodes and have them join the cluster to make room for more containers.

> By splitting the control plane and compute node components of different machines, you are making your cluster highly available and scalable. Kubernetes was built with all of the cloud-native concerns in mind; its components are stateless, easy to scale, and built to be distributed across different hosts. The whole idea is to avoid having a single point of failure by grouping all the components on the same host.

Here is a simplified diagram of a full-featured Kubernetes cluster with all the components listed. In this chapter, we're going to explain all of the components listed in this diagram, their roles, and their responsibilities. Here, all of the control plane components are installed on a single master node machine:

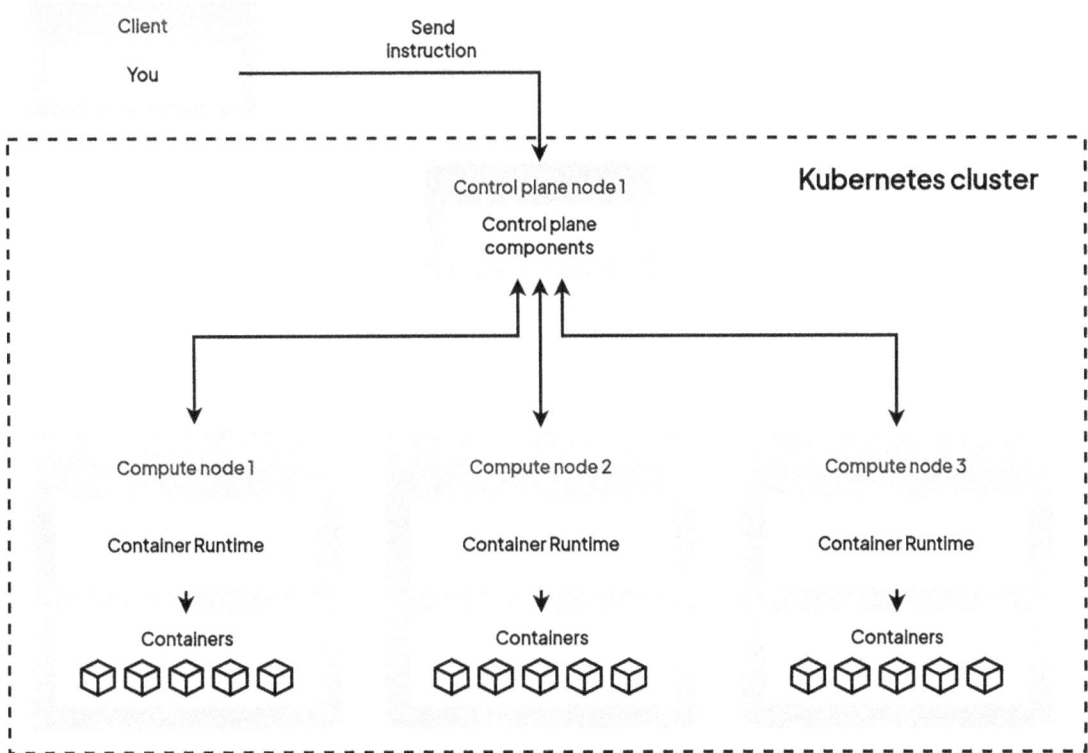

Figure 2.2: A full-featured Kubernetes cluster with one control plane node and three compute nodes

The preceding diagram displays a four-node Kubernetes cluster with all the necessary components.

Bear in mind that Kubernetes is modified and, therefore, can be modified to fit a given environment. When Kubernetes is deployed and used as part of a distribution such as Amazon EKS or Red Hat Open-Shift, additional components could be present, or the behavior of the default ones might differ. In this book, for the most part, we will discuss bare or vanilla Kubernetes. The components discussed in this chapter are the default ones and you will find them everywhere as they are the backbone of Kubernetes.

The following diagram shows the basic and core components of a Kubernetes cluster.

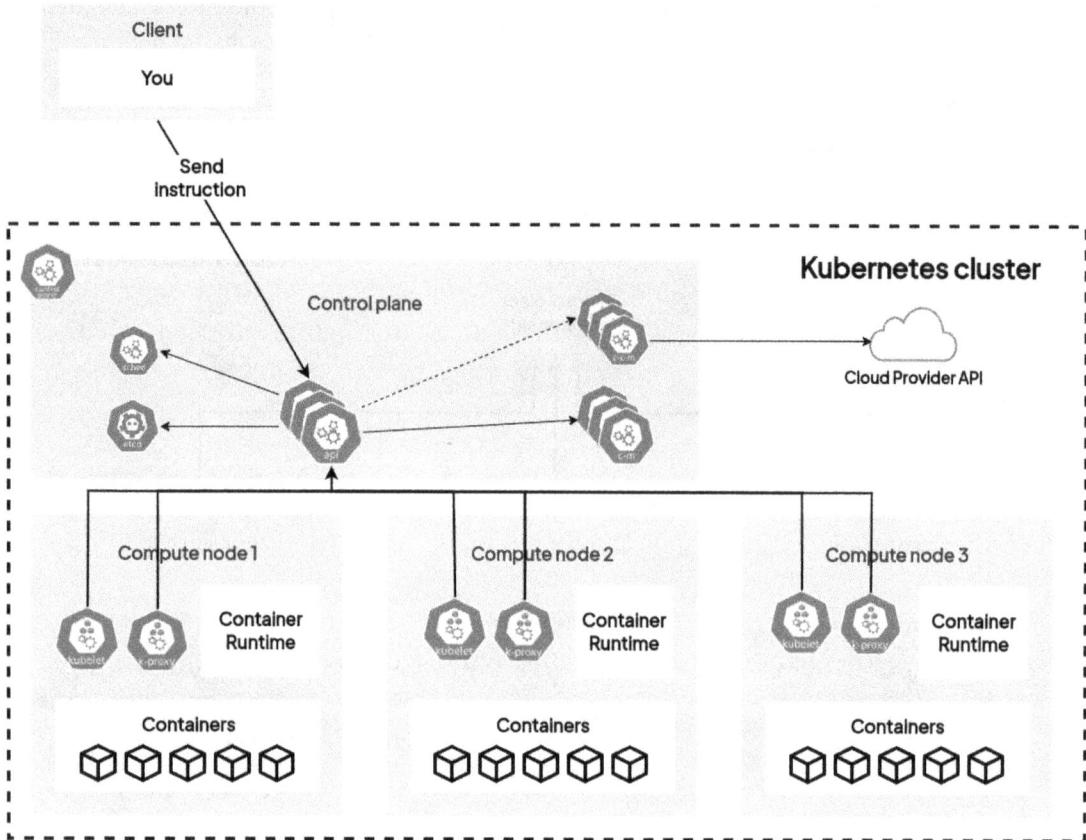

Figure 2.3: The components of a Kubernetes cluster (image source: https://kubernetes.io/docs/con-cepts/overview/components)

You might have noticed that most of these components have a name starting with kube: these are the components that are part of the Kubernetes project. Additionally, you might have noticed that there are two components with a name that does not start with kube. The other two components (etcd and Container Engine) are two external dependencies that are not strictly part of the Kubernetes project, but which Kubernetes needs to work:

- etcd is a third-party data store used by the Kubernetes project. Don't worry; you won't have to master it to use Kubernetes.
- The container engine is also a third-party engine.

Rest assured, you will not have to install and configure these components all by yourself. Almost no one bothers with managing the components by themselves, and, in fact, it's super easy to get a working Kubernetes without having to install the components.

For development purposes, you can use **minikube**, which is a tool that enables developers to run a single-node Kubernetes cluster locally on their machine. It's a lightweight and easy-to-use solution for testing and developing Kubernetes applications without the need for a full-scale cluster. minikube is absolutely NOT recommended for production.

For production deployment, cloud offerings like Amazon EKS or Google GKE provide an integrated, scalable Kubernetes cluster. Alternatively, **kubeadm**, a Kubernetes installation utility, is suitable for platforms without cloud access.

For educational purposes, a renowned tutorial known as *Kubernetes the Hard Way* by *Kelsey Hightower* guides users through manual installations, covering PKI management, networking, and computing provisioning on bare Linux machines in Google Cloud. While this tutorial may feel difficult for beginners, it is still recommended to practice, offering a valuable opportunity to comprehend the internals of Kubernetes. Note that establishing and managing a production-grade Kubernetes cluster, as demonstrated in *Kubernetes the Hard Way*, is intricate and time-consuming. It's advised against using its results in a production environment. You will observe many references to this tutorial on the internet because it's very famous.

We will learn about the Kubernetes control plane and compute node components in the next section.

Control plane components

These components are responsible for maintaining the state of the cluster. They should be installed on a control plane node. These are the components that will keep the list of containers executed by your Kubernetes cluster or the number of machines that are part of the cluster. As an administrator, when you interact with Kubernetes, you interact with the control plane components and the following are the major components in the control plane:

- `kube-apiserver`
- `etcd`
- `kube-scheduler`
- `kube-controller-manager`
- `cloud-controller-manager`

Compute node components

These components are responsible for interacting with the container runtime in order to launch containers according to the instructions they receive from the control plane components. Compute node components must be installed on a Linux machine running a supported container runtime and you are not supposed to interact with these components directly. It's possible to have hundreds or thousands of compute nodes in a Kubernetes cluster. The following are the major component parts of the compute nodes:

- `kubelet`
- `kube-proxy`
- Container runtime

Add-on components

Add-ons utilize Kubernetes resources such as DaemonSet, Deployment, and others to implement cluster features. As these features operate at the cluster level, resources for add-ons that are namespaced are located within the kube-system namespace. The following are some of the add-on components you will see commonly in your Kubernetes clusters:

- DNS
- Web UI (dashboard)
- Container resource monitoring
- Cluster-level logging
- Network plugins

Control plane in managed Kubernetes clusters

In contrast to self-managed Kubernetes clusters, cloud services like Amazon EKS, Google GKE, and similar offerings handle the installation and configuration of most Kubernetes control plane components. They provide access to a Kubernetes endpoint, or optionally, the kube-apiserver endpoint, without exposing intricate details about the underlying machines or provisioned load balancers. This holds true for components such as kube-scheduler, kube-controller-manager, etcd, and others.

Here is a screenshot of a Kubernetes cluster created on the Amazon EKS service:

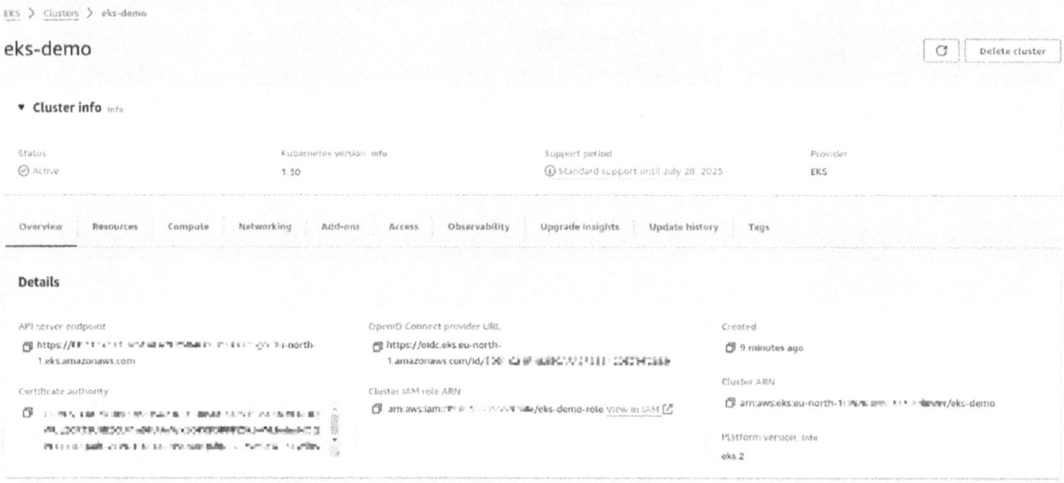

Figure 2.4: The UI console showing details of a Kubernetes cluster provisioned on Amazon EKS

We have detailed chapters to learn about EKS, GKE, and AKS later in this book.

We will learn about control plane components that are responsible for maintaining the state of the cluster in the next sections.

The Control Plane Components

In the following sections, let us explore the different control plane components and their responsibilities.

kube-apiserver

Kubernetes' most important component is a **Representational State Transfer** (**REST**) API called kube-apiserver, which exposes all the Kubernetes features. You will be interacting with Kubernetes by calling this REST API through the kubectl command-line tool, direct API calls, or the Kubernetes dashboard (Web UI) utilities.

The role of kube-apiserver

kube-apiserver is a part of the control plane in Kubernetes. It's written in Go, and its source code is open and available on GitHub under the Apache 2.0 license. To interact with Kubernetes, the process is straightforward. Whenever you want to instruct Kubernetes, you send an HTTP request to kube-apiserver. Whether it's creating, deleting, or updating a container, you always make these calls to the appropriate kube-apiserver endpoint using the right HTTP verb. This is the routine with Kubernetes—kube-apiserver serves as the sole entry point for all operations directed to the orchestrator. It's considered a good practice to avoid direct interactions with container runtimes (unless it is some troubleshooting activity).

kube-apiserver is constructed following the REST standard. REST proves highly efficient in showcasing functionalities through HTTP endpoints, accessible by employing different methods of the HTTP protocol like GET, POST, PUT, PATCH, and DELETE. When you combine HTTP methods and paths, you can perform various operations specified by the method on resources identified by the path.

The REST standard provides considerable flexibility, allowing easy extension of any REST API by adding new resources through the addition of new paths. Typically, REST APIs employ a datastore to manage the state of objects or resources.

Data retention in such an API can be approached in several ways, including the following:

REST API memory storage:

- Keeps data in its own memory.
- However, this results in a stateful API, making scaling impossible.

> Kubernetes uses etcd to store state and it is pronounced /ˈɛtsiːdiː/, which means distributed etc directory. The etcd is an open source distributed key-value store used to hold and manage the critical information that distributed systems need to keep running.

Database engine usage:

- Utilizes full-featured database engines like MariaDB or PostgreSQL.
- Delegating storage to an external engine makes the API stateless and horizontally scalable.

Any REST API can be easily upgraded or extended to do more than its initial intent. To sum up, here are the essential properties of a REST API:

- Relies on the HTTP protocol
- Defines a set of resources identified by URL paths
- Specifies a set of actions identified by HTTP methods
- Executes actions against resources based on a properly forged HTTP request
- Maintains the state of their resources on a datastore

In summary, `kube-apiserver` is nothing more than a REST API, which is at the heart of any Kubernetes cluster you will set up, no matter if it's local, on the cloud, or on-premises. It is also stateless; that is, it keeps the state of the resources by relying on a database engine called `etcd`. This means you can horizontally scale the `kube-apiserver` component by deploying it onto multiple machines and load balance request issues to it using a layer 7 load balancer without losing data.

As HTTP is supported almost everywhere, it is very easy to communicate with and issue instructions to a Kubernetes cluster. However, most of the time, we interact with Kubernetes via the command-line utility named `kubectl`, which is the HTTP client that is officially supported as part of the Kubernetes project. When you download `kube-apiserver`, you'll end up with a Go-compiled binary that is ready to be executed on any Linux machine. The Kubernetes developers defined a set of resources for us that are directly bundled within the binary. So, do expect the resources in `kube-apiserver` related to container management, networking, and computing in general.

A few of these resources are as follows:

- Pod
- ReplicaSet
- PersistentVolume
- NetworkPolicy
- Deployment

Of course, this list of resources is not exhaustive. If you want a full list of the Kubernetes components, you can access it from the official Kubernetes documentation API reference page at `https://kubernetes.io/docs/reference/kubernetes-api/`.

You might be wondering why there are no *container* resources here. As mentioned in *Chapter 1, Kubernetes Fundamentals*, Kubernetes makes use of a resource called a Pod to manage the containers. For now, you can think of pods as though they were containers.

> Although pods can hold multiple containers, it's common to have a pod with just one container inside. If you're interested in using multiple containers within a pod, we'll explore patterns like `sidecar` and `init containers` in *Chapter 5, Using Multi-Container Pods and Design Patterns*.

We will learn a lot about them in the coming chapters. Each of these resources is associated with a dedicated URL path, and changing the HTTP method when calling the URL path will have a different effect. All of these behaviors are defined in kube-apiserver. Note that these behaviors are not something you have to develop; they are directly implemented as part of kube-apiserver.

After the Kubernetes objects are stored on the etcd database, other Kubernetes components will *convert* these objects into raw container instructions.

Remember, kube-apiserver is the central hub and the definitive source for the entire Kubernetes cluster. All actions in Kubernetes revolve around it. Other components, including administrators, interact with kube-apiserver via HTTP, avoiding direct interaction with cluster components in most cases.

This is because kube-apiserver not only manages the cluster's state but also incorporates numerous mechanisms for authentication, authorization, and HTTP response formatting. Consequently, manual interventions are strongly discouraged due to the complexity of these processes.

How do you run kube-apiserver?

In *Chapter 3, Installing Your First Kubernetes Cluster,* we will focus on how to install and configure a Kubernetes cluster locally.

Essentially, there are two ways to run kube-apiserver (and other components), as follows:

- By running kube-apiserver as a container image
- By downloading and installing kube-apiserver and running it using a systemd unit file

Since the recommended method is to run the containerized kube-apisever, let's put aside the systemd method. Depending on the Kubernetes cluster deployment mechanisms, kube-apiserver and other components will be configured as containers by downloading the appropriate images from the container registry (e.g., registry.k8s.io).

Where do you run kube-apiserver?

kube-apiserver should be run on the control plane node(s) as it is part of the control plane. Ensure that the kube-apiserver component is installed on a robust machine solely dedicated to the control plane operations. This component is crucial, and if it becomes inaccessible, your containers will persist but lose connectivity with Kubernetes. They essentially turn into "orphan" containers on isolated machines, no longer under Kubernetes management.

Also, the other Kubernetes components from all cluster nodes constantly send HTTP requests to kube-apiserver to understand the state of the cluster or to update it. And the more compute nodes you have, the more HTTP requests will be issued against kube-apiserver. That's why kube-apiserver should be independently scaled as the cluster itself scales out.

As mentioned earlier, kube-apiserver is a stateless component that does not directly maintain the state of the Kubernetes cluster itself and relies on a third-party database to do so. You can scale it horizontally by hosting it on a group of machines that are behind a load balancer such as an HTTP API. When using such a setup, you interact with kube-apiserver by calling your API load balancer endpoint.

In the next section, we will learn how Kubernetes stores the cluster and resource information using etcd.

The etcd datastore

We explained that kube-apiserver can be scaled horizontally. We also mentioned that to store the state of the cluster status and details, kube-apiserver uses etcd, an open source, distributed key-value store. Strictly speaking, etcd is not a part of the Kubernetes project but a separate project that is maintained by the etcd-io community.

> While etcd is the commonly used datastore for Kubernetes clusters, some distributions like **k3s** leverage alternatives by default, such as SQLite or even external databases like MySQL or PostgreSQL (https://docs.k3s.io/datastore).
>
> etcd is also an open source project (written in Go just like Kubernetes), which is available on GitHub (https://github.com/etcd-io/etcd) under license Apache 2.0. It's also a project incubated (in 2018 and graduated in 2020) by the **Cloud Native Computing Foundation** (**CNCF**), which is the organization that maintains Kubernetes.

When you call kube-apiserver, each time you implement a read or write operation by calling the Kubernetes API, you will read or write data from or to etcd.

Let's zoom into what is inside the master node now:

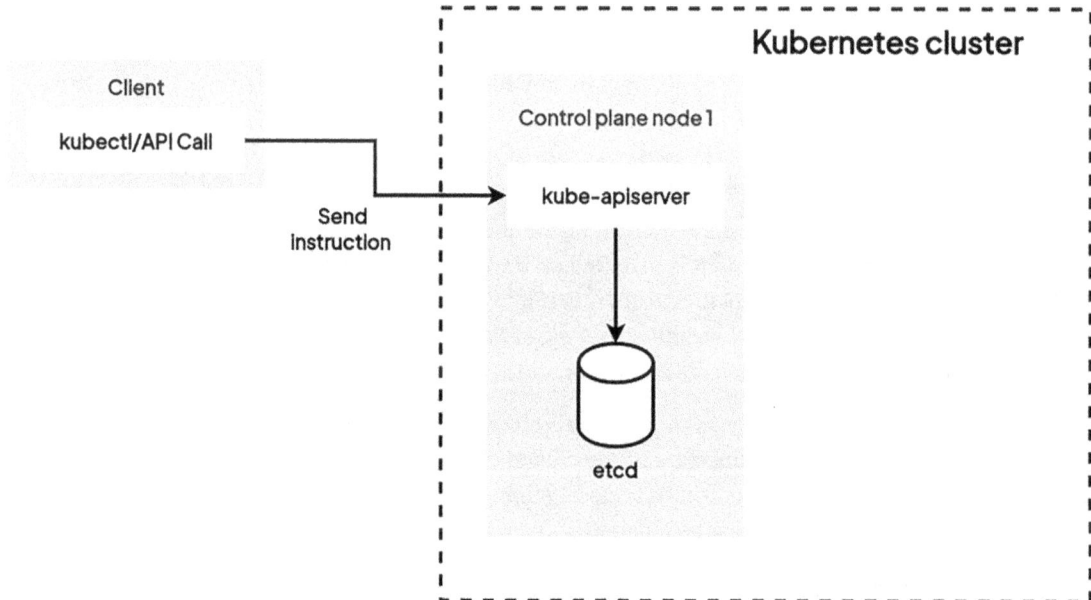

Figure 2.5: The kube-apiserver component is in front of the etcd datastore and acts as a proxy in front of it; kube-apiserver is the only component that can read or write from and to etcd

etcd is like the heart of your cluster. If you lose the data in etcd, your Kubernetes cluster won't work anymore. It's even more crucial than kube-apiserver. If kube-apiserver crashes, you can restart it. But if etcd data is lost or messed up without a backup, your Kubernetes cluster is done for.

Fortunately, you do not need to master etcd in depth to use Kubernetes. It is even strongly recommended that you do not touch it at all if you do not know what you are doing. This is because a bad operation could corrupt the data stored in etcd and, therefore, the state of your cluster.

Remember, the general rule in Kubernetes architecture says that every component has to go through kube-apiserver to read or write in etcd. This is because, from a technical point of view, kubectl authenticates itself against kube-apiserver through a TLS client certificate that only kube-apiserver has. Therefore, it is the only component of Kubernetes that has the right to read or write in etcd. This is a very important notion in the architecture of Kubernetes. All of the other components won't be able to read or write anything to or from etcd without calling the kube-apiserver endpoints through HTTP.

> Please note that etcd is also designed as a REST API. By default, it listens to port 2379.

Let's explore a simple kubectl command, as follows:

```
$ kubectl run nginx --restart Never --image nginx
```

When you execute the preceding command, the kubectl tool will forge an HTTP POST request that will be executed against the kube-apiserver component specified in the kubeconfig file. kube-apiserver will write a new entry in etcd, which will be persistently stored on disk.

At that point, the state of Kubernetes changes: it will then be the responsibility of the other Kubernetes components to reconcile the actual state of the cluster to the desired state of the cluster (that is, the one in etcd).

Unlike Redis or Memcached, etcd is not in-memory storage. If you reboot your machine, you do not lose the data because it is kept on disk.

Where do you run etcd?

In a self-managed Kubernetes setup, you can operate etcd either within a container or as part of a systemd unit file. etcd can naturally expand horizontally by distributing its dataset across several servers, making it an independent clustering solution.

Also, you have two places to run etcd for Kubernetes, as follows:

- etcd can be deployed together with kube-apiserver (and other control plane components) on the control plane nodes – this is the default and simple setup (in most Kubernetes clusters, components like etcd and kube-apiserver are initially deployed using static manifests. We'll explore this approach and alternatives in more detail later in the book).

- You can configure to use a dedicated etcd cluster – this is a more complex approach but more reliable if your environment is demanding for such reliability.

Operating etcd clusters for Kubernetes

The details about single-node or multi-node dedicated etcd clusters can be found in the official Kubernetes documentation at https://kubernetes.io/docs/tasks/administer-cluster/configure-upgrade-etcd/.

Learning more about etcd

If you are interested in learning how etcd works and want to play with the etcd dataset, there is a free playground available online. Visit http://play.etcd.io/play and learn how to manage etcd clusters and data inside.

Let us explore and learn about kube-scheduler in the next section.

kube-scheduler

kube-scheduler is responsible for electing a worker node out of those available to run a newly created pod.

Upon creation, pods are unscheduled, indicating that no worker node has been designated for their execution. An unscheduled pod is recorded in etcd without any assigned worker node. Consequently, no active kubelet will be informed of the need to launch this pod, leading to the non-execution of any container outlined in the pod specification.

Internally, the pod object, as it is stored in etcd, has a property called nodeName. As the name suggests, this property should contain the name of the worker node that will host the pod. When this property is set, we say that the pod has been *scheduled*; otherwise, the pod is *pending* for schedule.

kube-scheduler queries kube-apiserver at regular intervals in order to list the pods that have not been *scheduled* or with an empty nodeName property. Once it finds such pods, it will execute an algorithm to elect a worker node. Then, it will update the nodeName property in the pod by issuing an HTTP request to the kube-apiserver component. While electing a worker node, the kube-scheduler component will take into account some configuration values that you can pass:

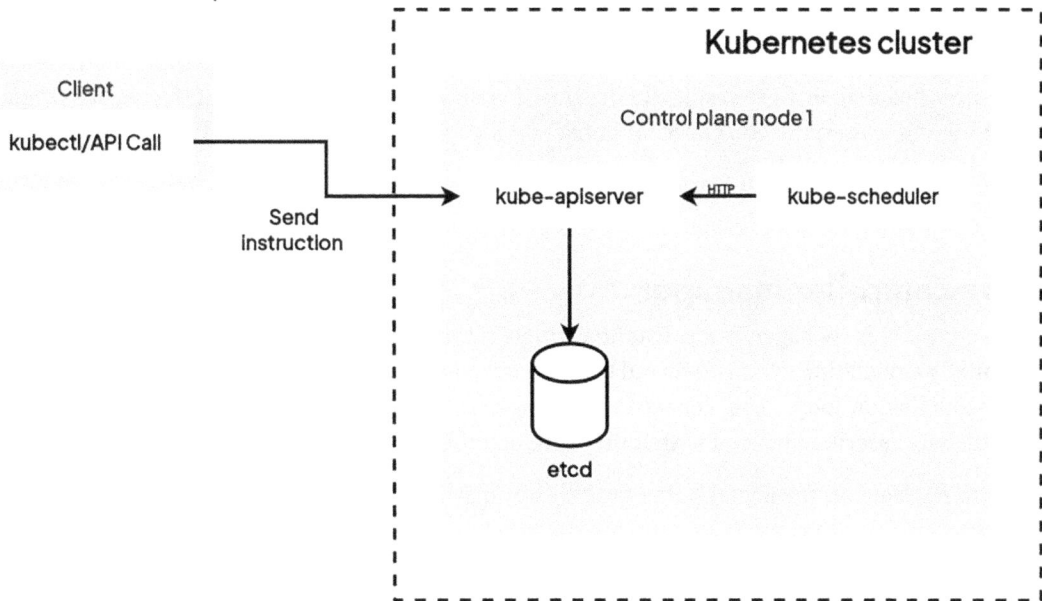

Figure 2.6: The kube-scheduler component polls the kube-apiserver component to find unscheduled pods

The kube-scheduler component will take into account some configuration values that you can pass optionally. By using these configurations, you can precisely control how the kube-scheduler component will elect a worker node. Here are some of the features to bear in mind when scheduling pods on your preferred node:

- Node selector
- Node affinity and anti-affinity
- Taint and toleration

There are also advanced techniques for scheduling that will completely bypass the kube-scheduler component. We will examine these features later.

> The kube-scheduler component can be replaced by a custom one. You can implement your own kube-scheduler component with your custom logic to select a node and use it on your cluster. It's one of the strengths of the distributed nature of Kubernetes components.

Where do you install kube-scheduler?

You can choose to install kube-scheduler on a dedicated machine or the same machine as kube-apiserver. It's a short process and won't consume many resources, but there are some things to pay attention to.

The `kube-scheduler` component should be highly available. That's why you should install it on more than one machine. If your cluster does not have a working `kube-scheduler` component, new pods won't be scheduled, and the result will be a lot of pending pods. Also note that if no `kube-scheduler` component is present, it won't have an impact on the already scheduled pods.

In the next section, we will learn about another important control plane component called `kube-controller-manager`.

kube-controller-manager

`kube-controller-manager` is a substantial single binary that encompasses various functionalities, essentially embedding what is referred to as a controller. It is the component that runs what we call the reconciliation loop. `kube-controller-manager` tries to maintain the actual state of the cluster with the one described in `etcd` so that there are no differences between the states.

In certain instances, the actual state of the cluster may deviate from the desired state stored in `etcd`. This discrepancy can result from pod failures or other factors. Consequently, the `kube-controller-manager` component plays a crucial role in reconciling the actual state with the desired state. As an illustration, consider the replication controller, one of the controllers operating within the `kube-controller-manager` component. In practical terms, Kubernetes allows you to specify and maintain a specific number of pods across different compute nodes. If, for any reason, the actual number of pods varies from the specified count, the replication controller initiates requests to the `kube-apiserver` component. This aims to recreate a new pod in `etcd`, thereby replacing the failed one on a compute node.

Here is a list of a few controllers that are part of `kube-controller-manager`:

- **Node Controller:** Handles the life cycle of nodes, overseeing their addition, removal, and updates within the cluster

- **Replication Controller:** Ensures that the specified number of replicas for a pod specification is consistently maintained

- **Endpoints Controller:** Populates the endpoints objects for services, reflecting the current pods available for each service

- **Service Account Controller:** Oversees the management of ServiceAccounts within namespaces, ensuring the presence of a ServiceAccount named `default` in each currently active namespace

- **Namespace Controller:** Manages the lifecycle of namespaces, encompassing creation, deletion, and isolation

- **Deployment Controller:** Manages the lifecycle of deployments, ensuring that the desired pod count for each deployment is maintained

- **StatefulSet Controller:** Manages the lifecycle of stateful sets, preserving the desired replica count, pod order, and identity

- **DaemonSet Controller:** Manages the lifecycle of daemon sets, guaranteeing that a copy of the daemon pod is active on each cluster node

- **Job Controller:** Manages the lifecycle of jobs, ensuring the specified pod count for each job is maintained until job completion

- **Horizontal Pod Autoscaler (HPA) Controller:** Dynamically scales the number of replicas for a deployment or stateful set based on resource utilization or other metrics
- **Pod Garbage Collector:** Removes pods no longer under the control of an owner, such as a replication controller or deployment

As you can gather, the kube-controller-manager component is quite big. But essentially, it's a single binary that is responsible for reconciling the actual state of the cluster with the desired state of the cluster that is stored in etcd.

Where do you run kube-controller-manager?

The kube-controller-manager component can run as a container or a systemd service similar to kube-apiserver on the control plane nodes. Additionally, you can decide to install the kube-controller-manager component on a dedicated machine. Let's now talk about cloud-controller-manager.

cloud-controller-manager

cloud-controller-manager is a component in the Kubernetes control plane that manages the interactions between Kubernetes and the underlying cloud infrastructure. cloud-controller-manager handles the provisioning and administration of cloud resources, including nodes and volumes, to facilitate Kubernetes workloads. It exclusively operates controllers tailored to your cloud provider. In cases where Kubernetes is self-hosted, within a learning environment on a personal computer, or on-premises, the cluster does not feature a cloud controller manager.

Similar to kube-controller-manager, cloud-controller-manager consolidates multiple logically independent control loops into a unified binary, executed as a single process. Horizontal scaling, achieved by running multiple copies, is an option to enhance performance or enhance fault tolerance.

Controllers with potential cloud provider dependencies include:

- **Node Controller:** Verifies if a node has been deleted in the cloud after it stops responding
- **Route Controller:** Establishes routes in the underlying cloud infrastructure
- **Service Controller:** Manages the creation, updating, and deletion of cloud provider load balancers

Where do you run cloud-controller-manager?

The cloud-controller-manager component can run as a container or a systemd service similar to kube-apiserver on the control plane nodes.

In the next sections, we will discuss the component parts of the compute nodes (also known as worker nodes) in the Kubernetes cluster.

The compute node components

We will dedicate this part of the chapter to explaining the anatomy of a compute node by explaining the three components running on it:

- Container engine and container runtime

- kubelet
- The kube-proxy component

> kubelet, kube-proxy, and container runtime are essential components for both control plane (master) nodes and worker nodes. We'll cover them in this section to highlight their functionalities in both contexts.

Container engine and container runtime

A **container engine** is a software platform designed to oversee the creation, execution, and lifecycle of containers. It offers a more abstract layer compared to a **container runtime**, streamlining container management and enhancing accessibility for developers. Well-known container engines are Podman, Docker Engine, and CRI-O. In contrast, **container runtime** is a foundational software component responsible for the creation, execution, and administration of containers in the backend when instructed by a container engine or container orchestrator. It furnishes essential functionality for container operation, encompassing tasks such as image loading, container creation, resource allocation, and container lifecycle management. Containerd, runc, dockerd, and Mirantis Container Runtime are some of the well-known container runtimes.

> The terms "container engine" and "container runtime" can sometimes be used interchangeably, leading to confusion. Container runtime (low-level) is the core engine responsible for executing container images, managing their lifecycles (start, stop, pause), and interacting with the underlying operating system. Examples include runc and CRI-O (when used as a runtime). Container engine (high-level) builds upon the container runtime, offering additional features like image building, registries, and management tools. Think Docker, Podman, or CRI-O (when used with Kubernetes). Remember, the key is understanding the core functionalities: low-level runtimes handle container execution, while high-level engines add a layer of management and user-friendliness.

Docker was the default option for running containers in the backend of Kubernetes in earlier days. But Kubernetes is not limited to Docker now; it can utilize several other container runtimes such as containerd, CRI-O (with runc), Mirantis Container Runtime, etc. However, in this book, we will be using Kubernetes with containerd or CRI-O for several reasons, including the following:

- **Focus and Flexibility:** containerd and CRI-O specialize in container runtime functionality, making them more lightweight and potentially more secure compared to Docker's broader feature set. This focus also allows for seamless integration with container orchestration platforms like Kubernetes. Unlike Docker, you don't require additional components like cri-dockerd for Kubernetes compatibility.
- **Alignment with Kubernetes:** Kubernetes is actively moving away from Docker as the default runtime. Previously (pre-v1.24), Docker relied on a component called dockershim for integration with Kubernetes.

However, this approach has been deprecated, and Kubernetes now encourages the use of runtimes adhering to the **Container Runtime Interface (CRI)** standard specifically designed for the platform. By choosing `containerd` or CRI-O, you ensure a more native and efficient integration with your Kubernetes environment.

- **Kubernetes-Centric Design:** CRI-O, in particular, is designed as a lightweight container runtime specifically for Kubernetes. It closely follows Kubernetes release cycles with respect to its minor versions (e.g., 1.x.y), simplifying version management. When a Kubernetes release reaches its end of life, the corresponding CRI-O version can likely be considered deprecated as well, streamlining the decision-making process for maintaining a secure and up-to-date Kubernetes environment.

Container Runtime Interface

Kubernetes employs a container runtime to execute containers within Pods. By default, Kubernetes utilizes the CRI to establish communication with the selected container runtime. The CRI was first introduced in Kubernetes version 1.5, released in December 2016.

The CRI serves as a plugin interface, empowering the kubelet to seamlessly integrate with a diverse range of container runtimes. This flexibility enables the selection of an optimal container runtime tailored to specific environmental requirements, such as `containerd`, Docker Engine, or CRI-O.

Within the CRI, a set of defined APIs allows the kubelet to engage with the container runtime efficiently. These APIs cover essential operations like creating, starting, stopping, and deleting containers, along with managing pod sandboxes and networking.

The following table shows the known endpoints for Linux machines.

Runtime	Path to Unix domain socket
`containerd`	`unix:///var/run/containerd/containerd.sock`
CRI-O	`unix:///var/run/crio/crio.sock`
Docker Engine (using `cri-dockerd`)	`unix:///var/run/cri-dockerd.sock`

Table 2.1: Known container runtime endpoints for Linux machines

Refer to the documentation (`https://kubernetes.io/docs/setup/production-environment/tools/kubeadm/install-kubeadm`) to learn more.

Kubernetes and Docker

In Kubernetes releases prior to v1.24, there was a direct integration with Docker Engine facilitated by a component called **dockershim**. However, this specific integration has been discontinued, and its removal was communicated with the v1.20 release. The deprecation of Docker as the underlying runtime is underway, and Kubernetes is now encouraging the use of runtimes aligned with the CRI designed for Kubernetes.

Despite these changes, Docker-produced images will persistently function in your cluster with any runtime, ensuring compatibility as it has been previously. Refer to `https://kubernetes.io/blog/2020/12/02/dont-panic-kubernetes-and-docker/` to learn more.

Therefore, any Linux machine running `containerd` can be used as a base on which to build a Kubernetes worker node. (We will discuss Windows compute nodes in the later chapters of this book.)

Open Container Initiative

The **Open Container Initiative** (OCI) is an open-source initiative that defines standards for container images, containers, container runtimes, and container registries. This effort aims to establish interoperability and compatibility across container systems, ensuring consistent container execution in diverse environments. Additionally, the CRI collaborates with OCI, providing a standardized interface for the `kubelet` to communicate with container runtimes. The OCI defines standards for container images and runtimes supported by the CRI, fostering efficient container management and deployment in Kubernetes.

Container RuntimeClass

Kubernetes **RuntimeClass** allows you to define and assign different container runtime configurations to Pods. This enables balancing performance and security for your applications. Imagine high-security workloads scheduled with a hardware virtualization runtime for stronger isolation, even if it means slightly slower performance. RuntimeClass also lets you use the same runtime with different settings for specific Pods. To leverage this, you'll need to configure the CRI on your nodes (installation varies) and create corresponding RuntimeClass resources within Kubernetes.

In the next section, we will learn about the `kubelet` agent, another important component of a Kubernetes cluster node.

kubelet

The `kubelet` is the most important component of the compute node since it is the one that will interact with the local container runtime installed on the compute node.

The `kubelet` functions solely as a system daemon and cannot operate within a container. Its execution is mandatory directly on the host system, often facilitated through `systemd`. This distinguishes the `kubelet` from other Kubernetes components, emphasizing its exclusive requirement to run on the host machine.

When the kubelet gets started, by default, it reads a configuration file located at `/etc/kubernetes/kubelet.conf`.

This configuration specifies two values that are really important for the `kubelet` to work:

- The endpoint of the `kube-apiserver` component
- The local container runtime Unix socket

Once the compute node has joined the cluster, the `kubelet` will act as a bridge between `kube-apiserver` and the local container runtime. The `kubelet` is constantly running HTTP requests against `kube-apiserver` to retrieve information about pods it has to launch.

By default, **every 20 seconds**, the kubelet runs a GET request against the kube-apiserver component to list the pods created on etcd that are destined to it.

Once it receives a pod specification in the body of an HTTP response from kube-apiserver, it can convert this into a container specification that will be executed against the specified UNIX socket. The result is the creation of your containers on your compute node using the local container runtime (e.g., containerd).

> Remember that, like any other Kubernetes components, kubelet does not read directly from etcd; rather it interacts with kube-apiserver, which exposes what is inside the etcd data layer. The kubelet is not even aware that an etcd server runs behind the kube-apiserver it polls.

The polling mechanisms, called **watch** mechanisms in Kubernetes terminology, are precisely to define how Kubernetes proceeds to run and delete containers against your worker nodes at scale. There are two things to pay attention to here:

- The kubelet and kube-apiserver must be able to communicate with each other through HTTP. That's why HTTPS port 6443 must be opened between the compute and control plane nodes.
- As they are running on the same machine, the kubelet, CRI, and container runtimes are interfaced through the usage of UNIX sockets.

Each worker node in the Kubernetes cluster needs its own kubelet, causing heightened HTTP polling against kube-apiserver with additional nodes. In larger clusters, particularly those with hundreds of machines, this increased activity can adversely affect kube-apiserver's performance and potentially lead to a situation that may impact API availability. Efficient scaling is essential to ensure the high availability of the kube-apiserver and other control plane components.

Also note that you can completely bypass Kubernetes and create containers on your worker nodes without having to use the kubelet, and the sole job of the kubelet is that its local container runtime reflects the configuration that is stored in etcd. So, if you create containers manually on a worker node, the kubelet won't be able to manage it. However, exposing the container runtime socket to containerized workloads is a security risk. It bypasses Kubernetes' security mechanisms and is a common target for attackers. A key security practice is to prevent containers from mounting this socket, safeguarding your Kubernetes cluster.

> Please note that the container engine running on the worker node has no clue that it is managed by Kubernetes through a local kubelet agent. A compute node is nothing more than a Linux machine running a container runtime with a kubelet agent installed next to it, executing container management instructions.

We will learn about the kube-proxy component in the next section.

The kube-proxy component

An important part of Kubernetes is networking. We will have the opportunity to dive into networking later; however, you need to understand that Kubernetes has tons of mechanics when it comes to exposing pods to the outside world or exposing pods to one another in the Kubernetes cluster.

These mechanics are implemented at the kube-proxy level; that is, each worker node requires an instance of a running kube-proxy so that the pods running on them are accessible. We will explore a Kubernetes feature called **Service**, which is implemented at the level of the kube-proxy component. Just like the kubelet, the kube-proxy component also communicates with the kube-apiserver component.

Several other sub-components or extensions operate at the compute node level, such as **cAdvisor** or **Container Network Interface** (**CNI**). However, they are advanced topics that we will discuss later.

Now we have learned about the different Kubernetes components and concepts, let us learn about the kubectl client utility and how it interacts with the Kubernetes API in the next section.

Exploring the kubectl command-line tool and YAML syntax

kubectl is the official command-line tool used to manage the Kubernetes platform. This is an HTTP client that is fully optimized to interact with Kubernetes and allows you to issue commands to your Kubernetes cluster.

> **Kubernetes and Linux-Based Learning Environment**
>
> For effective learning in Linux containers and related topics, it's best to use workstations or lab machines with a Linux OS. A good understanding of Linux basics is essential for working with containers and Kubernetes. Using a Linux OS on your workstation automatically places you in the Linux environment, making your learning experience better. You can choose the Linux distribution you prefer, like Fedora, Ubuntu, or another. We're committed to inclusivity and will offer alternative steps for Windows and macOS users when needed, ensuring a diverse and accessible learning experience for everyone. However, it is not mandatory to have a Linux OS-installed workstation to learn Kubernetes. If you are using a Windows machine, then you can use alternatives such as **Windows Subsystem for Linux** (**WSL**) (https://learn.microsoft.com/en-us/windows/wsl/).

Installing the kubectl command-line tool

The kubectl command-line tool can be installed on your Linux, Windows, or macOS workstations. You need to ensure that your kubectl client version stays within one minor version of your Kubernetes cluster for optimal compatibility. This means a v1.30 kubectl can manage clusters at v1.29, v1.30, and v1.31. Sticking to the latest compatible version helps avoid potential issues.

Since you are going to need the kubectl utility in the coming chapter, you can install it right now, as explained in the following sections.

Kubernetes Legacy Package Repositories

As of January 2024, the legacy Linux package repositories – namely, apt.kubernetes.io and yum. kubernetes.io (also known as packages.cloud.google.com) – have been frozen since September 13, 2023, and are no longer available. Users are advised to migrate to the new community-owned package repositories for Debian and RPM packages at pkgs.k8s.io, which were introduced on August 15, 2023. These repositories serve as replacements for the now-deprecated Google-hosted repositories (apt. kubernetes.io and yum.kubernetes.io). This change impacts users directly installing upstream versions of Kubernetes and those who have installed kubectl using the legacy package repositories. For further details, refer to the official announcement: Legacy Package Repository Deprecation (https:// kubernetes.io/blog/2023/08/31/legacy-package-repository-deprecation/).

Installing kubectl on Linux

To install kubectl on Linux, you need to download the kubectl utility and copy it to an executable path as follows:

```
$ curl -LO "https://dl.k8s.io/release/$(curl -L -s https://dl.k8s.io/release/
stable.txt)/bin/linux/amd64/kubectl"
$ chmod +x ./kubectl
$ sudo mv ./kubectl /usr/local/bin/kubectl
$ kubectl version
Client Version: v1.30.0
Kustomize Version: v5.0.4-0.20230601165947-6ce0bf390ce3
The connection to the server localhost:8080 was refused - did you specify the
right host or port?
```

Ignore the connection error here as you do not have a Kubernetes cluster configured to access using kubectl.

> The path /usr/local/bin/kubectl could be different in your case. You need to ensure appropriate **PATH** variables are configured to ensure the kubectl utility is under a detectable path. You can also use /etc/profile to configure the kubectl utility path.

To download a particular version, substitute the $(curl -L -s https://dl.k8s.io/release/stable. txt) section of the command with the desired version.

For instance, if you wish to download version 1.28.4 on Linux x86-64, enter:

```
$ curl -LO https://dl.k8s.io/release/v1.30.0/bin/linux/amd64/kubectl
```

This command will download the specific version of the kubectl utility, and you can copy it to your preferred path.

Let us learn how to install the kubectl utility on macOS now.

Installing kubectl on macOS

Installation is pretty similar on macOS except for the different kubectl packages for Intel and Apple versions:

```
# Intel
$ curl -LO "https://dl.k8s.io/release/$(curl -L -s
 https://dl.k8s.io/release/stable.txt)/bin/darwin/amd64/kubectl"
# Apple Silicon
$ curl -LO "https://dl.k8s.io/release/$(curl -L -s https://dl.k8s.io/release/
stable.txt)/bin/darwin/arm64/kubectl"
$ chmod +x ./kubectl
$ sudo mv ./kubectl /usr/local/bin/kubectl
$ sudo chown root: /usr/local/bin/kubectl
```

Installing kubectl on Windows

Download the kubectl.exe (https://dl.k8s.io/release/v1.28.4/bin/windows/amd64/kubectl.exe) using the browser or using curl (if you have curl or an equivalent command tool installed on Windows):

```
$ curl.exe -LO https://dl.k8s.io/release/v1.28.4/bin/windows/amd64/kubectl.exe
```

Finally, append or prepend the kubectl binary folder to your **PATH** environment variable and test to ensure the version of kubectl matches the downloaded one.

> You can also install kubectl using the native package manager, such as apt-get, yum, Zypper, brew (macOS), or Chocolatey (Windows). Refer to the documentation (https://kubernetes.io/docs/tasks/tools/install-kubectl-linux) to learn more.

We will learn about the usage of the kubectl command in the next section.

The role of kubectl

Since kube-apiserver is nothing more than an HTTP API, any HTTP client will work to interact with a Kubernetes cluster. You can even use curl to manage your Kubernetes cluster, but of course, there is a better way to do that.

So, why would you want to use such a client and not go directly with curl calls? Well, the reason is simplicity. Indeed, kube-apiserver manages a lot of different resources and each of them has its own URL path.

Calling kube-apiserver constantly through curl would be possible but extremely time-consuming. This is because remembering the path of each resource and how to call it is not user-friendly. Essentially, curl is not the way to go since kubectl also manages different aspects related to authentication against the Kubernetes authentication layer, managing cluster contexts, and more.

You would have to constantly go to the documentation to remember the URL path, HTTP header, or query string. kubectl will do that for you by letting you call kube-apiserver through commands that are easy to remember, secure, and entirely dedicated to Kubernetes management.

When you call kubectl, it reads the parameters you pass to it and, based on them, will create and issue HTTP requests to the kube-apiserver component of your Kubernetes cluster:

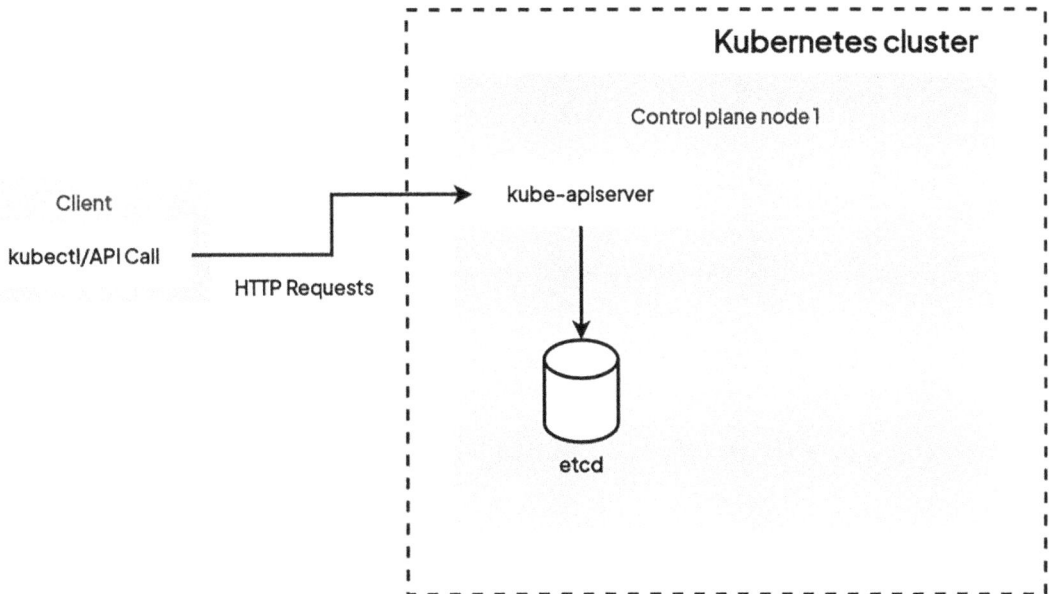

Figure 2.7: The kubectl command line will call kube-apiserver with the HTTP protocol; you'll interact with your Kubernetes cluster through kubectl all of the time

Once the kube-apiserver component receives a valid HTTP request coming from you, it will read or update the state of the cluster in etcd based on the request you submitted. If it's a write operation – for example, to update the image of a running container – kube-apiserver will update the state of the cluster in etcd. Then, the components running on the worker node where said container is being hosted will issue the proper container management commands in which to launch a new container based on the new image. This is so that the actual state of the container reflects what's in etcd.

Given that you won't have to interact with the container engine by yourself, or with etcd, we can say that the mastery of Kubernetes is largely based on your knowledge of the kubectl commands. To be effective with Kubernetes, you must master the Kubernetes API and details as much as possible. You won't have to interact with any other components than kube-apiserver and the kubectl command-line tool that allows you to call it.

As the kube-apiserver component is reachable via the HTTP(S) protocol, you can engage with any Kubernetes cluster using an HTTP-based library or programmatically with your preferred programming language. Numerous alternatives to kubectl are available, but kubectl, recognized as the official tool of the Kubernetes project, is consistently demonstrated in the documentation. The majority of examples you encounter will utilize kubectl.

How does kubectl work?

When you call the kubectl command, it will try to read a configuration file called kubeconfig from the default location $HOME/.kube/config. The kubeconfig file should contain the following information so that kubectl can use it and authenticate against kube-apiserver:

- The URL of the kube-apiserver endpoint and the port

- The user account

- Client certificates (if any) used to authenticate against kube-apiserver

- User-to-cluster mapping, known as context

> It is also possible to pass the details (such as cluster information, user authentication details, etc.) to the kubectl command as arguments but this is not a handy method when you have an environment with multiple clusters to manage.

A typical kubeconfig file and details are depicted in the following diagram.

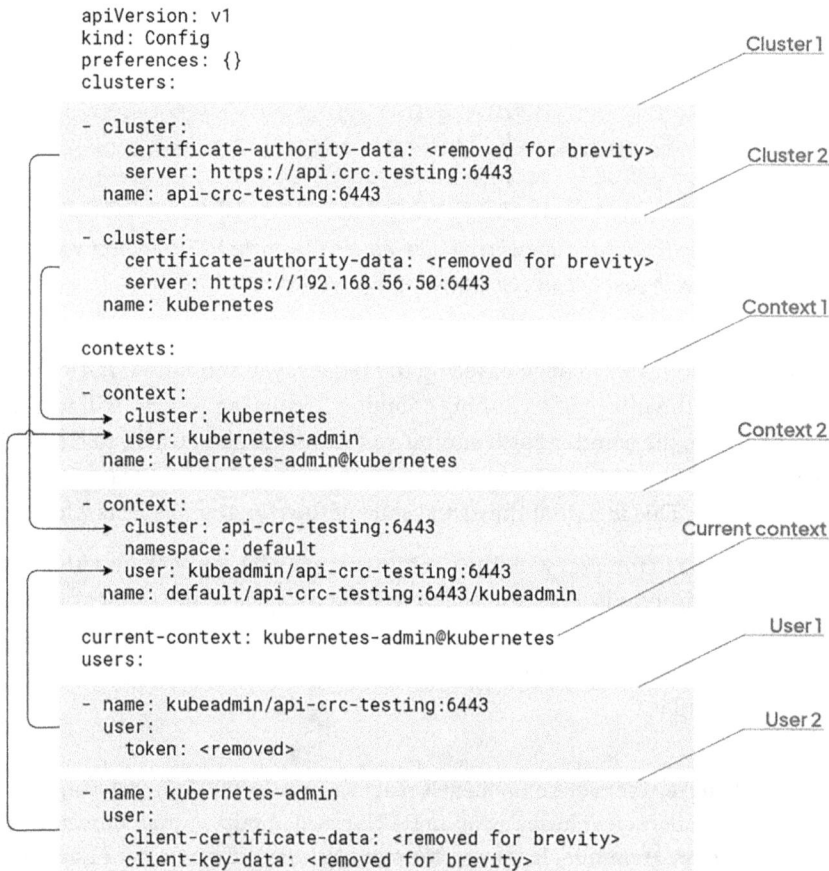

```
apiVersion: v1
kind: Config                                                    Cluster 1
preferences: {}
clusters:

- cluster:
    certificate-authority-data: <removed for brevity>          Cluster 2
    server: https://api.crc.testing:6443
  name: api-crc-testing:6443

- cluster:
    certificate-authority-data: <removed for brevity>
    server: https://192.168.56.50:6443
  name: kubernetes                                             Context 1

contexts:

- context:
    cluster: kubernetes
    user: kubernetes-admin                                     Context 2
  name: kubernetes-admin@kubernetes

- context:
    cluster: api-crc-testing:6443
    namespace: default                                         Current context
    user: kubeadmin/api-crc-testing:6443
  name: default/api-crc-testing:6443/kubeadmin

current-context: kubernetes-admin@kubernetes                   User 1
users:

- name: kubeadmin/api-crc-testing:6443
  user:                                                        User 2
    token: <removed>

- name: kubernetes-admin
  user:
    client-certificate-data: <removed for brevity>
    client-key-data: <removed for brevity>
```

Figure 2.8: kubeconfig context and structure

In the preceding diagram, there are multiple clusters, users, and contexts configured:

- **Clusters:** This section defines the Kubernetes clusters you can interact with. It contains information like the server address, API version, and certificate authority details for each cluster, allowing kubectl to connect and send commands to them.

- **Users:** This section stores your credentials for accessing the Kubernetes clusters. It typically includes a username and a secret (like a token or client certificate) used for authentication with the API server. kubeconfig files can also reference certificates in the user section to securely authenticate users with the Kubernetes API server. This two-way verification ensures that only authorized users with valid certificates can access the cluster, preventing unauthorized access and potential security breaches.

- **Contexts:** This section acts as a bridge between clusters and users. Each context references a specific cluster and a specific user within that cluster. By choosing a context, you define which cluster and user credentials kubectl will use for subsequent commands.

With multiple clusters, users, and contexts configured inside the kubeconfig file, it is easy to switch to different Kubernetes clusters with different user credentials.

The path of kubeconfig can be overridden on your system by setting an environment variable, called KUBECONFIG, or by using the --kubeconfig parameter when calling kubectl:

```
$ export KUBECONFIG="/custom/path/.kube/config"
$ kubectl --kubeconfig="/custom/path/.kube/config"
```

Each time you run a kubectl command, the kubectl command-line tool will look for a kubeconfig file in which to load its configuration in the following order:

1. First, it checks whether the --kubeconfig parameter has been passed and loads the config file.
2. At that point, if no kubeconfig file is found, kubectl looks for the KUBECONFIG environment variable.
3. Ultimately, it falls back to the default one in $HOME/.kube/config.

To view the config file currently used by your local kubectl installation, you can run this command:

```
$ kubectl config view
```

Then, the HTTP request is sent to kube-apiserver, which produces an HTTP response that kubectl will reformat in a human-readable format and output to your Terminal.

The following command is probably one that you'll type almost every day when working with Kubernetes:

```
$ kubectl get pods
```

This command lists the *Pods*. Essentially, it will issue a GET request to kube-apiserver to retrieve the list of containers (Pods) on your cluster. Internally, kubectl associates the Pods parameter passed to the command to the /api/v1/pods URL path, which is the path that kube-apiserver uses to expose the pod resource.

Here is another command:

```
$ kubectl run nginx --restart Never --image nginx
```

This one is slightly trickier because run is not an HTTP method. This command will issue a **POST** request against the kube-apiserver component, which will result in the creation of a container called nginx, based on the nginx image hosted on the container registry (e.g., Docker Hub or quay.io).

> In fact, this command won't create a container but a Pod. We will discuss the pod resource extensively in *Chapter 4, Running Your Containers in Kubernetes*. Let's try not to talk about containers anymore; instead, let's move on to pods and familiarize ourselves with Kubernetes concepts and wordings. From now on, if you come across the word *container*, it means a real container from a container perspective. Additionally, pods refer to the Kubernetes resource.

We will learn how to enable kubectl completion in the next section.

kubectl auto-completion

kubectl offers a built-in auto-completion feature for various shells, saving you precious time and frustration. kubectl supports autocompletion for popular shells like:

- Bash
- Zsh
- Fish
- PowerShell

To enable the autocompletion in Linux Bash, as an example:

```
# Install Bash Completion (if needed)
$ sudo yum install bash-completion  # For RPM-based systems
$ sudo apt install bash-completion  # For Debian/Ubuntu-based systems


# Add the source line to your shell configuration
$ echo 'source <(kubectl completion bash)' >>~/.bashrc
```

Now, when you start typing a kubectl command, magic happens! kubectl will suggest completions based on available resources and options. Simply press *Tab* to accept suggestions or keep typing to narrow down the options.

> The process for enabling autocompletion might differ slightly for other shells like Zsh or Fish. Refer to the official kubectl documentation for specific instructions: https:// kubernetes.io/docs/reference/kubectl/generated/kubectl_completion/

This setup ensures that autocompletion works every time you open a new Terminal session.

In the next section, we will start with the kubectl command and how to use imperative and declarative syntaxes.

The imperative syntax

Almost every instruction that you send to kube-apiserver through kubectl can be written using two types of syntax: **imperative** and **declarative**. The imperative syntax focuses on issuing commands that directly modify the state of the cluster based on the arguments and parameters you passed to the kubectl command.

Let us see some of the imperative style operations, as follows:

```
# Creates a pod, called my-pod, based on the busybox:latest container image:
$ kubectl run my-pod --restart Never --image busybox:latest
#  list all the ReplicaSet resources in the my-namespace namespace created on
the Kubernetes cluster:
$ kubectl get rs -n my-namespace
# Delete a pod, called my-pod, in the default namespace:
$ kubectl delete pods my-pod
```

The imperative syntax has multiple benefits. If you already understand what kind of instructions to send to Kubernetes and the proper command to achieve this, you are going to be incredibly fast. The imperative syntax is easy to type, and you can do a lot with just a few commands. Some operations are only accessible with the imperative syntax, too. For example, listing existing resources in the cluster is only possible with the imperative syntax.

However, the imperative syntax has a big problem. It is very complicated having to keep records of what you did previously in the cluster. If, for some reason, you were to lose the state of your cluster and need to recreate it from scratch, it's going to be incredibly hard to remember all of the imperative commands that you typed in earlier to bring your cluster back to the state you want. You could read your .bash_history file but, of course, there is a better way to do this, and we will learn about that declarative method in the next section.

The declarative syntax

"Declarative" is exactly what the name suggests. We "declare" the state we want the cluster to be, and then Kubernetes creates the required resources to achieve that state. Both JSON and YAML formats are supported; however, by convention, Kubernetes users prefer YAML syntax because of its simplicity.

YAML ("YAML Ain't Markup Language" or "Yet Another Markup Language") is a human-readable data serialization format widely used in Kubernetes. It allows you to define configuration for Kubernetes resources like Deployments, Services and Pods in a clear and concise way. This format makes it easy to manage and version control your Kubernetes configurations, promoting collaboration and repeatability. Also note, YAML is not a programming language and there is no real logic behind it. It's simply a kind of key:value configuration syntax that is used by a lot of projects nowadays, and Kubernetes is one of them.

Each key:value pair represents the configuration data that you want to set to the Kubernetes resource you want to create.

The following is the imperative command that created the pod named my-pod using the busybox:latest container image we used earlier:

```
$ kubectl run my-pod --restart Never --image busybox:latest
```

We will now do the same but with the declarative syntax instead:

```yaml
apiVersion: v1
kind: Pod
metadata:
  name: my-pod
spec:
  containers:
  - name: busybox-container
    image: busybox:latest
```

Let's say this file is saved with the name pod.yaml. To create the actual pod, you'll need to run the following command:

```
$ kubectl create -f pod.yaml
```

This result will be the equivalent of the previous command.

Each YAML file that is created for Kubernetes must contain four mandatory keys:

- apiVersion: This field tells you in which API version the resource is declared. Each resource type has an apiVersion key that must be set in this field. The pod resource type is in API version v1.
- kind: This field indicates the resource type the YAML file will create. Here, it is a **pod** that is going to be created.
- metadata: This field tells Kubernetes about the name of the actual resource. Here, the pod is named my-pod. This field describes the Kubernetes resource, not the container one. This metadata is for Kubernetes, not for container engines like Docker Engine or Podman.
- spec: This field tells Kubernetes what the object is made of. In the preceding example, the pod is made of one container that will be named busybox-container based on the busybox:latest container image. These are the containers that are going to be created in the backend container runtime.

Another important aspect of the declarative syntax is that it enables you to declare multiple resources in the same file using three dashes as a separator between the resources. Here is a revised version of the YAML file, which will create two pods:

```yaml
apiVersion: v1
kind: Pod
```

```
metadata:
  name: my-pod
spec:
  containers:
  -  name: busybox-container
     image: busybox:latest
---
apiVersion: v1
kind: Pod
metadata:
  name: my-second-pod
spec:
  containers:
  - name: nginx-container
    image: nginx:latest
```

You should be able to read this file by yourself and understand it; it just creates two pods. The first one uses the busybox image, and the second one uses the nginx image.

Of course, you don't have to memorize all of the syntaxes and what value to set for each key. You can always refer to the official Kubernetes documentation for the sample declaration YAML files. If the documentation is not enough or does not explain particular details, you can use the kubectl explain command to understand the resource details, as follows:

```
$ kubectl explain pod.spec.containers.image
KIND:        Pod
VERSION:     v1

FIELD: image <string>
DESCRIPTION:
    Container image name. More info:
    https://kubernetes.io/docs/concepts/containers/images
```

You will get a very clear explanation and field information from the kubectl explain output.

The declarative syntax offers a lot of benefits, too. With it, you'll be slower because writing these YAML files is a lot more time-consuming than just issuing a command in an imperative way. However, it offers two major benefits:

- **Infrastructure as Code (IaC) management**: You'll be able to keep the configuration stored somewhere and use Git (source code management) to version your Kubernetes resources, just as you would do with IaC. If you were to lose the state of your cluster, keeping the YAML files versioned in Git will enable you to recreate it cleanly and effectively.

- **Create multiple resources at the same time**: Since you can declare multiple resources in the same YAML file, you can have entire applications and all of their dependencies in the same place. Additionally, you get to create and recreate complex applications with just one command. Later, you'll discover a tool called Helm that can achieve templating on top of the Kubernetes YAML files.

There is no *better* way to use kubectl; these are just two ways to interact with it, and you need to master both. This is because some features are not available with the imperative syntax, while others are not available with the declarative syntax. Remember that, in the end, both call the kube-apiserver component by using the HTTP protocol.

kubectl should be installed on any machine that needs to interact with the cluster.

From a technical point of view, you must install and configure a kubectl command-line tool whenever and wherever you want to interact with a Kubernetes cluster.

Of course, it can be your local machine or a server from where you are accessing the Kubernetes cluster. However, in larger projects, it's also a good idea to install kubectl in the agent/runner of your continuous integration platform.

Indeed, you will probably want to automate maintenance or deployment tasks to run against your Kubernetes cluster, and you will probably use a **continuous integration** (CI) platform such as GitLab CI, Tekton, or Jenkins to do that.

If you want to be able to run Kubernetes commands in a CI pipeline, you will need to install kubectl on your CI agents and have a properly configured kubeconfig file written on the CI agent filesystem. This way, your CI/CD pipelines will be able to issue commands against your Kubernetes cluster and update the state of your cluster, too.

Just to add, kubectl should not be seen as a Kubernetes client for *human* users only. It should be viewed as a generic tool to communicate with Kubernetes: install it wherever you want to communicate with your cluster.

How to make Kubernetes highly available

As you've observed earlier, Kubernetes is a clustering solution. Its distributed nature allows it to run on multiple machines. By splitting the different components across different machines, you'll be able to make your Kubernetes cluster highly available. Next, we will have a brief discussion on the different Kubernetes setups.

The single-node cluster

Installing all Kubernetes components on the same machine is the worst possible idea if you want to deploy Kubernetes in production. However, it is perfectly fine for testing your development. The single-node way consists of grouping all of the different Kubernetes components on the same host or a virtual machine:

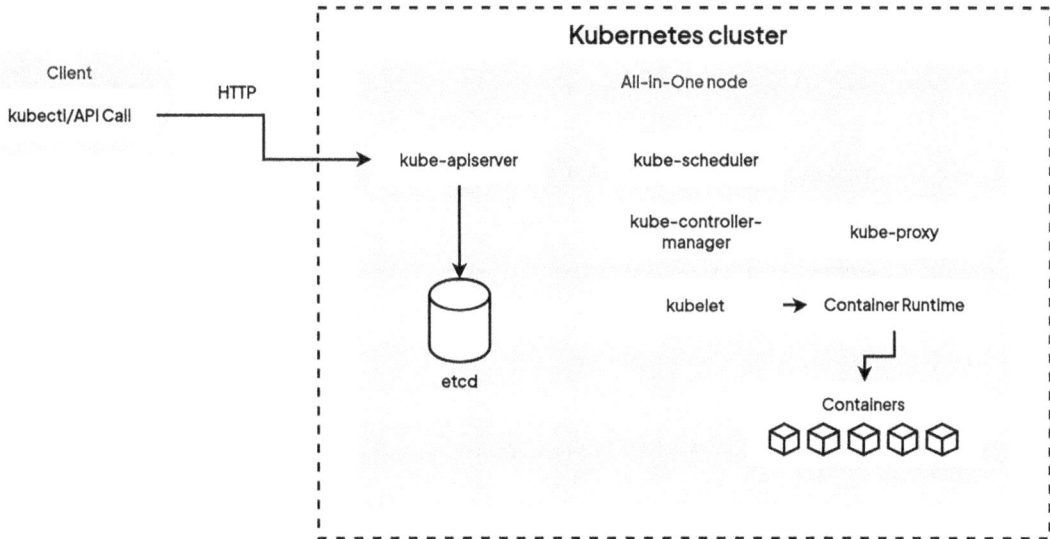

Figure 2.9: All of the components are working on the same machine

Typically, this arrangement is seen as a solid beginning for getting into Kubernetes through local testing. There's a tool called minikube that makes it easy to set up single-node Kubernetes on your computer. It runs a virtual machine with all the necessary components already configured. While minikube is handy for local tests and running minikube as a multi-node cluster is possible, keep in mind that minikube is definitely not recommended for production. The following table provides some of the pros and cons of using single-node Kubernetes clusters.

Pros	Cons
Good for testing	Impossible to scale
Easy to set up locally	Not highly available
Supported natively by minikube	Not recommended for production

Table 2.2: Pros and cons of single-node kubernetes clusters

Single-node Kubernetes is a well-suited option for resource-constrained edge environments. It offers a lightweight footprint while still enabling robust deployments with disaster recovery strategies in place.

The single-master cluster

This setup consists of having one node executing all of the control plane components with as many compute nodes as you want:

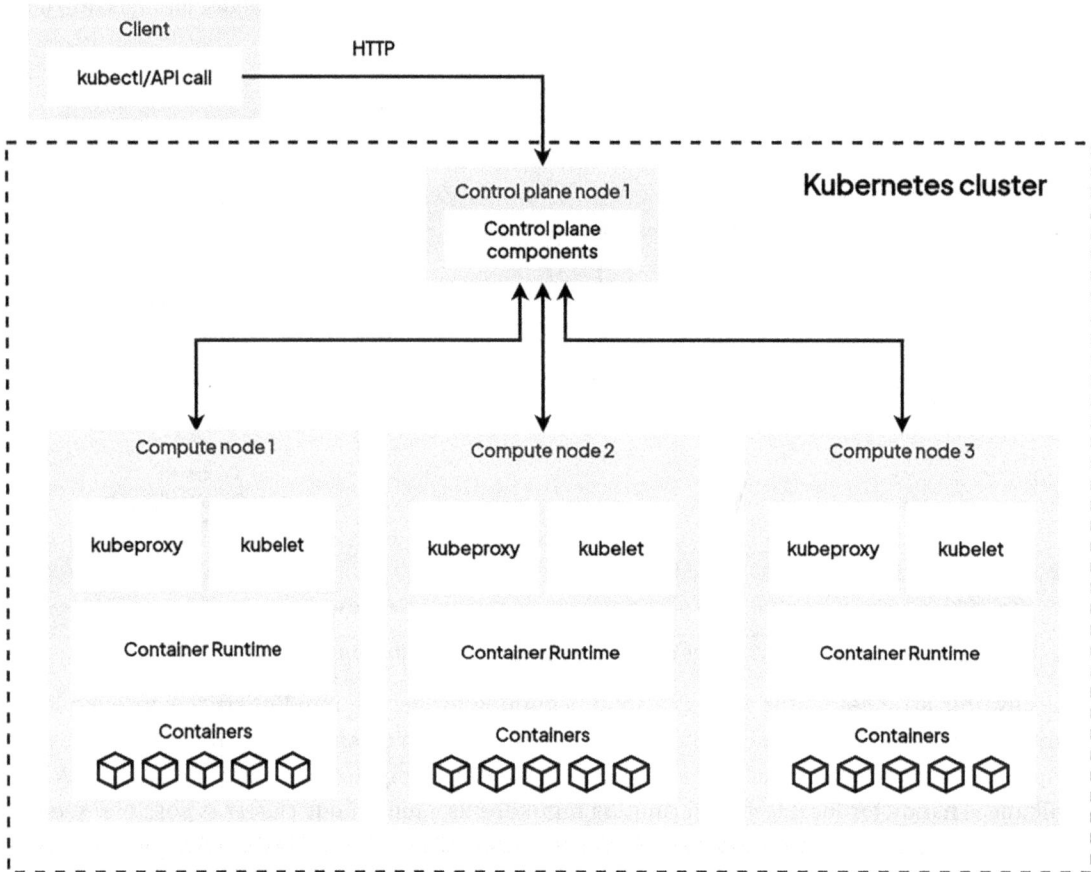

Figure 2.10: A single control plane node rules all of the compute nodes (here, it is three)

This setup is quite good compared to single-node clusters and the fact that there are multiple compute nodes will enable high availability for your containerized application. However, there is still room for improvement:

- There is a single point of failure since there is only one control plane node. If this single node fails, you won't be able to manage your running containers in a Kubernetes way anymore. Your containers will become orphans, and the only way to stop/update them would be to SSH on the worker node and run plain old container management commands (e.g., `ctr`, `crictl`, or Docker commands depending on the container runtime you are using).

- Also, there is a major problem here: by using a single `etcd` instance, there is a huge risk that you'll lose your dataset if the control plane node gets corrupted. If this happens, your cluster will be impossible to recover.

- Lastly, your cluster will encounter an issue if you start scaling your worker nodes. Each compute node brings its own kubelet agent, and periodically, the kubelet polls `kube-apiserver` every 20 seconds. If you start adding dozens of servers, you might impact the availability of your `kube-apiserver`, resulting in an outage of your control plane. Remember that your control plane must be able to scale and handle such traffic.

Pros	Cons
It has high-availability compute nodes	The control plane is a single point of failure
It supports multi-node features	A single `etcd` instance is running
It is possible to run it locally with projects such as kind or minikube but it is not perfect	It cannot scale effectively

Table 2.3: Pros and cons of a single-controller multi-compute Kubernetes cluster

Overall, this setup will always be better than single-node Kubernetes; however, it's still not highly available.

The multi-master multi-node cluster

This is the best way to achieve a highly available Kubernetes cluster. Both your running containers and your control plane are replicated to avoid a single point of failure.

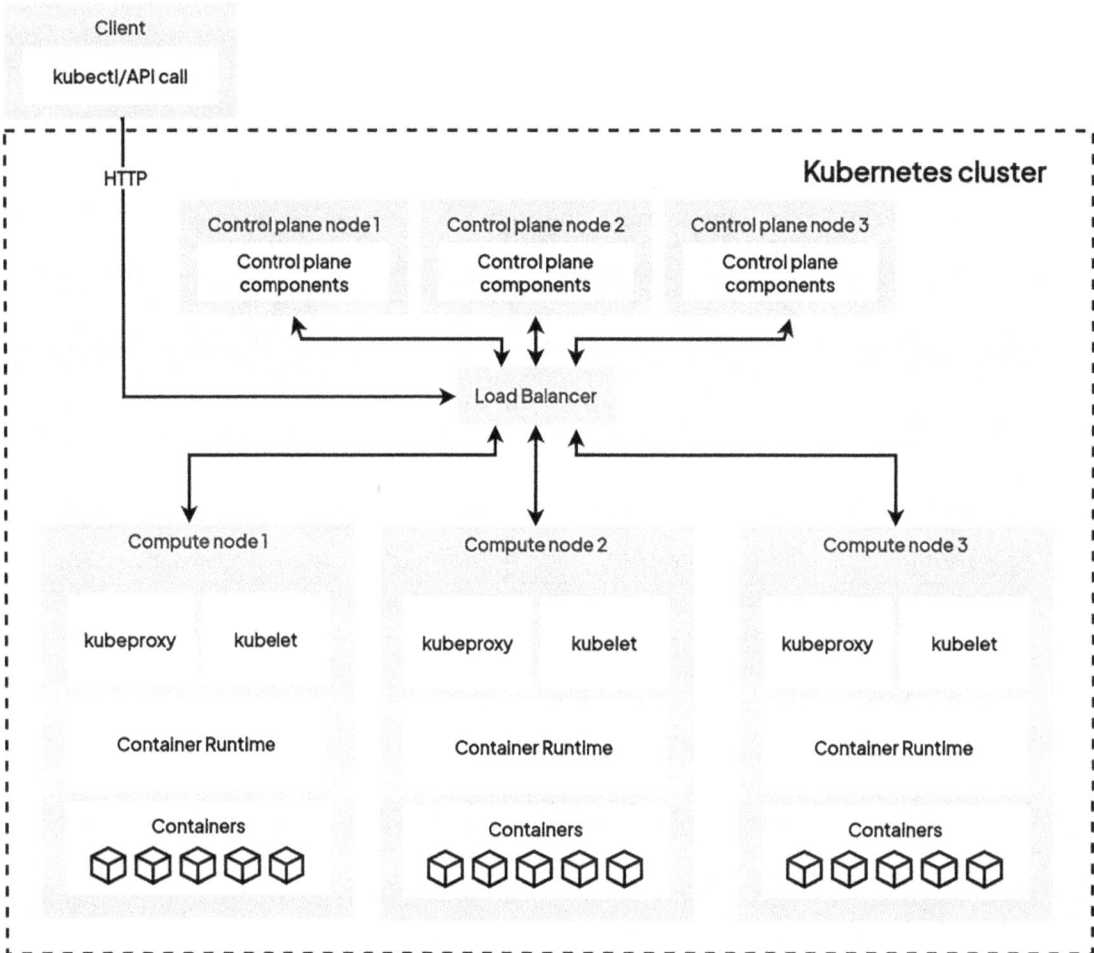

Figure 2.11: Multi-control plane node Kubernetes cluster

By using such a cluster, you are eliminating many of the risks we learned in the earlier cluster architectures because you are running multiple instances of your compute nodes and your control plane nodes. You will need a load balancer on top of your kube-apiserver instances in order to spread the load evenly between all of them, which will require a little bit more planning. Cloud providers such as Amazon EKS or Google GKE are provisioning Kubernetes clusters that are multi-controller and multi-compute clusters. If you wish to take it a step further, you can also split all of the different control plane components across a dedicated host. It's better but not mandatory, though. The cluster described in the preceding diagram is perfectly fine.

Managing etcd in Kubernetes with multiple control plane nodes

In a multi-control plane cluster, each control plane node runs an `etcd` instance. This ensures that the cluster has a high availability `etcd` store, even if some of the control plane nodes are unavailable. The `etcd` instances in a multi-control plane Kubernetes cluster will form an `etcd` cluster internally. This means that the `etcd` instances will communicate with each other to replicate the cluster state and ensure that all of the instances have the same data. The `etcd` cluster will use a consensus algorithm, known as Raft, to ensure that there is a single leader at all times. The leader is responsible for accepting writes to the cluster state and replicating the changes to the other instances. If the leader becomes unavailable, the other instances will elect a new leader.

We will learn about `etcd` member management and `etcd` backup/restore mechanisms in the later chapters of this book.

Before we end this chapter, we would like to sum up all the Kubernetes components. The following table will help you to memorize all of their responsibilities:

Component name	Communicates with	Role
`kube-apiserver`	`kubectl clients, etcd, kube-scheduler, kube-controller-manager, kubelet, kube-proxy`	The HTTP REST API. It reads and writes the state stored in `etcd`. The only component that is able to communicate with `etcd` directly.
`etcd`	`kube-apiserver`	This stores the state of the Kubernetes cluster.
kube-scheduler	`kube-apiserver`	This reads the API every 20 seconds to list unscheduled pods (an empty nodeName property), elects a worker node, and updates the nodeName property in the pod entry by calling `kube-apiserver`.
kube-controller-manager	`kube-apiserver`	This polls the API and runs the reconciliation loops.
kubelet	`kube-apiserver` and container runtime	This reads the API every 20 seconds to get pods scheduled to the node it's running on and translates the pod specs into running containers by calling the local container runtime operations.
kube-proxy	`kube-apiserver`	This implements the networking layer of Kubernetes.
Container engine	kubelet	This runs the containers by receiving instructions from the local kubelet.

Table 2.4: Kubernetes components and connectivity

These components are the default ones and are officially supported as part of the Kubernetes project. Remember that other Kubernetes distributions might bring additional components, or they might change the behavior of these.

These components are the strict minimum that you need to have a working Kubernetes cluster.

Summary

This was quite a big chapter, but at least you now have a list of all the Kubernetes components. Everything we will do later will be related to these components: they are the core of Kubernetes. This chapter was full of technical details too, but it was still relatively theoretical. Don't worry if things are still not very clear to you. You will gain a better understanding through practice.

The good news is that you are now completely ready to install your first Kubernetes cluster locally, and things are going to be a lot more practical from now on. That is the next step, and that's what we will do in the next chapter. After the next chapter, you'll have a Kubernetes cluster running locally on your workstation, and you will be ready to run your first pods using Kubernetes!

Further reading

To learn more about the topics that were covered in this chapter, take a look at the following resources:

- Kubernetes components: `https://kubernetes.io/docs/concepts/overview/components/`
- Cloud controller manager: `https://kubernetes.io/docs/concepts/architecture/cloud-controller/`
- Installing Kubernetes utilities (kubectl, kind, kubeadm, and minikube): `https://kubernetes.io/docs/tasks/tools/`
- Kubernetes legacy package repository changes on September 13, 2023: `https://kubernetes.io/blog/2023/08/31/legacy-package-repository-deprecation/`
- `pkgs.k8s.io`: Introducing Kubernetes Community-Owned Package Repositories: `https://kubernetes.io/blog/2023/08/15/pkgs-k8s-io-introduction/`
- Operating etcd clusters for Kubernetes: `https://kubernetes.io/docs/tasks/administer-cluster/configure-upgrade-etcd/`
- kubectl completion: `https://kubernetes.io/docs/reference/kubectl/generated/kubectl_completion/`
- Runtime class: `https://kubernetes.io/docs/concepts/containers/runtime-class/`

Join our community on Discord

Join our community's Discord space for discussions with the authors and other readers:

`https://packt.link/cloudanddevops`

3

Installing Your First Kubernetes Cluster

In the previous chapter, we had the opportunity to explain what Kubernetes is, its distributed architecture, the anatomy of a working cluster, and how it can manage your Docker containers on multiple Linux machines. Now, we are going to get our hands dirty because it's time to install Kubernetes. The main objective of this chapter is to get you a working Kubernetes installation for the coming chapters. This is so that you have your own cluster to work on, practice with, and learn about while reading this book.

Installing Kubernetes means that you have to get the different components to work together. Of course, we won't do that the hard way of setting up individual cluster components; instead, we will use automated tools. These tools have the benefit of launching and configuring all of the components for us locally. This automated Kubernetes cluster setup is particularly beneficial for DevOps teams rapidly testing changes to YAML, developers wanting a local environment to test applications, and security teams rapidly testing changes to Kubernetes object YAML definitions.

If you don't want to have a Kubernetes cluster on your local machine, we're also going to set up minimalist yet full-featured production-ready Kubernetes clusters on **Google Kubernetes Engine (GKE)**, **Amazon Elastic Kubernetes Service (EKS)**, and **Azure Kubernetes Service (AKS)** in later chapters of this book. These are cloud-based and production-ready solutions. In this way, you can practice and learn on a real-world Kubernetes cluster hosted on the cloud.

Whether you want to go local or on the cloud, it is your choice. You'll have to choose the one that suits you best by considering each solution's benefits and drawbacks. In both cases, however, you'll require a working `kubectl` installed on your local workstation to communicate with the resulting Kubernetes cluster. Installation instructions for `kubectl` are available in the previous chapter, *Chapter 2*, *Kubernetes Architecture – from Container Images to Running Pods*.

In this chapter, we're going to cover the following main topics:

- Installing a Kubernetes cluster with `minikube`
- Multi-node Kubernetes cluster with `kind`
- Alternative Kubernetes learning environments

- Production-grade Kubernetes clusters

Technical requirements

To follow along with the examples in this chapter, you will require the following:

- `kubectl` installed on your local machine
- A workstation with 2 CPUs or more, 2 GB of free memory, and 20 GB of free disk space. (You will need more resources if you want to explore the multi-node cluster environments.)
- A container or virtual machine manager installed on the workstation, such as Docker, QEMU, Hyperkit, Hyper-V, KVM, Parallels, Podman, VirtualBox, or VMware Fusion/Workstation
- Reliable internet access

You can download the latest code samples for this chapter from the official GitHub repository at `https://github.com/PacktPublishing/The-Kubernetes-Bible-Second-Edition/tree/main/Chapter03`

Installing a Kubernetes cluster with minikube

In this section, we are going to learn how to install a local Kubernetes cluster using `minikube`. It's probably the easiest way to get a working Kubernetes installation locally. By the end of this section, you're going to have a working single-node Kubernetes cluster installed on your local machine.

`minikube` is easy to use and is completely free. It's going to install all of the Kubernetes components on your local machine and configure all of them. Uninstalling all of the components through `minikube` is easy too, so you won't be stuck with it if, one day, you want to destroy your local cluster.

`minikube` has one big advantage compared to full-fledged production cluster deployment methods: it's a super useful tool for testing the Kubernetes scenarios quickly. If you do not wish to use `minikube`, you can completely skip this section and choose other methods described in this chapter.

While `minikube` is a popular choice for local Kubernetes development, it comes with some trade-offs in resource usage and feature fidelity compared to a full-blown production cluster:

- **Resource strain:** Running `minikube` alongside other processes on your local machine can be resource-intensive. It requires a good amount of CPU and RAM when you want to create larger Kubernetes clusters, potentially impacting the performance of other applications.
- **Networking discrepancies:** Unlike a production Kubernetes cluster, `minikube`'s default network setup may not fully mimic real-world networking environments. This can introduce challenges when replicating or troubleshooting network-related issues that might occur in production.
- **Compatibility considerations:** Certain Kubernetes features or third-party tools might require a more complete Kubernetes setup than what `minikube` offers, leading to compatibility issues during development.
- **Persistent storage challenges:** Managing persistent storage for applications within `minikube` can be cumbersome due to limitations in its persistent volume support compared to a full Kubernetes cluster.

We will learn how to install `minikube` and deploy and develop a Kubernetes cluster in the next section.

Installing minikube

Here, we will see how the `minikube` tool can be installed on Linux, macOS, and Windows. Installing `minikube` using the binary or package manager method is a straightforward task, as explained in the following sections.

> You can install `minikube` using the native package manager such as `apt-get`, yum, Zypper, Homebrew (macOS), or Chocolatey (Windows). Refer to the documentation (`https://minikube.sigs.k8s.io/docs/start`) to learn more.

Installing minikube on Linux

On Linux, `minikube` can be installed using the Debian package, the RPM package, or the binary, as explained below:

```
$ curl -LO https://storage.googleapis.com/minikube/releases/latest/minikube-linux-amd64
$ sudo install minikube-linux-amd64 /usr/local/bin/minikube
# Verify minikube command and path
$ which minikube
/usr/local/bin/minikube
```

Please note, the path can be different in your workstation depending on the operating system. You need to ensure the path is included in the **PATH** environment variables so that `minikube` command will work without any issues.

Installing minikube on macOS

On macOS, `minikube` can be installed with the binary, as explained below:

```
$ curl -LO https://storage.googleapis.com/minikube/releases/latest/minikube-darwin-amd64
$ sudo install minikube-darwin-amd64 /usr/local/bin/minikube
# Verify minikube command and path
$ which minikube
```

It is also possible to install `minikube` on macOS using the package manager, Homebrew.

Installing minikube on Windows

Like macOS and Linux, it is possible to install `minikube` on Windows using multiple methods, as follows:

```
# Using Windows Package Manager (if installed)
$ winget install minikube
# Using Chocolatey
$ choco install minikube
# Via .exe download and setting the PATH
```

```
# 1. Download minikube: https://storage.googleapis.com/minikube/releases/
latest/minikube-installer.exe
# 2. Set PATH
```

Once you have configured `minikube`, then you can create different types of Kubernetes clusters using `minikube`, as explained in the next sections.

minikube configurations

The `minikube` utility comes with minimal but effective customizations required for a development environment.

For example, the default specification of the Kubernetes cluster created by `minikube` will be 2 CPUs and 2 GB memory. It is possible to adjust this value using the following command if you need a bigger Kubernetes cluster node:

```
$ minikube config set cpus 4
  These changes will take effect upon a minikube delete and then a minikube
start
$ minikube config set memory 16000
  These changes will take effect upon a minikube delete and then a minikube
start
$ minikube config set container-runtime containerd
  These changes will take effect upon a minikube delete and then a minikube
start
```

As you see on the screen, you need to delete and recreate the `minikube` cluster to apply the settings.

Drivers for minikube

`minikube` acts as a simple and lightweight way to run a local Kubernetes cluster on your development machine. To achieve this, it leverages **drivers** – the workhorses behind managing the cluster's lifecycle. These drivers interact with different virtualization and containerization technologies, allowing `minikube` to create, configure, and control the underlying infrastructure for your local Kubernetes environment. `minikube`'s driver flexibility empowers you to deploy your cluster as a virtual machine, a container, or even directly on the bare metal of your development machine, tailoring the setup to your specific needs and preferences:

- **Container drivers:** For a containerized approach, `minikube` can leverage a local Podman or Docker installation. This allows you to run `minikube` directly within a container on your development machine, potentially offering a more lightweight and resource-efficient setup.

- **Virtual machine (VM) drivers:** If you prefer a VM approach, `minikube` can launch VMs on your machine. These VMs will then house and wrap the necessary Kubernetes components, providing a more isolated environment for your local cluster.

The choice between container and VM drivers depends on your specific needs and preferences, as well as your development environment's capabilities.

Refer to the `minikube` driver documentation (`https://minikube.sigs.k8s.io/docs/drivers/`) to learn about available and supported `minikube` drivers and supported operating systems.

It is also possible to set the default driver for `minikube` using the following command:

```
$  minikube config set driver docker
   These changes will take effect upon a minikube delete and then a minikube
start
# or set the VirtualBox as driver
$  minikube config set driver virtualbox
   These changes will take effect upon a minikube delete and then a minikube
start$  minikube config view driver
- driver: docker
```

Also, the driver can be set while creating the `minikube` cluster as follows:

```
$ minikube start --driver=docker
```

Prerequisites depend on the individual `minikube` drivers and must be installed and prepared. These may include an installation of Docker, Podman, or VirtualBox with permissions granted on a specific operating system. Installation and configuration instructions can be found in the `minikube` driver-specific documentation (`https://minikube.sigs.k8s.io/docs/drivers`).

Let us learn how to launch our first Kubernetes cluster using `minikube` in the next section.

Launching a single-node Kubernetes cluster using minikube

The main purpose of `minikube` is to launch the Kubernetes components on your local system and have them communicate with each other. In the following sections, we will learn how to deploy `minikube` clusters using the VirtualBox driver and Docker.

Setting up minikube using VMs

The VM method requires you to install a hypervisor on top of your workstation as follows:

- Linux: KVM2 (preferred), VirtualBox, QEMU
- Windows: Hyper-V (preferred), VirtualBox, VMware Workstation, QEMU
- macOS: Hyperkit, VirtualBox, Parallels, VMware Fusion, QEMU

Then, `minikube` will wrap all of the Kubernetes components into a VM that will be launched.

In the following example, we are using Fedora 39 as our workstation and VirtualBox as our hypervisor software as it is available for Linux, macOS, and Windows.

Refer to `https://www.virtualbox.org/wiki/Downloads` to download and install VirtualBox for your workstation. You are free to use your own choice of virtualization software and always follow the documentation (`https://minikube.sigs.k8s.io/docs/drivers/`) to see the supported virtualization software.

Do not confuse the `minikube` version and the deployed Kubernetes version. For example, `minikube 1.32` uses Kubernetes 1.28 for stability and compatibility reasons. This allows for thorough testing, broader tool support, controlled rollouts, and longer-term support for older versions. Users still have the flexibility to run different versions of Kubernetes independently. This balance between stability and flexibility makes `minikube` a reliable and versatile platform for developers.

On your workstation where you have installed `minikube` and VirtualBox, execute the following command:

```
$  minikube start --driver=virtualbox --memory=8000m --cpus=2
```

If you are using a particular version of `minikube` but want to install a different version of Kubernetes, then you can mention the specific version, as follows:

```
$ minikube start --driver=virtualbox --memory=8000m --cpus=2 --kubernetes-
version=1.29.0
```

You will see that `minikube` is starting the Kubernetes deployment process including the VM image downloading, as follows:

```
😄   minikube v1.32.0 on Fedora 39
❗   Specified Kubernetes version 1.29.0 is newer than the newest supported
version: v1.28.3. Use `minikube config defaults kubernetes-version` for
details.
❗   Specified Kubernetes version 1.29.0 not found in Kubernetes version list
💸   Searching the internet for Kubernetes version...
✅   Kubernetes version 1.29.0 found in GitHub version list
👍   Using the virtualbox driver based on user configuration
👍   Starting control plane node minikube in cluster minikube
🔥   Creating virtualbox VM (CPUs=2, Memory=8000MB, Disk=20000MB) ...
🐳   Preparing Kubernetes v1.29.0 on Docker 24.0.7 ...
    ▪ Generating certificates and keys ...
    ▪ Booting up control plane ...
    ▪ Configuring RBAC rules ...
🔗   Configuring bridge CNI (Container Networking Interface) ...
    ▪ Using image gcr.io/k8s-minikube/storage-provisioner:v5
```

You may also see the below information based on your workstation's operating system and the virtualization software as a recommendation:

```
|    You have selected "virtualbox" driver, but there are better options!
|    For better performance and support consider using a different driver:
|         - kvm2
|         - qemu2
```

```
|
|     To turn off this warning run:
|            $ minikube config set WantVirtualBoxDriverWarning false
|     To learn more about on minikube drivers checkout https://minikube.sigs.
k8s.io/docs/drivers/    |
|     To see benchmarks checkout https://minikube.sigs.k8s.io/docs/benchmarks/
cpuusage/                |
```

Finally, you will see the following success message from `minikube`:

```
🔎  Verifying Kubernetes components...
🌟  Enabled addons: storage-provisioner, default-storageclass
🏄  Done! kubectl is now configured to use "minikube" cluster and "default"
namespace by default
```

Yes, you have deployed a fully working Kubernetes cluster within a minute and are ready to deploy your application.

Now verify the Kubernetes cluster status using the `minikube` command, as follows:

```
$  minikube status
minikube
type: Control Plane
host: Running
kubelet: Running
apiserver: Running
kubeconfig: Configured
```

You can also see the new `minikube` VM in your VirtualBox UI, as follows:

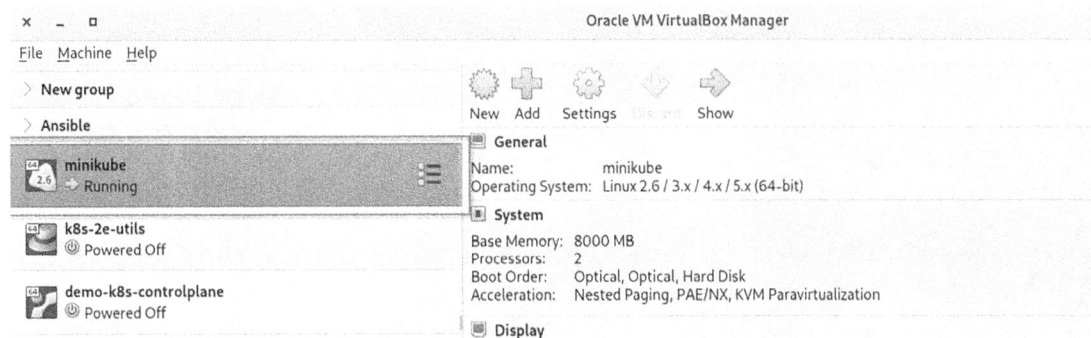

Figure 3.1: The minikube VM on the VirtualBox UI

In the next section, we will learn how to deploy Kubernetes clusters using `minikube` and containers.

Setting up minikube using a container

The container method is simpler. Instead of using a VM, `minikube` uses a local Docker Engine instance or Podman to launch the Kubernetes components inside a big container. To use the container-based `minikube`, make sure that you install Docker or Podman by following the instructions for your workstation operating system on which you are installing `minikube`; `minikube` will not install Podman or Docker for you. If the provided driver is missing or if the `minikube` cannot find the driver on the system, you may get an error, as shown below:

```
$ minikube start --driver=podman
😄  minikube v1.32.0 on Fedora 39 (hyperv/amd64)
✨  Using the podman driver based on user configuration
💣  Exiting due to PROVIDER_PODMAN_NOT_FOUND: The 'podman' provider was not
found: exec: "podman": executable file not found in $PATH
```

The Docker installation process is easy, but the steps can vary depending on your operating system, and you can take a look at the documentation (`https://docs.docker.com/engine/install/`) for more information. Similarly, the Podman installation steps are available at `https://podman.io/docs/installation` for different operating system flavors.

> If you are using a Windows workstation and Hyper-V-based VM for your hands-on lab, remember to disable Dynamic Memory for the VM in which you are installing `minikube` and the container engine.
>
> When running with the Podman driver, `minikube` performs a check of the available memory when it starts, and will report the "in-use" memory (set dynamically). So, you need to ensure enough memory is available or configure memory requirements for the Kubernetes node.

In the following example, we are using Fedora 39 as our workstation and Docker as the container engine:

```
$ minikube start --driver=docker --kubernetes-version=1.29.0
😄  minikube v1.32.0 on Fedora 39
❗  Specified Kubernetes version 1.29.0 is newer than the newest supported
version: v1.28.3. Use `minikube config defaults kubernetes-version` for
details.
❗  Specified Kubernetes version 1.29.0 not found in Kubernetes version list
🤔  Searching the internet for Kubernetes version...
✅  Kubernetes version 1.29.0 found in GitHub version list
✨  Using the docker driver based on user configuration
👍  Using Docker driver with root privileges
👍  Starting control plane node minikube in cluster minikube
🚜  Pulling base image ...
🔥  Creating docker container (CPUs=2, Memory=8000MB) ...
🐳  Preparing Kubernetes v1.29.0 on Docker 24.0.7 ...
```

```
    ▪ Generating certificates and keys ...
    ▪ Booting up control plane ...
    ▪ Configuring RBAC rules ...
  🔗 Configuring bridge CNI (Container Networking Interface) ...
    ▪ Using image gcr.io/k8s-minikube/storage-provisioner:v5
  🔍 Verifying Kubernetes components...
  🌟 Enabled addons: storage-provisioner, default-storageclass
  🏄 Done! kubectl is now configured to use "minikube" cluster and "default"
namespace by default
```

We can also use Podman as the container engine and create a Kubernetes cluster using `minikube` with the following command:

```
$ minikube start --driver=podman
```

Now we have the Kubernetes cluster created using `minikube`, in the next section let us learn how to access and manage the cluster using `kubectl`.

Accessing the Kubernetes cluster created by minikube

Now, we need to create a `kubeconfig` file for our local `kubectl` CLI to be able to communicate with this new Kubernetes installation. The good news is that `minikube` also generated one on the fly for us when we launched the `minikube start` command. The `kubeconfig` file generated by `minikube` is pointing to the local `kube-apiserver` endpoint, and your local `kubectl` was configured to call this cluster by default. So, essentially, there is nothing to do: the `kubeconfig` file is already formatted and in the proper location.

By default, this configuration is in `~/.kube/config`, and you should be able to see that a `minikube` context is now present:

```
$ cat ~/.kube/config
...<removed for brevity>..
- context:
    cluster: minikube
    extensions:
    - extension:
        last-update: Mon, 03 Jun 2024 13:06:44 +08
        provider: minikube.sigs.k8s.io
        version: v1.33.1
      name: context_info
    namespace: default
    user: minikube
  name: minikube
...<removed for brevity>..
```

Use the following command to display the current `kubeconfig` file. You should observe a cluster, named `minikube`, that points to a local IP address:

```
$ kubectl config view
```

Following this, run the following command, which will show the Kubernetes cluster that your `kubectl` is pointing to right now:

```
$ kubectl config current-context
minikube
```

Now, let's try to issue a real `kubectl` command to list the nodes that are part of our `minikube` cluster. If everything is okay, this command should reach the `kube-apiserver` component launched by `minikube`, which will return only one node since `minikube` is a single-node solution. Let's list the nodes with the following command:

```
$ kubectl get nodes
NAME       STATUS   ROLES          AGE      VERSION
minikube   Ready    control-plane  3m52s    v1.29.0
```

If you don't view any errors when running this command, it means that your `minikube` cluster is ready to be used and is fully working!

This is the very first real `kubectl` command you ran as part of this book. Here, a real `kube-apiserver` component received your API call and answered back with an HTTP response containing data coming from a real `etcd` data store. In our scenario, this is the list of the nodes in our cluster.

> Since `minikube` creates a single-node Kubernetes cluster by default, this command only outputs one node. This node is both a control plane node and a compute node at the same time. It's good for local testing, but do not deploy such a setup in production.

What we can do now is list the status of the control plane components so that you can start familiarizing yourself with `kubectl`:

```
$ kubectl get componentstatuses
Warning: v1 ComponentStatus is deprecated in v1.19+
NAME                 STATUS    MESSAGE   ERROR
controller-manager   Healthy   ok
scheduler            Healthy   ok
etcd-0               Healthy   ok
```

This command should output the status of the control plane components. You should see the following:

- A running `etcd` datastore
- A running `kube-scheduler` component
- A running `kube-controller-manager` component

In the next section, we will learn how to housekeep your Kubernetes learning environment by stopping and deleting the `minikube` Kubernetes cluster.

Stopping and deleting the local minikube cluster

You might want to stop or delete your local `minikube` installation. To proceed, do not kill the VM or container directly, but rather, use the `minikube` command-line utility. Here are the two commands to do so:

```
$ minikube stop
     Stopping node "minikube"  ...
     1 node stopped.
```

The preceding command will stop the cluster. However, it will continue to exist; its state will be kept, and you will be able to resume it later using the following `minikube start` command again. You can check it by calling the `minikube status` command again:

```
$ minikube status
minikube
type: Control Plane
host: Stopped
kubelet: Stopped
apiserver: Stopped
kubeconfig: Stopped
```

It is also possible to pause the cluster instead of stopping so that you can quickly re-start the Kubernetes cluster:

```
$  minikube pause
     Pausing node minikube ...
     Paused 14 containers in: kube-system, kubernetes-dashboard, storage-
gluster, istio-operator

$  minikube status
minikube
type: Control Plane
host: Running
kubelet: Stopped
apiserver: Paused
kubeconfig: Configured
```

And later, you can resume the cluster as follows:

```
$ minikube unpause
```

If you want to destroy the cluster, use the following command:

```
$ minikube delete
    Deleting "minikube" in docker ...
    Deleting container "minikube" ...
    Removing /home/gmadappa/.minikube/machines/minikube ...
    Removed all traces of the "minikube" cluster.
```

If you use this command, the cluster will be completely destroyed. Its state will be lost and impossible to recover.

Now that your minikube cluster is operational, it's up to you to decide whether you want to use it to follow the next chapters or pick another solution.

Multi-node Kubernetes cluster using minikube

It is also possible to create multi-node kubernetes clusters using minikube. In the following demonstration, we are creating a three-node Kubernetes cluster using minikube:

> You need to ensure that, your workstation has enough resources to create multiple Kubernetes nodes (either VMs or containers) when you create multi-node clusters. Also, note that minikube will spin up nodes with the same vCPU and memory you mentioned in settings or arguments.

```
$ minikube start --driver=podman --nodes=3
```

Once the cluster is provisioned, check the node details and find all the nodes as follows:

```
$ kubectl get nodes
NAME            STATUS    ROLES           AGE    VERSION
minikube        Ready     control-plane   93s    v1.29.0
minikube-m02    Ready     <none>          74s    v1.29.0
minikube-m03    Ready     <none>          54s    v1.29.0
```

minikube created a three-node cluster (--nodes=3) with the first node as the control plane node (or master node) and the remaining two nodes as compute nodes (you will need to assign appropriate labels later; we will learn about this in later chapters).

Multi-master Kubernetes cluster using minikube

There might be situations where you want to deploy and test Kubernetes clusters with a high availability control plane with multiple control plane nodes. You can implement the same using minikube using the following command:

```
$ minikube start \
    --driver=virtualbox \
    --nodes 5 \
```

```
    --ha true \
    --cni calico \
    --cpus=2 \
    --memory=2g \
    --kubernetes-version=v1.30.0 \
    --container-runtime=containerd
$ kubectl get nodes
NAME            STATUS    ROLES            AGE       VERSION
minikube        Ready     control-plane    6m28s     v1.30.0
minikube-m02    Ready     control-plane    4m36s     v1.30.0
minikube-m03    Ready     control-plane    2m45s     v1.30.0
minikube-m04    Ready     <none>           112s      v1.30.0
minikube-m05    Ready     <none>           62s       v1.30.0
```

minikube will create a five-node cluster (--nodes 5) and configure the first three nodes as control plane nodes (--ha true).

Again, remember to ensure you have enough resources on your workstation to create such multi-node clusters.

Multiple Kubernetes clusters using minikube

As we learned, minikube is meant for the development and testing of Kubernetes environments. There might be situations where you want to simulate the environment with multiple Kubernetes clusters. In that case, you can use minikube again as it is possible to create multiple Kubernetes clusters using minikube. But remember to give different names (--profile) for your different Kubernetes clusters, as explained below:

```
# Start a minikube cluster using VirtualBox as driver.
$  minikube start --driver=virtualbox --kubernetes-version=1.30.0 --profile
cluster-vbox
# Start another minikube cluster using Docker as driver
$ minikube start --driver=docker --kubernetes-version=1.30.0 --profile cluster-
docker
```

You can list the minikube clusters and find the details, as shown in the below image:

```
● ● ●
$  minikube profile list
|----------------|------------|-----------|----------------|------|---------|---------|-------|----------------|--------------------|
|    Profile     | VM Driver  |  Runtime  |       IP       | Port | Version | Status  | Nodes | Active Profile | Active Kubecontext |
|----------------|------------|-----------|----------------|------|---------|---------|-------|----------------|--------------------|
| cluster-docker | docker     | containerd | 192.168.49.2  | 8443 | v1.30.0 | Running |   1   |                |                    |
| cluster-vbox   | virtualbox | containerd | 192.168.59.145 | 8443 | v1.30.0 | Running |   1   |                |         *          |
|----------------|------------|-----------|----------------|------|---------|---------|-------|----------------|--------------------|
```

Figure 3.2: minikube profile list showing multiple Kubernetes clusters

```
# Stop cluster with profile name
$ minikube stop --profile cluster-docker

# Remove the cluster with profile name
$ minikube delete --profile cluster-docker
```

We have learned how to create different types and sizes of Kubernetes clusters; now let's examine another tool for setting up a local Kubernetes cluster, called kind.

Multi-node Kubernetes cluster with kind

In this section, we are going to discuss a tool called kind, which is also designed to run a Kubernetes cluster locally, just like minikube.

The whole idea behind kind is to use Docker or Podman containers as Kubernetes nodes thanks to the **Docker-in-Docker** (**DinD**) or **Containers-in-Container** model. By launching containers, which themselves contain the container engines and the kubelet, it is possible to make them behave as Kubernetes worker nodes.

The following diagram shows the high-level architecture of kind cluster components:

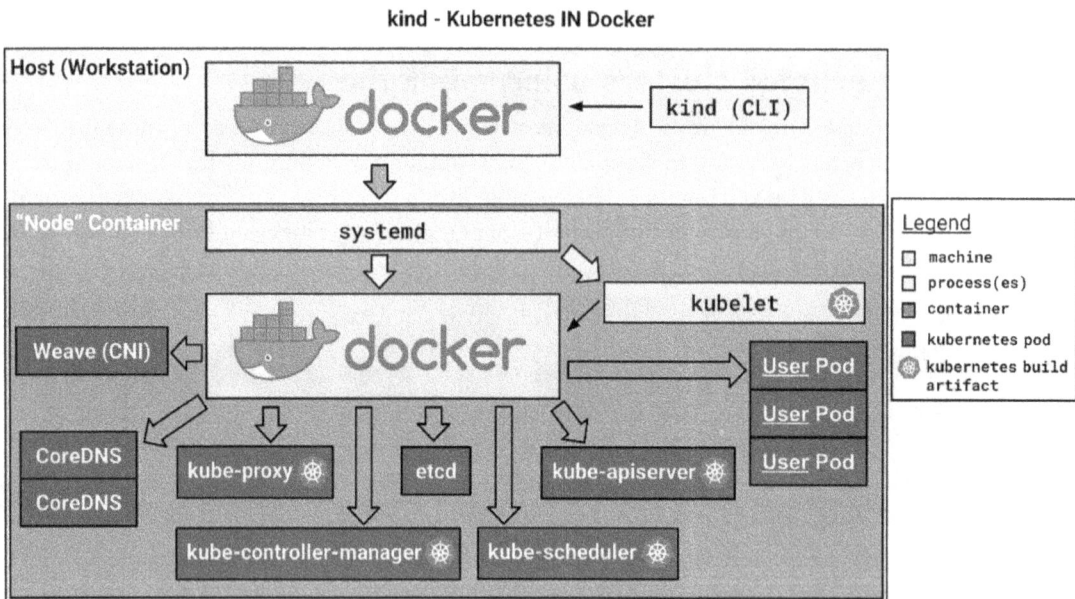

Figure 3.3: kind cluster components (image source: https://kind.sigs.k8s.io/docs/design/initial)

This is exactly the same as when you use the Docker driver for minikube, except that there, it will not be done in a single container but in several. The result is a local multi-node cluster. Similar to minikube, kind is a free open-source tool.

Similar to minikube, kind is a tool that is used for local development and testing. Please never use it in production because it is not designed for it.

Installing kind onto your local system

Since kind is a tool entirely built around Docker and Podman, you need to have either of these container engines installed and working on your local system.

Since the Docker and Podman installation instructions are available as documentation, we will skip those steps here (refer to the earlier section, *Setting up minikube using a container*, for the details).

Refer to the kind release page for the kind version information and availability (https://github.com/kubernetes-sigs/kind/releases).

Again, the process of installing kind will depend on your operating system:

- Use the following commands for Linux:

```
$ curl -Lo ./kind https://kind.sigs.k8s.io/dl/v0.23.0/kind-linux-amd64
$ chmod +x ./kind
$ mv ./kind /usr/local/bin/kind
# Check version
$ kind version
kind v0.23.0 go1.21.10 linux/amd64
```

- Use the following commands for macOS:

```
$ curl -Lo ./kind https://kind.sigs.k8s.io/dl/v0.20.0/kind-darwin-amd64
$ chmod +x ./kind
$ mv ./kind /usr/local/bin/kind
```

- You can also install it with Homebrew:

```
$ brew install kind
```

- Use the following commands for Windows PowerShell:

```
$ curl.exe -Lo kind-windows-amd64.exe https://kind.sigs.k8s.io/dl/v0.23.0/kind-windows-amd64
$ Move-Item .\kind-windows-amd64.exe c:\some-dir-in-your-PATH\kind.exe
```

- You can also install it with Chocolatey:

```
$ choco install kind
```

- If you have Go language installed, then you can use the following command:

```
$ go install sigs.k8s.io/kind@v0.22.0 && kind create cluster
```

Refer to the documentation (`https://kind.sigs.k8s.io/docs/user/quick-start#installation`) to learn other installation methods for your system.

Let us learn how to create a Kubernetes cluster using `kind` in the next section.

Creating a Kubernetes cluster with kind

Once `kind` has been installed on your system, you can immediately proceed to launch a new Kubernetes cluster using the following command:

```
$ kind create cluster --name test-kind
Creating cluster "kind" ...
```

When you run this command, `kind` will start to build a Kubernetes cluster locally by pulling a container image containing all the control plane components. The result will be a single-node Kubernetes cluster with a Docker container acting as a *control plane node*.

Podman can be used as the provider for the `kind` cluster if you prefer, as follows:

```
$ KIND_EXPERIMENTAL_PROVIDER=podman kind create cluster
```

We do not want this setup since we can already achieve it with `minikube`. What we want is a multi-node cluster with `kind` where we can customize the cluster and nodes. To do this, we must write a very small configuration file and tell `kind` to use it as a template to build the local Kubernetes cluster. So, let's get rid of the single-node `kind` cluster that we just built, and let's rebuild it as a multi-node cluster:

1. Run this command to delete the cluster:

   ```
   $ kind delete cluster
   Deleting cluster "kind" ...
   ```

2. Then, we need to create a `config` file that will serve as a template for `kind` to build our cluster. Simply copy the following content to a local file in this directory, for example, `~/.kube/kind_cluster`:

   ```
   kind: Cluster
   apiVersion: kind.x-k8s.io/v1alpha4
   nodes:
   - role: control-plane
   - role: worker
   - role: worker
   - role: worker
   ```

   Please note that this file is in YAML format. Pay attention to the `nodes` array, which is the most important part of the file. This is where you tell `kind` how many nodes you want in your cluster. The role key can take two values: control plane and worker.

Depending on which role you choose, a different node will be created.

3. Let's relaunch the `kind create` command with this `config` file to build our multi-node cluster. For the given file, the result will be a one-master, three-worker Kubernetes cluster:

```
$ kind create cluster --config ~/.kube/kind_cluster
```

It is also possible to build a specific version of Kubernetes by using the appropriate image details while creating the `kind` cluster, as follows:

```
$ kind create cluster \
    --name my-kind-cluster \
    --config ~/.kube/kind_cluster \
    --image kindest/node:v1.29.0@
sha256:eaa1450915475849a73a9227b8f201df25e55e268e5d619312131292e324d570
```

A new Kubernetes cluster will be deployed and configured by `kind` and you will receive the messages related to cluster access at the end, as follows:

```
Creating cluster "my-kind-cluster" ...
✓ Ensuring node image (kindest/node:v1.29.0) 🖼
✓ Preparing nodes 📦 📦 📦
✓ Writing configuration 📜
✓ Starting control-plane 🕹
✓ Installing CNI 🔌
✓ Installing StorageClass 💾
✓ Joining worker nodes 🚜
Set kubectl context to "kind-my-kind-cluster"
You can now use your cluster with:

kubectl cluster-info --context kind-my-kind-cluster

Thanks for using kind! 😊
```

Following this, you should have four new Docker containers: one running as a master node and the other three as worker nodes of the same Kubernetes cluster.

Now, as always with Kubernetes, we need to write a `kubeconfig` file for our `kubectl` utility to be able to interact with the new cluster. And guess what, `kind` has already generated the proper configuration and appended it to our `~/.kube/config` file, too. Additionally, `kind` set the current context to our new cluster, so there is essentially nothing left to do. We can immediately start querying our new cluster. Let's list the nodes using the `kubectl get nodes` command. If everything is okay, we should view four nodes:

```
$ kubectl get nodes
NAME                            STATUS   ROLES           AGE    VERSION
my-kind-cluster-control-plane   Ready    control-plane   4m     v1.29.0
```

```
my-kind-cluster-worker          Ready     <none>        3m43s   v1.29.0
my-kind-cluster-worker2         Ready     <none>        3m42s   v1.29.0
```

Everything seems to be perfect. Your kind cluster is working!

Just as we did with minikube, you can also check for the component's statuses using the following command:

```
$ kubectl cluster-info
Kubernetes control plane is running at https://127.0.0.1:42547
CoreDNS is running at https://127.0.0.1:42547/api/v1/namespaces/kube-system/
services/kube-dns:dns/proxy
```

To further debug and diagnose cluster problems, use 'kubectl cluster-info dump':

```
$  kubectl get --raw='/readyz?verbose'
[+]ping ok
[+]log ok
[+]etcd ok
...<removed for brevity>...
[+]poststarthook/apiservice-openapiv3-controller ok
[+]shutdown ok
readyz check passed
```

As part of the development and learning environment housekeeping, we need to learn how to stop and delete Kubernetes clusters created using kind. Let us learn how to do this in the next section.

Stopping and deleting the local kind cluster

You might want to stop or remove everything kind created on your local system to clean the place after your practice. To do so, you can use the following command:

```
$ kind stop
```

This command will stop the Docker containers that kind is managing. You will achieve the same result if you run the Docker stop command on your containers manually. Doing this will stop the containers but will keep the state of the cluster. That means your cluster won't be destroyed, and simply relaunching it using the following command will get the cluster back to its state before you stop it.

If you want to completely remove the cluster from your system, use the following command. Running this command will result in removing the cluster and its state from your system. You won't be able to recover the cluster:

```
$ kind delete cluster
Deleting cluster "kind" ...
```

Now that your kind cluster is operational, it's up to you to decide whether you want to use it to practice while reading the coming chapters. You can also decide whether to pick another solution described in the following sections of this chapter. kind is particularly nice because it's free to use and allows you to install a multi-node cluster. However, it's not designed for production and remains a development and testing solution for a non-production environment. kind makes use of Docker containers to create *Kubernetes nodes*, which, in the real world, are supposed to be Linux machines.

Let us learn about some of the alternative Kubernetes learning and testing environments in the next section.

Alternative Kubernetes learning environments

You can also utilize some of the available zero-configuration learning environments, designed to make your Kubernetes journey smooth and enjoyable.

Play with Kubernetes

This interactive playground (labs.play-with-k8s.com), brought to you by **Docker** and **Tutorius**, provides a simple and fun way to experiment with Kubernetes. Within seconds, you'll be running your own Kubernetes cluster directly in your web browser.

The environment comes with the following features:

- Free Alpine Linux VM: Experience a realistic VM environment without leaving your browser.
- DinD: This technology creates the illusion of multiple VMs, allowing you to explore distributed systems concepts.

Killercoda Kubernetes playground

Killercoda (https://killercoda.com/playgrounds/scenario/kubernetes) is a zero-configuration playground that offers a temporary Kubernetes environment accessible through your web browser. Stay on top of the latest trends with their commitment to providing the newest kubeadm Kubernetes version just a few weeks after release.

The environment comes with the following features:

- Ephemeral environment: Get started quickly with a preconfigured cluster that vanishes once you're done. This makes it perfect for quick experimentation without commitment.
- Empty kubeadm cluster with two nodes: Dive into the core functionalities of Kubernetes with a readily available two-node cluster.
- Control plane node with scheduling ability: Unlike some playgrounds, this one lets you schedule workloads on the control plane node, providing more flexibility for testing purposes.

We will explore some of the production-grade Kubernetes options in the next section.

Production-grade Kubernetes clusters

We have been talking about the Kubernetes environments for development and learning purposes so far. How do you build a production-grade Kubernetes environment that meets your specific needs? Next, we'll see some of the well-known options adopted by Kubernetes users.

In the following section, let us understand the managed Kubernetes services offered by the major **Cloud Service Providers (CSPs)**.

Managed Kubernetes clusters using cloud services

If you prefer to have your Kubernetes environment using managed services, then there are several options available, such as GKE, AKS, EKS, and so on:

- **Google Kubernetes Engine (GKE)**: Offered by **Google Cloud Platform (GCP)**, GKE is a fully managed Kubernetes service. It takes care of the entire cluster lifecycle, from provisioning and configuration to scaling and maintenance. GKE integrates seamlessly with other GCP services, making it a great choice for existing GCP users.

- **Azure Kubernetes Service (AKS)**: Part of Microsoft Azure, AKS is another managed Kubernetes offering. Similar to GKE, AKS handles all aspects of cluster management, allowing you to focus on deploying and managing your containerized applications. AKS integrates well with other Azure services, making it a natural fit for Azure users.

- **Amazon Elastic Kubernetes Service (EKS)**: Offered by **Amazon Web Services (AWS)**, EKS provides a managed Kubernetes service within the AWS ecosystem. Like GKE and AKS, EKS takes care of cluster management, freeing you to focus on your applications. EKS integrates with other AWS services, making it a strong option for AWS users.

These managed Kubernetes services provide a convenient and scalable way to deploy and manage your containerized applications without the complexities of self-managed Kubernetes clusters.

We have detailed chapters to learn how to deploy and manage such clusters as follows:

- *Chapter 15, Kubernetes Clusters on Google Kubernetes Engine*
- *Chapter 16, Launching a Kubernetes Cluster on Amazon Web Services with Amazon Elastic Kubernetes Service*
- *Chapter 17, Kubernetes Clusters on Microsoft Azure with Azure Kubernetes Service*

If you do not have a local Kubernetes setup, as we explained in the previous sections of this chapter, you can create one using the managed Kubernetes service on your choice of cloud platform by referring to the respective chapter. But it is a requirement to have a working Kubernetes cluster before you start reading the next part of this book, *Part 2*: *Diving into Kubernetes Core Concepts*.

We will learn about the Kubernetes distributions and platforms in the next section.

Kubernetes distributions

Kubernetes distributions are essentially pre-packaged versions of Kubernetes that include additional features and functionalities beyond the core Kubernetes offering. They act like value-added packages, catering to specific needs and simplifying deployments for users. For a more feature-rich experience, consider these Kubernetes distributions:

- **Red Hat OpenShift:** This enterprise-grade distribution extends Kubernetes with developer tools (image builds and CI/CD pipelines), multi-cluster management, security features (RBAC and SCC), and built-in scaling for complex deployments (`https://www.redhat.com/en/technologies/cloud-computing/openshift`).

- **Rancher:** A complete container management platform, Rancher goes beyond Kubernetes. It offers multi-cluster management across diverse environments, workload management for various orchestration platforms, and a marketplace for preconfigured applications (`https://www.rancher.com/`).

- **VMware Tanzu:** Designed for the VMware ecosystem, Tanzu integrates seamlessly for infrastructure provisioning, security, and hybrid cloud deployments. It provides lifecycle management tools for containerized applications within the VMware environment (`https://tanzu.vmware.com/platform`).

Please note, some of the above-listed Kubernetes distributions are subscription-based or license-based products and are not freely available.

Let us learn about some of the Kubernetes deployment tools in the next section.

Kubernetes installation tools

The following tools provide flexibility and control over the Kubernetes cluster setup. Of course, you need to add more automation using other third-party tools and platforms to manage your Kubernetes environment:

- **kubeadm:** This official Kubernetes tool provides a user-friendly way to set up Kubernetes clusters, making it suitable for both testing and production environments. Its simplicity allows for quick cluster deployment but may require additional configuration for production-grade features like high availability (`https://kubernetes.io/docs/reference/setup-tools/kubeadm/`).

- **kops:** For managing robust Kubernetes clusters in production, kops is an official Kubernetes project offering command-line control. It streamlines the creation, upgrading, and maintenance of highly available clusters, ensuring the reliable operation of your containerized applications (`https://kops.sigs.k8s.io/`).

- **Kubespray:** Looking to deploy Kubernetes on bare metal or VMs? Kubespray leverages the power of Ansible automation. It combines Ansible playbooks with Kubernetes resources, allowing for automated cluster deployment on your preferred infrastructure (`https://github.com/kubespray`).

- **Terraform:** This tool allows you to define and manage your Kubernetes cluster infrastructure across various cloud providers. The code-driven approach ensures consistency and repeatability when deploying clusters in different environments (`https://www.terraform.io/`).

- **Pulumi:** Similar to Terraform, Pulumi provides infrastructure-as-code capabilities. It allows you to define and manage your Kubernetes cluster infrastructure using programming languages like Python or Go. This approach offers greater flexibility and customization compared to purely declarative configuration languages (`https://www.pulumi.com/`).

If the Kubernetes landscape is very large with several Kubernetes clusters, then you need to consider hybrid-multi-cluster management solutions; let us learn about those in the next section.

Hybrid and multi-cloud solutions

Managing Kubernetes clusters across diverse environments requires powerful tools, and there are a few offering such multi-cluster management features:

- **Anthos** (Google): This hybrid and multi-cloud platform facilitates managing Kubernetes clusters across diverse environments. Anthos allows organizations to leverage a consistent approach for deploying and managing containerized applications on-premises, in the cloud, or at the edge (`https://cloud.google.com/anthos`).

- **Red Hat Advanced Cluster Management (RHACM) for Kubernetes:** Red Hat also offers a solution for managing Kubernetes clusters across hybrid and multi-cloud environments. Their Advanced Cluster Management platform provides a centralized control plane for consistent deployment, management, and governance of your containerized workloads (`https://www.redhat.com/en/technologies/management/advanced-cluster-management`).

- **VMware Tanzu Mission Control:** This centralized management tool simplifies the process of overseeing Kubernetes clusters across various environments. From a single console, you can provision, monitor, and manage clusters regardless of their location, be that on-premises, cloud, or hybrid (`https://docs.vmware.com/en/VMware-Tanzu-Mission-Control/index.html`).

How do you choose your Kubernetes platform and management solutions? Let's explore some of the key points in the next section.

Choosing the right environment

The best production-grade Kubernetes environment depends on several factors:

- **Level of control:** Do you need complete control over the cluster configuration, or are you comfortable with preconfigured managed services?

- **Existing infrastructure:** Consider your existing infrastructure (cloud provider, bare metal) when choosing a deployment method.

- **Scalability needs:** How easily do you need to scale your cluster up or down to meet changing demands?

- **Team expertise:** Evaluate your team's experience with Kubernetes and cloud infrastructure to determine which solution best suits their skills.

By carefully considering these factors and exploring the various options available, you can build a production-grade Kubernetes environment that delivers optimal performance and scalability for your containerized applications.

In the next section we will discuss some of the cluster maintenance tasks.

Running Kubernetes On-Premises: Challenges and Considerations

Running Kubernetes in an on-premises environment provides more control over infrastructure but also demands careful management. Compared to cloud-managed solutions, maintaining an on-premises Kubernetes cluster requires handling all aspects, from provisioning to upgrades, manually. Below, we explore key considerations and challenges that arise when managing Kubernetes on-premises.

- **Infrastructure Provisioning**: Setting up infrastructure for Kubernetes on-premises means automating the provisioning of nodes. Tools like Rancher's cloud controllers or Terraform help streamline this process by ensuring consistency. Packer can also be used to create VM images, enabling smoother upgrades by deploying updated images across nodes.

- **Cluster Setup and Maintenance**: Deploying a Kubernetes cluster on-premises involves using tools such as kubeadm. This process is often more involved than in cloud-managed environments. Cluster maintenance tasks include renewing certificates, managing nodes, and handling high availability setups, which add further complexity.

- **Load Balancing and Access**: Providing external access to applications in on-premises environments can be challenging. Standard Kubernetes options like **NodePort** and **LoadBalancer** services may not be enough. **MetalLB** can offer a load balancing solution for bare-metal setups but comes with limitations, such as not being able to load balance the API server in high availability environments.

- **Persistent Storage**: Persistent storage is critical for running production workloads. Kubernetes relies on **PersistentVolumeClaims** (PVCs) and **PersistentVolumes** (PVs), which require integration with physical storage systems. Tools like Longhorn allow dynamic provisioning of volumes and replication across nodes, providing flexibility in on-prem setups.

- **Upgrades and Scalability**: Kubernetes releases frequent updates, which means managing upgrades on-premises can be tricky. It's essential to test new versions before rolling them out to production. Tools like Packer and Terraform can assist in scaling by simplifying node additions and upgrades.

- **Networking**: On-premises Kubernetes networking depends on your data center configuration. Manual management of DNS, load balancers, and network settings is necessary. Monitoring tools such as Prometheus, alongside solutions like MetalLB for load balancing, can help, though they require integration and constant monitoring.

- **Monitoring and Management**: Monitoring on-premises clusters is essential for ensuring the system's health. Tools like Prometheus and Grafana can be used to monitor resource usage. Additionally, logging and alerting systems should be set up to detect and resolve issues swiftly, helping to minimize downtime.

- **Tooling and Automation:** Automating tasks such as node management and upgrades are vital in on-premises clusters. Enterprise Kubernetes platforms like Rancher or OpenShift help reduce manual intervention, providing a more streamlined and manageable Kubernetes environment.

- **Security and Compliance:** Security is crucial in enterprise Kubernetes setups. Including **FIPS** (**Federal Information Processing Standards**) support from the beginning can help meet compliance needs and maintain a secure environment as the system evolves.

- In summary, managing Kubernetes on-premises provides more flexibility but demands careful attention to infrastructure, networking, and storage setups. With the right tools and strategies, organizations can effectively scale and maintain a robust Kubernetes environment on their own infrastructure.

Summary

This chapter was quite intense! You require a Kubernetes cluster to follow this book, and so we examined five ways in which to set up Kubernetes clusters on different platforms. You learned about `minikube`, which is the most common way to set up a cluster on a local machine. You also discovered `kind`, which is a tool that can set up multi-node local clusters, which is a limitation of `minikube`.

We learned about some of the Kubernetes learning environments and also explored the production-grade Kubernetes environments including three major Kubernetes cloud services, GKE, Amazon EKS, and AKS. These three services allow you to create a Kubernetes cluster on the cloud for you to practice and train with. This was just a quick introduction to these services, and we will have the opportunity to dive deeper into these services later. For the moment, simply pick the solution that is the best for you.

In the next chapter, we are going to dive into Kubernetes by exploring the concept of Pods. The Pod resource is the most important resource that Kubernetes manages. We will learn how to create, update, and delete Pods. Additionally, we will look at how to provision them, how to get information from them, and how to update the containers they are running.

We will deploy an NGINX Pod on a Kubernetes cluster and examine how we can access it from the outside. By the end of the next chapter, you will be capable of launching your first containers on your Kubernetes cluster through the usage of Pods. The cluster that you installed here will be very useful when you follow the real-world examples that are coming in the next chapter.

Further reading

- Installing `minikube`: https://minikube.sigs.k8s.io/docs/start/
- `minikube` drivers: https://minikube.sigs.k8s.io/docs/drivers/
- Installing Docker: https://docs.docker.com/engine/install/
- Installing Podman: https://podman.io/docs/installation
- Multi-node Kubernetes using `minikube`: https://minikube.sigs.k8s.io/docs/tutorials/multi_node/
- Installing `kind`: https://kind.sigs.k8s.io/docs/user/quick-start#installation

Join our community on Discord

Join our community's Discord space for discussions with the authors and other readers:

`https://packt.link/cloudanddevops`

4

Running Your Containers in Kubernetes

This chapter is probably the most important one in this book. Here, we are going to discuss the concept of **Pods**, which are the objects Kubernetes uses to launch your application containers. Pods are at the heart of Kubernetes and mastering them is essential.

In *Chapter 3, Installing your First Kubernetes Cluster*, we said that the Kubernetes API defines a set of resources representing a computing unit. Pods are resources that are defined in the Kubernetes API that represent one or several containers. We never create containers directly with Kubernetes, but we always create Pods, which will be *converted* into containers on a compute node in our Kubernetes cluster.

At first, it can be a little difficult to understand the connection between Kubernetes Pods and containers, which is why we are going to explain what Pods are and why we use them rather than containers directly. A Kubernetes Pod can contain one or more application containers. In this chapter, however, we will focus on Kubernetes Pods that contain only one container. We will then have the opportunity to discover Pods that contain several containers in the next chapter.

We will create, delete, and update Pods using the **BusyBox** image, which is a Linux-based image containing many utilities useful for running tests. We will also launch a Pod based on the NGINX container image to launch an HTTP server. We will explore how to access the default NGINX home page via a feature that kubectl exposes called port forwarding. It's going to be useful to access and test the Pods running on your Kubernetes cluster from your web browser.

Then, we will discover how to label and annotate our Pods to make them easily accessible. This will help us organize our Kubernetes cluster so that it's as clean as possible. Finally, we will discover two additional resources, which are **Jobs** and **CronJobs**. By the end of this chapter, you will be able to launch your first containers managed by Kubernetes, which is the first step in becoming a Kubernetes master!

In this chapter, we're going to cover the following main topics:

- Let's explain the notion of Pods
- Launching your first Pods

- Labeling and annotating the Pods
- Launching your first Job
- Launching your first CronJob

Technical requirements

To follow along with the examples in this chapter, you will require the following:

- A properly configured Kubernetes cluster so that you can practice the commands shown as you read. Whether it's a minikube, Kind, GKE, EKS, or AKS cluster is not important.
- A working kubectl installation on your local machine. You can have more than one node if you want, but at least one Ready node is required to have a working Kubernetes setup.

You can download the latest code samples for this chapter from the official GitHub repository at https://github.com/PacktPublishing/The-Kubernetes-Bible-Second-Edition/tree/main/Chapter04

Let's explain the notion of Pods

In this section, we will explain the concept of Pods from a theoretical point of view. Pods have certain peculiarities that must be understood if you wish to master them well.

What are Pods?

When you want to create, update, or delete a container through Kubernetes, you do so through a Pod. A Pod is a group of one or more containers that you want to launch on the same machine, in the same Linux namespace. That's the first rule to understand about Pods: they can be made up of one or more containers but all the containers that belong to the same Pod will be launched on the same worker node. A Pod cannot and *won't ever* span across multiple worker nodes: that's an absolute rule.

But why do we bother delegating the management of our containers to this intermediary resource? After all, Kubernetes could have a container resource that would just launch a single container. The reason is that containerization invites you to think in terms of Linux processes rather than in terms of virtual machines. You may already know about the biggest and most recurrent container anti-pattern, which consists of using containers as virtual machine replacements: in the past, you used to install and deploy all your processes on top of a virtual machine. But containers are no virtual machine replacements, and they are not meant to run multiple processes.

Container technology invites you to follow one golden rule: *there should be a one-to-one relationship between a container and a Linux process*. That being said, modern applications are often made up of multiple processes, not just one, so in most cases, using only one container won't suffice to run a full-featured microservice. This implies that the processes, and thus the containers, should be able to communicate with each other by sharing file systems, networking, and so on. That's what Kubernetes Pods offer you: the ability to group your containers logically. All the containers/processes that make up an application should be grouped in the same Pod. That way, they'll be launched together and benefit from all the features when it comes to facilitating inter-process and inter-container communications.

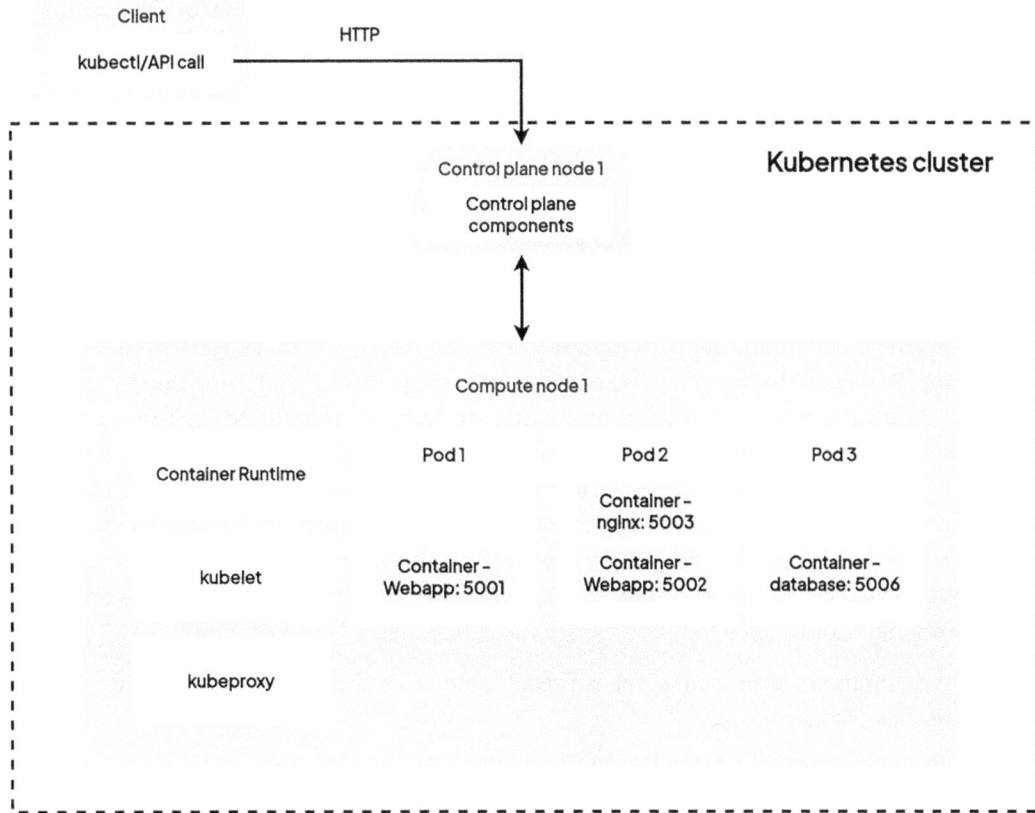

Figure 4.1: Containers and Pods

To help you understand this, imagine you have a working WordPress blog on a virtual machine and you want to convert that virtual machine into a WordPress Pod to deploy your blog on your Kubernetes cluster. WordPress is one of the most common pieces of software and is a perfect example to illustrate the need for Pods. This is because WordPress requires multiple processes to work properly.

WordPress is a PHP application that requires both a web server and a PHP interpreter to work. Let's list what Linux processes WordPress needs to work on:

- **An NGINX HTTP server:** It's a web application, so it needs an HTTP server running as a process to receive and serve server blog pages. NGINX is a good HTTP server that will do the job perfectly.
- **The PHP-FastCGI-Process-Manager (FPM) interpreter:** It's a blog engine written in PHP, so it needs a PHP interpreter to work.

NGINX and PHP-FPM are two processes: they are two binaries that you need to launch separately, but they need to be able to work together. On a virtual machine, the job is simple: you just install NGINX and **PHP-FPM** on the virtual machine and have both of them communicate through Unix sockets. You can do this by telling NGINX that the Linux socket PHP-FPM is accessible thanks to the /etc/nginx. config configuration file.

In the container world, things become harder because running these two processes in the same container is an anti-pattern: you have to run two containers, one for each process, and you must have them communicate with each other and share a common directory so that they can both access the application code. To solve this problem, you have to use the Docker networking layer to have the NGINX container be able to communicate with the PHP-FPM one. Then, you must use a volume mount to share the WordPress code between the two containers. You can do this with some Docker commands but imagine it now in production at scale, on multiple machines, on multiple environments, and so on. Achieving inter-process communication is possible with bare Docker, but that's difficult to achieve at scale while keeping all the production-related requirements in mind. With tons of microservices to manage and spread on different machines, it would become a nightmare to manage all these Docker networks, volume mounts, and so on. As you can imagine, that's the kind of problem the Kubernetes Pod resource solves. Pods are very useful because they wrap multiple containers and enable easy inter-process communication. The following are the core benefits Pods bring you:

- All the containers in the same Pod can reach each other through localhost as they share the same network namespace.
- All the containers in the same Pod share the same port space.
- You can attach a volume to a Pod, and then mount the volume to underlying containers, allowing them to share directories and file locations.

With the benefits Kubernetes brings you, it would be super easy to provision your WordPress blog as you can create a Pod that will run two containers: NGINX and PHP-FPM. Since they both can access each other on localhost, having them communicate is super easy. You can then use a volume to expose WordPress's code to both containers.

The most complex applications will forcibly require several containers, so it's a good idea to group them in the same Pod to have Kubernetes launch them together. Keep in mind that the Pod is here for only one reason: to ease inter-container (or inter-process) communications at scale.

> That being said, it is not uncommon at all to have Pods that are only made up of one container. But in any case, the Pod is the lowest level of abstraction provided by the Kubernetes APIs and the one you will interact with.

Lastly, please note that a container that was launched manually on a machine managed by a Kubernetes cluster won't be seen by Kubernetes as a container it manages. It becomes a kind of *orphan* container outside of the scope of the orchestrator. Kubernetes only manages the container it has launched through its Pod API.

Each Pod gets an IP address

Containers inside a single Pod are capable of communicating with each other through localhost, but Pods are also capable of communicating with each other. At launch time, each Pod gets a private IP address. Each Pod can communicate with any other Pod in the cluster by calling it through its IP address.

Kubernetes uses a flat network model that is implemented by a component called **Container Network Interface (CNI)**. CNI acts as a standardized bridge between containerized applications and the underlying network infrastructure within Kubernetes clusters. This eliminates the need for custom networking configurations for each container, streamlining communication and data flow.

CNI leverages a flexible plugin-based architecture. These plugins, written in various languages, communicate with the container runtime using standard input/output. The plugin specification defines a clear interface for network configuration, IP address provisioning, and maintaining connections across multiple hosts. Container runtimes call upon these plugins, enabling dynamic management and updates to container networks within the Kubernetes environment. This approach ensures seamless and adaptable networking for your containerized applications.

The following diagram shows the high-level communication flow between Pods and containers.

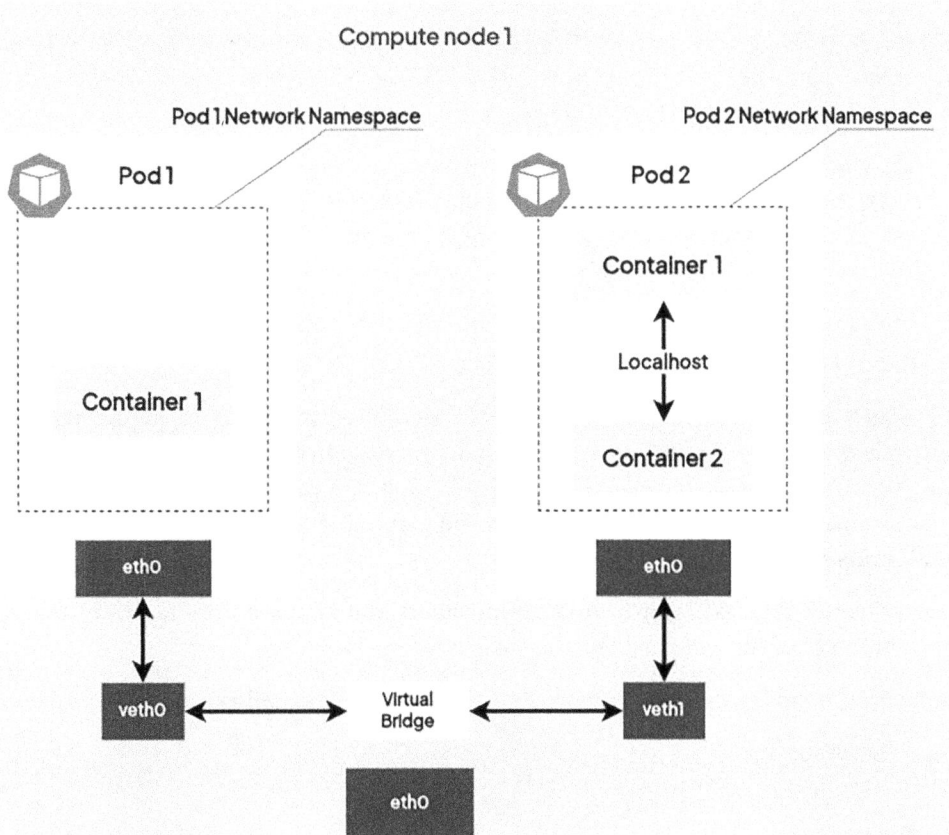

Figure 4.2: Container and Pod communication

How should you design your Pods?

While understanding Pods is crucial, in the real world of Kubernetes, most teams leverage a more powerful construct: Deployment. Deployments provide a higher-level abstraction for managing Pods. They automate tasks like scaling and restarting Pods in case of failures, ensuring a more robust and manageable application experience. We'll delve deeper into deployments in a moment, but for now, let's explore the Pods API to solidify your understanding of these foundational building blocks.

So, here is the second golden rule about Pods: they are meant to be destroyed and recreated easily. Pods can be destroyed voluntarily or not. For example, if a given worker node running four Pods were to fail, each of the underlying containers would become inaccessible. Because of this, you should be able to destroy and recreate your Pods at will, without it affecting the stability of your application. The best way to achieve this is to respect two simple design rules when building your Pods:

- A Pod should contain everything required to launch an application.
- A Pod should store any kind of state outside of the Pod using external storage (PersistentVolume).

When you start designing Pods on Kubernetes, it's hard to know exactly what a Pod should and shouldn't contain. It's pretty straightforward to explain: a Pod has to contain an application or a microservice. Take the example of our WordPress Pod, which we mentioned earlier: the Pod should contain the NGINX and PHP-FPM containers, which are required to launch WordPress. If such a Pod were to fail, our WordPress would become inaccessible, but recreating the Pod would make WordPress accessible again because the Pod contains everything necessary to run WordPress.

That being said, every modern application stores its state outside by utilizing external storage, database storage, such as Redis or MySQL, or by calling another microservice application to store the state. WordPress on its own does that too – it uses MySQL (or MariaDB) to store and retrieve your post. So, you'll also have to run a MySQL container somewhere. Two solutions are possible here:

- You run the MySQL container as part of the WordPress Pod.
- You run the MySQL container as part of a dedicated MySQL Pod.

Both solutions can be used, but the second is preferred. It's a good idea to decouple your application (here, this is WordPress, but tomorrow, it could be a microservice) from its database or logic layer by running them in two separate Pods. Remember that Pods are capable of communicating with each other. You can benefit from this by dedicating a Pod to running MySQL and giving its Pod IP address to your WordPress blog.

By separating the database layer from the application, you improve the stability of the setup: the application Pod crashing will not affect the database.

To summarize, grouping the application layers in the same Pods would cause three problems:

- Data durability
- Availability
- Stability

That's why it is recommended to keep your application Pods stateless as much as possible, by storing their states in an independent Pod. By treating the data layer as a separate application with its own development and management life cycle, we can achieve a decoupled architecture. This separation allows for independent scaling, updates, and testing of the data layer without impacting the application code itself.

Stateful monolithic applications

Despite some niche possibilities, running fast-moving monolithic stateful workloads on Kubernetes in 2024 is generally discouraged due to the complexity of managing monolithic applications within containers, potential inefficiencies for frequent updates in fast-paced environments, and increased management overhead for persistent storage needs compared to traditional deployments.

Accessing Pods via an IP address

You can access Pods using their IP addresses; however, this is not the recommended method for interacting with running applications. In upcoming chapters, we will delve into the Service resource, which plays a crucial role in mapping IP addresses to Pods. Stay tuned for a detailed explanation of how Services enhance Pod accessibility and application communication.

Now, let's launch our first Pod. Creating a WordPress Pod would be too complex for now, so let's start easy by launching some NGINX Pods and see how Kubernetes manages the container.

Launching your first Pods

In this section, we will explain how to create our first Pods in our Kubernetes cluster. Pods have certain peculiarities that must be understood to master them well.

We are not going to create a resource on your Kubernetes cluster at the moment; instead, we are simply going to explain what Pods are. In the next section, we'll start building our first Pods.

Creating a Pod with imperative syntax

In this section, we are going to create a Pod based on the NGINX image. We need two parameters to create a Pod:

- The Pod's name, which is arbitrarily defined by you
- The container images to build its underlying containers

As with almost everything on Kubernetes, you can create Pods using either of the two syntaxes available: the imperative syntax and the declarative syntax, which you have learned about in *Chapter 2, Kubernetes Architecture – from Container Images to Running Pods*. As a reminder, the imperative syntax is to run `kubectl` commands directly from a terminal, while with declarative syntax, you must write a YAML file containing the configuration information for your Pod, and then apply it with the `kubectl apply -f` command.

To create a Pod on your Kubernetes cluster, you have to use the `kubectl run` command. That's the simplest and fastest way to get a Pod running on your Kubernetes cluster. Here is how the command can be called:

```
$ kubectl run nginx-pod --image nginx:latest
```

In this command, the Pod's name is set to `nginx-pod`. This name is important because it is a pointer to the Pod: when you need to run the `update` or `delete` command on this Pod, you'll have to specify that name to tell Kubernetes which Pod the action should run on. The `--image` flag will be used to mention the container that this Pod will run. Once the Pod is created by the cluster, the status can be checked as follows:

```
$ kubectl get pods
NAME         READY   STATUS    RESTARTS   AGE
nginx-pod    1/1     Running   0          79s
```

> Standing up a Pod isn't instantaneous. Kubernetes might need to pull the container image from a registry if it's not available locally and configure the Pod's environment. To track this process in real time, use `kubectl get po -w`, which shows Pod information and refreshes automatically.

Here, you are telling Kubernetes to build a Pod based on the `nginx:latest` container image hosted on Docker Hub. This `nginx-pod` Pod contains only one container based on this `nginx:latest` image: you cannot specify multiple images here; this is a limitation of the imperative syntax.

If you want to build a Pod containing multiple containers built from several different container images, then you will have to go through the declarative syntax and write a YAML file.

Tags versus digests — ensuring image consistency

While creating Pods, you might encounter references to tags and digests. Both are used to identify container images, but with a key difference:

- **Tags:** Think of tags as human-readable names for image versions. They can be changed to point to different versions of the same image, potentially causing unexpected behavior.
- **Digests:** These are unique fingerprints of an image, ensuring you always reference the exact desired version. This is crucial for security and reproducibility, especially in light of potential software supply chain attacks.

For example, instead of using `nginx:latest` (tag), you might use `nginx@sha256:1445eb9c6dc5e9 619346c836ef6fbd6a95092e4663f27dcfce116f051cdbd232` (digest). You can fetch the digest information for the container image from the registry itself or by using the `podman manifest inspect nginx:latest` command.

Figure 4.3: Fetching image digest from a container registry

This guarantees you're deploying the specific image version with the unique `abcd1234` hash. This practice is becoming increasingly important for secure and reliable deployments.

Let us learn how to create Pods using YAML declarations in the next section.

Creating a Pod with declarative syntax

Creating a Pod with declarative syntax is simple too. You have to create a YAML file containing your Pod definition and apply it against your Kubernetes cluster using the `kubectl apply -f` command.

Remember that Kubernetes cannot run two Pods with the same name in the same namespace (e.g., the `default` namespace in our case): the Pod's name is the unique identifier and is used to identify the Pods within a namespace. You need to delete the existing Pod that you created in the previous step before you create a new Pod with the same name in the same namespace:

```
$ kubectl delete pod nginx-pod
pod "nginx-pod" deleted
```

Here is the content of the `nginx-pod.yaml` file, which you can create on your local workstation:

```
apiVersion: v1
kind: Pod
metadata:
  name: nginx-pod
spec:
  containers:
    - name: nginx-container
      image: nginx:latest
```

Try to read this file and understand its content. YAML files are only key-value pairs. The Pod's name is `nginx-Pod`, and then we have an array of containers in the `spec:` part of the file containing only one container created from the `nginx:latest` image. The container itself is named `nginx-container`.

Once the `nginx-Pod.yaml` file has been saved, run the following command to create the Pod:

```
$ kubectl apply -f nginx-pod.yaml
pod/nginx-pod created
```

If a Pod called `nginx-pod` already exists in your cluster, this command will fail. Try to edit the YAML file to update the Pod's name and then apply it again.

Namespaces in Kubernetes

If you omit to specify a namespace during resource creation, it defaults to the default namespace. Stay tuned for *Chapter 6, Namespaces, Quotas, and Limits for Multi-Tenancy in Kubernetes*, where we'll delve into the significance of Kubernetes namespaces.

Reading the Pod's information and metadata

At this point, you should have a running Pod on your Kubernetes cluster. Here, we are going to try to read its information. At any time, we need to be able to retrieve and read information regarding the resources that were created on your Kubernetes cluster; this is especially true for Pods. Reading the Kubernetes cluster can be achieved using two `kubectl` commands: `kubectl get` and `kubectl describe`. Let's take a look at them:

- `kubectl get`: The `kubectl get` command is a list operation; you use this command to list a set of objects. Do you remember when we listed the nodes of your cluster after all the installation procedures described in the previous chapter? We did this using `kubectl get nodes`. The command works by requiring you to pass the object type you want to list. In our case, it's going to be the `kubectl get pods` operation. In the upcoming chapters, we will discover other objects, such as `configmaps` and `secrets`. To list them, you'll have to type `kubectl get configmaps`; the same goes for the other object types. For example, the `nginx-pod` can be listed as follows:

```
$ kubectl get pods
```

- `kubectl describe`: The `kubectl describe` command is quite different. It's intended to retrieve a complete set of information for one specific object that's been identified from both its kind and object name. You can retrieve the information of our previously created Pod by using `kubectl describe pods nginx-pod`. Calling this command will return a full set of information available about that specific Pod, such as its IP address. To see the details of `nginx-pod`, the following command can be used:

```
$ kubectl describe pod nginx-pod
Name:           nginx-pod
Namespace:      default
...<removed for brevity>...
Containers:
  nginx-container:
    Container ID:
containerd://3afbbe30b51b77994df69f4c4dbefb02fc304efb2bf0f5bdb65a65
```

```
1154a8e311
    Image:              nginx:latest
...<removed for brevity>...
Conditions:
  Type                          Status
  PodReadyToStartContainers     True
  Initialized                   True
...<removed for brevity>...
Events:
  Type      Reason      Age    From            Message
  ----      ------      ----   ----            -------
...<removed for brevity>...
  Normal    Created     112s   kubelet         Created container nginx-
container
  Normal    Started     112s   kubelet         Started container nginx-
container
```

From the preceding command output, you can read a lot of information, including the following items:

- **Pod name and namespace:** This identifies the specific Pod you requested information on (e.g., nginx-pod).
- **Container details:** This lists information about the containers within the Pod, including image name, resource requests/limits, and current state.
- **Pod conditions:** This shows the current operational state of the Pod (e.g., Running, Pending, CrashLoopBackOff).
- **Events:** This provides a history of relevant events related to the Pod's life cycle, including creation, restarts, or errors.

Now, let's look at some more advanced options for listing and describing objects in Kubernetes.

Listing the objects in JSON or YAML

The -o or --output option is one of the most useful options offered by the kubectl command line. This one has some benefits you must be aware of. This option allows you to customize the output of the kubectl command line. By default, the kubectl get pods command will return a list of the Pods in your Kubernetes cluster in a formatted way so that the end user can see it easily. You can also retrieve this information in JSON format or YAML format by using the -o option:

```
$ kubectl get pods --output yaml # In YAML format
$ kubectl get pods --output json # In JSON format
```

If you know the Pod name, you can also get a specific Pod:

```
$ kubectl get pods <POD_NAME> -o yaml
# OR
$ kubectl get pods <POD_NAME> -o json
```

This way, you can retrieve and export data from your Kubernetes cluster in a scripting-friendly format.

Backing up your resource using the list operation

You can also use these flags to back up your Kubernetes resources. Imagine a situation where you created a Pod using the imperative way, so you don't have the YAML declaration file stored on your computer. If the Pod fails, it's going to be hard to recreate it. The -o option helps us retrieve the YAML declaration file of a resource that's been created in Kubernetes, even if we created it using the imperative way. To do this, run the following command:

```
$ kubectl get pods/nginx-pod -o yaml > nginx-pod-output.yaml
```

This way, you have a YAML backup of the `nginx-pod` resource as it is running on your cluster. You can always compare the output file with the original YAML declaration and analyze the differences using the `diff` command or other utilities:

```
$ diff nginx-pod.yaml nginx-pod-output.yaml
```

> There are tools available to clean up the YAML and get a clean output of a usable declaration. For example, `kube-neat` is such a utility that will help to clean the unwanted information from the detailed output. Refer to `https://github.com/itaysk/kubectl-neat` to learn more.

If something goes wrong, you'll be able to recreate your Pod easily. Pay attention to the `nginx-pod` section of this command. To retrieve the YAML declaration, you need to specify which resource you are targeting. By redirecting the output of this command to a file, you get a nice way to retrieve and back up the configuration of the object inside your Kubernetes cluster.

Getting more information from the list operation

It's also worth mentioning the -o wide format, which is going to be very useful for you: using this option allows you to expand the default output to add more data. By using it on the `Pods` object, for example, you'll get the name of the worker node where the Pod is running:

```
$ kubectl get pods -o wide
NAME          READY    STATUS    RESTARTS    AGE    IP           NODE
NOMINATED NODE    READINESS GATES
nginx-pod     1/1      Running   0           15m    10.244.0.4   minikube    <none>
<none>
```

Keep in mind that the -o option can take a lot of different parameters and that some of them are much more advanced, such as `jsonpath`, which allows you to directly execute sorting operations on top of a JSON body document to retrieve only specific information, just like the `jq` library you used previously if you have already written some bash scripts that deal with JSON parsing.

Accessing a Pod from the outside world

At this point, you should have a Pod containing an NGINX HTTP server on your Kubernetes cluster. You should now be able to access it from your web browser. However, this is a bit complicated.

By default, your Kubernetes cluster does not expose the Pod it runs to the outside world. For that, you will need to use another resource called a service, which we will cover in more detail in *Chapter 8, Exposing Your Pods with Services*. However, kubectl does offer a command for quickly accessing a running container on your cluster called kubectl port-forward. This is how you can use it:

```
$ kubectl port-forward pod/nginx-pod 8080:80
Forwarding from 127.0.0.1:8080 -> 80
Forwarding from [::1]:8080 -> 80
```

This command is quite easy to understand: we are telling kubectl to forward port 8080 on my local machine (the one running kubectl) to port 80 on the Pod identified by pod/nginx-Pod.

Kubectl then outputs a message, telling you that it started to forward your local 8080 port to the 80 one of the Pod. If you get an error message, it's probably because your local port 8080 is currently being used. Try to set a different port or simply remove the local port from the command to let kubectl choose a local port randomly:

```
$ kubectl port-forward pod/nginx-pod 8080:80
Forwarding from 127.0.0.1:8080 -> 80
Forwarding from [::1]:8080 -> 80
```

Now, you can launch your browser and try to reach the http://localhost:<localport> address, which in your case is http://localhost:8080:

Welcome to nginx!

If you see this page, the nginx web server is successfully installed and working. Further configuration is required.

For online documentation and support please refer to nginx.org. Commercial support is available at nginx.com.

Thank you for using nginx.

Figure 4.4: The NGINX default page running in a Pod and accessible on localhost, which indicates the port-forward command worked

Once you have finished the testing, use the *Ctrl + C* command to end the port forwarding task.

Entering a container inside a Pod

When a Pod is launched, you can access the Pods it contains. Under Docker, the command to execute a command in a running container is called `docker exec`. Kubernetes copies this behavior via a command called `kubectl exec`. Use the following command to access our NGINX container inside `nginx-pod`, which we launched earlier:

```
$ kubectl exec -it nginx-pod -- bash
root@nginx-pod:/# hostname
nginx-pod
```

After running this command, you will be inside the NGINX container. You can do whatever you want here, just like with any other container. The preceding command assumes that the `bash` binary is installed in the container you are trying to access. Otherwise, the `sh` binary is generally installed on a lot of containers and might be used to access the container. Don't be afraid to take a full binary path, like so:

```
$ kubectl exec -it nginx-pod -- /bin/bash
```

Once you have finished testing, exit from the bash shell of the container using the `exit` command:

```
root@nginx-pod:/# exit
exit
```

> **IMPORTANT: Security and Non-Root Users in Containers**
>
> It's generally recommended to run containers with a non-root user. You need to limit potential damage from vulnerabilities. If a vulnerability is exploited, a non-root user has less access to the system, reducing the impact. Also, follow the principle of least privilege, which grants only the necessary permissions for the container to function, minimizing its attack surface. We will explore the security context in *Chapter 18, Security in Kubernetes*.

Now, let's discover how to delete a Pod from a Kubernetes cluster.

Deleting a Pod

Deleting a Pod is super easy. You can do so using the `kubectl delete` command. You need to know the name of the Pod you want to delete. In our case, the Pod's name is `nginx-pod`. Run the following command:

```
$ kubectl delete pods nginx-pod
# or...
$ kubectl delete pods/nginx-pod
```

If you do not know the name of the Pod, remember to run the `kubectl get pods` command to retrieve the list of the Pods and find the one you want to delete.

There is also something you must know: if you have built your Pod with declarative syntax and you still have its YAML configuration file, you can delete your Pod without having to know the name of the container because it is contained in the YAML file.

Run the following command to delete the Pod using the declarative syntax:

```
$ kubectl delete -f nginx-pod.yaml
```

After you run this command, the Pod will be deleted in the same way.

> Remember that all containers belonging to the Pod will be deleted. The container's life cycle is bound to the life cycle of the Pod that launched it. If the Pod is deleted, the containers it manages will be deleted. Remember to always interact with the Pods and not with the containers directly.

With that, we have reviewed the most important aspects of Pod management, such as launching a Pod with the imperative or declarative syntax, deleting a Pod, and listing and describing them. Now, we will introduce one of the most important aspects of Pod management in Kubernetes: labeling and annotating.

Labeling and annotating the Pods

We will now discuss another key concept of Kubernetes: labels and annotations. Labels are key-value pairs that you can attach to your Kubernetes objects. Labels are meant to tag your Kubernetes objects with key-value pairs defined by you. Once your Kubernetes objects have been labeled, you can build a custom query to retrieve specific Kubernetes objects based on the labels they hold. In this section, we are going to discover how to interact with labels through kubectl by assigning some labels to our Pods.

What are labels and why do we need them?

What label you define for your objects is up to you – there is no specific rule regarding this. These labels are attributes that will allow you to organize your objects in your Kubernetes cluster. To give you a very concrete example, you could attach a label called environment = prod to some of your Pods, and then use the kubectl get pods command to list all the Pods within that environment. So, you could list all the Pods that belong to your production environment in one command:

```
$ kubectl get pods --label "environment=production"
```

As you can see, it can be achieved using the --label parameter, which can be shortened using its -l equivalent:

```
$ kubectl get pods --label "environment=production"
```

This command will list all the Pods holding a label called environment with a value of production. Of course, in our case, no Pods will be found since none of the ones we created earlier are holding this label. You'll have to be very disciplined about labels and not forget to set them every time you create a Pod or another object, and that's why we are introducing them quite early in this book: not only Pods but almost every object in Kubernetes can be labeled, and you should take advantage of this feature to keep your cluster resources organized and clean.

You use labels not only to organize your cluster but also to build relationships between your different Kubernetes objects: you will notice that some Kubernetes objects will read the labels that are carried by certain Pods and perform certain operations on them based on the labels they carry. If your Pods don't have labels or they are misnamed or contain the wrong values, some of these mechanisms might not work as you expect.

On the other hand, using labels is completely arbitrary: there is no particular naming rule, nor any convention Kubernetes expects you to follow. Thus, it is your responsibility to use the labels as you wish and build your convention. If you are in charge of the governance of a Kubernetes cluster, you should enforce the usage of mandatory labels and build some monitoring rules to quickly identify non-labeled resources.

Keep in mind that labels are limited to 63 characters; they are intended to be short. Here are some label ideas you could use:

- `environment` (prod, dev, uat, and so on)
- `stack` (blue, green, and so on)
- `tier` (frontend and backend)
- `app_name` (wordpress, magento, mysql, and so on)
- `team` (business and developers)

Labels are not intended to be unique between objects. For example, perhaps you would like to list all the Pods that are part of the production environment. Here, several Pods with the same label key-value pair can exist in the cluster at the same time without posing any problem – it's even recommended if you want your list query to work. For example, if you want to list all the resources that are part of the prod environment, a label environment such as = `prod` should be created on multiple resources. Now, let's look at annotations, which are another way we can assign metadata to our Pods.

What are annotations and how do they differ from labels?

Kubernetes also uses another type of metadata called **annotations**. Annotations are very similar to labels as they are also key-value pairs. However, annotations do not have the same use as labels. Labels are intended to identify resources and build relationships between them, while annotations are used to provide contextual information about the resource that they are defined on.

For example, when you create a Pod, you could add an annotation containing the email of the support team to contact if this app does not work. This information has its place in an annotation but has nothing to do with a label.

While it is highly recommended that you define labels wherever you can, you can omit annotations: they are less important to the operation of your cluster than labels. Be aware, however, that some Kubernetes objects or third-party applications often read annotations and use them as configuration. In this case, their usage of annotations will be explained explicitly in their documentation.

Adding a label

In this section, we will learn how to add and remove labels and annotations from Pods. We will also learn how to modify the labels of a Pod that already exists on a cluster.

Let's take the Pod based on the NGINX image that we used earlier. We will recreate it here with a label called `tier`, which will contain the `frontend` value. Here is the `kubectl` command to run for that:

```
$ kubectl run nginx-pod --image nginx --labels "tier=frontend"
```

As you can see, a label can be assigned using the `--labels` parameter. You can also add multiple labels by using the `--labels` parameter and comma-separated values, like this:

```
$ kubectl run nginx-pod --image nginx
--labels="app=myapp,env=dev,tier=frontend"
```

Here, the `nginx` Pod will be created with two labels.

The `--labels` flag has a short version called `-l`. You can use this to make your command shorter and easier to read. Labels can be appended to a YAML Pod definition. Here is the same Pod, holding the two labels we created earlier, but this time, it's been created with the declarative syntax:

```
# labelled_pod.yaml
apiVersion: v1
kind: Pod
metadata:
  name: nginx-pod
  labels:
    environment: prod
    tier: frontend
spec:
  containers:
    - name: nginx-container
      image: nginx:latest
```

Consider the file that was created at `~/labelled_pod.yaml`. The following `kubectl` command would create the Pod the same way as it was created previously:

```
$ kubectl apply -f ~/labelled_pod.yaml
```

This time, running the command we used earlier should return at least one Pod – the one we just created:

```
$ kubectl get pod -l environment=prod
NAME        READY   STATUS    RESTARTS   AGE
nginx-pod   1/1     Running   0          31m
```

Now, let's learn how we can list the labels attached to our Pod.

Listing labels attached to a Pod

There is no dedicated command to list the labels attached to a Pod, but you can make the output of `kubectl get pods` a little bit more verbose. By using the `--show-labels` parameter, the output of the command will include the labels attached to the Pods:

```
$ kubectl get pods --show-labels
NAME           READY    STATUS     RESTARTS    AGE    LABELS
nginx-pod      1/1      Running    0           56s    environment=prod,tier=frontend
```

This command does not run any kind of query based on the labels; instead, it displays the labels themselves as part of the output.

Adding or updating a label to/of a running Pod

Now that we've learned how to create Pods with labels, we'll learn how to add labels to a running Pod. You can add, create, or modify the labels of a resource at any time using the `kubectl label` command. Here, we are going to add another label to our `nginx` Pod. This label will be called `stack` and will have a value of `blue`:

```
$ kubectl label pod nginx-pod stack=blue
pod/nginx-pod labeled
```

This command only works if the Pod has no label called `stack`. When the command is executed, it can only add a new tag and not update it. This command will update the Pod by adding a label called `stack` with a value of `blue`. Run the following command to see that the change was applied:

```
$ kubectl get pods nginx-pod --show-labels
NAME           READY    STATUS     RESTARTS    AGE    LABELS
nginx-pod      1/1      Running    0           38m
environment=prod,stack=blue,tier=frontend
```

To update an existing label, you must append the `--overwrite` parameter to the preceding command. Let's update the `stack=blue` label to make it `stack=green`; pay attention to the `overwrite` parameter:

```
$ kubectl label pod nginx-pod stack=green --overwrite
pod/nginx-pod labeled
```

Here, the label should be updated. The `stack` label should now be equal to `green`. Run the following command to show the Pod and its labels again:

```
$ kubectl get pods nginx-pod --show-labels
NAME           READY    STATUS     RESTARTS    AGE    LABELS
nginx-pod      1/1      Running    0           41m
environment=prod,stack=green,tier=frontend
```

> Adding or updating labels using the `kubectl label` command might be dangerous. As we mentioned earlier, you'll build relationships between different Kubernetes objects based on labels. By updating them, you might break some of these relationships and your resources might start to behave not as expected. That's why it's better to add labels when a Pod is created and keep your Kubernetes configuration immutable. It's always better to destroy and recreate rather than update an already running configuration.

The last thing we must do is learn how to delete a label attached to a running Pod.

Deleting a label attached to a running Pod

Just like we added and updated the labels of a running Pod, we can also delete them. The command is a little bit trickier. Here, we are going to remove the label called `stack`, which we can do by adding a minus symbol (-) right after the label name:

```
$ kubectl label pod nginx-pod stack-
pod/nginx-pod unlabeled
```

Adding that minus symbol at the end of the command might be quite strange, but running `kubectl get pods --show-labels` again should show that the `stack` label is now gone:

```
$ kubectl get pods nginx-pod --show-labels
```

Now, let's learn about annotations in Kubernetes in the next section.

Adding an annotation

Kubernetes annotations are key-value pairs that you can attach to various Kubernetes objects such as Pods, Deployments, and Services. They allow you to add extra information to these objects without changing their core functionality. Unlike labels, which are used for identification and selection, annotations are designed to store additional data that can be used for human readability or by external tools. Annotations can include details such as configuration information or the creator's name.

Let's learn how to add annotations to a Pod:

```
# annotated_pod.yaml
apiVersion: v1
kind: Pod
metadata:
  annotations:
    tier: webserver
  name: nginx-pod
  labels:
    environment: prod
    tier: frontend
spec:
```

```
containers:
  - name: nginx-container
    image: nginx:latest
```

Here, we simply added the `tier: webserver` annotation, which can help us identify that this Pod is running an HTTP server. Just keep in mind that it's a way to add additional metadata.

When you apply this new configuration, you can use the `kubectl replace -f` command to replace the existing Pod configuration.

> `kubectl replace` is a command used to update or replace existing Kubernetes resources using a manifest file. It offers a more forceful approach compared to `kubectl apply -f`. The `kubectl replace` command replaces the existing resource definition with the one specified in the manifest file. This essentially overwrites the existing resource configuration. Unlike `kubectl apply`, which might attempt to merge changes, `kubectl replace` aims for a complete replacement. This command can be helpful when you want to ensure a specific configuration for a resource, regardless of its current state. It's also useful for situations where the resource definition might have become corrupted and needs to be replaced entirely.

The name of an annotation can be prefixed by a DNS name. This is the case for Kubernetes components such as `kube-scheduler`, which must indicate to cluster users that this component is part of the Kubernetes core. The prefix can be omitted completely, as shown in the preceding example.

You can see the annotations by using the `kubectl describe` Pod, `kubectl get po -o yaml`, or with the `jq` utility as follows:

```
$ kubectl get pod nginx-pod -o json | jq '.metadata.annotations'
{
  "cni.projectcalico.org/containerID":
"666d12cd2fb7d6ffe09add73d8466db218f01e7c7ef5315ef0187a675725b5ef",
  "cni.projectcalico.org/podIP": "10.244.151.1/32",
  "cni.projectcalico.org/podIPs": "10.244.151.1/32",
  "kubectl.kubernetes.io/last-applied-configuration": "{\"apiVersion\":\"v1\",\
"kind\":\"Pod\",\"metadata\":{\"annotations\":{},\"name\":\"nginx-pod\",\"names
pace\":\"default\"},\"spec\":{\"containers\":[{\"image\":\"nginx:latest\",\"nam
e\":\"nginx-container\"}]}}\n"
}
```

Let us learn about Jobs in Kubernetes in the next section of this chapter.

Launching your first Job

Now, let's discover another Kubernetes resource that is derived from Pods: the Job resource. In Kubernetes, a computing resource is a Pod, and everything else is just an intermediate resource that manipulates Pods.

This is the case for the Job object, which is an object that will create one or multiple Pods to complete a specific computing task, such as running a Linux command.

What are Jobs?

A Job is another kind of resource that's exposed by the Kubernetes API. In the end, a job will create one or multiple Pods to execute a command defined by you. That's how jobs work: they launch Pods. You have to understand the relationship between the two: jobs are not independent of Pods, and they would be useless without Pods. In the end, the two things they are capable of are launching Pods and managing them. Jobs are meant to handle a certain task and then exit. Here are some examples of typical use cases for a Kubernetes Job:

- Taking a backup of a database
- Sending an email
- Consuming some messages in a queue

These are tasks you do not want to run forever. You expect the Pods to be terminated once they have completed their task. This is where the Jobs resource will help you.

But why bother using another resource to execute a command? After all, we can create one or multiple Pods directly that will run our command and then exit.

This is true. You can use a Pod based on a container image to run the command you want and that would work fine. However, jobs have mechanisms implemented at their level that allow them to manage Pods in a more advanced way. Here are some things that jobs are capable of:

- Running Pods multiple times
- Running Pods multiple times in parallel
- Retrying to launch the Pods if they encountered any errors
- Killing a Pod after a specified number of seconds

Another good point is that a job manages the labels of the Pods it will create so that you won't have to manage the labels on those Pods directly.

All of this can be done without using jobs, but this would be very difficult to manage and that is the reason we have the Jobs resource in Kubernetes.

Creating a job with restartPolicy

Since creating a job might require some advanced configurations, we are going to focus on declarative syntax here. This is how you can create a Kubernetes job through YAML. We are going to make things simple here; the job will just echo `Hello world`:

```yaml
# hello-world-job.yaml
apiVersion: batch/v1
kind: Job
metadata:
  name: hello-world-job
```

```
spec:
  template:
    metadata:
      name: hello-world-job
    spec:
      restartPolicy: OnFailure
      containers:
      - name: hello-world-container
        image: busybox
        command: ["/bin/sh", "-c"]
        args: ["echo 'Hello world'"]
```

Pay attention to the kind resource, which tells Kubernetes that we need to create a job and not a Pod, as we did previously. Also, notice apiVersion:, which also differs from the one that's used to create the Pod.

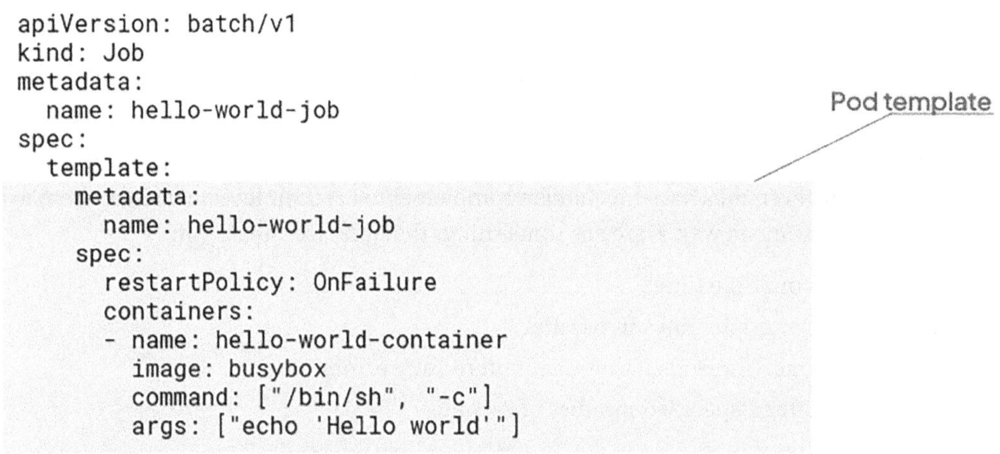

```
apiVersion: batch/v1
kind: Job
metadata:
  name: hello-world-job
spec:
  template:
    metadata:
      name: hello-world-job
    spec:
      restartPolicy: OnFailure
      containers:
      - name: hello-world-container
        image: busybox
        command: ["/bin/sh", "-c"]
        args: ["echo 'Hello world'"]
```

Pod template

Figure 4.5: Job definition details and Pod template

You can create the job with the following command:

```
$ kubectl apply -f hello-world-job.yaml
```

As you can see, this job will create a Pod based on the busybox container image:

```
$ kubectl get jobs
NAME              COMPLETIONS   DURATION   AGE
hello-world-job   1/1           9s         8m46s
```

This will run the echo 'Hello World' command. Lastly, the restartPolicy option is set to OnFailure, which tells Kubernetes to restart the Pod or the container in case it fails. If the entire Pod fails, a new Pod will be relaunched. If the container fails (the memory limit has been reached or a non-zero exit code occurs), the individual container will be relaunched on the same node because the Pod will remain untouched, which means it's still scheduled on the same machine.

The restartPolicy parameter can take two options:

- Never
- OnFailure

Setting it to Never will prevent the job from relaunching the Pods, even if it fails. When debugging a failing job, it's a good idea to set restartPolicy to Never to help with debugging. Otherwise, new Pods might be recreated over and over, making your life harder when it comes to debugging.

In our case, there is little chance that our job was not successful since we only want to run a simple Hello world. To make sure that our job worked well, we can read the job logs as follows:

```
$kubectl logs jobs/hello-world-job
Hello world
```

We can also retrieve the name of the Pod the job created using the kubectl get pods command. Then, we can use the kubectl logs command, as follows:

```
$ kubectl logs pods/hello-world-job-2qh4d
Hello world
```

Here, we can see that our job has worked well since we can see the Hello world message displayed in the log of our Pod. However, what if it had failed? Well, this depends on restartPolicy – if it's set to Never, then nothing would happen, and Kubernetes wouldn't try to relaunch the Pods.

However, if restartPolicy was set to OnFailure, Kubernetes would try to restart the job after 10 seconds and then double that time on each new failure. 10 seconds, then 20 seconds, then 40 seconds, then 80 seconds, and so on. After 6 minutes, Kubernetes would give up.

Understanding the job's backoffLimit

By default, the Kubernetes job will try to relaunch the failing Pod six times during the next six minutes after its failure. You can change this limitation by changing the backoffLimit option. Here is the updated YAML file:

```yaml
# hello-world-job-2.yaml
apiVersion: batch/v1
kind: Job
metadata:
  name: hello-world-job-2
spec:
  backoffLimit: 3
  template:
    metadata:
      name: hello-world-job-2
    spec:
      restartPolicy: OnFailure
      containers:
```

```
    - name: hello-world-container
      image: busybox
      command: ["/bin/sh", "-c"]
      args: ["echo 'Hello world'"]
```

This way, the job will only try to relaunch the Pods twice after its failure.

Running a task multiple times using completions

You can also instruct Kubernetes to launch a job multiple times using the Job object. You can do this by using the completions option to specify the number of times you want a command to be executed. The number of completions will create 10 different Pods that will be launched one after the other in the following example. Once one Pod has finished, the next one will be started. Here is the updated YAML file:

```
# hello-world-job-3.yaml
apiVersion: batch/v1
kind: Job
metadata:
  name: hello-world-job-3
spec:
  backoffLimit: 3
  completions: 10
  template:
    metadata:
      name: hello-world-job-3
    spec:
      restartPolicy: OnFailure
      containers:
      - name: hello-world-container
        image: busybox
        command: ["/bin/sh", "-c"]
        args: ["echo 'Hello world'"]
```

The completions option was added here. Also, please notice that the args section was updated by us by adding the sleep 3 option. Using this option will make the task sleep for three seconds before completing, giving us enough time to notice the next Pod being created. Once you've applied this configuration file to your Kubernetes cluster, you can run the following command:

```
$ kubectl get pods --watch
$ kubectl get jobs -w


NAME                 STATUS      COMPLETIONS   DURATION    AGE
hello-world-job-2    Complete    1/1           10s         52s
```

```
hello-world-job-3    Running    5/10         49s       49s
hello-world-job-3    Running    5/10         51s       51s
hello-world-job-3    Running    5/10         52s       52s
hello-world-job-3    Running    6/10         52s       52s
hello-world-job-3    Running    6/10         56s       56s
<removed for brevity>
```

The watch (-w or –watch) mechanism will update your kubectl output when something new arrives, such as the creation of the new Pods being managed by your Kubernetes. If you want to wait for the job to finish, you'll see 10 Pods being created with a 3-second delay between each.

Running a task multiple times in parallel

The completions option ensures that the Pods are created one after the other. You can also enforce parallel execution using the parallelism option. If you do that, you can get rid of the completions option. Kubernetes Jobs can leverage parallelism to significantly speed up execution. By running multiple Pods concurrently, you distribute workload across your cluster, leading to faster completion times and improved resource utilization, especially for large or complex tasks. Here is the updated YAML file:

```yaml
# hello-world-job-4.yaml
apiVersion: batch/v1
kind: Job
metadata:
  name: hello-world-job-4
spec:
  backoffLimit: 3
  parallelism: 5
  template:
    metadata:
      name: hello-world-job-4
    spec:
      restartPolicy: OnFailure
      containers:
      - name: hello-world-container
        image: busybox
        command: ["/bin/sh", "-c"]
        args: ["echo 'Hello world'; sleep 3"]
```

Please notice that the completions option is now gone and that we replaced it with parallelism. The job will now launch five Pods at the same time and will have them run in parallel:

```
$ kubectl get pods -w
NAME                     READY   STATUS              RESTARTS   AGE
hello-world-job-4-9dspk  0/1     ContainerCreating   0          7s
hello-world-job-4-n6qv9  0/1     Completed           0          7s
```

```
hello-world-job-4-pv754     0/1        ContainerCreating    0          7s
hello-world-job-4-ss4g8     1/1        Running              0          7s
hello-world-job-4-v78cj     1/1        Running              0          7s
...<removed for brevity>...
```

In the next section, we will learn how to terminate a Job automatically after a particular time period.

Terminating a job after a specific amount of time

You can also decide to terminate a Pod after a specific amount of time. This can be very useful when you are running a job that is meant to consume a queue, for example. You could poll the messages for one minute and then automatically terminate the processes. You can do that using the activeDeadlineSeconds parameter. Here is the updated YAML file:

```yaml
# hello-world-job-5.yaml
apiVersion: batch/v1
kind: Job
metadata:
  name: hello-world-job-5
spec:
  backoffLimit: 3
  activeDeadlineSeconds: 60
  template:
    metadata:
      name: hello-world-job-5
    spec:
      restartPolicy: OnFailure
      containers:
      - name: hello-world-container
        image: busybox
        command: ["/bin/sh", "-c"]
        args: ["echo 'Hello world'"]
```

Here, the job will terminate after 60 seconds, no matter what happens. It's a good idea to use this feature if you want to keep a process running for an exact amount of time and then terminate it.

What happens if a job succeeds?

If your job is completed, it will remain created in your Kubernetes cluster and will not be deleted automatically: that's the default behavior. The reason for this is that you can read its logs a long time after its completion. However, keeping your jobs created on your Kubernetes cluster that way might not suit you. You can delete the jobs automatically, and the Pods they created, by using the ttlSecondsAfterFinished option. Here is the updated YAML file for implementing this solution:

```yaml
# hello-world-job-6.yaml
apiVersion: batch/v1
kind: Job
metadata:
  name: hello-world-job-6
spec:
  backoffLimit: 3
  ttlSecondsAfterFinished: 30
  template:
    metadata:
      name: hello-world-job-6
    spec:
      restartPolicy: OnFailure
      containers:
      - name: hello-world-container
        image: busybox
        command: ["/bin/sh", "-c"]
        args: ["echo 'Hello world'"]
```

Here, the jobs are going to be deleted 30 seconds after their completion.

Deleting a job

Keep in mind that the Pods that are created are bound to the life cycle of their parent. Deleting a job will result in deleting the Pods they manage.

Start by getting the name of the job you want to destroy. In our case, it's hello-world-job. Otherwise, use the kubectl get jobs command to retrieve the correct name. Then, run the following command:

```
$ kubectl delete jobs hello-world-job
```

If you want to delete the jobs but not the Pods it created, you need to add the --cascade=false parameter to the delete command:

```
$ kubectl delete jobs hello-world-job --cascade=false
```

Thanks to this command, you can get rid of all the jobs that will be kept on your Kubernetes cluster once they've been completed. Now, let's move on to launching the first CronJob.

Launching your first CronJob

To close this first chapter on Pods, we will look at another Kubernetes resource called **CronJob**.

What are CronJobs?

The name **CronJob** can mean two different things and it is important that we do not get confused with these:

- The Unix cron feature
- The Kubernetes CronJob resource

Historically, CronJobs are commands scheduled using the cron Unix feature, which is the most robust way to schedule the execution of a command in Unix systems. This idea was later taken up in Kubernetes.

Be careful because even though the two ideas are similar, they don't work the same at all. On Unix and other derived systems such as Unix, you schedule commands by editing a file called Crontab, which is usually found in /etc/crontab or related paths. In the world of Kubernetes, things are different: you are not going to schedule the execution of commands but the execution of Job resources, which themselves will create Pod resources. Keep in mind that the CronJob object you'll create will create Job objects.

Think of it as a kind of wrapper around the Job resource: in Kubernetes, we call that a controller. CronJob can do everything the Job resource is capable of because it is nothing more than a wrapper around the Job resource, according to the cron expression specified.

The good news is that the Kubernetes CronJob resource is using the cron format inherited from Unix. So, if you have already written some CronJobs on a Linux system, mastering Kubernetes CronJobs will be super straightforward.

But first, why would you want to execute a Pod? The answer is simple; here are some concrete use cases:

- Taking database backups regularly every Sunday at 1 A.M.
- Clearing cached data every Monday at 4 P.M.
- Sending a queued email every 5 minutes.
- Various maintenance operations to be executed regularly.

The use cases of Kubernetes CronJobs do not differ much from their Unix counterparts – they are used to answer the same need, but they do provide the massive benefit of allowing you to use your already configured Kubernetes cluster to schedule regular jobs using your container images and your already existing Kubernetes cluster.

Preparing your first CronJob

It's time to create your first CronJob. Let's do this using declarative syntax. First, let's create a cronjob.yaml file and place the following YAML content into it:

```
# hello-world-cronjob.yaml
apiVersion: batch/v1beta1
kind: CronJob
metadata:
```

```
    name: hello-world-cronjob
spec:
  schedule: "* * * * *"
  jobTemplate:
    spec:
      template:
        metadata:
          name: hello-world-cronjob
        spec:
          restartPolicy: OnFailure
          containers:
          - name: hello-world-container
            image: busybox
            imagePullPolicy: IfNotPresent
            command: ["/bin/sh", "-c"]
            args: ["echo 'Hello world'"]
```

Before applying this file to the Kubernetes cluster, let's start to explain it. There are two important things to notice here:

- The schedule key, which lets you input the cron expression
- The jobTemplate section, which is exactly what you would input in a job YAML manifest

Let's explain these two keys quickly before applying the file.

Understanding the schedule key

The schedule key allows you to insert an expression in a cron format such as Linux. Let's explain how these expressions work; if you already know these expressions, you can skip these explanations:

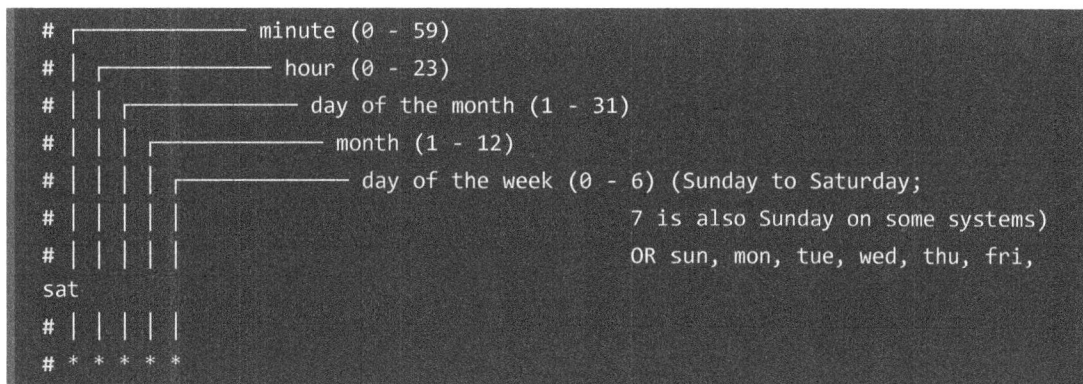

```
# ┌───────────── minute (0 - 59)
# │ ┌───────────── hour (0 - 23)
# │ │ ┌───────────── day of the month (1 - 31)
# │ │ │ ┌───────────── month (1 - 12)
# │ │ │ │ ┌───────────── day of the week (0 - 6) (Sunday to Saturday;
# │ │ │ │ │                                      7 is also Sunday on some systems)
# │ │ │ │ │                                      OR sun, mon, tue, wed, thu, fri,
sat
# │ │ │ │ │
# * * * * *
```

A cron expression is made up of five entries separated by white space. From left to right, these entries correspond to the following:

- Minutes
- Hour

- Day of the month
- Month
- Day of the week

Each entry can be filled with an asterisk, which means *every*. You can also set several values for one entry by separating them with a ,. You can also use a – to input a range of values. Let me show you some examples:

- "10 11 * * *" means "At 11:10 every day of every month."
- "10 11 * 12 *" means "At 11:10 every day of December."
- "10 11 * 12 1" means "At 11:10 of every Monday of December."
- "10 11 * * 1,2" means "At 11:10 of every Monday and Tuesday of every month."
- "10 11 * 2-5 *" means "At 11:10 every day from February to May."

Here are some examples that should help you understand how cron works. Of course, you don't have to memorize the syntax: most people help themselves with documentation or cron expression generators online such as crontab.cronhub.io and crontab.guru. If this is too complicated, feel free to use this kind of tool; it can help you confirm that your syntax is valid before you deploy the object to Kubernetes.

Understanding the role of the jobTemplate section

If you've been paying attention to the structure of the YAML file, you may have noticed that the jobTemplate key contains the definition of a Job object. When we use the CronJob object, we are simply delegating the creation of a Job object to the CronJob object.

```
apiVersion: batch/v1beta1
kind: CronJob
metadata:
  name: hello-world-cronjob                                    Cronjob Spec
spec:
  schedule: "* * * * *"                                        Job template spec
  jobTemplate:
    spec:
      template:                                                Pod template spec
        metadata:
          name: hello-world-cronjob
        spec:
          restartPolicy: OnFailure
          containers:
          - name: hello-world-container
            image: busybox
            imagePullPolicy: IfNotPresent
            command: ["/bin/sh", "-c"]
            args: ["echo 'Hello world'"]
```

Figure 4.6: CronJob YAML architecture

Therefore, the CronJob object is a resource that only manipulates another resource.

Later, we will discover many objects that will allow us to create Pods so that we don't have to do it ourselves. These special objects are called **controllers**: they manipulate other Kubernetes resources by obeying their own logic. Moreover, when you think about it, the Job object is itself a controller since, in the end, it only manipulates Pods by providing them with its own features, such as the possibility of running Pods in parallel.

In a real context, you should always try to create Pods using these intermediate objects as they provide additional and more advanced management features.

Try to remember this rule: the basic unit in Kubernetes is a Pod, but you can delegate the creation of Pods to many other objects. In the rest of this section, we will continue to discover *naked* Pods. Later, we will learn how to manage their creation and management via controllers.

Controlling the CronJob execution deadline

For some reason, a CronJob may fail to execute. In this case, Kubernetes cannot execute the Job at the moment it is supposed to start. If jobs surpass their configured deadline, Kubernetes considers them unsuccessful.

The optional .spec.startingDeadlineSeconds field establishes a deadline (in complete seconds) for initiating the Job in case it misses its scheduled time due to any reason. Once the deadline is missed, the Cronjob skips that specific instance of the Job, though future occurrences are still scheduled.

Managing the history limits of jobs

After the completion of a Cronjob, regardless of its success status, your Kubernetes cluster retains a history. The history setting can be configured at the CronJob level, allowing you to determine whether to preserve the history for each CronJob. If you choose to keep it, you can specify the number of entries to retain for both succeeded and failed jobs using the optional .spec.successfulJobsHistoryLimit and .spec.failedJobsHistoryLimit fields.

Creating a CronJob

If you already have the YAML manifest file, creating a CronJob object is easy:

```yaml
# hello-world-cronjob.yaml
apiVersion: batch/v1
kind: CronJob
metadata:
  name: hello-world-cronjob
spec:
  schedule: "*/1 * * * *"
  # Run every minute
  successfulJobsHistoryLimit: 5
  startingDeadlineSeconds: 30
  jobTemplate:
    spec:
      template:
```

```
    metadata:
      name: hello-world-cronjob
    spec:
      restartPolicy: OnFailure
      containers:
        - name: hello-world-container
          image: busybox
          imagePullPolicy: IfNotPresent
          command: ["/bin/sh", "-c"]
          args: ["echo 'Hello world'"]
```

See the details in the preceding YAML sample.

successfulJobsHistoryLimit: 5, instructs the CronJob controller to retain the 5 most recent successful job runs. Older successful jobs will be automatically deleted.

You can create the CronJob using the kubectl apply -f command as follows:

```
$ kubectl apply -f hello-world-cronjob.yaml
cronjob.batch/hello-world-cronjob created
```

With that, the CronJob has been created on your Kubernetes cluster. It will launch a scheduled Pod, as configured in the YAML file; in this case, every minute:

```
$ kubectl get cronjobs
NAME                    SCHEDULE      TIMEZONE     SUSPEND    ACTIVE    LAST SCHEDULE
AGE
hello-world-cronjob     */1 * * * *   <none>       False      0         37s
11m
$ kubectl get jobs
NAME                            COMPLETIONS   DURATION   AGE
hello-world-cronjob-28390196    1/1           3s         4m47s
hello-world-cronjob-28390197    1/1           3s         3m47s
hello-world-cronjob-28390198    1/1           3s         2m47s
hello-world-cronjob-28390199    1/1           3s         107s
hello-world-cronjob-28390200    1/1           4s         47s


$ kubectl get pods
NAME                                  READY   STATUS      RESTARTS   AGE
hello-world-cronjob-28390196-fpmc6    0/1     Completed   0          4m52s
hello-world-cronjob-28390197-vkzw2    0/1     Completed   0          3m52s
hello-world-cronjob-28390198-tj6qv    0/1     Completed   0          2m52s
hello-world-cronjob-28390199-dd666    0/1     Completed   0          112s
hello-world-cronjob-28390200-kn89r    0/1     Completed   0          52s
```

Since you have configured `successfulJobsHistoryLimit:` 5, only the last 5 jobs or Pods will be visible.

Deleting a CronJob

Like any other Kubernetes resource, deleting a `CronJob` can be achieved through the `kubectl delete` command. Like before, if you have the YAML manifest, it's easy:

```
$ kubectl delete -f ~/cronjob.yaml
cronjob/hello-world-cronjob deleted
```

With that, the `CronJob` has been destroyed by your Kubernetes cluster. No scheduled jobs will be launched anymore.

Summary

We have come to the end of this chapter on Pods and how to create them; we hope you enjoyed it. You've learned how to use the most important objects in Kubernetes: Pods.

The knowledge you've developed in this chapter is part of the essential basis for mastering Kubernetes: all you will do in Kubernetes is manipulate Pods, label them, and access them. But also remember, in the actual Kubernetes environment, you will not be creating or modifying resources directly, instead using other methods to deploy your application Pods and other resources. In addition, you saw that Kubernetes behaves like a traditional API, in that it executes CRUD operations to interact with the resources on the cluster. In this chapter, you learned how to launch containers on Kubernetes, how to access these containers using `kubectl` port forwarding, how to add labels and annotations to Pods, how to delete Pods, and how to launch and schedule jobs using the `CronJob` resource.

Just remember this rule about container management: any container that will be launched in Kubernetes will be launched through the object. Mastering this object is like mastering most of Kubernetes: everything else will consist of automating things around the management of Pods, just like we did with the `CronJob` object; you have seen that the `CronJob` object only launches Job objects that launch Pods. If you've understood that some objects can manage others, but in the end, all containers are managed by Pods, then you've understood the philosophy behind Kubernetes, and it will be very easy for you to move forward with this orchestrator.

Also, we invite you to add labels and annotations to your Pods, even if you don't see the need for them right away. Know that it is essential to label your objects well to keep a clean, structured, and well-organized cluster.

However, you still have a lot to discover when it comes to managing Pods because, so far, we have only seen Pods that are made up of only one Docker container. The greatest strength of Pods is that they allow you to manage multiple containers at the same time, and of course, to do things properly, there are several design patterns that we can follow to manage our Pods when they are made of several containers.

In the next chapter, we will learn how to manage Pods that are composed of several containers. While this will be very similar to the Pods we've seen so far, you'll find that some little things are different and that some are worth knowing. First, you will learn how to launch multi-container Pods using kubectl (hint: kubectl will not work), and then how to get the containers to communicate with each other. After that, you will learn how to access a specific container in a multi-container Pod, as well as how to access logs from a specific container. Finally, you will learn how to share volumes between containers in the same Pod.

As you read the next chapter, you will learn about the rest of the fundamentals of Pods in Kubernetes. So, you'll get an overview of Pods while we keep moving forward by discovering additional objects in Kubernetes that will be useful for deploying applications in our clusters.

Further reading

- **Managing Pods:** https://kubernetes.io/docs/concepts/workloads/pods/
- **Kubernetes Jobs:** https://kubernetes.io/docs/concepts/workloads/controllers/job/
- **Well-Known Labels, Annotations and Taints:** https://kubernetes.io/docs/reference/labels-annotations-taints/
- **Kubernetes CronJob:** https://kubernetes.io/docs/concepts/workloads/controllers/cron-jobs/

Join our community on Discord

Join our community's Discord space for discussions with the authors and other readers:

https://packt.link/cloudanddevops

5

Using Multi-Container Pods and Design Patterns

Running complex applications on Kubernetes will require that you run not one but several containers in the same Pods. The strength of Kubernetes also lies in its ability to create Pods made up of several containers. We will focus on those Pods in this chapter by studying the different aspects of hosting several containers in the same Pod, as well as having these different containers communicate with each other.

So far, we've only created Pods running a single container: those were the simplest forms of Pods, and you'll use these Pods to manage the simplest of applications. We also discovered how to update and delete them by running simple **create, read, update, and delete** (CRUD) operations against those Pods using the kubectl command-line tool.

Besides mastering the basics of CRUD operations, you have also learned how to access a running Pod inside a Kubernetes cluster.

While single-container Pods are more common, there are situations where using multiple containers in a single Pod is beneficial. For example, using a dedicated container to handle log gathering with the main container inside a Pod or another dedicated container to enable proxy communication between services. In this chapter, we will push all of this one step forward and discover how to manage Pods when they are meant to launch not one but several containers. The good news is that everything you learned previously will also be valid for multi-container Pods. Things won't differ much in terms of raw Pod management because updating and deleting Pods is not different, no matter how many containers the Pod contains.

Besides those basic operations, we are also going to cover how to access a specific container inside a multi-container Pod and how to access its logs. When a given Pod contains more than one container, you'll have to run some specific commands with specific arguments to access it, and that's something we are going to cover in this chapter.

We will also discover some important design patterns such as Ambassador, Sidecar, and Adapter containers. You'll need to learn these architectures to effectively manage multi-container Pods. You'll also learn how to deal with volumes from Kubernetes. Docker also provides volumes, but in Kubernetes, they are used to share data between containers launched by the same Pod, and this is going to be an important part of this chapter. After this chapter, you're going to be able to launch complex applications inside Kubernetes Pods.

In this chapter, we're going to cover the following main topics:

- Understanding what multi-container Pods are
- Sharing volumes between containers in the same Pod
- The Ambassador Design Pattern
- The Sidecar Design Pattern
- The Adapter Design Pattern
- Sidecars versus Kubernetes Native Sidecars

Technical requirements

You will require the following prerequisites for this chapter:

- A working `kubectl` command-line utility.
- A local or cloud-based Kubernetes cluster to practice with.

You can download the latest code samples for this chapter from the official GitHub repository: `https://github.com/PacktPublishing/The-Kubernetes-Bible-Second-Edition/tree/main/Chapter05`.

Understanding what multi-container Pods are

Multi-container Pods are a way to package tightly coupled applications together in Kubernetes. This allows multiple containers to share resources and easily communicate with each other, which is ideal for scenarios like sidecars and service mesh. In this section, we'll learn about the core concepts of Pods for managing multiple containers at once by discussing some concrete examples of multi-container Pods.

Concrete scenarios where you need multi-container Pods

You should group your containers into a Pod when they need to be tightly linked. More broadly, a Pod must correspond to an application or a process running in your Kubernetes cluster. If your application requires multiple containers to function properly, then those containers should be launched and managed through a single Pod.

When the containers are supposed to work together, you should group them into a single Pod. Keep in mind that a Pod cannot span across multiple compute nodes. So, if you create a Pod containing several containers, then all these containers will be created on the same compute node. To understand where and when to use multi-container Pods, take the example of two simple applications:

- **A log forwarder**: In this example, imagine that you have deployed a web server such as NGINX that stores its logs in a dedicated directory. You might want to collect and forward these logs. For that, you could deploy something like a Splunk forwarder as a container within the same Pod as your NGINX server.

These log forwarding tools are used to forward logs from a source to a destination location, and it is very common to deploy agents such as Splunk, Fluentd, or Filebeat to grab logs from a container and forward them to a central location such as an Elasticsearch cluster. In the Kubernetes world, this is generally achieved by running a multi-container Pod with one container dedicated to running the application, and another one dedicated to grabbing the logs and sending them elsewhere. Having these two containers managed by the same Pod would ensure that they are launched on the same node as the log forwarder and at the same time.

• **A proxy server:** Imagine an NGINX reverse proxy container in the same Pod as your main application, efficiently handling traffic routing and security with custom rules. This concept extends to **service mesh**, where a dedicated proxy like Envoy can be deployed alongside your application container, enabling features like load balancing and service discovery within a microservices architecture. (We will learn about service mesh in detail in *Chapter 8*, *Exposing Your Pods with Services*.) By bundling the two containers in the same Pod, you'll get two Pods running in the same node. You could also run a third container in the same Pod to forward the logs that are emitted by the two others to a central logging location! This is because Kubernetes has no limit on the number of containers you can have in the same Pod, as long as you have enough computing resources to run them all.

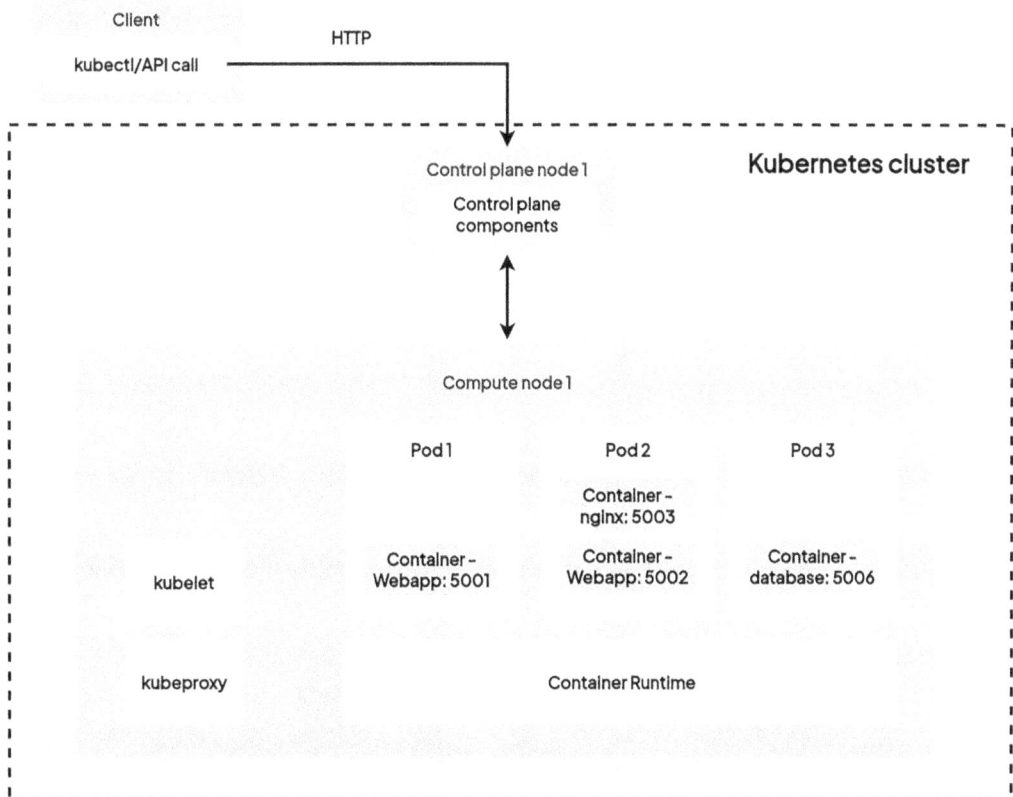

Figure 5.1: Sample multi-container Pod scenario

In general, every time several of your containers work together and are tightly coupled, you should have them in a multi-container Pod.

Now, let's discover how to create multi-container Pods.

Creating a Pod made up of two containers

In the previous chapter, we discovered two syntaxes for manipulating Kubernetes:

- The imperative syntax
- The declarative syntax

Most of the Kubernetes objects we are going to discover in this book can be created or updated using these two methods, but unfortunately, this is not the case for multi-container Pods.

When you need to create a Pod containing multiple containers, you will need to go through the declarative syntax. This means that you will have to create a YAML file containing the declaration of your Pods and all the containers it will manage, and then apply it through `kubectl apply -f file.yaml`.

Consider the following YAML manifest file stored in `~/multi-container-pod.yaml`:

```yaml
# multi-container-pod.yaml
apiVersion: v1
kind: Pod
metadata:
  name: multi-container-pod
spec:
  restartPolicy: Never
  containers:
    - name: nginx-container
      image: nginx:latest
    - name: debian-container
      image: debian
      command: ["/bin/sh"]
      args: ["-c", "while true; do date;echo debian-container; sleep 5 ; done"]
```

This YAML manifest will create a Kubernetes Pod made up of two containers: one based on the `nginx:latest` image and the other one based on the `debian` image.

To create it, use the following command:

```
$ kubectl apply -f multi-container-pod.yaml
pod/multi-container-pod created
```

This will result in the Pod being created. The kubelet on the elected node will have the container runtime (e.g., containerd, CRI-O, or Docker daemon) to pull both images and instantiate two containers.

To check whether the Pod was correctly created, we can run `kubectl get pods`:

```
$ kubectl get pods
NAME                 READY   STATUS    RESTARTS   AGE
multi-container-pod  2/2     Running   0          2m7s
```

Do you remember the role of `kubelet` from *Chapter 2, Kubernetes Architecture – from Container Images to Running Pods*? This component runs on each node that is part of your Kubernetes cluster and is responsible for converting Pod manifests received from `kube-apiserver` into actual containers.

> All the containers that are declared in the same Pod will be scheduled, or launched, on the same node and Pods cannot span multiple machines.
>
> Containers in the same Pod are meant to live together. If you terminate a Pod, all its containers will be killed together, and when you create a Pod, the kubelet will, at the very least, attempt to create all its containers together.
>
> High availability is generally achieved by replicating multiple Pods over multiple nodes, which you will learn about later in this book.

From a Kubernetes perspective, applying this file results in a fully working multi-container Pod made up of two containers, and we can make sure that the Pod is running with the two containers by running a standard `kubectl get pods` command to fetch the Pod list from `kube-apiserver`.

Do you see the column that states 2/2 in the previous `kubectl` command output? This is the number of containers inside the Pod. Here, this is saying that the two containers that are part of this Pod were successfully launched! We can see the logs from different containers as follows.

```
$ kubectl logs multi-container-pod -c debian-container
Mon Jan  8 01:33:23 UTC 2024
debian-container
Mon Jan  8 01:33:28 UTC 2024
debian-container
...<removed for brevity>...


$ kubectl logs multi-container-pod -c nginx-container
/docker-entrypoint.sh: /docker-entrypoint.d/ is not empty, will attempt to
perform configuration
...<removed for brevity>...
2024/01/08 01:33:20 [notice] 1#1: start worker process 39
2024/01/08 01:33:20 [notice] 1#1: start worker process 40
```

We learned how to create and manage multi-container Pods, and in the next section, we will learn how to troubleshoot when a multi-container Pod fails.

What happens when Kubernetes fails to launch one container in a Pod?

Kubernetes keeps track of all the containers that are launched in the same Pod. But it often happens that a specific container cannot be launched. Let's introduce a typo in the YAML manifest to demonstrate how Kubernetes reacts when some containers of a specific Pod cannot be launched.

In the following example, we have defined a container image that does not exist at all for the NGINX container; note the `nginx:i-do-not-exist` tag:

```yaml
# failed-multi-container-pod.yaml
apiVersion: v1
kind: Pod
metadata:
  name: multi-container-pod
spec:
  restartPolicy: Never
  containers:
    - name: nginx-container
      image: nginx:i-do-not-exist
    - name: debian-container
      image: debian
      command: ["/bin/sh"]
      args: ["-c", "while true; do date;echo debian-container; sleep 5 ; done"]
```

Now, we can apply the following container using the `kubectl apply -f failed-multi-container-pod.yaml` command:

```
$ kubectl apply -f failed-multi-container-pod.yaml
pod/failed-multi-container-pod created
```

Here, you can see that the Pod was effectively created. This is because even if there's a non-existent image, the YAML remains valid from a Kubernetes perspective. So, Kubernetes simply creates the Pod and persists the entry into `etcd`, but we can easily imagine that the kubelet will encounter an error when it launches the container to retrieve the image from the container registry (e.g., Docker Hub).

Let's check the status of the Pod using `kubectl get pod`:

```
$ kubectl get pod
NAME                         READY   STATUS            RESTARTS   AGE
failed-multi-container-pod   1/2     ImagePullBackOff  0          93s
```

As you can see, the status of the Pod is `ImagePullBackOff`. This means that Kubernetes is trying to launch the Pod but failing with an image access issue. To find out why it's failing, you have to describe the Pod using the `kubectl describe pod failed-multi-container-pod` command as follows:

```
$ kubectl describe pod failed-multi-container-pod
Name:              failed-multi-container-pod
Namespace:         default
...<removed for brevity>...
Events:
  Type      Reason    Age                    From           Message
  ----      ------    ----                   ----           -------
...<removed for brevity>...
  Warning   Failed    5m23s (x3 over 6m13s)  kubelet        Error:
ErrImagePull
  Warning   Failed    4m55s (x5 over 6m10s)  kubelet        Error:
ImagePullBackOff
  Normal    Pulling   4m42s (x4 over 6m17s)  kubelet        Pulling image
"nginx:i-do-not-exist"
  Warning   Failed    4m37s (x4 over 6m13s)  kubelet        Failed to pull
image "nginx:i-do-not-exist": Error response from daemon: manifest for nginx:i-
do-not-exist not found: manifest unknown: manifest unknown
  Normal    BackOff   75s (x19 over 6m10s)   kubelet        Back-off
pulling image "nginx:i-do-not-exist"
```

It's a little bit hard to read, but by following this log, you can see that debian-container is okay since kubelet has succeeded in creating it, as shown by the last line of the preceding output. But there's a problem with the other container; that is, nginx-container.

Here, you can see that the output error is ErrImagePull and, as you can guess, it's saying that the container cannot be launched because the image pull fails to retrieve the nginx:i-do-not-exist image tag.

So, Kubernetes does the following:

1. First, it creates the entry in etcd if the Pod of the YAML file is valid.
2. Then, it simply tries to launch the container.
3. If an error is encountered, it will try to launch the failing container again and again.

If any other container works properly, it's fine. However, your Pod will never enter the Running status because of the failed container. After all, your app certainly needs the failing container to work properly; otherwise, that container should not be there at all!

Now, let's learn how to delete a multi-container Pod.

Deleting a multi-container Pod

When you want to delete a Pod containing multiple containers, you have to go through the kubectl delete command, just like you would for a single-container Pod.

Then, you have two choices:

* You can specify the path to the YAML manifest file that's used by using the -f option.
* You can delete the Pod without using its YAML path if you know its name.

The first way consists of specifying the path to the YAML manifest file. You can do so using the following command:

```
$ kubectl delete -f multi-container-pod.yaml
```

Otherwise, if you already know the Pod's name, you can do this as follows:

```
$ kubectl delete pods/multi-pod
$ # or equivalent
$ kubectl delete pods multi-pod
```

To figure out the name of the Pods, you can use the kubectl get commands:

```
$ kubectl get pod
NAME                          READY    STATUS            RESTARTS    AGE
failed-multi-container-pod    1/2      ImagePullBackOff  0           13m
```

When we ran them, only failed-multi-container-pod was created in the cluster, so that's why you can just see one line in the output.

Here is how you can delete failed-multi-container-pod imperatively without specifying the YAML file that created it:

```
$ kubectl delete -f failed-multi-container-pod.yaml
pod "failed-multi-container-pod" deleted
```

After a few seconds, the Pod is removed from the Kubernetes cluster, and all its containers are removed from the container daemon and the Kubernetes cluster node.

The amount of time that's spent before the command is issued and the Pod's name is deleted and released is called the **grace period**. Let's discover how to deal with it!

Understanding the Pod deletion grace period

One important concept related to deleting Pods is what is called the grace period. Both single-container Pods and multi-container Pods have this grace period, which can be observed when you delete them. This grace period can be ignored by passing the --grace-period=0 --force option to the kubectl delete command.

During the deletion of a Pod, certain kubectl commands display its status as Terminating. Notably, this Terminating status is not categorized within the standard Pod phases. Pods are allocated a designated grace period for graceful termination, typically set to 30 seconds. To forcefully terminate a Pod, the --force flag can be employed. When the deletion is forced by setting --grace-period=0 with the --force flag, the Pod's name is immediately released and becomes available for another Pod to take it. During an unforced deletion, the grace period is respected, and the Pod's name is released after it is effectively deleted.

```
$ kubectl delete pod failed-multi-container-pod --grace-period=0 --force
Warning: Immediate deletion does not wait for confirmation that the running
resource has been terminated. The resource may continue to run on the cluster
indefinitely.
pod "failed-multi-container-pod" force deleted
```

> This command should be used carefully if you don't know what you are doing. Forcefully deleting a Pod shouldn't be seen as the norm because, as the output states, you cannot be sure that the Pod was effectively deleted. If, for some reason, the Pod cannot be deleted, it might run indefinitely, so do not run this command if you are not sure of what to do.

Now, let's discover how to access a specific container that is running inside a multi-container Pod.

Accessing a specific container inside a multi-container Pod

When you have several containers in the same Pod, you can access each of them individually. Here, we will access the NGINX container of our multi-container Pods. Let's start by recreating it because we deleted it in our previous example:

```
$ kubectl apply -f multi-container-pod.yaml
pod/multi-container-pod created
```

To access a running container, you need to use the kubectl exec command, just like you need to use docker exec to launch a command in an already created container when using Docker without Kubernetes.

This command will ask for two important parameters:

- The Pod that wraps the container you want to target
- The name of the container itself, as entered in the YAML manifest file

We already know the name of the Pod because we can easily retrieve it with the kubectl get command. In our case, the Pod is named multi-container-pod.

However, we don't have the container's name because there is no kubectl get containers command that would allow us to list the running containers. This is why we will have to use the kubectl describe pods/multi-container-pod command to find out what is contained in this Pod:

```
$ kubectl describe pods/multi-container-pod
```

This command will show the names of all the containers contained in the targeted Pod. Here, we can see that our Pod is running two containers, one called debian-container and another called nginx-container.

Additionally, the following is a command for listing all the container names contained in a dedicated Pod:

```
$ kubectl get pod/multi-container-pod -o jsonpath="{.spec.containers[*].name}"
nginx-container debian-container
```

This command will spare you from using the `describe` command. However, it makes use of `jsonpath`, which is an advanced feature of `kubectl`: this command might look strange but it mostly consists of a sort filter that's applied against the command. The `jsonpath` expression `{.spec.containers[*].name}` can be used with the `kubectl get pod` command to retrieve the names of all containers within a specific Pod. The `.` denotes the entire response object, while `spec.containers` targets the containers section within the Pod specification. The `[*]` operator instructs `jsonpath` to iterate through all elements within the containers list, and `.name` extracts the `name` property from each container object. Essentially, this expression provides a comma-separated list of container names within the specified Pod.

`jsonpath` filters are not easy to get right, so, feel free to add this command as a bash alias or note it somewhere because it's a useful one.

In any case, we can now see that we have these two containers inside the `multi-container-pod` Pod:

- `nginx-container`
- `busybox-container`

Now, let's access `nginx-container`. You have the name of the targeted container in the targeted Pod, so use the following command to access the Pod:

```
$ kubectl exec -it multi-container-pod --container nginx-container -- /bin/bash
root@multi-container-pod:/# hostname
multi-container-pod
root@multi-container-pod:/#
```

After running this command, you will find yourself inside `nginx-container`. Let's explain this command a little bit. `kubectl exec` does the same as `docker exec`:

- `kubectl exec`: This is the command to execute commands in a Kubernetes container.
- `-it`: These are options for the execution. `-t` allocates a pseudo-TTY, and `-i` allows interaction with the container.
- `multi-container-pod`: This is the name of the Pod in which you want to execute the command.
- `--container nginx-container`: This specifies the container within the Pod where the command should be executed. In Pods with multiple containers, you need to specify the container you want to interact with.
- `-- /bin/bash`: This is the actual command that will be executed in the specified container. It launches a Bash shell (`/bin/bash`) in interactive mode, allowing you to interact with the container's command line.

When you run this command, you get the shell of the container, inside the multi-container Pod, at a point at which you will be ready to run commands inside this very specific container on your Kubernetes cluster.

The main difference from the single container Pod situation is the `--container` option (the `-c` short option works, too). You need to pass this option to tell `kubectl` what container you want to reach.

Now, let's discover how to run commands in the containers running in your Pods!

Running commands in containers

One powerful aspect of Kubernetes is that you can, at any time, access the containers running on your Pods to execute some commands. We did this previously, but did you know you can also execute any command you want directly from the kubectl command-line tool?

First, we are going to recreate the multi-container Pod:

```
$ kubectl apply -f multi-container-pod.yaml
pod/multi-container-pod created
```

To run a command in a container, you need to use kubectl exec, just like we did previously. But this time, you have to remove the -ti parameter to prevent kubectl from attaching to your running terminal session.

Here, we are running the ls command to list files in nginx-container from the multi-container-pod Pod:

```
$ kubectl exec pods/multi-container-pod -c nginx-container -- ls
bin
boot
dev
docker-entrypoint.d
docker-entrypoint.sh
...<removed for brevity>
```

You can omit the container name but if you do so, kubectl will use the default first one.

Next, we will discover how to override the commands that are run by your containers.

Overriding the default commands run by the containers

Overriding default commands is important in multi-container Pods because it lets you control each container's behavior individually. This means you can customize how each container works within the Pod. For example, a web server container might normally run a start server command, but you could override a sidecar container's command to handle logging instead. This approach can also help with resource management. If a container usually runs a heavy process, you can change it to a lighter one in the Pod to ensure other containers have enough resources. Finally, it helps with managing dependencies. For instance, a database container might typically start right away, but you could override its command to wait until a related application container is ready.

When using Docker, you have the opportunity to write files called Dockerfiles to build container images. Dockerfiles make use of two keywords to tell us what commands and arguments the containers that were built with this image will launch when they're created using the docker run command. These two keywords are ENTRYPOINT and CMD:

- ENTRYPOINT is the main command the container will launch.
- CMD is used to replace the parameters that are passed to the ENTRYPOINT command.

For example, a classic `Dockerfile` that should be launched to run the `sleep` command for 30 seconds would be written like this:

```
# ~/Dockerfile
FROM busybox:latest
ENTRYPOINT ["sleep"]
CMD ["30"]
```

Containerfile and Podman

A Containerfile acts as a recipe for building a container image similar to Dockerfiles. It contains a series of instructions specifying the operating system, installing dependencies, copying application code, and configuring settings. Podman, a tool similar to Docker, can interpret this Containerfile and construct the image based on the instructions.

The previous code snippet is just plain old Docker and you should be familiar with these concepts. The `CMD` argument is what you can pass to the `docker run` command. If you build this image with this `Dockerfile` using the `docker build` command, you'll end up with a BusyBox image that just runs the `sleep` command (`ENTRYPOINT`) when the `docker run` command is run for 30 seconds (the `CMD` argument).

Thanks to the `CMD` instruction, you can override the default 30 seconds like so:

```
$ docker run my-custom-ubuntu:latest 60
$ docker run my-custom-ubuntu:latest # Just sleep for 30 seconds
```

Kubernetes, on the other hand, allows us to override both `ENTRYPOINT` and `CMD` thanks to YAML Pod definition files. To do so, you must append two optional keys to your YAML configuration file: `command` and `args`.

This is a very big benefit that Kubernetes brings you because you can decide to append arguments to the command that's run by your container's `Dockerfile`, just like the `CMD` arguments do with bare Docker, or completely override `ENTRYPOINT`!

Here, we are going to write a new manifest file that will override the default `ENTRYPOINT` and `CMD` parameters of the `busybox` image to make the `busybox` container sleep for 60 seconds. Here is how to proceed:

```
# nginx-debian-with-custom-command-and-args
apiVersion: v1
kind: Pod
metadata:
  name: nginx-debian-with-custom-command-and-args
spec:
  restartPolicy: Never
  containers:
```

```
- name: nginx-container
  image: nginx:latest
- name: debian-container
  image: debian
  command: ["sleep"] # Corresponds to the ENTRYPOINT
  args: ["60"] # Corresponds to CMD
```

This is a bit tricky to understand because what `Dockerfile` calls `ENTRYPOINT` corresponds to the command argument in the YAML manifest file, and what `Dockerfile` calls `CMD` corresponds to the `args` configuration key in the YAML manifest file.

What if you omit one of them? Kubernetes will default to what is inside the container image. If you omit the `args` key in the YAML, then Kubernetes will go for the `CMD` provided in the `Dockerfile`, while if you omit the command key, Kubernetes will go for the `ENTRYPOINT` declared in the `Dockerfile`. Most of the time, or at least if you're comfortable with your container's `ENTRYPOINT`, you're just going to override the `args` file (the `CMD` `Dockerfile` instruction).

When we create the Pod, we can check the output as follows:

```
$ kubectl apply -f nginx-debian-with-custom-command-and-args.yaml
pod/nginx-debian-with-custom-command-and-args created

$ kubectl get po -w
NAME                                          READY   STATUS            RESTARTS   AGE
nginx-debian-with-custom-command-and-args     0/2     ContainerCreating  0
2s
nginx-debian-with-custom-command-and-args     2/2     Running            0
6s
nginx-debian-with-custom-command-and-args     1/2     NotReady           0
66s
```

Therefore, overriding default commands offers granular control over container behavior within multi-container Pods. This enables tailored functionality, resource optimization, and dependency management for seamless Pod operation. In this section, we learned that Kubernetes allows overriding defaults through the command and args fields in Pod YAML definitions.

Now, let's look at another feature: `initContainers`! In the next section, you'll see another way to execute some additional side containers in your Pod to configure the main ones.

Introducing initContainers

`initContainers` is a feature provided by Kubernetes Pods to run setup scripts before the actual containers start. You can think of them as additional side containers you can define in your Pod YAML manifest file: they will first run when the Pod is created. Then, once they complete, the Pod starts creating its main containers.

You can execute not one but several initContainers in the same Pod, but when you define lots of them, keep in mind that they will run one after another, not in parallel. Once an initContainer completes, the next one starts, and so on. In general, initContainers are used for preparation tasks; some of them are outlined in the following list:

- **Database initialization:** Set up and configure databases before the main application starts.
- **Configuration file download:** Download essential configuration files from remote locations.
- **Package installation:** Install dependencies required by the main application.
- **Waiting for external service:** Ensure external services are available before starting the main container.
- **Running pre-checks:** Perform any necessary checks or validations before starting the main application.
- **Secret management:** Download and inject secrets securely into the main container's environment.
- **Data migration:** Migrate data to a database or storage system before the main application starts.
- **Customizing file permissions:** Set appropriate file permissions for the main application.

Since initContainers can have their own container images. You can offload some configuration to them by keeping the main container images as small as possible, thus increasing the whole security of your setup by removing unnecessary tools from your main container images. Here is a YAML manifest that introduces an initContainer:

```yaml
# nginx-with-init-container.yaml
---
apiVersion: v1
kind: Pod
metadata:
  name: nginx-with-init-container
  labels:
    environment: prod
    tier: frontend
spec:
  restartPolicy: Never
  volumes:
    - name: website-volume
      emptyDir: {}
  initContainers:
    - name: download-website
      image: busybox
      command:
        - sh
        - -c
```

```
        - |
          wget https://github.com/iamgini/website-demo-one-page/archive/refs/
heads/main.zip -O /tmp/website.zip && \
          mkdir /tmp/website && \
          unzip /tmp/website.zip -d /tmp/website && \
          cp -r /tmp/website/website-demo-one-page-main/* /usr/share/nginx/html
      volumeMounts:
        - name: website-volume
          mountPath: /usr/share/nginx/html
  containers:
    - name: nginx-container
      image: nginx:latest
      volumeMounts:
        - name: website-volume
          mountPath: /usr/share/nginx/html
```

As you can see from this YAML file, initContainer runs the BusyBox image, which will download the application (in this case, simple website content from https://github.com/iamgini/website-demo-one-page) and copy the same application to a shared volume called website-volume. (You will learn about volumes and persistent storage in *Chapter 9, Persistent Storage in Kubernetes*, later in this book.) The same volume is also configured to mount under the NGINX container so that NGINX will use it as the default website content. Once the execution of initContainer is complete, Kubernetes will create the nginx-container container.

> Keep in mind that if initContainer fails, Kubernetes won't initiate the primary containers. It's crucial not to perceive initContainer as an optional component or one that can afford to fail. If included in the YAML manifest file, they are mandatory and their failure prevents the launch of the main containers!

Let's create the Pod. After this, we will run the kubectl get Pods -w command for kubectl to watch for a change in the Pod list. The output of the command will be updated regularly, showing the change in the Pod's status. Please take note of the status column, which says that an initContainer is running:

```
$ kubectl apply -f nginx-with-init-container.yaml
pod/nginx-with-init-container created

$ kubectl get po -w
NAME                        READY    STATUS          RESTARTS    AGE
nginx-with-init-container   0/1      Init:0/1        0           3s
nginx-with-init-container   0/1      Init:0/1        0           4s
nginx-with-init-container   0/1      PodInitializing 0           19s
nginx-with-init-container   1/1      Running         0           22s
```

As you can see, `Init:0/1` indicates that `initContainer` is being launched. After its completion, the `Init:` prefix disappears for the next statuses, indicating that we are done with `initContainer` and that Kubernetes is now creating the main container – in our case, the NGINX one!

If you want to explore this further, you can expose the Pod with a NodePort service as follows:

```
$ kubectl expose pod nginx-with-init-container --port=80 --type=NodePort
service/nginx-with-init-container exposed
```

Now, start a port forwarding service using `kubectl port-forward` command as follows so that we can access the service outside of the cluster:

```
$ kubectl port-forward pod/nginx-with-init-container 8080:80
Forwarding from 127.0.0.1:8080 -> 80
Forwarding from [::1]:8080 -> 80
```

Now, access `http://localhost:8080` and you will see a one-page website with content copied from `https://github.com/iamgini/website-demo-one-page`. We will learn about exposing services in *Chapter 8*, *Exposing Your Pods with Services*. Also remember to stop the port forward by pressing the *ctrl + c* to in your console before proceeding to the next section.

Use `initContainer` wisely when you're building your Pods! You are not forced to use `init` containers, but they can be really helpful for running configuration scripts or pulling something from external servers before you launch your actual containers!

Now, let's learn how to access the logs of a specific container inside a running Pod!

Accessing the logs of a specific container

When using multiple containers in a single Pod, you can retrieve the logs of a dedicated container inside the Pod. The proper way to proceed is by using the `kubectl logs` command.

The most common way a containerized application exposes its logs is by sending them to `stdout`. The `kubectl logs` command is capable of streaming the `stdout` property of a dedicated container in a dedicated Pod and retrieving the application logs from the container. For it to work, you will need to know the name of both the precise container and its parent Pod, just like when we used `kubectl exec` to access a specific container.

Please read the previous section, *Accessing a specific container inside a multi-container Pod*, to discover how to do this:

```
$ kubectl logs -f pods/multi-container-pod --container nginx-container
```

Please note the `--container` option (the `-c` short option works, too), which specifies the container you want to retrieve the logs for. Note that it also works the same for `initContainers`: you have to pass its name to this option to retrieve its logs.

Remember that if you do not pass the `--container` option, you will retrieve all the logs from all the containers that have been launched inside the Pod. Not passing this option is useful in the case of a single-container Pod, but you should consider this option every time you use a multi-container Pod.

There are other multiple useful options you need to be aware of when it comes to accessing the logs of a container in a Pod. You can decide to retrieve the logs written in the last two hours by using the following command:

```
$ kubectl logs --since=2h pods/multi-container-pod --container nginx-container
```

Also, you can use the `--tail` option to retrieve the most recent lines of a log's output. Here's how to do this:

```
$ kubectl logs --tail=30 pods/multi-container-pod --container nginx-container
```

Here, we are retrieving the 30 most recent lines in the log output of `nginx-container`.

Now, you are ready to read and retrieve the logs from your Kubernetes Pods, regardless of whether they are made up of one or several containers!

In this section, we discovered how to create, update, and delete multi-container Pods. We also discovered how to force the deletion of a Pod. We then discovered how to access a specific container in a Pod, as well as how to retrieve the logs of a specific container in a Pod. Though we created an NGINX and a Debian container in our Pod, they are relatively poorly linked since they don't do anything together. To remediate that, we will now learn how to deal with volumes so that we can share files between our two containers.

Sharing volumes between containers in the same Pod

In this section, we'll learn what volumes are from a Kubernetes point of view and how to use them. Docker also has volumes, but they differ from Kubernetes volumes: they answer the same need but they are not the same.

We will discover what Kubernetes volumes are, why they are useful, and how they can help us when it comes to Kubernetes volumes.

What are Kubernetes volumes?

We are going to answer a simple problem. Our multi-container Pods are currently made up of two containers: an NGINX one and a Debian one. We are going to try sharing the log directory in the NGINX container with the Debian container by mounting the log directory of NGINX in the directory of the Debian container. This way, we will create a relationship between the two containers to have them share a directory.

Kubernetes has two kinds of volumes:

- Volumes, which we will discuss here.
- PersistentVolume, which is a more advanced feature we will discuss later, in *Chapter 9, Persistent Storage in Kubernetes*.

Keep in mind that these two are not the same. PersistentVolume is a resource of its own, whereas "volumes" is a Pod configuration. As the name suggests, PersistentVolume is persistent, whereas volumes are not supposed to be. But keep in mind that this is not always the case!

In straightforward terms, volumes in Kubernetes are intricately linked to the life cycle of a Pod. When you instantiate a Pod, you can define and connect volumes to the containers within it. Essentially, volumes represent storage tied to the existence of the Pod. Once the Pod is removed, any associated volumes are also deleted.

While volumes serve a broader range of purposes beyond this scenario, it's worth noting that this description doesn't universally apply. However, you can view volumes as an especially effective method for facilitating the sharing of directories and files among containers coexisting within a Pod.

> Remember that volumes are bound to the Pod's life cycle, not the container's life cycle. If a container crashes, the volume will survive because if a container crashes, it won't cause its parent Pod to crash, and thus, no volume will be deleted. So long as a Pod is alive, its volumes are, too.

Volumes are a core concept for managing data in containerized applications. They provide a way to persist data independent of the container's life cycle. Kubernetes supports various volume types, including those mounted from the host file system, cloud providers, and network storage systems.

However, Kubernetes expanded on this by introducing support for various drivers, enabling integration of Pod volumes with external solutions. For instance, an AWS EBS (Elastic Block Store) volume seamlessly serves as a Kubernetes volume. Some widely utilized solutions include the following:

- hostPath
- emptyDir
- nfs
- persistentVolumeClaim (when you need to use a PersistentVolume, which is outside the scope of this chapter)

Please note that some of the old volume types are removed or deprecated in the latest Kubernetes version; refer to the documentation to learn more (https://kubernetes.io/docs/concepts/storage/volumes/). Refer to the following list for examples:

- azureDisk (removed)
- gcePersistentDisk (removed)
- glusterfs (removed)
- azureFile (deprecated)

Please note that using external solutions to manage the Kubernetes volumes will require you to follow those external solutions' requirements. For example, using an AWS EBS volume as a Kubernetes volume will require your Pods to be executed on a Kubernetes worker node, which would be an EC2 instance. The reason for this is that AWS EBS volumes can only be attached to EC2 instances. Thus, a Pod exploiting such a volume would need to be launched on an EC2 instance. Refer to `https://kubernetes.io/docs/concepts/storage/volumes/` to learn more.

The following diagram shows the high-level idea about the `hostPath` and `emptyDir` volumes.

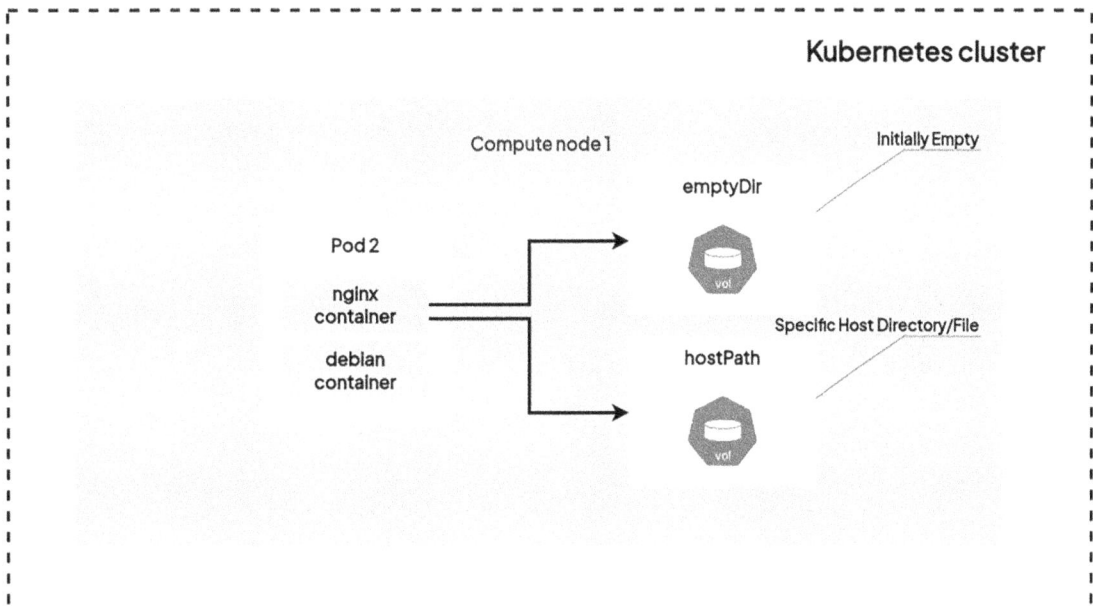

Figure 5.2: hostPath and emptyDir volumes

We are going to discover the two basic volume drivers here: `emptyDir` and `hostPath`. We will also talk about `persistentVolumeClaim` because this one is going to be a little special in comparison to the other volumes and will be fully discovered in *Chapter 9, Persistent Storage in Kubernetes*.

Now, let's start discovering how to share files between containers in the same Pod using volumes with the `emptyDir` volume type!

Creating and mounting an emptyDir volume

As the name suggests, it is simply an empty directory that is initialized at Pod creation that you can mount to the location of each container running in the Pod.

It is certainly the easiest and simplest way to have your containers share data between them. Let's create a Pod that will manage two containers.

In the following example, we are creating a Pod that will launch two containers, and just like we had previously, it's going to be an NGINX container and a Debian container. We are going to override the command that's run by the Debian container when it starts to prevent it from completing. That way, we will get it running indefinitely as a long process and we will be able to launch additional commands to check whether our emptyDir has been initialized correctly.

Both containers will have a common volume mounted at /var/i-am-empty-dir-volume/, which will be our emptyDir volume, initialized in the same Pod. Here is the YAML file for creating the Pod:

```yaml
# multi-container-with-emptydir-pod.yaml
---
apiVersion: v1
kind: Pod
metadata:
  name: multi-container-with-emptydir-pod
spec:
  containers:
    - name: nginx-container
      image: nginx:latest
      volumeMounts:
      - mountPath: /var/i-am-empty-dir-volume
        name: empty-dir-volume
    - name: debian-container
      image: debian
      command: ["/bin/sh"]
      args: ["-c", "while true; do sleep 30; done;"] # Prevents container from
exiting after completion
      volumeMounts:
      - mountPath: /var/i-am-empty-dir-volume
        name: empty-dir-volume
  volumes:
  - name: empty-dir-volume # name of the volume
    emptyDir: {} # Initialize an empty directory # The path on the worker node.
```

Note that the object we will create in our Kubernetes cluster will become more complex as we go through this example, and, as you can imagine, most complex things cannot be achieved with just imperative commands. That's why you are going to see more and more examples relying on the YAML manifest file: you should start a habit of trying to read them to figure out what they do.

We can now apply the manifest file using the following kubectl apply -f command:

```
$ kubectl apply -f multi-container-with-emptydir-pod.yaml
pod/multi-container-with-emptydir-pod created
```

Now, we can check that the Pod is successfully running by issuing the kubectl get Pods command:

```
$ kubectl get po
NAME                                   READY   STATUS    RESTARTS   AGE
multi-container-with-emptydir-pod      2/2     Running   0          25s
```

Now that we are sure the Pod is running and that both the NGINX and Debian containers have been launched, we can check that the directory can be accessed in both containers by issuing the ls command.

If the command is not failing, as we saw previously, we can run the ls command in the containers by simply running the kubectl exec command. As you may recall, the command takes the Pod's name and the container's name as arguments. We are going to run it twice to make sure the volume is mounted in both containers:

```
$ kubectl exec multi-container-with-emptydir-pod -c debian-container -- ls /var
backups
cache
i-am-empty-dir-volume
lib
local
lock
log
mail
opt
run
spool
tmp
$ kubectl exec multi-container-with-emptydir-pod -c nginx-container  -- ls /var
backups
cache
i-am-empty-dir-volume
lib
local
lock
log
mail
opt
run
spool
tmp
```

As you can see, the ls /var command is showing the name in both containers! This means that emptyDir was initialized and mounted in both containers correctly.

Now, let's create a file in one of the two containers. The file should be immediately visible in the other container, proving that the volume mount is working properly!

In the following command, we are simply creating a .txt file called hello-world.txt in the mounted directory:

```
$ kubectl exec multi-container-with-emptydir-pod -c debian-container -- bin/sh
-c "echo 'hello world' >> /var/i-am-empty-dir-volume/hello-world.txt"
$ kubectl exec multi-container-with-emptydir-pod -c nginx-container -- cat /
var/i-am-empty-dir-volume/hello-world.txt
hello world
$ kubectl exec multi-container-with-emptydir-pod -c debian-container -- cat /
var/i-am-empty-dir-volume/hello-world.txt
hello world
```

As you can see, we used debian-container to create the /var/i-am-empty-dir-volume/hello-world.txt file, which contains the hello-world string. Then, we simply used the cat command to access the file from both containers; you can see that the file is accessible in both cases. Again, remember that emptyDir volumes are completely tied to the life cycle of the Pod. If the Pod declares it is destroyed, then the volume is destroyed, too, along with all its content, and it will become impossible to recover!

Now, we will discover another volume type: the hostPath volume. As you can imagine, it's going to be a directory that you can mount on your containers that is backed by a path on the host machine – the Kubernetes node running the Pod!

Creating and mounting a hostPath volume

As the name suggests, hostPath will allow you to mount a directory in the host machine to containers in your Pod! The host machine is the Kubernetes compute node (or controller node) executing the Pod. Here are some examples:

- If your cluster is based on minikube (a single-node cluster), the host is your local machine.
- On Amazon EKS, the host machine will be an EC2 instance.
- In a kubeadm cluster, the host machine is generally a standard Linux machine.

The host machine is the machine running the Pod, and you can mount a directory from the file system of the host machine to the Kubernetes Pod!

In the following example, we will be working on a Kubernetes cluster based on minikube, so hostPath will be a directory that's been created on your computer that will then be mounted in a Kubernetes Pod.

> Using the hostPath volume type can be useful, but in the Kubernetes world, you can consider it an anti-pattern. The hostPath volume type, while convenient, is discouraged in Kubernetes due to reduced portability and potential security risks. It can also be incompatible with advanced security features like SELinux **multi-category security** (**MCS**), now supported by many Kubernetes distributions. For a more portable, secure, and future-proof approach, leverage **persistent volumes** (**PVs**) and **persistent volume claims** (**PVCs**) to manage persistent data within your containerized applications.

The whole idea behind Pods is that they are supposed to be easy to delete and reschedule on another worker node without problems. Using hostPath will create a tight relationship between the Pod and the worker node, and that could lead to major issues if your Pod were to fail and be rescheduled on a node where the required path on the host machine is not present.

Now, let's discover how to create hostPath.

Let's imagine that we have a file on the worker node on worker-node/nginx.conf and we want to mount it on /var/config/nginx.conf on the nginx container.

Here is the YAML file to create the setup. As you can see, we declared a hostPath volume at the bottom of the file that defines a path that should be present on the host machine. Now, we can mount it on any container that needs to deal with the volume in the containers block:

```yaml
# multi-container-with-hostpath.yaml
---
apiVersion: v1
kind: Pod
metadata:
  name: multi-container-with-hostpath
spec:
  containers:
    - name: nginx-container
      image: nginx:latest
      volumeMounts:
      - mountPath: /foo
        name: my-host-path-volume
    - name: debian-container
      image: debian
      command: ["/bin/sh"]
      args: ["-c", "while true; do sleep 30; done;"] # Prevents container from exiting after completion
  volumes:
  - name: my-host-path-volume
    hostPath:
      path: /tmp # The path on the worker node.
      type: Directory
```

As you can see, mounting the value is just like what we did with the emptyDir volume in the previous section regarding the emptyDir volume type. By using a combination of volumes at the Pod level and volumeMounts at the container level, you can mount a volume on your containers.

> In the previous YAML snippet, we mentioned `type: Directory`, which means the directory already exists on the host machine. If you want to create the directory or file on the host machine, then use `DirectoryOrCreate` and `FileOrCreate` respectively.

You can also mount the directory on the Debian container so that it gets access to the directory on the host.

Before running the YAML manifest file, though, you need to create the path on your host and create the necessary file:

```
$ echo "Hello World" >> /tmp/hello-world.txt
```

If you are using a minikube cluster, remember to do this step inside the minikube VM as follows:

```
$ minikube ssh
docker@minikube:~$ echo "Hello World" > /tmp/hello-world.txt
docker@minikube:~$ exit
Logout
```

If your minikube cluster is created using Podman containers (e.g., `minikube start --profile cluster2-podman --driver=podman`), then log in to the minikube Pod and create the file:

```
$ sudo podman exec -it minikube /bin/bash
root@minikube:/# cat /tmp/hello-world.txt
```

Now that the path exists on the host machine, we can apply the YAML file to our Kubernetes cluster and, immediately after, launch a `kubectl get Pod` command to check that the Pod was created correctly:

```
$ kubectl apply -f multi-container-with-hostpath.yaml
pod/multi-container-with-hostpath created
$ kubectl get pod
NAME                             READY   STATUS    RESTARTS   AGE
multi-container-with-hostpath    2/2     Running   0          11
```

Everything seems good! Now, let's echo the file that should be mounted at `/foo/hello-world.txt`:

```
$ kubectl exec multi-container-with-hostpath -c nginx-container -- cat /foo/
hello-world.txt
Hello World
```

We can see the local file (on the Kubernetes node) is available inside the container via the hostPath volume mount.

At the beginning of this chapter, we discovered the different aspects of multi-container Pods! We discovered how to create, update, and delete multi-container Pods, as well as how to use `initContainers`, access logs, override command-line arguments passed to containers directly from the Pod's resources, and share directories between containers using the two basic volumes.

Now, we are going to put a few architecting principles together and discover some notions related to multi-container Pods called "patterns."

The ambassador design pattern

When designing a multi-container Pod, you can decide to follow some architectural principles to build your Pod. Some typical needs are answered by these design principles, and the ambassador pattern is one of them.

Here, we are going to discover what the ambassador design pattern is, learn how to build an ambassador container in Kubernetes Pods, and look at a concrete example of them.

What is the ambassador design pattern?

In essence, the ambassador design pattern applies to multi-container Pods. We can define two containers in the same Pod:

- The first container will be called the main container.
- The other container will be called the ambassador container.

In this design pattern, we assume that the main container might have to access external services to communicate with them. For example, you can have an application that must interact with an SQL database that is living outside of your Pod, and you need to reach this database to retrieve data from it.

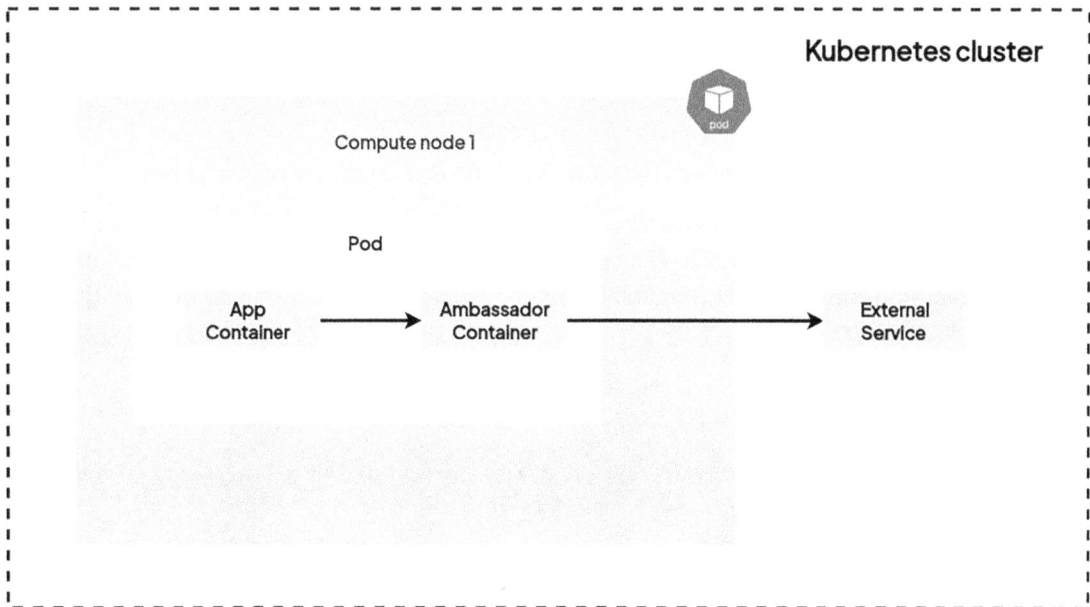

Figure 5.3: Ambassador design pattern in Kubernetes

This is the typical use case where you can deploy an adapter container alongside the main container, next to it, in the same Pod. The whole idea is to get the ambassador container to proxy the requests run by the main container to the database server.

In this case, the ambassador container will be essentially an SQL proxy. Every time the main container wants to access the database, it won't access it directly but rather create a connection to the ambassador container that will play the role of a proxy.

> Running an ambassador container is fine, but only if the external API is not living in the same Kubernetes cluster. To run requests on another Pod, Kubernetes provides strong mechanics called Services. We will have the opportunity to discover them in *Chapter 8, Exposing Your Pods with Services*.

But why would you need a proxy to access external databases? Here are some concrete benefits this design pattern can bring you:

- Offloading SQL configuration
- Management of **Secure Sockets Layer/Transport Layer Security (SSL/TLS)** certificates

Please note that having an ambassador proxy is not limited to an SQL proxy but this example is demonstrative of what this design pattern can bring you. Note that the ambassador proxy is only supposed to be called for outbound connections from your main container to something else, such as data storage or an external API. It should not be seen as an entry point to your cluster! Now, let's quickly discover how to create an ambassador SQL proxy with a YAML file.

Ambassador multi-container Pod – an example

Now that we know about ambassador containers, let's learn how to create one with Kubernetes. The following YAML manifest file creates a Pod that creates two containers:

- `nginx-app`, derived from the `nginx:latest` image
- `sql-ambassador-proxy`, created from the `mysql-proxy:latest` container image

The following example is only to demonstrate the concept of the ambassador SQL proxy. If you want to test the full functionality, you should have a working AWS **RDS (Relational Database Service)** instance reachable from your Kubernetes cluster and a proper application to test the database operations.

```
# ~/ nginx-with-ambassador.yaml
apiVersion: v1
kind: Pod
metadata:
  name: nginx-with-ambassador
spec:
  containers:
    - name: mysql-proxy-ambassador-container
      image: mysql-proxy:latest
      ports:
          - containerPort: 3306
      env:
      - name: DB_HOST
```

```
          value: mysql.xxx.us-east-1.rds.amazonaws.com
    - name: nginx-container
      image: nginx:latest
```

As you can imagine, it's going to be the developer's job to get the application code running in the NGINX container to query the ambassador instead of the Amazon RDS endpoint. As the ambassador container can be configured from environment variables, it's going to be easy for you to just input the configuration variables in `ambassador-container`.

> Do not get tricked by the order of the containers in the YAML file. The fact that the ambassador container appears first does not make it the *main* container of the Pod. This notion of the *main* container does not exist at all from a Kubernetes perspective – both are plain containers that run in parallel with no concept of a hierarchy between them. Here, we just access the Pod from the NGINX container, which makes it the most important one.

Remember that the ambassador running in the same Pod as the NGINX container makes it accessible from NGINX on `localhost:3306`.

In the next section, we will learn about the sidecar pattern, another important concept in multi-container Pods.

The sidecar design pattern

The sidecar design pattern is good when you want to extend the features of your main container with features it would normally not be able to achieve on its own.

Just like we did for the ambassador container, we're going to explain exactly what the sidecar design pattern is by covering some examples.

What is the sidecar design pattern?

Think of the sidecar container as an extension or a helper for your main container. Its main purpose is to extend the main container to bring it a new feature, but without changing anything about it. Unlike the ambassador design pattern, the main container may even not be aware of the presence of a sidecar.

Just like the ambassador design pattern, the sidecar design pattern makes use of a minimum of two containers:

- The main container – the one that is running the application
- The sidecar container – the one that is bringing something additional to the first one

You may have already guessed, but this pattern is especially useful when you want to run monitoring or log forwarder agents. The following figure shows a simple sidecar design with a main app container and a sidecar container to collect the application logs.

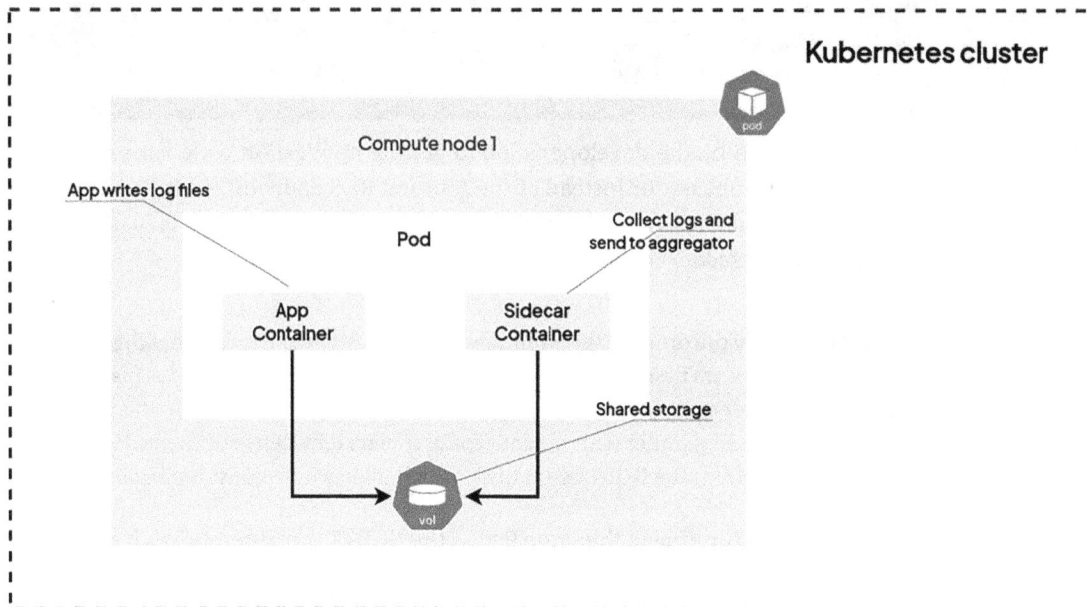

Figure 5.4: Sidecar design pattern in Kubernetes

There are three things to understand when you want to build a sidecar that is going to forward your logs to another location:

- You must locate the directory where your main containers write their data (e.g., logs).
- You must create a volume to make this directory accessible to the sidecar container (e.g., a log forwarder sidecar).
- You must launch the sidecar container with the proper configuration.

Based on these concepts, the main container remains unchanged, and even if the sidecar fails, it wouldn't have an impact on the main container, which could continue to work.

When to use a Sidecar design pattern?

When considering the usage of sidecar containers, they prove particularly beneficial in the following scenarios:

- **Network proxies:** Network proxies can be configured to initialize before other containers in the Pod, ensuring their services are available immediately. The Istio "Envoy" proxy is a great example of a sidecar container used as a proxy.
- **Enhanced logging:** Log collection sidecars can start early and persist until the Pod terminates, capturing logs reliably even in case of Pod crashes.
- **Jobs:** Sidecars can be deployed alongside Kubernetes Jobs without affecting Job completion. No additional configuration is required for sidecars to run within Jobs.
- **Credential management:** Many third-party credential management platforms utilize sidecar Pods to inject and manage credentials within workloads. They can also facilitate secure credential rotation and revocation.

Sidecar multi-container Pod — an example

Just like the ambassador design pattern, the sidecar makes use of multi-container Pods. We will define two containers in the same Pod as follows – an NGINX container, which acts as the application container, and a `Fluentd` container, which acts as the sidecar to collect the logs from the NGINX web server:

```yaml
# nginx-with-fluentd-sidecar.yaml
apiVersion: v1
kind: Pod
metadata:
  name: nginx-with-sidecar
spec:
  containers:
    - name: nginx-container
      image: nginx:latest
      ports:
        - containerPort: 80
      volumeMounts:
        - name: log-volume
          mountPath: /var/log/nginx
    - name: fluentd-sidecar
      image: fluent/fluentd:v1.17
      volumeMounts:
        - name: log-volume
          mountPath: /var/log/nginx
  volumes:
    - name: log-volume
      emptyDir: {}
```

Please note, for the Fluentd to work properly, we need to pass the configuration via ConfigMap; a typical configuration can be found in the following code (you will learn more about ConfigMaps in *Chapter 7, Configuring your Pods using ConfigMaps and Secrets)*:

```yaml
---
apiVersion: v1
kind: ConfigMap
metadata:
  name: fluentd-config-map
  namespace: default
data:
  fluentd.conf: |
    <source>
      @type tail
```

```
    path /var/log/nginx/*.log
    pos_file /var/log/nginx/nginx.log.pos
    tag nginx
    <parse>
      @type nginx
    </parse>
  </source>

  <match nginx.**>
    @type elasticsearch
    host elastic.lab.example.com
    port 9200
    logstash_format true
    logstash_prefix fluentd
    logstash_dateformat %Y.%m.%d
  </match>
```

> **Fluentd** is a popular open-source log collection and forwarding agent, often used as a sidecar container in Kubernetes deployments. It efficiently collects logs from various sources, parses them for structure, and forwards them to centralized logging platforms like Elasticsearch, Google Cloud Logging, or Amazon CloudWatch Logs. This allows for streamlined log management, improved observability, and easier analysis of application health and performance. While this example demonstrates sending logs to a dummy Elasticsearch server (e.g., `elastic.lab.example.com`), Fluentd offers flexibility to integrate with various external logging solutions depending on your specific needs.

In the following section of this chapter, we will discuss the adapter design pattern.

The adapter design pattern

As its name suggests, the adapter design pattern is going to *adapt* an entry from a source format to a target format.

As with the ambassador and sidecar design patterns, this one expects that you run at least two containers:

- The first one is the main container.
- The second one is the adapter container.

This design pattern is helpful and should be used whenever the main containers emit data in a format, A, that should be sent to another application that is expecting the data in another format, B. As the name suggests, the adapter container is here to *adapt*.

Again, this design pattern is especially well suited for log or monitoring management. Imagine a Kubernetes cluster where you have dozens of applications running; they are writing logs in Apache format, which you need to convert into JSON so that they can be indexed by a search engine. This is exactly where the adapter design pattern comes into play. Running an adapter container next to the application containers will help you get these logs adapted to the source format before they are sent somewhere else.

Just like for the sidecar design pattern, this one can only work if both the containers in your Pod are accessing the same directory using volumes.

Adapter multi-container Pod — an example

In this example, we are going to use a Pod that uses an adapter container with a shared directory mounted as a Kubernetes volume.

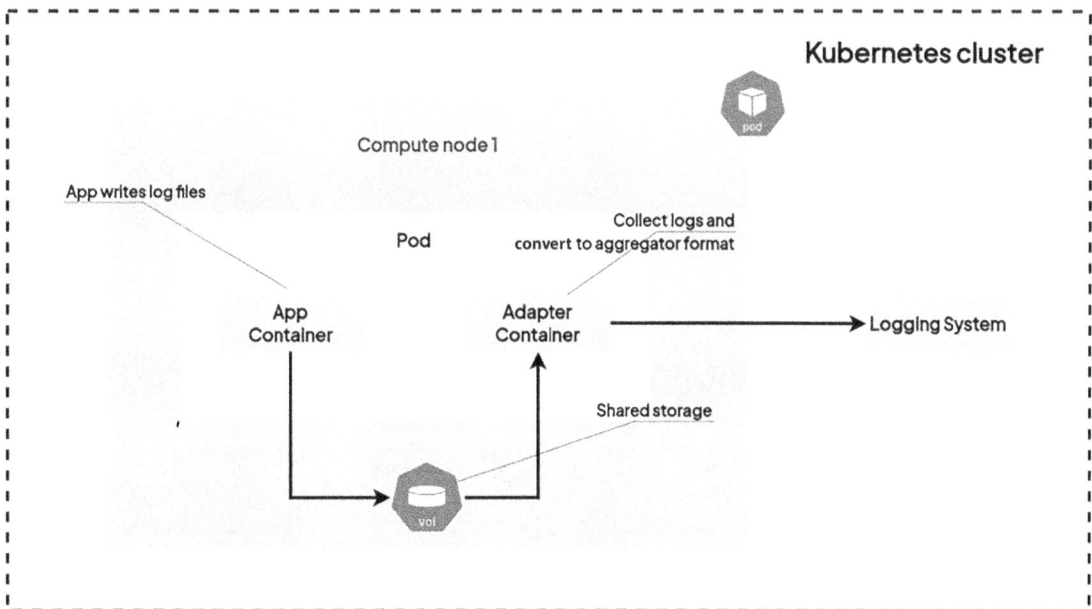

Figure 5.5: Adapter design pattern in Kubernetes

This Pod is going to run two containers:

- `alpine-writer`: Main app container, which writes logs to /var/log/app.
- `log-adapter`: Adapter container, which will read the log and convert it to another format (e.g., append a `PROCESSED` string at the end of each log).

The following YAML file contains the definition for an adapter multi-container Pod with multiple containers:

```
# alpine-with-adapter.yaml
apiVersion: v1
kind: Pod
```

```
metadata:
  name: pod-with-adapter
spec:
  containers:
    - name: alpine-writer
      image: alpine:latest
      command: [ "sh", "-c", "i=1; while true; do echo \"$(date) - log $i\" >>
/var/log/app/app.log; i=$(((i+1)); sleep 5; done" ]
      volumeMounts:
        - name: log-volume
          mountPath: /var/log/app
      # adapter container
    - name: log-adapter
      image: alpine:latest
      command: [ "sh", "-c", "while true; do cat /logs/app.log | sed 's/$/
PROCESSED/' > /logs/processed_app.log; cat /logs/processed_app.log; sleep 10;
done" ]
      volumeMounts:
        - name: log-volume
          mountPath: /logs
  volumes:
    - name: log-volume
      emptyDir: {}
```

Apply the YAML and check the Pod status as follows:

```
$ kubectl apply -f nginx-with-adapter.yaml
pod/pod-with-adapter created
```

Once the Pod is created, the logs will be generated, and we can verify the logs from both containers. The following command will display the logs by the `alpine-writer` container:

```
$ kubectl exec -it pod-with-adapter -c alpine-writer -- head -5 /var/log/app/
app.log
Sun Jun 30 15:05:26 UTC 2024 - log 1
Sun Jun 30 15:05:31 UTC 2024 - log 2
Sun Jun 30 15:05:36 UTC 2024 - log 3
Sun Jun 30 15:05:41 UTC 2024 - log 4
Sun Jun 30 15:05:46 UTC 2024 - log 5
```

We can also check the converted logs by using `log-adapter`:

```
$ kubectl exec -it pod-with-adapter -c log-adapter -- head -5 /logs/processed_
app.log
Sun Jun 30 15:05:26 UTC 2024 - log 1 PROCESSED
```

```
Sun Jun 30 15:05:31 UTC 2024 - log 2 PROCESSED
Sun Jun 30 15:05:36 UTC 2024 - log 3 PROCESSED
Sun Jun 30 15:05:41 UTC 2024 - log 4 PROCESSED
Sun Jun 30 15:05:46 UTC 2024 - log 5 PROCESSED
```

By using the adapter containers, it is possible to handle complex operations without modifying your original application containers.

Before we conclude the chapter, in the next section, let us learn about one more feature in Kubernetes related to multi-container Pods.

Sidecars versus Kubernetes Native Sidecars

Traditionally, sidecars in Kubernetes have been regular containers deployed alongside the main application in a Pod, as we learned in the previous sections. This approach offers additional functionalities but has limitations. For instance, sidecars might continue running even after the main application exits, wasting resources. Additionally, Kubernetes itself isn't inherently aware of sidecars and their relationship with the primary application.

To address these limitations, Kubernetes v1.28 introduced a new concept: native sidecars. These leverage existing `init` containers with special configurations. This allows you to define a `restartPolicy` for containers within the Pod's `initContainers` section. These special sidecar containers can be independently started, stopped, or restarted without affecting the main application or other init containers, offering more granular control over their life cycle.

The following Deployment definition file explains how native sidecar containers can be configured in Kubernetes:

```
...<removed for brevity>...
    spec:
      containers:
        - name: myapp
          image: alpine:latest
          command: ['sh', '-c', 'while true; do echo "logging" >> /opt/logs.
txt; sleep 1; done']
          volumeMounts:
            - name: data
              mountPath: /opt
      initContainers:
        - name: logshipper
          image: alpine:latest
          restartPolicy: Always
          command: ['sh', '-c', 'tail -F /opt/logs.txt']
          volumeMounts:
            - name: data
              mountPath: /opt
      volumes:
```

```
    - name: data
      emptyDir: {}
```

This approach ensures synchronized startup and shutdown of the sidecar with the main container, optimizing resource usage. More importantly, Kubernetes gains awareness of the sidecar's role in the Pod, potentially enabling future features for tighter integration.

By leveraging multi-container Pods with init containers, sidecars, and the adapter or ambassador patterns, Kubernetes empowers you to build complex applications as modular units. This streamlines deployments and promotes efficient resource utilization within your containerized environment.

Summary

This chapter was quite a long one, but you should now have a good understanding of what Pods are and how to use them, especially when it comes to managing multiple containers in the same Pod.

We recommend that you focus on mastering the declarative way of creating Kubernetes resources. As you have noticed in this chapter, the key to achieving the most complex things with Kubernetes resides in writing YAML files. One example is that you simply cannot easily create a multi-container Pod without writing YAML files.

This chapter complements the previous one: *Chapter 4, Running Your Containers in Kubernetes*. You need to understand that everything we will do with Kubernetes will be Pod management because everything in Kubernetes revolves around them. Keep in mind that containers are never created directly, but always through a Pod object, and that all the containers within the same Pod are created on the same Kubernetes node. If you understand that, then you can continue to the next chapter!

In the next chapter, we're going to cover another important aspect of Kubernetes called namespaces.

Further reading

- How Pods manage multiple containers: `https://kubernetes.io/docs/concepts/workloads/pods/#how-pods-manage-multiple-containers`
- Kubernetes volumes: `https://kubernetes.io/docs/concepts/storage/volumes/`
- Sidecar containers: `https://kubernetes.io/docs/concepts/workloads/pods/sidecar-containers/`

Join our community on Discord

Join our community's Discord space for discussions with the authors and other readers:

`https://packt.link/cloudanddevops`

6

Namespaces, Quotas, and Limits for Multi-Tenancy in Kubernetes

So far, we've learned about Kubernetes' key concepts by launching objects into our clusters and observing their behavior. You may have noticed that, in the long run, it would be difficult to maintain a cleanly organized cluster. As your clusters grow, it will become more and more difficult to maintain the ever-increasing number of resources managed in your cluster. That's when Kubernetes namespaces come into play.

In this chapter, we will learn about **namespaces**. They help us keep our clusters well organized by grouping our resources by application or environment. Kubernetes namespaces are another key aspect of Kubernetes management, and it's really important to master them!

In this chapter, we're going to cover the following main topics:

- Introduction to Kubernetes namespaces
- How namespaces impact your resources and services
- Configuring ResourceQuota and Limits at the namespace level

Technical requirements

For this chapter, you will need the following:

- A working Kubernetes cluster (local or cloud-based, but this is not important)
- A working kubectl CLI configured to communicate with the cluster

If you do not have these technical requirements, please read *Chapter 2*, *Kubernetes Architecture – from Container Images to Running Pods*, and *Chapter 3*, *Installing Your First Kubernetes Cluster*, to get them.

You can download the latest code samples for this chapter from the official GitHub repository at https://github.com/PacktPublishing/The-Kubernetes-Bible-Second-Edition/tree/main/Chapter06.

Introduction to Kubernetes namespaces

The more applications you deploy on your Kubernetes clusters, the greater the need to keep your cluster resources organized. You can use labels and annotations to manage the objects within your cluster, but you can take organization further by using **namespaces**. Namespaces in Kubernetes allow you to logically isolate parts of your cluster, helping you manage resources more effectively. However, to enforce resource allocation and limits, additional objects like `ResourceQuotas` are required. Once namespaces have been created, you can launch Kubernetes objects such as Pods, which will only exist in that namespace. So all the operations that are run against the cluster with `kubectl` will be scoped to that individual namespace, where you can perform as many operations as possible while eliminating the risk of impacting resources that are in another namespace.

Let's start by finding out what exactly namespaces are and why they were created.

> For advanced multi-cluster and multi-tenancy scenarios in Kubernetes, projects like **Capsule** (https://capsule.clastix.io/) and **HyperShift** (https://github.com/openshift/hypershift) offer robust solutions. Capsule enables secure, multi-tenant Kubernetes clusters by allowing different teams or tenants to manage their own isolated namespaces. HyperShift simplifies the management of multiple clusters by providing a lightweight and scalable approach to securely isolate and manage Kubernetes resources across different environments.

Now, let's move on to the next section on the importance of namespaces in Kubernetes.

The importance of namespaces in Kubernetes

As we mentioned previously, namespaces in Kubernetes are a way to help the cluster administrator keep everything clean and organized, while providing resource isolation.

The biggest Kubernetes clusters can run hundreds or even thousands of applications. When everything is deployed in the same namespace, it can become very complex to know which particular resource belongs to which application.

If, by misfortune, you update or modify the wrong resource, you might end up breaking an app running in your cluster. To resolve that, you can use labels and selectors, but even then, as the number of resources grows, managing the cluster will quickly become chaotic if you don't start using namespaces.

We learned the basics of creating namespaces in *Chapter 8, Exposing Your Pods with Services*, but we didn't learn that in much detail. Let's now learn in detail how the namespaces are used to keep everything clean and organized while providing resource isolation.

How namespaces are used to split resources into chunks

Right after you've installed Kubernetes, when your cluster is brand new, it is created with a few namespaces for the cluster components. So even if you didn't notice previously, you are already using namespaces as follows:

```
$ kubectl get namespaces
NAME              STATUS   AGE
default           Active   8d
kube-node-lease   Active   8d
kube-public       Active   8d
kube-system       Active   8d
```

The main concept is to deploy your Pods and other objects in Kubernetes while specifying a namespace of your preference. This practice helps to keep your cluster tidy and well structured. It's worth noting that Kubernetes comes with a default namespace, which is used if you don't specify one explicitly.

The following image illustrates the namespaces and isolation at a high level.

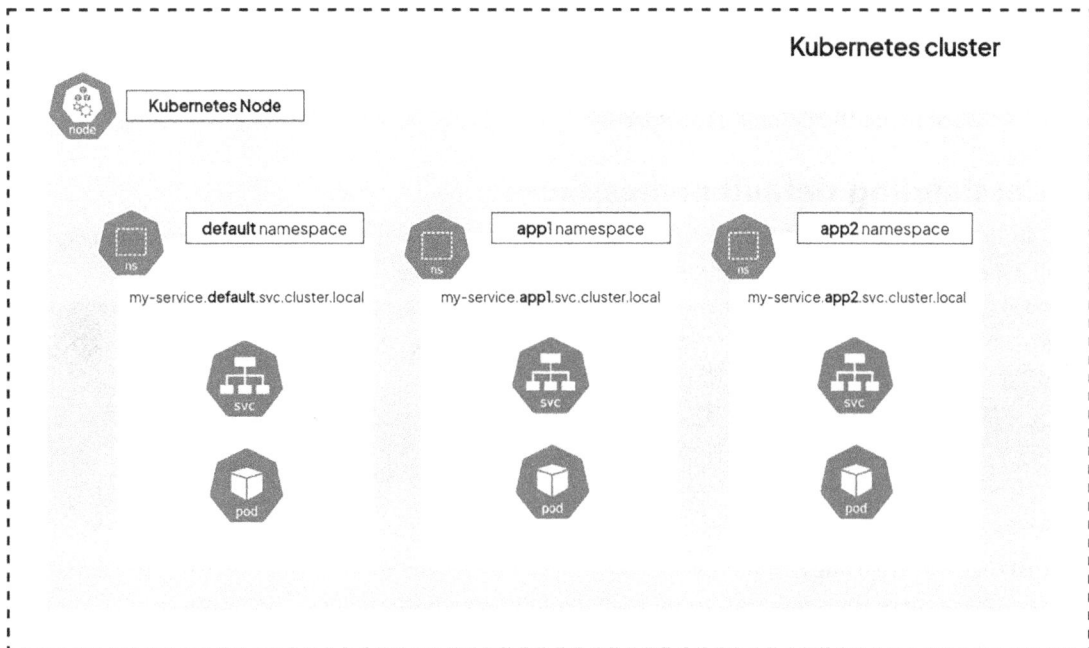

Figure 6.1: Kubernetes namespaces and resource isolation

In a broader sense, Kubernetes namespaces serve several purposes for administrators, including:

- Resource isolation
- Scoping resource names
- Hardware allocation and consumption limitation
- Permissions and access control with Role-Based Access Control

We recommend that you create one namespace per microservice or application, and then deploy all the resources that belong to a microservice within its namespace. However, be aware that Kubernetes does not impose any specific rules on you. For example, you could decide to use namespaces in these ways:

- **Differentiating between environments:** For example, one namespace is for a production environment, and the other one is for a development environment.

- **Differentiating between the tiers:** One namespace is for databases, one is for application Pods, and another is for middleware deployment.

- **Using the default namespace:** For the smallest clusters that only deploy a few resources, you can go for the simplest setup and just use one big default namespace, deploying everything into it.

Either way, keep in mind that even though two Pods are deployed in different namespaces and exposed through services, they can still interact and communicate with each other. Even though Kubernetes services are created in a given namespace, they'll receive a **fully qualified domain name (FQDN)** that will be accessible on the whole cluster. So even if an application running on namespace A needs to interact with an application in namespace B, it will have to call the service exposing app B by its FQDN. You don't need to worry about cross-namespace communication, as this is allowed by default and can be controlled via network policies.

Now, let's learn about the default namespaces.

Understanding default namespaces

Most Kubernetes clusters are created with a few namespaces by default. You can list your namespaces using kubectl get namespaces (or kubectl get ns), as follows:

```
$ kubectl get namespaces
NAME              STATUS   AGE
default           Active   8d
kube-node-lease   Active   8d
kube-public       Active   8d
kube-system       Active   8d
```

For instance, we are using a minikube cluster. By reading this command's output, we can see that the cluster we are currently using was set up with the following namespaces from the start:

- default: Kubernetes automatically provides this namespace, allowing you to begin using your new cluster without the need to create one manually. This namespace has been the default location to create all your resources thus far and is also utilized as the default namespace when no other is specified.

- kube-public: This namespace is accessible to all clients, including those without authentication. Primarily designated for cluster-wide purposes, it ensures certain resources are publicly visible and readable across an entire cluster. However, it's important to note that the public aspect of this namespace is more of a convention than a strict requirement. Currently unused, you can safely leave it as is.

- kube-system: This namespace is reserved for objects created by the Kubernetes system itself. It's where Kubernetes deploys the necessary objects for its operation. In typical Kubernetes setups, essential components like kube-scheduler and kube-apiserver are deployed as Pods within this namespace.

These components are vital for the proper functioning of the Kubernetes cluster they serve. Therefore, it's advisable to refrain from making changes to this namespace, as any alterations could potentially disrupt the cluster's functionality.

- `kube-node-lease`: The purpose of this namespace is to store Lease objects linked to individual nodes. These node leases enable the kubelet to transmit heartbeats, facilitating the detection of node failures by the control plane.

> Depending on which Kubernetes distribution you use, the pre-existing set of namespaces can change. But most of the time, these namespaces will be created by default.

Let's leave this namespace aside for now because we are going to get to the heart of the matter and start creating namespaces. We will look at the impact that these can have on your Pods, particularly at the level of the DNS resolution of your services.

How namespaces impact your resources and services

In this section, we will learn how to create, update, and delete namespaces, as well as the impacts that namespaces have on services and Pods.

We will also learn how to create resources by specifying a custom namespace so that we don't rely on the default one.

Listing namespaces inside your cluster

We saw this in the previous section, *Understanding default namespaces*, but in this section, we will learn how to list and explore the namespaces that have been created in your Kubernetes cluster:

```
$ kubectl get namespaces
NAME              STATUS   AGE
default           Active   8d
kube-node-lease   Active   8d
kube-public       Active   8d
kube-system       Active   8d
```

Keep in mind that all the commands that make use of the namespaces resource kind can also use the ns alias to benefit from a shorter format.

Retrieving the data of a specific namespace

Retrieving the data of a specific namespace can be achieved using the kubectl describe command, as follows:

```
$ kubectl describe namespaces default
Name:         default
Labels:       kubernetes.io/metadata.name=default
```

```
Annotations:   <none>
Status:        Active

No resource quota.

No LimitRange resource.
```

You can also use the get command and redirect the YAML format to a file to get the data from a specific namespace:

```
$ kubectl get namespaces default -o yaml > default-ns.yaml
```

Please note that a namespace can be in one of two states:

- **Active:** The namespace is active; it can be used to place new objects into it.
- **Terminating:** The namespace is being deleted, along with all its objects. It can't be used to host new objects while in this status.

Now, let's learn how to create a new namespace imperatively.

Creating a namespace using imperative syntax

To imperatively create a namespace, you can use the kubectl create namespaces command by specifying the name of the namespace to create. Here, we are going to create a new namespace called custom-ns. Please notice that all the operations related to namespaces in kubectl can be written with the shorter ns alias:

```
$ kubectl create ns custom-ns
namespace/custom-ns created
```

The new namespace, called custom-ns, should now be created in your cluster. You can check it by running the kubectl get command once more:

```
$ kubectl get ns custom-ns
NAME           STATUS   AGE
custom-ns      Active   35s
```

As you can see, the namespace has been created and is in the Active state. We can now place resources in it.

> Please avoid naming your cluster with a name starting with the kube- prefix, as this is the terminology for Kubernetes' objects and system namespaces.

Now, let's learn how to create another namespace using declarative syntax.

Creating a namespace using declarative syntax

Let's discover how to create a namespace using declarative syntax. As always, you must use a YAML (or JSON) file. Here is a basic YAML file to create a new namespace in your cluster. Please pay attention to the kind: Namespace in the file:

```
# custom-ns-2.yaml
apiVersion: v1
kind: Namespace
metadata:
  name: custom-ns-2
```

Apply the definition using the kubectl create command, by defining the YAML file path:

```
$ kubectl apply -f custom-ns-2.yaml
namespace/custom-ns-2 created
```

With that, we have created two custom namespaces. The first one, which was created imperatively, is called custom-ns, while the second one, which was created declaratively, is called custom-ns-2.

Now, let's learn how to remove these two namespaces using kubectl.

Deleting a namespace

You can delete a namespace using kubectl delete, as follows:

```
$ kubectl delete namespaces custom-ns
namespace "custom-ns" deleted
```

Please note this can also be achieved using declarative syntax. Let's delete the custom-ns-2 namespace that was created using the previous YAML file:

```
$ kubectl delete -f custom-ns-2.yaml
namespace "custom-ns-2" deleted
```

Running this command will take the namespace out of the Active status; it will enter the Terminating status. Right after the command, the namespace will be unable to host new objects, and after a few moments, it should completely disappear from the cluster.

> We have to warn you about using the kubectl delete namespace command, as it is extremely dangerous. Deleting a namespace is permanent and definitive. All the resources that were created in the namespace will be destroyed. If you need to use this command, be sure to have YAML files to recreate the destroyed resources and even the destroyed namespace.

Now, let's discover how to create resources inside a specific namespace.

Creating a resource inside a namespace

The following code shows how to create an NGINX Pod by specifying a custom namespace. Here, we are going to recreate a new `custom-ns` namespace and launch an NGINX Pod in it:

```
$ kubectl create ns custom-ns
$ kubectl run nginx --image nginx:latest -n custom-ns
Pod/nginx created
```

Pay attention to the `-n` option, which, in its long form, is the `--namespace` option. This is used to enter the name of the namespace where you want to create the resource (or get the details from). This option is supported by all the `kind` resources that can be scoped in a namespace.

Here is another command to demonstrate this. The following command will create a new `configmap` in the `custom-ns` namespace:

```
$ kubectl create configmap configmap-custom-ns --from-literal=Lorem=Ipsum -n
custom-ns
configmap/configmap-custom-ns created
```

You can also specify a namespace when using declarative syntax. Here is how to create a Pod in a specific namespace with declarative syntax:

```
# pod-in-namespace.yaml
apiVersion: v1
kind: Pod
metadata:
  name: nginx2
  namespace: custom-ns
spec:
  containers:
  - name: nginx
    image: nginx:latest
```

Please note the `namespace` key under the `metadata` section, just under the Pod's name, which says to create the Pod in the `custom-ns` namespace. Now, we can apply this file using `kubectl`:

```
$ kubectl apply -f pod-in-namespace.yaml
pod/nginx2 created
```

At this point, we should have a namespace called `custom-ns` that contains two `nginx` Pods, as well as a `configmap` called `configmap-custom-ns`.

> When you're using namespaces, you should always specify the `-n` flag to target the specific namespace of your choosing. Otherwise, you might end up running operations in the wrong namespace.

Now, let's move on to list the resources inside specific namespaces.

Listing resources inside a specific namespace

To be able to list the resources within a namespace, you must add the -n option, just like when creating a resource. Use the following command to list the Pods in the custom-ns namespace:

```
$ kubectl get pods -n custom-ns
NAME      READY   STATUS    RESTARTS   AGE
nginx     1/1     Running   0          9m23s
nginx2    1/1     Running   0          94s
```

Here, you can see that the nginx Pod that we created earlier is present in the namespace. From now on, all the commands that target this particular Pod should contain the -n custom-ns option.

The reason for this is that the Pod does not exist in the default namespace, and if you omit passing the -n option, then the default namespace will be requested. Let's try to remove -n custom-ns from the get command. We will see that the nginx Pod is not here anymore:

```
$ kubectl get pods
No resources found in default namespace.
```

Now, we can also run the get configmap command to check whether configmap is listed in the output. As you can see, the behavior is the same as when trying to list Pods. If you omit the -n option, the list operation will run against the default namespace:

```
$ kubectl get cm
NAME                 DATA    AGE
kube-root-ca.crt     1       9d

$ kubectl get cm -n custom-ns
NAME                  DATA    AGE
configmap-custom-ns   1       70m
kube-root-ca.crt      1       76m
```

The most important point to remember from all that we have discussed so far in this section is to never forget to add the -n option when working on a cluster that has multiple namespaces. This little carelessness could waste your time because, if you forget it, everything you do will be done on the default namespace.

Instead of passing the namespace information in the command line every time, it is possible to set it in the kubeconfig context. In the next section, we will learn how to set the working namespace in the current context of kubeconfig.

Setting the current namespace using kubectl config set-context

It is also possible to set your current namespace in some situations. For example, if you are working on a specific project and using a specific namespace for your application and other resources, then you can set the namespace context as follows:

```
$ kubectl config set-context --current --namespace=custom-ns
Context "minikube" modified.
```

We can also check if any namespace is configured in the context as follows:

```
$ kubectl config view --minify --output 'jsonpath={..namespace}'
custom-ns
```

Now, we can get the details of the application or apply configuration without mentioning the `-n <namespace>` option:

```
$ kubectl get pods
NAME     READY   STATUS    RESTARTS   AGE
nginx    1/1     Running   0          79m
nginx2   1/1     Running   0          71m
```

Running the kubectl config command and sub-commands will only trigger modification or read operations against the `~/.kube/config` file, which is the configuration file that kubectl uses.

When you're using the kubectl config set-context command, you're just updating that file to make it point to another namespace.

Knowing how to switch between namespaces with kubectl is important, but before you run any write operations such as kubectl delete or kubectl create, make sure that you are in the correct namespace. Otherwise, you should continue to use the -n flag. As this switching operation might be executed a lot of times, Kubernetes users tend to create Linux aliases to make them easier to use. Do not hesitate to define a Linux alias if you think it can be useful to you.

For example, you can set an alias in your ~/.bashrc file (assuming that you are using Linux or macOS) as follows:

```
alias kubens='kubectl config set-context --current --namespace'
```

And use this alias next time, as follows:

```
$ kubens custom-ns
Context "minikube" modified.
$ kubectl config view --minify --output 'jsonpath={..namespace}'
custom-n
```

```
# change to another namespace
$ kubens default
Context "minikube" modified.
$ kubectl config view --minify --output 'jsonpath={..namespace}'
default
```

But again, it is highly recommended to use the -n namespace option to avoid any accidents in Kubernetes operations.

Before we continue our chapter and hands-on tutorials, let's set the namespace in context back to normal by running the following command:

```
$ kubectl config set-context --current --namespace=default
Context "minikube" modified.
```

Now, let's discover how to list all the resources inside a specific namespace.

Listing all the resources inside a specific namespace

If you want to list all the resources in a specific namespace, there is a very useful command that you can use called kubectl get all -n custom-ns:

```
$ kubectl get all -n custom-ns
```

As you can see, this command can help you retrieve all the resources that are created in the namespace specified in the -n flag.

Recognizing how names are scoped within a namespace

It's important to understand that namespaces offer an additional advantage: defining the scope for the names of the resources they contain.

Take the example of Pod names. When you work without namespaces, you interact with the default namespace, and when you create two Pods with the same name, you get an error because Kubernetes uses the names of the Pods as their unique identifiers to distinguish them.

Let's try to create two Pods in the default namespace. Both will be called nginx. Here, we can simply run the same command twice in a row:

```
$ kubectl run nginx --image nginx:latest
Pod/nginx created
$ kubectl run nginx --image nginx:latest
Error from server (AlreadyExists): Pods "nginx" already exists
```

The second command produces an error, saying that the Pod already exists, which it does. If we run kubectl get pods, we can see that only one Pod exists:

```
$ kubectl get pods
NAME    READY    STATUS     RESTARTS    AGE
nginx   1/1      Running    0           64s
```

Now, let's try to list the Pods again but, this time, in the `custom-ns` namespace:

```
$ kubectl get Pods --namespace custom-ns
NAME     READY   STATUS    RESTARTS   AGE
nginx    1/1     Running   0          89m
nginx2   1/1     Running   0          23m
```

As you can see, this namespace also has a Pod called `nginx`, and it's not the same Pod that is contained in the `default` namespace. This is one of the major advantages of namespaces. By using them, your Kubernetes cluster can now define multiple resources with the same names, so long as they are in different namespaces. You can easily duplicate microservices or applications by using the namespace element.

Also, note that you can override the key to the namespaces of the resources that you create declaratively. By adding the `-n` option to the `kubectl create` command, you force a namespace as the context for your command; `kubectl` will take the namespace that was passed in the command into account, not the one present in the YAML file. By doing this, it becomes very easy to duplicate your resources between different namespaces – for example, a production environment in a `production` namespace and a test environment in a `test` namespace.

Understanding that not all resources are in a namespace

In Kubernetes, not all objects belong to a namespace. This is the case, for example, with nodes, which are represented at the cluster level by an entry of the `Node` kind but that does not belong to any particular namespace. You can list resources that do not belong to a namespace using the following command:

```
$ kubectl api-resources --namespaced=false
NAME                        SHORTNAMES   APIVERSION
NAMESPACED     KIND
componentstatuses           cs           v1
false          ComponentStatus
namespaces                  ns           v1
false          Namespace
nodes                       no           v1
false          Node
persistentvolumes           pv           v1
false          PersistentVolume
...<removed for brevity>...
```

You can also list all the resources that belong to a namespace by passing `--namespaced` to `true`:

```
$ kubectl api-resources --namespaced=true
```

Now, let's learn how namespaces affect a service's DNS.

Resolving a service using namespaces

As we discovered in *Chapter 8*, *Exposing Your Pods with Services*, Pods can be exposed through a type of object called Services. When created, services are assigned a DNS record that allows Pods in the cluster to access them.

However, when a Pod tries to call a service through DNS, it can only reach it if the service is in the same namespace as the Pod, which is limiting. Namespaces have a solution to this problem. When a service is created in a particular namespace, the name of its service is added to its DNS:

```
<service_name>.<namespace_name>.svc.cluster.local
```

By querying this domain name, you can easily query any service that is in any namespace in your Kubernetes cluster. So you are not limited to that level. Pods are still capable of achieving inter-communication, even if they don't run in the same namespace.

In the following section, we will explore some of the best practices for Kubernetes namespaces.

Best practices for Kubernetes namespaces

Even though there are no strict rules on namespace creation and management, let us learn some of the industry best practices related to namespaces.

- Organization and separation

    - **Logical Grouping:** Put together apps, services, and resources that go together based on what they do, where they are in development (like dev, test, and prod), or who owns them (e.g., different teams). This keeps things organized and makes managing resources easier.

    - **Isolation:** Use namespaces to keep deployments separate. This means apps in one namespace won't mess with stuff in another, which reduces conflicts. You can also increase the security and isolation by applying appropriate **Role-Based Access Control (RBAC)** network policies.

- Naming rules

    - **Clear and Descriptive:** Give your namespaces names that say what they're for. This makes it easier to keep track of them, especially in big setups. Stick to common naming tricks like `dev-`, `test-`, or `prod-` for different environments.

    - **Stay Consistent:** Use the same naming style across your cluster. This makes it easier for your team to talk about and understand what's going on.

- Managing resources

    - **Resource Limits:** Set limits on how much stuff a namespace can use. This stops one deployment from hogging everything and makes sure everyone gets their fair share. Please remember that it is also possible to set the resource limits at the Pod level for increased control.

    - **Limits for All:** Make rules for how much stuff each Pod and container in a namespace can ask for or use. This gives you more control over how resources are used.

- Controlling access

  - **Role-Based Access:** Use RBAC to control who can do what in each namespace. Give people and services the right permissions to manage stuff in their namespace.

- Keeping an eye out

  - **Watch Things:** Keep an eye on how much stuff is used and if your apps are healthy in each namespace. This helps you spot problems and use resources better.

  - **Lifecycle of a Namespace:** Check up on your namespaces regularly. Get rid of the ones that you don't use anymore to keep things tidy and avoid security risks. Think about automating how you create and delete namespaces.

- Other stuff to think about

  - **Not Perfect Isolation:** Even though namespaces help keep things separate, they're not foolproof. You might need network rules for extra safety.

  - **Clusters versus Namespaces:** If your setup is complicated and needs a lot of separation, think about using different Kubernetes clusters instead of just namespaces.

By following these best practices, you can keep your Kubernetes setup organized, safe, and easy to handle using namespaces. Just remember to tweak things to fit your own setup for the best results. With that, we're done with the basics of namespaces in Kubernetes. We have learned what namespaces are, how to create and delete them, how to use them to keep a cluster clean and organized, and how to update the kubeconfig context to make kubectl point to a specific namespace.

Now, we'll look at a few more advanced options related to namespaces. It is a good time to introduce ResourceQuota and Limit, which you can use to limit the computing resources that an application deployed on Kubernetes can access!

Configuring ResourceQuota and Limit at the namespace level

In this section, we're going to discover that namespaces can not only be used to sort resources in a cluster but also to limit the computing resources that Pods can access.

Using ResourceQuota and Limits with namespaces, you can create limits regarding the computing resources your Pods can access. We're going to learn how to proceed and exactly how to use these new concepts. In general, defining ResourceQuota and Limits is considered good practice for production clusters – that's why you should use them wisely.

Understanding the need to set ResourceQuotas

Just like applications or systems, Kubernetes Pods will require a certain amount of computing resources to work properly. In Kubernetes, you can configure two types of computing resources:

- CPU
- Memory

All your nodes (compute and controller) work together to provide CPU and memory, and in Kubernetes, adding more CPU and memory simply consists of adding more compute (or worker) nodes to make room for more Pods. Depending on whether your Kubernetes cluster is based on-premises or in the cloud, adding more compute nodes can be achieved by purchasing the hardware and setup to do so, or by simply calling the cloud API to create additional virtual machines.

Understanding how Pods consume these resources

When you launch a Pod on Kubernetes, a control plane component, known as `kube-scheduler`, will elect a compute node and assign the Pods to it. Then, the `kubelet` on the elected compute node will attempt to launch the containers defined in the Pod.

This process of compute node election is called **Pod scheduling** in Kubernetes.

When a Pod gets scheduled and launched on a compute node, it has, by default, access to all the resources that the compute node has. Nothing prevents it from accessing more and more CPU and memory as the application is used, and ultimately, if the Pods run out of memory or CPU resources to work properly, then the application simply crashes.

This can become a real problem because compute nodes can be used to run multiple applications – and, therefore, multiple Pods – at the same time. So if 10 Pods are launched on the same compute node but one of them consumes all the computing resources, then this will have an impact on all 10 Pods running on the compute node.

This problem means that you have two things you must consider:

- Each Pod should be able to require some computing resources to work.
- The cluster should be able to restrict the Pod's consumption so that it doesn't take all the resources available, sharing them with other Pods too.

It is possible to address these two problems in Kubernetes, and we will discover how to use two options that are exposed to the Pod object. The first one is called **resource requests**, which is the option that's used to let a Pod indicate what amount of computing resources it needs, while the other one is called **resource limit** and will be used to indicate the maximum computing resources the Pod will have access to.

Let's explore these options now.

Understanding how Pods can require computing resources

The `request` and `limit` options will be declared within the YAML definition file of a Pod resource, or you can apply it to the running deployment using the `kubectl set resource` command. Here, we're going to focus on the `request` option.

The resource request is simply the minimal amount of computing resources a Pod will need to work properly, and it is a good practice to always define a `request` option for your Pods, at least for those that are meant to run in production.

Let's say that you want to launch an NGINX Pod on your Kubernetes cluster. By filling in the request option, you can tell Kubernetes that your NGINX Pod will need, at the bare minimum, 512 MiB of memory and 25% of a CPU core to work properly.

Here is the YAML definition file that will create this Pod:

```
# pod-in-namespace-with-request.yaml
apiVersion: v1
kind: Pod
metadata:
  name: nginx-with-request
  namespace: custom-ns
spec:
  containers:
    - name: nginx
      image: nginx:latest
      resources:
        requests:
          memory: "512Mi"
          cpu: "250m"
```

As you can see in the above YAML snippet, you can define request at the container level and set different ones for each container.

There are three things to note about this Pod:

- It is created inside the custom-ns namespace.
- It requires 512Mi of memory.
- It requires 250m of CPU.

But what do these metrics mean?

Memory is expressed in bytes (one MiB is equal to 1,048,576 bytes), whereas CPU is expressed in **millicores** and allows fractional values. If you want your Pod to consume one entire CPU core, you can set the cpu key to 1000m. If you want two cores, you must set it to 2000m; for half of a core, it will be 500m or 0.5; and so on. However, to request a full CPU core, it's simpler and more common practice to use the whole number (e.g., 2) instead of 2000m. So the preceding YAML definition says that the NGINX Pod we will create will forcibly need 512 MiB of memory, since memory is expressed in bytes, and one-quarter of a CPU core of the underlying compute node. There is nothing related to the CPU or memory frequency here.

When you apply this YAML definition file to your cluster, the scheduler will look for a compute node that is capable of launching your Pods. This means that you need a compute node where there is enough room in terms of available CPU and memory to meet your Pods' requests.

But what if no compute node is capable of fulfilling these requirements? Here, the Pod will never be scheduled and never be launched. Unless you remove some running Pods to make room for this one or add a compute node that is capable of launching this Pod, it won't ever be launched.

> Keep in mind that Pods cannot span multiple nodes. So if you set 8000m (which represents eight CPU cores) but your cluster is made up of two compute nodes with four cores each, then no compute node will be able to fulfill the request, and your Pod won't be scheduled.

So use the `request` option wisely – consider it as the minimum amount of compute resources the Pod will need to work. You have the risk that your Pod will never be scheduled if you set too high a request, but on the other hand, if your Pod is scheduled and launched successfully, this amount of resources is guaranteed.

Now, let's see how we can limit resource consumption.

Understanding how you can limit resource consumption

When you write a YAML definition file, you can define resource limits regarding what a Pod will be able to consume.

Setting a resource request won't suffice to do things properly. You should set a limit each time you set a resource. Setting a limit will tell Kubernetes to let the Pod consume resources up to that limit, and never above. This way, you ensure that your Pod won't take all the resources of the compute for itself.

However, be careful – Kubernetes won't behave the same, depending on what kind of limit is reached. If the Pod reaches its CPU limit, it is going to be throttled, and you'll notice performance degradation. But if your Pod reaches its memory limit, then it might be terminated. The reason for this is that memory is not something that can be throttled, and Kubernetes still needs to ensure that other applications are not impacted and remain stable. So be aware of that.

Without a limit, the Pod will be able to consume all the resources of the compute node. Here is an updated YAML file corresponding to the NGINX Pod we saw earlier, but now, it has been updated to define a limit on memory and CPU.

Here, the Pod will be able to consume up to 1 GiB of memory and up to 1 entire CPU core of the underlying compute node:

```yaml
# pod-in-namespace-with-request-and-limit.yaml
apiVersion: v1
kind: Pod
metadata:
  name: nginx-with-request-and-limit
  namespace: quota-ns
spec:
  containers:
    - name: nginx
```

```
    image: nginx:latest
    resources:
      requests:
        memory: "512Mi"
        cpu: "250m"
      limits:
        memory: "1Gi"
        cpu: "1000m"
```

So when you set a request, set a limit too. Let us try this request and limit it in our next hands-on lab.

For this exercise, let us check current system resource availability. Since we are using the minikube cluster for our demonstration, let us enable metrics for detailed resource usage information. You will use metrics in *Chapter 10, Running Production-Grade Kubernetes Workloads*:

```
$ minikube addons enable metrics-server
```

Wait for the metrics server Pods to enter a `Running` state before you continue:

```
$ kubectl get po -n kube-system | grep metrics
metrics-server-7c66d45ddc-82ngt            1/1       Running    0
113m
```

Let us check the cluster usage using the metrics information:

```
$ kubectl top node
NAME        CPU(cores)   CPU%   MEMORY(bytes)   MEMORY%
minikube    181m         1%     806Mi           2%
```

In this case, we have about 800Mi memory and 180m CPU available to consume in our Kubernetes cluster.

> If you are using minikube with Podman or the Docker driver, the minikube will show the actual host CPU and memory, not the memory of the minikube Kubernetes cluster node. In such cases, you can try another minikube cluster using VirtualBox (`minikube start --profile cluster2-vb --driver=virtualbox`) so that it will use the minikube VM CPU and memory resource.

Let us create a new namespace for the resource request and limit demonstration:

```
$ kubectl create ns quota-ns
namespace/quota-ns created
```

Now, let us create a new YAML file where we have non-realistic memory requests such as `100Gi` `resources.requests.memory`, as follows:

```
# pod-with-request-and-limit-1.yaml
apiVersion: v1
```

```
kind: Pod
metadata:
  name: nginx-with-request-and-limit-1
  namespace: quota-ns
spec:
  containers:
    - name: nginx
      image: nginx:latest
      resources:
        requests:
          memory: "100Gi"
          cpu: "100m"
        limits:
          memory: "100Gi"
          cpu: "500m"
```

Create the Pod using kubectl apply, as follows:

```
$ kubectl apply -f pod-with-request-and-limit-1.yaml
pod/nginx-with-request-and-limit-1 created
```

Check the Pod status to see the Pod creation:

```
$ kubectl get pod -n quota-ns
NAME                              READY   STATUS    RESTARTS   AGE
nginx-with-request-and-limit-1    0/1     Pending   0          45s
```

This says Pending and that the Pod is not running yet. Let us check the Pod details using the kubectl describe command, as follows:

```
$ kubectl describe po nginx-with-request-and-limit-1 -n quota-ns
Name:              nginx-with-request-and-limit-1
Namespace:         quota-ns
...<removed for brevity>...
Status:            Pending
IP:
IPs:               <none>
Containers:
  nginx:
...<removed for brevity>...
    Limits:
      cpu:      500m
      memory:   100Gi
```

```
   Requests:
     cpu:         100m
     memory:      100Gi
   Environment:  <none>
...<removed for brevity>...
  Warning  FailedScheduling  105s  default-scheduler  0/1 nodes are available:
1 Insufficient memory. preemption: 0/1 nodes are available: 1 No preemption
victims found for incoming pod.
```

You will see an error, as the scheduler cannot find any nodes in the cluster to accommodate your Pod with the memory requests, which means that the Pod will be scheduled until the Kubernetes cluster has a suitable node to place the Pod.

Now, we will update the YAML with reasonable memory, `1Gi resources.requests.memory`, as follows:

```yaml
# pod-with-request-and-limit-2.yaml
apiVersion: v1
kind: Pod
metadata:
  name: nginx-with-request-and-limit-2
  namespace: quota-ns
spec:
  containers:
    - name: nginx
      image: nginx:latest
      resources:
        requests:
          memory: "1Gi"
          cpu: "100m"
        limits:
          memory: "2Gi"
          cpu: "500m"
```

Let us create the Pod now:

```
$ kubectl apply -f pod-with-request-and-limit-2.yaml
pod/nginx-with-request-and-limit-2 created
$ kubectl get po -n quota-ns
NAME                             READY   STATUS    RESTARTS   AGE
nginx-with-request-and-limit-1   0/1     Pending   0          8m24s
nginx-with-request-and-limit-2   1/1     Running   0          20s
```

Now, the Kubernetes scheduler can find the suitable node, based on the resource request, and the Pod has already started running as usual.

So now that you are aware of this request and limit consideration, don't forget to add it to your Pods!

Understanding why you need ResourceQuota

You can entirely manage your Pod's consumptions by relying entirely on its request and limit options. All the applications in Kubernetes are just Pods, so setting these two options provides you with a strong and reliable way to manage resource consumption on your cluster, given that you never forget to set these.

It is easy to forget these two options and deploy a Pod on your cluster that won't define any request or limit. Maybe it will be you, or maybe a member of your team, but the risk of deploying such a Pod is high because everyone can forget about these two options. And if you do so, the risk of application instability is high, as a Pod without a limit can eat all the resources on the compute node it is launched on.

Kubernetes provides a way to mitigate this issue, thanks to two objects called ResourceQuota and LimitRange. These two objects are extremely useful because they can enforce these constraints at the namespace level.

ResourceQuota is another resource kind, just like a Pod or ConfigMap. The workflow is quite simple and consists of two steps:

1. You must create a new namespace.
2. You must create a ResourceQuota and a LimitRange object inside that namespace.

Then, all the Pods that are launched in that namespace will be constrained by these two objects.

These quotas are used, for example, to ensure that all the containers that are accumulated in a namespace do not consume more than 4 GiB of RAM.

Therefore, it is possible and even recommended to set restrictions on what can and cannot run within Pods. It is strongly recommended that you always define a ResourceQuota and a LimitRange object for each namespace you create in your cluster!

Without these quotas, the deployed resources could consume as much CPU or RAM as they want, which would ultimately make your cluster and all the applications running on it unstable, given that the Pods don't hold requests and limits as part of their respective configurations.

In general, ResourceQuota is used to do the following:

*   Limit CPU consumption within a namespace
*   Limit memory consumption within a namespace
*   Limit the absolute number of objects such as Pods, Services, ReplicationControllers, Replicas, Deployments, etc. operating within a namespace
*   Limit consumption of storage resources based on the associated storage class

There are a lot of use cases, and you can discover all of them directly in the Kubernetes documentation. Now, let's learn how to define ResourceQuota in a namespace.

Creating a ResourceQuota

To demonstrate the usefulness of ResourceQuota, we are going to create one ResourceQuota object for the namespace quota-ns. This ResourceQuota will be used to create requests and limits that all the Pods within this namespace combined will be able to use. Here is the YAML file that will create ResourceQuota; please note the resource kind:

```yaml
# resourcequota.yaml
apiVersion: v1
kind: ResourceQuota
metadata:
  name: resourcequota
  namespace: quota-ns
spec:
  hard:
    requests.cpu: "1000m"
    requests.memory: "1Gi"
    limits.cpu: "2000m"
    limits.memory: "2Gi"
```

Keep in mind that the ResourceQuota object is scoped to one namespace.

This one is stating that, in this namespace, the following will occur:

- All the Pods combined won't be able to request more than one CPU core.
- All the Pods combined won't be able to request more than 1 GiB of memory.
- All the Pods combined won't be able to consume more than two CPU cores.
- All the Pods combined won't be able to consume more than 2 GiB of memory.

Create the ResourceQuota by applying the YAML configuration:

```
$ kubectl apply -f resourcequota.yaml
resourcequota/resourcequota created

$ kubectl get quota -n quota-ns
NAME            AGE    REQUEST                                              LIMIT
resourcequota   4s     requests.cpu: 100m/1, requests.memory: 1Gi/1Gi      limits.
cpu: 500m/2, limits.memory: 2Gi/2Gi
```

Let's check the current resources in the quota-ns namespace as follows:

```
$ kubectl get po -n quota-ns
NAME                            READY   STATUS    RESTARTS        AGE
nginx-with-request-and-limit-2  1/1     Running   1 (5h41m ago)   7h18m

$ kubectl top pod -n quota-ns
NAME                            CPU(cores)    MEMORY
```

There is an nginx Pod (if you haven't deleted the previous demo Pod) and usage is very low.

Now, we have a new Pod YAML file but we request 3Gi memory for the Pod as follows:

```yaml
# pod-with-request-and-limit-3.yaml
apiVersion: v1
kind: Pod
metadata:
  name: nginx-with-request-and-limit-3
  namespace: quota-ns
spec:
  containers:
    - name: nginx
      image: nginx:latest
      resources:
        requests:
          memory: "3Gi"
          cpu: "100m"
        limits:
          memory: "4Gi"
          cpu: "500m"
```

Now, let us try to create this Pod and see what the result will be:

```
$ kubectl apply -f pod-with-request-and-limit-3.yaml
Error from server (Forbidden): error when creating "pod-with-request-and-
limit-3.yaml": pods "nginx-with-request-and-limit-3" is forbidden: exceeded
quota: resourcequota, requested: limits.memory=4Gi,requests.memory=3Gi, used:
limits.memory=2Gi,requests.memory=1Gi, limited: limits.memory=2Gi,requests.
memory=1Gi
```

Yes, the error is very clear; we are requesting more resources than the quota, and Kubernetes will not allow the creation of a new Pod.

You can have as many Pods and containers in the namespace, so long as they respect these constraints. Most of the time, ResourceQuotas are used to enforce constraints on requests and limits, but they can also be used to enforce these limits per namespace.

> While setting up ResourceQuotas at the namespace level, it's crucial to prevent any single namespace from consuming all cluster resources; it's also important to apply resource requests and limits at the Pod or Deployment level. This dual-layered approach ensures that resource-hogging is contained both within individual namespaces and at the application level. By enforcing these limits, you create a more predictable and stable environment, preventing any one component from disrupting the entire system.

In the following example, the preceding ResourceQuota has been updated to specify that the namespace where it is created cannot hold more than 10 ConfigMaps and 5 services, which is pointless but a good example to demonstrate the different possibilities with ResourceQuota:

```yaml
# resourcequota-with-object-count.yaml
apiVersion: v1
kind: ResourceQuota
metadata:
  name: my-resourcequota
  namespace: quota-ns
spec:
  hard:
    requests.cpu: "1000m"
    requests.memory: "1Gi"
    limits.cpu: "2000m"
    limits.memory: "2Gi"
    configmaps: "10"
    services: "5"
```

> When applying a ResourceQuota YAML definition, ensure that the ResourceQuota is assigned to the correct namespace. If the namespace isn't specified within the YAML file, remember to use the --namespace flag to specify where the ResourceQuota should be applied.

Create the ResourceQuota as follows:

```
$ kubectl apply -f resourcequota-with-object-count.yaml
```

In the following section, we will learn about the storage ResourceQuota in Kubernetes.

Storage resource quotas

In Kubernetes, resource quotas allow you to control the total storage resources requested within a namespace. You can set limits on both the sum of storage requests across all persistent volume claims and the number of persistent volume claims allowed. Additionally, quotas can be defined based on specific storage classes, enabling separate limits for different storage class types. For example, you can set quotas for gold and bronze storage classes separately. Starting from release 1.8, quotas also support local ephemeral storage, allowing you to limit the sum of local ephemeral storage requests and limits across all Pods within a namespace.

Now, let's learn how to list ResourceQuotas.

Listing ResourceQuota

ResourceQuota objects can be accessed through kubectl using the quota's resource name option. The kubectl get command will do this for us:

```
$ kubectl get resourcequotas -n quota-ns
NAME              AGE    REQUEST                                         LIMIT
resourcequota     15m    requests.cpu: 100m/1, requests.memory: 1Gi/1Gi  limits.
cpu: 500m/2, limits.memory: 2Gi/2Gi
```

Now, let's learn how to delete ResourceQuota from a Kubernetes cluster.

Deleting ResourceQuota

To remove a ResourceQuota object from your cluster, use the kubectl delete command:

```
$ kubectl delete -f resourcequota-with-object-count.yaml
resourcequota "my-resourcequota" deleted
```

Now, let's introduce the notion of LimitRange.

Introducing LimitRange

LimitRange is another object that is similar to ResourceQuota, as it is created at the namespace level. The LimitRange object is used to enforce default requests and limit values to individual containers. Even by using the ResourceQuota object, you could create one object that consumes all the available resources in the namespace, so the LimitRange object is here to prevent you from creating too small or too large containers within a namespace.

Here is a YAML file that will create LimitRange:

```
# limitrange.yaml
apiVersion: v1
kind: LimitRange
metadata:
  name: my-limitrange
  namespace: quota-ns
spec:
  limits:
    - default: # this section defines default limits
        cpu: 500m
        memory: 256Mi
      defaultRequest: # this section defines default requests
        cpu: 500m
        memory: 128Mi
      max: # max and min define the limit range
        cpu: "1"
        memory: 1000Mi
      min:
        cpu: 100m
        memory: 128Mi
      type: Container
```

As you can see, the `LimitRange` object consists of four important keys that all contain `memory` and `cpu` configuration. These keys are as follows:

- `default`: This helps you enforce default values for the `memory` and `cpu` limits of containers if you forget to apply them at the Pod level. Each container that is set up without limits will inherit these default ones from the `LimitRange` object.
- `defaultRequest`: This is the same as `default` but for the `request` option. If you don't set a `request` option to one of your containers in a Pod, the ones from this key in the `LimitRange` object will be automatically used by default.
- `max`: This value indicates the maximum limit (not request) container that a Pod can set. You cannot configure a Pod with a limit value that is higher than this one. It is the same as the `default` value in that it cannot be higher than the one defined here.
- `min`: This value works like `max` but for requests. It is the minimum amount of computing resources that a Pod can request, and the `defaultRequest` option cannot be lower than this one.

Finally, note that if you omit the `default` and `defaultRequest` keys, then the `max` key will be used as the `default` key, and the `min` key will be used as the `default` key.

Defining `LimitRange` is a good idea if you want to protect yourself from forgetting to set requests and limits on your Pods. At least with `LimitRange`, these objects will have default limits and requests!

```
$ kubectl apply -f limitrange.yaml
```

Now, let's learn how to list `LimitRanges`.

Listing LimitRanges

The `kubectl` command line will help you list your `LimitRanges`. Don't forget to add the `-n` flag to scope your request to a specific namespace:

```
$ kubectl get limitranges -n quota-ns
NAME             CREATED AT
my-limitrange    2024-03-10T16:13:00Z

$ kubectl describe limitranges my-limitrange -n quota-ns
Name:         my-limitrange
Namespace:    quota-ns
Type         Resource  Min    Max     Default Request  Default Limit  Max Limit/
Request Ratio
----         --------  ---    ---     ---------------  -------------  ----------
------------
Container    memory    128Mi  1000Mi  128Mi            256Mi          -
Container    cpu       100m   1       500m             500m           -
```

Now, let's learn how to delete `LimitRange` from a namespace.

Deleting LimitRange

Deleting LimitRange can be achieved using the kubectl command-line tool. Here is how to proceed:

```
$ kubectl delete limitranges my-limitrange -n quota-ns
limitrange "my-limitrange" deleted
```

As always, don't forget to add the -n flag to scope your request to a specific namespace; otherwise, you may target the wrong one!

Summary

This chapter introduced you to namespaces, which are extremely important in Kubernetes. You cannot manage your cluster effectively without using namespaces because they provide logical resource isolation in your cluster. Most people use production and development namespaces, for example, or one namespace for each application. It is generally not rare to see clusters where hundreds of namespaces are created.

We discovered that most Kubernetes resources are scoped to a namespace, although some are not. Keep in mind that, by default, Kubernetes is set up with a few preconfigured namespaces, such as kube-system, and that it is generally a bad idea to change the things that run in these namespaces, especially if you do not know what you are doing.

We also discovered that namespaces can be used to set quotas and limit the resources that Pods can consume, and it is a really good practice to set these quotas and limits at the namespace level, using the ResourceQuota and LimitRange objects, to prevent your Pods from consuming too many computing resources. By implementing these measures, you're laying the foundation for effective capacity management, a critical consideration for organizations aiming to maintain the stability and efficiency of all applications running on a cluster.

In the next chapter, we will learn how to handle configuration infomation and sensitive data in Kubernetes using ConfigMaps and Secrets.

Further reading

- *Initial namespaces*: https://kubernetes.io/docs/concepts/overview/working-with-objects/namespaces/#initial-namespaces
- *Resource Management for Pods and Containers*: https://kubernetes.io/docs/concepts/configuration/manage-resources-containers/
- *Resource Quotas*: https://kubernetes.io/docs/concepts/policy/resource-quotas/
- *Limit Ranges*: https://kubernetes.io/docs/concepts/policy/limit-range/
- *Resource metrics pipeline*: https://kubernetes.io/docs/tasks/debug/debug-cluster/resource-metrics-pipeline/
- *Configure Default Memory Requests and Limits for a Namespace*: https://kubernetes.io/docs/tasks/administer-cluster/manage-resources/memory-default-namespace/

Join our community on Discord

Join our community's Discord space for discussions with the authors and other readers:

`https://packt.link/cloudanddevops`

7

Configuring Your Pods Using ConfigMaps and Secrets

The previous chapters introduced you to launching application containers using Kubernetes. You now know that whenever you need to launch a container on Kubernetes, you will need to do so using Pods. This was the key concept for you to understand and assimilate. Kubernetes is a complex system managed through a RESTful API. The core component handling this is the Kubernetes API server, which provides the primary interface to interact with the cluster. When users create Kubernetes objects, such as Pods, through this API, the system responds by provisioning the necessary resources on cluster nodes. Among these resources, a Pod stands out, as its creation on a Kubernetes node leads to the instantiation of application containers.

In this chapter, we'll learn about two new Kubernetes objects: **ConfigMaps** and **Secrets**. Kubernetes leverages ConfigMaps and Secrets to decouple application configuration from the code itself. These objects provide a mechanism to manage configuration values independently, enhancing application portability and security. ConfigMaps store non-sensitive data as key-value pairs, while Secrets handle sensitive information like passwords or API keys. Both can be injected into Pods as environment variables or mounted as volumes, allowing applications to dynamically access configuration without hardcoding values. By separating configuration from an application, you create more flexible, resilient, and secure deployments within the Kubernetes ecosystem.

The following are the main topics that we're going to cover in this chapter:

- Understanding what ConfigMaps and Secrets are
- Configuring your Pods using ConfigMaps
- Managing sensitive configuration with the Secret object

Technical requirements

For this chapter, you will need the following:

- A working Kubernetes cluster (local or cloud-based, although this is not important)
- A working `kubectl` CLI configured to communicate with the Kubernetes cluster

You can get these two prerequisites by following *Chapter 2, Kubernetes Architecture – From Container Images to Running Pods*, and *Chapter 3, Installing Your First Kubernetes Cluster*, to get a working Kubernetes cluster and a properly configured `kubectl` client, respectively.

You can download the latest code samples for this chapter from the official GitHub repository: `https://github.com/PacktPublishing/The-Kubernetes-Bible-Second-Edition/tree/main/Chapter07`.

Understanding what ConfigMaps and Secrets are

Kubernetes environments are dynamic and constantly changing. This makes managing application configurations a complex challenge. Traditional methods often fall short in keeping up with the rapid pace of cloud-native development.

To address this, Kubernetes provides ConfigMaps and Secrets, specialized ways to handle different types of configuration data. By separating configuration from the application code, these resources significantly improve application portability, security, and manageability.

In the following sections, we'll dive into the details of ConfigMaps and Secrets, exploring how they work and how to best utilize them in your Kubernetes deployments.

Decoupling your application and your configuration

Containers are immutable by nature, meaning they cannot be changed once created. This presents a challenge when managing configuration. Embedding configuration within a container image requires rebuilding the entire image for every configuration change, a time-consuming and inefficient process. Relying solely on environment variables or command-line arguments for configuration can also be cumbersome, especially for complex setups, and doesn't guarantee persistence across container restarts. These limitations highlight the need for a more effective configuration management strategy in Kubernetes.

When we use Kubernetes, we want our applications to be as portable as possible. A good way to achieve this is to decouple the application from its configuration. Back in the old days, configuration and application were the same thing; since the application code was designed to work only on one environment, configuration values were often bundled within the application code itself, so the configuration and application code were tightly coupled.

Having both application code and configuration values treated as the same thing reduces the portability of an application. Now, things have changed a lot, and we must be able to update the application configuration because we want to make our application as portable as possible, enabling us to deploy applications in multiple environments flawlessly.

Consider the following problem:

1. You deploy a Java application to the development environment for testing.
2. After the tests, the app is ready for production, and you need to deploy it. However, the MySQL endpoint for production is different from the one in development.

There are two possibilities here:

- The configuration and application code are not decoupled, and the MySQL is hardcoded and bundled within the application code; you are stuck and need to rebuild the whole app after editing the application code.
- The configuration and application code are decoupled. That's good news for you, as you can simply override the MySQL endpoint as you deploy it to production.

That's the key to the concept of portability – the application code should be independent of the infrastructure it is running on. The best way to achieve this is to decouple the application code from its configuration.

In the following image, we have a common application container image and different configurations for the different environments.

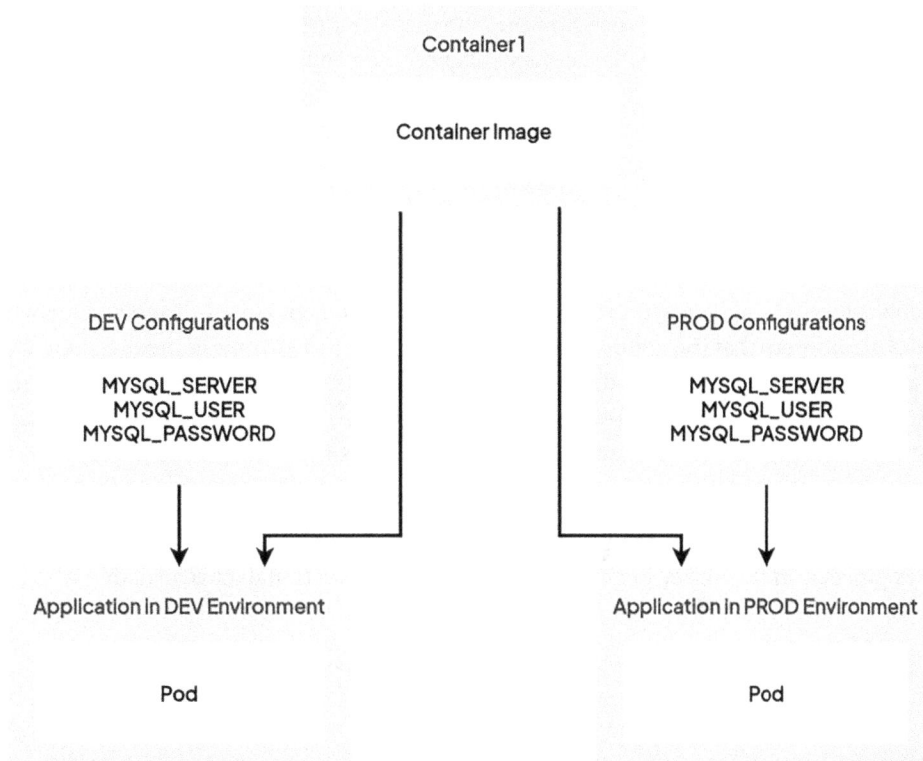

Figure 7.1: Application configuration decoupled

Let's look at some typical examples of the types of configuration values you should decouple from an app:

- API keys to access an Amazon S3 bucket
- The password of the MySQL server used by your application
- The endpoint of a Redis cluster used by your application
- Pre-computed values such as JWT token private keys

All of these values are likely to change from one environment to another, and the applications that are launched in your Pods should be able to load a different configuration, depending on the environment they are launched on. This is why we will seek to systematically maintain a barrier between our applications and the configurations they consume. Doing this allows us to treat them as two completely different entities in our Kubernetes cluster. The best way to achieve this is by considering our application and its configurations as two different entities.

This is why Kubernetes suggests using the ConfigMaps and Secrets objects, which are designed to carry your configuration data. Then, you will need to attach these ConfigMaps and Secrets when you create your Pods.

> Please avoid including your configuration values as part of your container images, such as Docker images. Your Dockerfile (or Containerfile for Podman) should build your application but not configure it. By including the container configuration at build time, you create a strong relationship between your application and how it's configured, which reduces the portability of your container to different environments.

ConfigMaps are meant to hold non-sensitive configuration values, whereas Secrets are globally the same but are meant to hold sensitive configuration values, such as database passwords, API keys, and so on.

So, you can imagine a ConfigMap and Secret for each environment and for each application, which will contain the parameters that the application needs to function in a specific context and environment. The key point is that ConfigMaps and Secrets serve as storage mechanisms for key/value pairs, for configuration data within a Kubernetes cluster. These key/value pairs can contain plain values called literals or full configuration files, such as YAML, TOML, and so on. Then, on Pod creation, you can pick the name of a ConfigMap or a Secret and link it to your Pod so that the configuration values are exposed to the containers running into it.

You always proceed in this order to guarantee that the configuration data is available when the Pod starts:

1. Create a ConfigMap or a Secret with the configuration values.
2. Create a Pod referencing the ConfigMap or Secret.

By adopting this approach, you enhance your application's portability and maintainability, aligning with common DevOps best practices.

Now that we've explained why it is important to decouple application code and configuration values, it is time to explain why and how to achieve this in a Kubernetes-friendly way.

Understanding how Pods consume ConfigMaps and Secrets

Before diving into the specifics of ConfigMaps and Secrets, let's examine traditional methods of managing configuration in containerized environments. Outside of Kubernetes, modern containerized applications consume their configuration in multiple ways:

- As OS environment variables
- As configuration files
- Command-line arguments
- API access

This is because overriding an environment variable is super-easy with Docker or Podman, and all programming languages offer functions to easily read environment variables. Configuration files can easily be shared and mounted as volumes between containers.

> IMPORTANT
>
> While ConfigMaps and Secrets are the primary methods for managing configuration in Kubernetes, there are alternative approaches. Command-line arguments can be used to pass configuration directly to containers, but this method is less flexible and secure than ConfigMaps and Secrets. Directly accessing the Kubernetes API to fetch configuration is generally discouraged, due to security risks and added complexity.

Back in the Kubernetes world, ConfigMaps and Secrets follow these two methods. Once created in your Kubernetes cluster, ConfigMaps can be consumed in one of two ways:

- Included as environment variables in the container running in your Pods
- Mounted as Kubernetes volumes, just like any other volume

You can choose to inject one value from a ConfigMap or a Secret as an environment variable, or inject all the values from a ConfigMap or a Secret as environment variables.

> IMPORTANT
>
> Using environment variables to expose Secrets is strongly discouraged due to potential security risks and the possibility of value truncation. Consider using the volume approach with file-based Secrets to leverage kubelet caching for dynamic updates and improved security.

ConfigMap and Secrets can also behave as volume mounts. When you mount a ConfigMap as a volume, you can inject all the values it contains in a directory into your container. If you store the full configuration files in your ConfigMap, using this feature to override a configuration directory becomes incredibly easy.

After this introduction to ConfigMap and Secrets, you should now understand why they are so important when it comes to configuring an application. Mastering them is crucial if you intend to work with Kubernetes cleanly and solidly. As we mentioned earlier in this chapter, ConfigMaps are used to store *unsecured* configuration values, whereas Secrets are used for more sensitive configuration data, such as hashes or database passwords.

The following table shows the high-level differences between ConfigMaps and Secrets in Kubernetes.

Feature	ConfigMaps	Secrets
Purpose	Store non-sensitive configuration data	Store sensitive information
Data format	Plain text	Base64-encoded
Security	Less secure	More secure
Common use cases	Application configuration, environment variables, and command-line arguments	Passwords, API keys, SSH keys, and TLS certificates
Handling sensitive data	Not recommended	Strongly recommended

Table 7.1: Differences between ConfigMaps and Secrets

Since the ConfigMaps and Secrets don't behave the same, let's look at them separately. First, we are going to discover how ConfigMaps work. We will discover Secrets after.

Configuring your Pods using ConfigMaps

In this section, we will learn how to list, create, delete, and read ConfigMaps. Also, we'll learn how to attach them to our Pods so that their values are injected into the Pods, in the form of environment variables or volumes.

In the following sections, we will learn how to list, create, and manage ConfigMaps in Kubernetes.

Listing ConfigMaps

Listing the ConfigMaps that were created in your cluster is fairly straightforward and can be accomplished using kubectl, just like any other object in Kubernetes. You can do this by using the full resource name, which is configmaps:

```
$ kubectl get configmaps
```

Alternatively, you can use the shorter alias, which is cm:

```
$ kubectl get cm
```

Both of these commands have the same effect. When executed, kubectl might display a few default ConfigMaps or issue an error, stating that no ConfigMaps were found. This discrepancy arises because certain cloud services generate default ConfigMaps for internal processes, while others do not. The presence or absence of these default ConfigMaps depends on the environment in which your Kubernetes cluster is deployed.

```
$ kubectl get configmaps -A
NAMESPACE          NAME                                        DATA
AGE
default            kube-root-ca.crt                            1
23d
kube-node-lease    kube-root-ca.crt                            1
23d
kube-public        cluster-info                                1
23d
kube-public        kube-root-ca.crt                            1
23d
...<removed for brevity>...
```

As you can see in the preceding output, there are multiple ConfigMaps in the default namespace and Kubernetes-managed namespaces, which are created during the cluster deployment. Now, let's learn how to create a new ConfigMap in the next section.

Creating a ConfigMap

Like other Kubernetes objects, ConfigMaps can be created imperatively or declaratively. You can decide to create an empty ConfigMap and then add values to it or create a ConfigMap directly with initial values. The following command will create an empty ConfigMap, called my-first-configmap, via the imperative method:

```
$ kubectl create configmap my-first-configmap
configmap/my-first-configmap created
```

Once this command has been executed, you can type the kubectl get cm command once again to see your new configmap:

```
$ kubectl get cm
NAME                DATA    AGE
my-first-configmap  0       42s
```

Now, we are going to create a new empty ConfigMap, but this time, we are going to create it with the declarative method. This way, we'll have to create a YAML file and apply it through kubectl.

The following content should be placed in a file called ~/my-second-configmap.yaml:

```
# ~/my-second-configmap.yaml
apiVersion: v1
kind: ConfigMap
metadata:
  name: my-second-configmap
```

Once this file has been created, you can apply it to your Kubernetes cluster using the `kubectl apply -f` command:

```
$ kubectl apply -f ~/my-second-configmap.yaml
configmap/my-second-configmap created
```

You can type the `kubectl get cm` command once more to see your new `configmap` added next to the one you created earlier:

```
$  kubectl get cm
NAME                     DATA    AGE
kube-root-ca.crt         1       23d
my-first-configmap       0       5s
my-second-configmap      0       2s
```

Please note that the output of the `kubectl get cm` command also returns the number of keys that each ConfigMap contains in the DATA column. For now, it's zero, but in the following examples, you'll see that we can fill a `configmap` when it's created, so DATA will reflect the number of keys we put in `configmap`.

Creating a ConfigMap from literal values

Having an empty ConfigMap is quite useless, so let's learn how to create a ConfigMap with values inside it. Let's do this imperatively: adding the `--from-literal` flag to the `kubectl create cm` command.

Here, we are going to create a ConfigMap called `my-third-configmap`, with a key named `color` and its value set to `blue`:

```
$ kubectl create cm my-third-configmap --from-literal=color=blue
configmap/my-third-configmap created
```

Also, be aware that you can create a ConfigMap with multiple parameters; you just need to add as much configuration data as you want to `configmap` by chaining as many from-literals as you need in your command:

```
$ kubectl create cm my-fourth-configmap --from-literal=color=blue --from-
literal=version=1 --from-literal=environment=prod
configmap/my-fourth-configmap created
```

Here, we create a ConfigMap with three configuration values inside it. Now, you can list your Config-Maps once more using this command. You should see the few additional ones you just created.

Please note that the DATA column in the return of `kubectl get cm` now reflects the number of configuration values inside each `configmap`:

```
$ kubectl get cm
NAME                     DATA    AGE
my-first-configmap       0       9m30s
my-fourth-configmap      3       6m23s
```

```
my-second-configmap     0        8m2s
my-third-configmap      1        7m9s
```

We can also see the details of the ConfigMap (or any other objects) by displaying a well-formatted output in the YAML or JSON format, as follows.

```
$ kubectl get cm my-fourth-configmap -o yaml
apiVersion: v1
data:
  color: blue
  environment: prod
  version: "1"
kind: ConfigMap
metadata:
  creationTimestamp: "2024-08-10T06:20:49Z"
  name: my-fourth-configmap
  namespace: default
  resourceVersion: "25647"
  uid: 3c8477dc-f3fe-4d69-b66a-403679a88450
```

This approach also facilitates object backup in the YAML or JSON formats, aligning with **Configuration as Code (CaC)** and **Infrastructure as Code (IaC)** best practices.

Now, it is also possible to create the same ConfigMap declaratively. Here is the declarative YAML configuration file that is ready to be applied against the cluster. Please note the new data YAML key, which contains all the configuration values:

```
# ~/my-fifth-configmap.yaml
apiVersion: v1
kind: ConfigMap
metadata:
  name: my-fifth-configmap
data:
  color: "blue"
  version: "1"
  environment: "prod"
```

Once you have created the file, you can create the ConfigMap using the kubectl apply command, as follows:

```
$ kubectl apply -f my-fifth-configmap.yaml
configmap/my-fifth-configmap created
```

A new ConfigMap object will be created based on the my-fifth-configmap.yaml file, which includes your data as per the YAML definition.

Now, let's learn how to store entire configuration files inside a ConfigMap in the next section of this chapter.

Storing entire configuration files in a ConfigMap

As we mentioned earlier, it's also possible to store complete files inside a ConfigMap – you are not restricted to literal values. The trick is to give the path of a file stored in your filesystem to the kubectl command line. The content of the file will then be taken by kubectl and used to populate a parameter in configmap.

Having the content of a configuration file stored in a ConfigMap is super-useful because you'll be able to mount your ConfigMaps in your Pods, just like you can do with volumes.

The good news is that you can mix literal values and files inside a ConfigMap. Literal values are meant to be short strings, whereas files are just treated as longer strings; they are not two different data types.

We have already seen such a sample ConfigMap in *Chapter 5*, in the *Sidecar multi-container Pod – an example* section, where we stored the Fluentd configuration content inside a ConfigMap as file content.

Here, a sixth ConfigMap is created with a literal value, just like it was previously, but now, we are also going to store the content of a file in it.

Let's create a file called configfile.txt in the $HOME/configfile.txt location with arbitrary content:

```
$ echo "I'm just a dummy config file" >> $HOME/configfile.txt
```

Here, that configuration file has the .txt extension, but it could be a .yaml, .toml, .rb, or any other configuration format that your application can use.

Now, we need to import that file into a ConfigMap, so let's create a brand-new ConfigMap to demonstrate this. You can do this using the --from-file flag, which can be used together with the --from-literal flag in the same command:

```
$ kubectl create cm my-sixth-configmap --from-literal=color=yellow --from-
file=$HOME/configfile.txt
configmap/my-sixth-configmap created
```

Let's run the kubectl get cm command once more to make sure that our sixth configmap is created. The command will show that it contains two configuration values – in our case, the one created from a literal and the other one created from the content of a file:

```
$ kubectl get cm my-sixth-configmap
NAME                  DATA   AGE
my-sixth-configmap    2      38s
```

As you can see, my-sixth-configmap contains two pieces of data: the literal and the file.

Now, let's create a seventh ConfigMap. Just like the sixth one, it's going to contain a literal and a file, but this time, we're going to create it declaratively.

The YAML format allows you to use multiple lines with the | symbol. We're using this syntax as part of our declaration file:

```
# ~/my-seventh-configmap.yaml
apiVersion: v1
kind: ConfigMap
metadata:
  name: my-seventh-configmap
data:
  color: "green"
  configfile.txt: |
    I'm another configuration file.
```

Let's apply this YAML file to create our `configmap` with the `kubectl apply` command:

```
$ kubectl apply -f my-seventh-configmap.yaml
configmap/my-seventh-configmap created
```

Let's list the ConfigMaps in our cluster using `kubectl get cm` to make sure our seventh `configmap` has been created and contains two values. So let's run the `kubectl get cm` command once more:

```
$ kubectl get configmap/my-seventh-configmap
NAME                     DATA   AGE
my-seventh-configmap     2      36s
```

Now, let's discover the last possible way to create a ConfigMap – that is, from an env file.

Creating a ConfigMap from an env file

As you might guess, you can create a ConfigMap from an env file imperatively using the `--from-env-file` flag.

An env file is a `key=value` format file where each key is separated by a line break. This is a configuration format that's used by some applications, so Kubernetes introduced a way to generate a ConfigMap from an existing env file. This is especially useful if you have an already existing application that you want to migrate into Kubernetes.

Here is a typical env file:

```
# ~/my-env-file.env
hello=world
color=blue
release=1.0
production=true
```

By convention, env files are named `.env`, but it's not mandatory. So long as the file is formatted correctly, Kubernetes will be able to generate a ConfigMap based on the parameters.

You can use the following command to import the configuration in the env file as a ConfigMap into your Kubernetes cluster:

```
$ kubectl create cm my-eight-configmap --from-env-file my-env-file.env
configmap/my-eight-configmap created
```

Lastly, let's list the ConfigMaps in our cluster to check that our new ConfigMap was created with three configuration values:

```
$ kubectl get cm my-eight-configmap
NAME                   DATA   AGE
my-eight-configmap     4      7s
```

As you can see, the new `configmap` is now available in the cluster, and it was created with the three parameters that were present in the env file. That's a solid way to import your env files into Kubernetes ConfigMaps.

> Remember that `ConfigMaps` are not meant to contain sensitive values. Data in ConfigMaps are not encoded, and that's why you can view them with just a `kubectl describe cm` command. For anything that requires privacy, you'll have to use the `Secret` object and not the `ConfigMap` one. It's important to note that while `Secrets` themselves are not inherently encrypted by default, they are stored in the base64 format, offering a basic level of obfuscation. Additionally, with Kubernetes versions 1.27 and later, you can leverage Kubernetes **Key Management System (KMS)** plugin providers to encrypt data within both Secrets and ConfigMaps. This provides a more robust security layer for sensitive information.

Now, let's discover how to read the values inside a ConfigMap.

Reading values inside a ConfigMap

So far, we've only listed the ConfigMaps to see the number of keys in them. Let's take this a little bit further: you can read actual data inside a ConfigMap, not just count the number of ConfigMaps. This is useful if you want to debug a ConfigMap or if you're not confident about what kind of data is stored in them.

The data in a ConfigMap is not meant to be sensitive, so you can read and retrieve it easily from `kubectl`; it will be displayed in the terminal's output.

You can read the value in a ConfigMap with the `kubectl describe` command. We will run this command against the `my-fourth-configmap` ConfigMap, since it's the one that contains the most data. The output is quite big, but as you can see, the two pieces of configuration data are displayed clearly:

```
$ kubectl describe cm my-fourth-configmap
Name:          my-fourth-configmap
Namespace:     default
Labels:        <none>
Annotations:   <none>
```

```
Data
====
color:
----
blue
environment:
----
prod
version:
----
1

BinaryData
====

Events:   <none>
```

The kubectl describe cm command returns these kinds of results. Expect to receive results similar to this one and not results in a computer-friendly format, such as JSON or YAML.

As the data is displayed clearly in the terminal output, keep in mind that any user of the Kubernetes cluster (with access to this ConfigMap) will be able to retrieve this data directly by typing the kubectl describe cm command, so be careful to not store any sensitive value in a ConfigMap.

Now, let's discover how we can inject ConfigMap data into running Pods as environment variables.

Linking ConfigMaps as environment variables

In this section, we're going to bring our ConfigMaps to life by linking them to Pods. First, we will focus on injecting ConfigMaps as environment variables. Here, we want the environment variables of a container within a Pod to come from the values of a ConfigMap.

You can do this in two different ways:

- **Injecting one or more given values in a given ConfigMap**: You can set the value of an environment variable based on the parameters contained in one or multiple ConfigMaps.
- **Injecting all the values contained in a given ConfigMap:** You take one ConfigMap and inject all the values it contains into an environment at once. This way is good if you are creating one ConfigMap per Pod specification or application so that each app has a ConfigMap ready to be deployed.

> Please note that it's impossible to link a ConfigMap to a Pod with the kubectl imperative method. The reason is that it's impossible to create a Pod referencing a ConfigMap directly from the kubectl run command. You will have to write declarative YAML files to use your ConfigMaps in your Pods.

Earlier in this chapter, we created a ConfigMap called `my-third-configmap` that contains a parameter called `color`, with a value of `blue`. In this example, we will create a Pod with the `quay.io/iamgini/my-flask-app:1.0` image, and we will link `my-third-configmap` to the Pod so that the flask application container is created with an environment variable called `COLOR`, with a value set to `blue`, according to what we have in the ConfigMap. Here is the YAML manifest to achieve that. Pay attention to the `env:` key in the `container spec`:

```yaml
# flask-pod-with-configmap.yaml
apiVersion: v1
kind: Pod
metadata:
  name: flask-pod-with-configmap
  labels:
    app: my-flask-app
spec:
  containers:
    - name: flask-with-configmap
      image: quay.io/iamgini/my-flask-app:1.0
      env:
        - name: COLOR # Any other name works here.
          valueFrom:
            configMapKeyRef:
              name: my-third-configmap
              key: color
```

Now, we can create this Pod using the `kubectl apply` command:

```
$ kubectl apply -f flask-pod-with-configmap.yaml
pod/flask-pod-with-configmap created
$ kubectl get pod
NAME                       READY   STATUS    RESTARTS   AGE
flask-pod-with-configmap   1/1     Running   0          5s
```

Now that our application Pod has been created, let's launch the env command inside the container to list all the environment variables that are available in the container. As you may have guessed, we will issue the env Linux command in this specific container by calling the `kubectl exec` command. Here is the command and the output to expect:

```
$ kubectl exec pods/flask-pod-with-configmap -- env
PATH=/usr/local/bin:/usr/local/sbin:/usr/local/bin:/usr/sbin:/usr/bin:/sbin:/bin
HOSTNAME=flask-pod-with-configmap
COLOR=blue
KUBERNETES_PORT_443_TCP=tcp://10.96.0.1:443
...<removed for brevity>...
```

You should see the COLOR environment variable in the output if your ConfigMap has been linked to your Pod correctly.

You can also check the application deployed by exposing the Pod using a service, as follows:

```
$ kubectl expose pod flask-pod-with-configmap --port=8081 --target-port=5000
--type=NodePort
```

Let's test the application over a browser using the kubectl port-forward method.

You can start a port-forward to test the application, as follows:

```
$ kubectl port-forward flask-pod-with-configmap 20000:5000
Forwarding from 127.0.0.1:20000 -> 5000
Forwarding from [::1]:20000 -> 5000
```

Using kubectl port-forward

kubectl port-forward creates a secure tunnel between your local machine and a specific Pod within a Kubernetes cluster. This allows you to access and interact with applications running inside the Pod as if they were running locally. It's a valuable tool for debugging, testing, and development, providing direct access to services without external exposure.

In the preceding code snippet, we forward the local port 20000 to the target port 5000 temporarily. You can now access the application in your local browser using the address 127.0.0.1:20000.

> While NodePort services are designed for external access, they can also be used in minikube environments. To access your application externally in this case, use the minikube service command to get the NodePort and corresponding IP address. For instance, the minikube service --url flask-pod-with-configmap might output http://192.168.49.2:31997, allowing you to access your Flask application at this URL.

Now, you can access the flask application over a web browser and can see the background with a BLUE color, as it was configured in the ConfigMap as COLOR=blue. You can change the value of COLOR in the ConfigMap and recreate the flask-pod-with-configmap Pod to see the changes.

We will learn more about Kubernetes services and DNS in *Chapter 8*, *Exposing Your Pods with Services*.

Please note that the kubectl port-forward will continue serving the forward until you end the command (e.g., by pressing the *Ctrl + C* keys).

Now, we are going to discover the second way of injecting ConfigMaps as environment variables.

In this demo, we will link another ConfigMap, the one called my-fourth-configmap. This time, we don't want to retrieve a single value in this ConfigMap but all the values inside of it instead. Here is the updated YAML Pod manifest. This time, we don't use individual env keys but an envFrom key in our container spec instead:

```
# flask-pod-with-configmap-all.yaml
apiVersion: v1
```

```
kind: Pod
metadata:
  name: flask-pod-with-configmap-all
spec:
  containers:
    - name: flask-with-configmap
      image: quay.io/iamgini/my-flask-app:1.0
      envFrom:
        - configMapRef:
            name: my-fourth-configmap
```

Once the manifest file is ready, you can recreate the NGINX Pod:

```
$ kubectl apply -f flask-pod-with-configmap-all.yaml
pod/ flask-pod-with-configmap-all created
```

Now, let's run the env command once more in the nginx container, using the kubectl exec command
to list the environment variables:

```
$ kubectl exec pods/flask-pod-with-configmap-all -- env
PATH=/usr/local/bin:/usr/local/sbin:/usr/local/bin:/usr/sbin:/usr/bin:/sbin:/
bin
HOSTNAME=flask-pod-with-configmap-all
environment=prod
version=1
color=blue
KUBERNETES_PORT_443_TCP_PORT=443
KUBERNETES_PORT_443_TCP_ADDR=10.96.0.1
...<removed for brevity>...
```

Note that the three parameters – color, version, and environment – that were declared in the my-
fourth-configmap have been set as environment variables in the container, but this time, you don't
have control over how the environment variables are named in the container. Their names are directly
inherited from the parameter key names in the ConfigMap.

Now, it's time to learn how to mount a ConfigMap as a volume in a container.

Mounting a ConfigMap as a volume mount

Earlier in this chapter, we created two ConfigMaps that store dummy configuration files. kubectl allows
you to mount a ConfigMap inside a Pod as a volume. This is especially useful when the ConfigMap
contains the content of a file that you want to inject into a container's filesystem.

Just like when we inject environment variables, we need to do this imperatively using a YAML manifest file. Here, we are going to mount a ConfigMap called `my-sixth-configmap` as a volume mount to a new Pod, `flask-pod-with-configmap-volume`, as follows:

```yaml
# flask-pod-with-configmap-volume.yaml
apiVersion: v1
kind: Pod
metadata:
  name: flask-pod-with-configmap-volume
spec:
  containers:
    - name: flask-with-configmap-volume
      image: quay.io/iamgini/my-flask-app:1.0
      volumeMounts:
        - name: configuration-volume # match the volume name
          mountPath: /etc/conf
  volumes:
    - name: configuration-volume
      configMap:
        name: my-sixth-configmap # Configmap name goes here
```

Here, we have declared a volume named `configuration volume` at the same level as the containers, and we have told Kubernetes that this volume was built from a ConfigMap. The referenced Config-Map (here, `my-sixth-configmap`) must be present in the cluster when we apply this file. Then, at the container level, we mounted the volume we declared earlier on `path /etc/conf:`. The parameter in the ConfigMap should be present at the specified location.

Let's apply this file to create a new ConfigMap, with the volume attached to our cluster:

```
$ kubectl apply -f flask-pod-with-configmap-volume.yaml
pod/flask-pod-with-configmap-volume created
```

Run the `ls` command in the container to make sure that the directory has been mounted:

```
$  kubectl exec pods/flask-pod-with-configmap-volume -- ls /etc/conf
color
configfile.txt
```

Here, the directory has been successfully mounted, and both parameters that were created in the ConfigMap are available in the directory as plain files.

Let's run the `cat` command to make sure that both files hold the correct values:

```
$ kubectl exec pods/flask-pod-with-configmap-volume -- cat /etc/conf/color
yellow
$ kubectl exec pods/flask-pod-with-configmap-volume -- cat /etc/conf/
```

```
configfile.txt
I'm just a dummy config file
```

Good! Both files contain the values that were declared earlier when we created the ConfigMap! For example, you could store a virtual host NGINX configuration file and have it mounted to the proper directory, allowing NGINX to serve your website based on the configuration values hosted in a ConfigMap. That's how you can override the default configuration and cleanly manage your app in Kubernetes. Now, you have a really strong and consistent interface to manage and configure the containers running in Kubernetes.

Next, we will learn how to delete and update a ConfigMap.

Deleting a ConfigMap

Deleting a ConfigMap is very easy. However, be aware that you can delete a ConfigMap even if its values are used by a container. Once the Pod has been launched, it's independent of the `configmap` object:

```
$ kubectl delete cm my-first-configmap
configmap "my-first-configmap" deleted
```

Regardless of whether the ConfigMap's values are used by the container, it will be deleted as soon as this command is entered. Note that the ConfigMap cannot be recovered, so please think twice before removing a ConfigMap you have created imperatively, since you won't be able to recreate it. Unlike declaratively created ConfigMaps, its content is not stored in any YAML file. But as we learned earlier, it is possible to collect the content of such resources in the YAML format by formatting the `kubectl get` commands – for example, `kubectl get cm <configmap-name> -o YAML > my-first-configmap.yaml`.

Also, we recommend that you are careful when removing your ConfigMaps, especially if you delete ConfigMaps that are used by running Pods. If your Pod were to crash, you wouldn't be able to relaunch it without updating the manifest file; the Pods would look for the missing ConfigMap you deleted.

Let's test this scenario with the `my-sixth-configmap` ConfigMap; delete the ConfigMap resource as follows:

```
$ kubectl delete cm my-sixth-configmap
configmap "my-sixth-configmap" deleted
```

Now, try relaunching the `flask-pod-with-configmap-volume` Pod (or recreate the Pod) to see the problem:

```
$ kubectl apply -f flask-pod-with-configmap-volume.yaml
pod/flask-pod-with-configmap-volume created
$ kubectl get pod
NAME                               READY   STATUS             RESTARTS   AGE
flask-pod-with-configmap-volume    0/1     ContainerCreating  0          61s
```

You will see the Pod is in `ContainerCreating` and not in a `Running` state. Let's check the details of the Pod as follows:

```
$ kubectl describe pod flask-pod-with-configmap-volume
Name:               flask-pod-with-configmap-volume
Namespace:          default
...<removed for brevity>...
Events:
  Type     Reason        Age                From              Message
  ----     ------        ----               ----              -------
  Normal   Scheduled     71s                default-scheduler  Successfully
assigned default/flask-pod-with-configmap-volume to minikube-m03
  Warning  FailedMount   7s (x8 over 71s)   kubelet            MountVolume.SetUp
failed for volume "configuration-volume" : configmap "my-sixth-configmap" not
found
```

The error in the previous code snippet clearly says that the required ConfigMap is missing and unable to mount the Volume; hence, the Pod is still in the `ContainerCreating` state.

Updating a ConfigMap

There are two primary methods to update a ConfigMap in Kubernetes. The first involves using the `kubectl apply` command with a modified ConfigMap definition file. This approach is ideal for version control and collaborative environments. Simply make the necessary changes to your ConfigMap YAML file and apply the updates using `kubectl apply`.

Alternatively, you can directly edit an existing ConfigMap using the `kubectl edit` command. This provides an interactive way to modify the ConfigMap's contents. However, be cautious when using this method, as it doesn't involve version control.

Immutable ConfigMaps

Kubernetes offers a feature called Immutable ConfigMaps to prevent accidental or intentional modifications to ConfigMap data. By marking a ConfigMap as immutable, you ensure that its contents remain unchanged.

This feature is particularly beneficial for clusters heavily reliant on ConfigMaps, as it safeguards against configuration errors causing application disruptions. Additionally, it enhances cluster performance by reducing the load on the `kube-apiserver`.

To create an immutable ConfigMap, simply set the immutable field to `true` within the ConfigMap definition, as follows:

```
# immutable-configmap.yaml
apiVersion: v1
kind: ConfigMap
metadata:
```

```
    name: immutable-configmap
data:
    color: "blue"
    version: "1.0"
    environment: "prod"
immutable: true
```

Once marked immutable, you cannot modify the ConfigMap's data. If changes are necessary, you must delete the existing ConfigMap and create a new one. It's essential to recreate any Pods referencing the deleted ConfigMap to maintain the correct configuration.

In the next section of this chapter, we will learn about Kubernetes Secrets.

Managing sensitive configuration with the Secret object

The Secret object is a resource that allows you to configure applications running on Kubernetes. Secrets are extremely similar to ConfigMaps and they can be used together. The difference is that Secrets are encoded and intended to store sensitive data such as passwords, tokens, or private API keys, while ConfigMaps are intended to host non-sensitive configuration data. Other than that, Secrets and ConfigMaps mostly behave the same.

To ensure the protection of sensitive information stored in Kubernetes Secrets, adhere to the following best practices:

- **Limit access:** Utilize **Role-Based Access Control (RBAC)** to restrict access to Secrets based on user roles and permissions. Grant only necessary privileges to individuals or services.
- **Avoid hardcoding secrets:** Never embed Secrets directly within your application code or configuration files.
- **Rotate secrets regularly:** Implement a regular rotation schedule for Secrets to mitigate the risk of unauthorized access.
- **Consider external secret management:** For advanced security requirements, explore dedicated secret management solutions like HashiCorp Vault or AWS Secrets Manager.
- **Leverage encryption:** Utilize KMS plugins to encrypt Secrets data at rest, providing an additional layer of protection.
- **Monitor and audit:** Regularly review access logs and audit trails to detect suspicious activity and potential security breaches.
- **Educate your team:** Foster a security-conscious culture by providing training on the best practices for handling and managing Secrets.

> While Secrets are used to store sensitive information and their data is encoded in base64, this encoding alone does not guarantee strong security. Base64 is a reversible encoding format, meaning that the original data can be recovered easily. For robust protection of sensitive information, consider additional security measures, such as encryption at rest using KMS plugins or external secret management solutions.

While Secrets are primarily used to store sensitive data, Kubernetes offers additional security measures. With the introduction of the **Key Management Service (KMS)** plugin providers, you can now encrypt data within both Secrets and ConfigMaps. This provides an extra layer of protection for sensitive information stored in ConfigMaps, making them more secure.

By utilizing KMS encryption for ConfigMaps, you can safeguard sensitive configuration data without resorting to Secrets. This approach simplifies configuration management while maintaining a high level of security.

> Even with KMS encryption, it's essential to carefully consider the sensitivity of the data stored in ConfigMaps. For highly confidential information, Secrets remain the recommended option.

Refer to the documentation (`https://kubernetes.io/docs/tasks/administer-cluster/kms-provider/`) to learn more about the KMS provider for data encryption.

Let's start by discovering how to list the Secrets that are available in your Kubernetes cluster.

Listing Secrets

Like any other Kubernetes resource, you can list secrets using the `kubectl get` command. The resource identifier is a Secret here:

```
$ kubectl get secret -A
```

Just like with ConfigMaps, the DATA column tells you the number of sensitive parameters that have been hashed and saved in your `secret`. When executed, `kubectl` might display a few default Secrets or issue an error, stating that no resources were found. This discrepancy arises because certain cloud services generate default Secrets for internal processes, while others do not. The presence or absence of these default ConfigMaps depends on the environment in which your Kubernetes cluster is deployed.

Creating a Secret imperatively with --from-literal

You can create a Secret imperatively or declaratively – both methods are supported by Kubernetes. Let's start by discovering how to create a Secret imperatively. Here, we want to store a database password, `my-db-password`, in a Secret object in our Kubernetes cluster. You can achieve that imperatively with `kubectl` by adding the `--from-literal` flag to the `kubectl create secret` command:

```
$ kubectl create secret generic my-first-secret --from-literal='db_password=my-db-password'
```

Now, run the `kubectl get secrets` command to retrieve the list of Secrets in your Kubernetes cluster. The new Secret should be displayed:

```
$ kubectl get secrets
NAME              TYPE      DATA    AGE
my-first-secret   Opaque    1       37s
```

Let's see the details of the Secret using the YAML output format, as shown below:

```
$ kubectl get secrets my-first-secret -o yaml
apiVersion: v1
data:
  db_password: bXktZGItcGFzc3dvcmQ=
kind: Secret
metadata:
  creationTimestamp: "2024-08-10T13:13:32Z"
  name: my-first-secret
  namespace: default
  resourceVersion: "36719"
  uid: 7ccf5120-d1c5-4874-ba4b-894274fd27e6
type: Opaque
```

You will see data with the `db_password: bXktZGItcGFzc3dvcmQ=` line, where the password is stored in an encoded format.

Now, let's figure out how to create a Secret declaratively.

Creating a Secret declaratively with a YAML file

Secrets can be created declaratively using YAML files. While it's possible to manually encode secret values as base64 for the `data` field, Kubernetes provides a more convenient approach. The `stringData` field allows you to specify secret values as plain text strings. Kubernetes automatically encodes these values into the base64 format when creating the Secret. This method simplifies the process and helps prevent accidental exposure of sensitive data in plain text configuration files.

> Note: While base64 encoding offers basic obfuscation, it's essential to remember that it's not a strong encryption method. For heightened security, consider using KMS plugins or external secret management solutions.

When you use `--from-literal`, Kubernetes will encode your strings in base64 itself, but when you create a Secret from a YAML manifest file, you will have to handle this step yourself.

So let's start by converting the `my-db-password` string into `base64`:

```
$ echo -n 'my-db-password' | base64
bXktZGItcGFzc3dvcmQ=
```

`bXktZGItcGFzc3dvcmQ=` is the `base64` representation of the `my-db-password` string, and that's what we will need to write in our YAML file. Here is the content of the YAML file to create the Secret object properly:

```
# ~/secret-from-file.yaml
apiVersion: v1
```

```
kind: Secret
metadata:
  name: my-second-secret
type: Opaque
data:
    db_password: bXktZGItcGFzc3dvcmQ=
```

Once this file has been stored on your system, you can create the `secret` using the `kubectl apply` command:

```
$ kubectl apply -f ~/secret-from-file.yaml
```

We can make sure that the `secret` has been created properly by listing the secrets, with details, in our Kubernetes cluster:

```
$ kubectl get secret my-second-secret -o yaml
apiVersion: v1
data:
  db_password: bXktZGItcGFzc3dvcmQ=
kind: Secret
...<removed for brevity>...
type: Opaque
```

Now, let's create the same Secret with `stringData` so that we do not need to encode it manually. Create a new YAML file as follows:

```
# ~/secret-from-file-stringData.yaml
apiVersion: v1
kind: Secret
metadata:
  name: my-secret-stringdata
type: Opaque
stringData:
  db_password: my-db-password
```

Note the password in plain text instead of encoded text.

Create the Secret from the file as follows:

```
kubectl apply -f secret-from-file-stringData.yaml
secret/my-secret-stringdata created
```

Verify the secret content and compare it with the Secret `my-second-secret` to see the encoded content:

```
$ kubectl get secrets my-secret-stringdata -o yaml
apiVersion: v1
data:
  db_password: bXktZGItcGFzc3dvcmQ=
```

```
kind: Secret
...<removed for brevity>...
  name: my-secret-stringdata
  namespace: default
  resourceVersion: "37229"
  uid: 5078124b-7318-44df-a2da-a30ea5088e3f
type: Opaque
```

You will notice that the `stringData` is already encoded and stored under the `data` section.

Now, let's discover another Kubernetes feature: the ability to create a Secret with values from a file.

Creating a Secret with content from a file

We can create a Secret with values from a file, the same as we did with ConfigMaps. We start by creating a file that will contain our secret value. Let's say that we have to store a password in a file and import it as a Secret object in Kubernetes:

```
$ echo -n 'mypassword' > ./password.txt
```

After running this command, we have a file called `password.txt` that contains a string called `mypassword`, which is supposed to be our Secret value. The `-n` flag is used here to ensure that `password.txt` does not contain any extra blank lines at the end of the text.

Now, let's run the `kubectl create secret` command by passing the location of `password.txt` to the `--from-file` flag. This will result in a new `secret`, containing a `base64` representation of the `mypassword` string being created:

```
$ kubectl create secret generic mypassword --from-file=./password.txt
secret/mypassword created
```

This new `secret` is now available in your Kubernetes cluster! Now, let's learn how to read a Kubernetes Secret.

Reading a Secret

As observed, Secrets are supposed to host sensitive data, and, as such, the `kubectl` output won't show you the secret decoded data to ensure confidentiality. You'll simply have to decode it yourself to understand. Why is this confidentiality maintained? Let's take a look:

- To prevent the secret from being accidentally opened by someone who shouldn't be able to open it.
- To prevent the secret from being displayed as part of a terminal output, which could result in it being logged somewhere.

> While base64 encoding obfuscates secret data, it doesn't provide strong encryption. Any user with API access to the Kubernetes cluster can retrieve and decode a Secret. To protect sensitive information, implement RBAC to restrict access to Secrets based on user roles and permissions. By carefully defining RBAC rules, you can limit who can view, modify, or delete Secrets, enhancing the overall security of your cluster.

Because of these securities, you simply won't be able to retrieve the actual content of a secret, but you can still grab information about its size, and so on.

You can do this using the `kubectl describe` command, just like we did earlier for ConfigMaps. As we mentioned previously, ConfigMaps and Secrets are very similar; they almost behave the same:

```
$ kubectl describe secret/mypassword
Name:          mypassword
Namespace:     default
Labels:        <none>
Annotations:   <none>


Type:  Opaque


Data
====
password.txt:  10 bytes
```

Do not get confused if your output is a little different than this one. If you receive something similar, it means that the new secret is available in your Kubernetes cluster and that you successfully retrieved its data!

However, also remember that the encoded data can be visible using the `kubectl` and YAML output format, as follows:

```
$ kubectl get secret my-second-secret -o yaml
apiVersion: v1
data:
  db_password: bXktZGItcGFzc3dvcmQ=
kind: Secret
metadata:
  creationTimestamp: "2024-02-13T04:15:56Z"
  name: my-second-secret
  namespace: default
  resourceVersion: "90372"
  uid: 94b7b529-baed-4844-8097-f6c2a001fa7b
type: Opaque
```

And anyone with access to this Secret can decode the data, as follows:

```
$ echo 'bXktZGItcGFzc3dvcmQ=' | base64 --decode
my-db-password
```

> **IMPORTANT**
>
> Base64 encoding is not encryption; it is simply a way to represent binary data in ASCII characters. While it makes the data look less readable, it's not encryption. Anyone with the knowledge and tools can easily decode it back to plain text. To safeguard Secrets effectively, combine base64 encoding with additional security controls like RBAC and encryption, or consider using external secret management solutions.

Now that we've explored the creation and management of Secrets, let's delve into how to make this sensitive information accessible to our applications within Pods.

Consuming a Secret as an environment variable

We've seen how we can inject the values of a ConfigMap into a Pod in the form of environment variables, and we can do the same with Secrets. Returning to the example with our NGINX container, we are going to retrieve the db_password value of the my-first-secret Secret and inject it as an environment variable into the Pod. Here is the YAML manifest. Again, everything occurs under the env: key:

```
# flask-pod-with-secret.yaml
apiVersion: v1
kind: Pod
metadata:
  name: flask-pod-with-secret
  labels:
    app: flask-with-secret
spec:
  containers:
    - name: flask-with-secret
      image: quay.io/iamgini/my-flask-app:1.0
      env:
        - name: PASSWORD_ENV_VAR # Name of env variable
          valueFrom:
            secretKeyRef:
              name: my-first-secret # Name of secret object
              key: db_password # Name of key in secret object
```

Now, you can apply this file using the kubectl apply command:

```
$ kubectl apply -f flask-pod-with-secret.yaml
```

Now, run the env command to list the environment variables in your container:

```
$ kubectl exec pods/flask-pod-with-secret -- env
PATH=/usr/local/bin:/usr/local/sbin:/usr/local/bin:/usr/sbin:/usr/bin:/sbin:/
bin
HOSTNAME=flask-pod-with-secret
PASSWORD_ENV_VAR=my-db-password
KUBERNETES_SERVICE_PORT=443
...<removed for brevity>...
```

As you can see, the my-db-password string is available as the environment variable PASSWORD_ENV_VAR.

Another way to find out details about a Secret is by using the envFrom YAML key. When using this key, you'll read all the values from a Secret and get them as environment variables in the Pod all at once. It works the same as for the ConfigMap object.

> For a Pod to start successfully, the referenced Secret must exist unless explicitly marked as optional within the Pod definition. Ensure that the Secret is created before deploying Pods that depend on it.

Create a Secret first by using the following YAML with the envFrom sample:

```
# secret-from-file-database.yaml
apiVersion: v1
kind: Secret
metadata:
  name: appdb-secret
type: Opaque
stringData:
  db_user: appadmin
  db_password: appdbpassword
```

```
$ kubectl apply -f secret-from-file-database.yaml
secret/appdb-secret created
```

Here is the preceding example of a Pod but updated with an envFrom key:

```
# flask-pod-with-secret-all.yaml
apiVersion: v1
kind: Pod
metadata:
  name: flask-pod-with-secret-all
  labels:
```

```
      app: flask-with-secret
spec:
  containers:
    - name: flask-with-secret
      image: quay.io/iamgini/my-flask-app:1.0
      envFrom:
        - secretRef:
            name: appdb-secret # Name of the secret object
```

Create the Pod using the kubectl apply command:

```
$ kubectl apply -f flask-pod-with-secret-all.yaml
pod/flask-pod-with-secret-all created
```

Using this, all the keys in the Secret object will be used as environment variables within the Pod. Let us verify the environment variables inside the Pod to ensure that the variables from the Secret object are loaded properly:

```
$ kubectl exec pods/flask-pod-with-secret-all -- env |grep -i app
db_user=appadmin
db_password=appdbpassword
```

> Note that if a key name cannot be used as an environment variable name, then it will be simply ignored!

Now, let's learn how to consume a secret as a volume mount in the next section.

Consuming a Secret as a volume mount

You can mount Secrets as a volume for your Pods, but you can only do so declaratively. So you'll have to write YAML files to do this successfully.

You must start from a YAML manifest file that will create a Pod. Here is a YAML file that mounts a Secret called mypassword in the /etc/passwords-mounted-path location:

```
# flask-pod-with-secret-volume.yaml
apiVersion: v1
kind: Pod
metadata:
  name: flask-pod-with-secret-volume
  labels:
    app: flask-with-secret-volume
spec:
  containers:
```

```
    - name: flask-with-secret
      image: quay.io/iamgini/my-flask-app:1.0
      volumeMounts:
        - name: mysecret-volume
          mountPath: '/etc/password-mounted-path'
          readOnly: true    # Setting readOnly to true to prevent writes to the
   secret
    volumes:
      - name: mysecret-volume
        secret:
          secretName: my-second-secret # Secret name goes here
```

Once you have created this file on your filesystem, you can apply the YAML file using kubectl:

```
$ kubectl apply -f flask-pod-with-secret-volume.yaml
```

Please make sure that the my-second-secret exists before you attempt to create the Secret.

Finally, you can run a command inside flask-with-secret, using the kubectl exec command, to check if the volume containing the Secret was set up correctly:

```
$ kubectl exec pods/flask-pod-with-secret-volume --  cat /etc/password-mounted-
path/db_password
my-db-password
```

As you can see, the my-db-password string is displayed correctly; the Secret was correctly mounted as a volume!

Now that we have learned how to create a Secret in Kubernetes and use it with a Pod in multiple methods, let's learn how to delete and update Secrets in the next section.

Deleting a Secret

Deleting a secret is very simple and can be done via the kubectl delete command:

```
$ kubectl delete secret my-first-secret
secret "my-first-secret" deleted
```

Now, let's learn how to update an existing secret in a Kubernetes cluster.

Updating a Secret

To update a Secret in Kubernetes, you can use the kubectl apply command with a modified Secret definition or by using the kubectl edit command.

Kubernetes Secrets provide a secure way to store and manage sensitive information like passwords, API keys, and certificates. Unlike ConfigMaps, Secrets are encoded to protect data confidentiality. This section explored how to create Secrets both imperatively and declaratively using YAML. You learned how to inject Secrets into Pods using environment variables and volume mounts, enabling applications to access sensitive data without exposing it in plain text.

Summary

This chapter delved into the fundamental concepts of managing configuration within Kubernetes. We explored the critical distinction between ConfigMaps and Secrets, understanding their respective roles in handling non-sensitive and sensitive data. By effectively utilizing these Kubernetes resources, you can significantly enhance application portability and security.

We learned how to create and manage both ConfigMaps and Secrets, employing both imperative and declarative approaches. You discovered how to inject configuration data into Pods using environment variables and volume mounts, ensuring seamless access to application settings.

To protect sensitive information, we emphasized the importance of implementing robust security measures beyond base64 encoding. By combining RBAC, encryption, and external secret management solutions, you can significantly strengthen the security posture of your Kubernetes environment.

By mastering the concepts presented in this chapter, you'll be well-equipped to build resilient and secure Kubernetes applications that are decoupled from their configuration, promoting flexibility and maintainability. In the next chapter, we will continue discovering Kubernetes by tackling another central concept of Kubernetes, which is Services. Services are Kubernetes objects that allow you to expose your Pods to not only each other but also the internet. This is a very important network concept for Kubernetes, and mastering it is essential to use the orchestrator correctly. Fortunately, mastering Services is not very complicated, and the next chapter will explain how to achieve this. You will learn how to associate the ports of a container with the ports of the worker node it is running on, and also how to associate a static IP with your Pods so that they can always be reached at the same address by other Pods in the cluster.

Further reading

- ConfigMaps: https://kubernetes.io/docs/concepts/configuration/configmap/
- Configure a Pod to Use a ConfigMap: https://kubernetes.io/docs/tasks/configure-pod-container/configure-pod-configmap/
- Managing Secrets using kubectl: https://kubernetes.io/docs/tasks/configmap-secret/managing-secret-using-kubectl/
- Managing Secrets using Configuration File: https://kubernetes.io/docs/tasks/configmap-secret/managing-secret-using-config-file/
- Using a KMS provider for data encryption: https://kubernetes.io/docs/tasks/administer-cluster/kms-provider/
- The security aspect of Secrets as well as compliance and DevOps practices: Kubernetes Secrets Handbook: Design, implement, and maintain production-grade Kubernetes Secrets management solutions: https://www.amazon.com/Kubernetes-Secrets-Handbook-production-grade-management/dp/180512322X/

Join our community on Discord

Join our community's Discord space for discussions with the authors and other readers:

`https://packt.link/cloudanddevops`

8

Exposing Your Pods with Services

After reading the previous chapters, you now know how to deploy applications on Kubernetes by building Pods, which can contain one container or multiple containers in the case of more complex applications. You also know that it is possible to decouple applications from their configuration by using Pods and **ConfigMaps** together and that Kubernetes is also capable of storing your sensitive configurations, thanks to **Secret** objects.

The good news is that with these three resources, you can start deploying applications on Kubernetes properly and get your first app running. However, you are still missing something important: you need to be able to expose Pods to end users or even to other Pods within the Kubernetes cluster. This is where Kubernetes **Services** comes in, and that's the concept we're going to discover now!

In this chapter, we'll learn about a new Kubernetes resource type called the Service. Since Kubernetes Services is a big topic with many things to cover, this chapter will be quite big with a lot of information. But after you master these Services, you're going to be able to expose your Pods and get your end users to your apps!

Services are also a key concept to master **high availability** (HA) and redundancy in your Kubernetes setup. Put simply, it is crucial to master them to be effective with Kubernetes!

In this chapter, we're going to cover the following main topics:

- Why would you want to expose your Pods?
- The `NodePort` Service
- The `ClusterIP` Service
- The `LoadBalancer` Service
- The `ExternalName` Service type
- Implementing Service readiness using probes

Technical requirements

To follow along with the examples in this chapter, make sure you have the following:

- A working Kubernetes cluster (whether this is local or cloud-based is of no importance)
- A working kubectl **command-line interface (CLI)** configured to communicate with the Kubernetes cluster

You can download the latest code samples for this chapter from the official GitHub repository at https://github.com/PacktPublishing/The-Kubernetes-Bible-Second-Edition/tree/main/Chapter08.

Why would you want to expose your Pods?

In the previous chapters, we discussed the microservice architecture, which exposes your functionality through **Representational State Transfer (REST) application programming interfaces (APIs)**. These APIs rely completely on **HyperText Transfer Protocol (HTTP)**, which means that your microservices must be accessible via the web, and thus via an **Internet Protocol (IP)** address on the network.

> While REST APIs are commonly used for communication in microservice architectures, it's important to evaluate other communication protocols such as **gRPC**, particularly in scenarios where performance and efficiency between services are critical. gRPC, built on **HTTP/2** and using binary serialization (Protocol Buffers), can offer significant advantages in distributed systems, such as faster communication, lower latency, and support for streaming. Before defaulting to REST, consider whether gRPC might be a better fit for your system's needs.

In the following section, we will learn about cluster networking in Kubernetes.

Cluster networking in Kubernetes

Networking is a fundamental aspect of Kubernetes, enabling communication between containers, Pods, Services, and external clients. Understanding how Kubernetes manages networking helps ensure seamless operation in distributed environments. There are four key networking challenges that Kubernetes addresses:

- **Container-to-container communication:** This is solved using Pods, allowing containers within the same Pod to communicate through localhost (i.e., internal communication).
- **Pod-to-pod communication:** Kubernetes enables communication between Pods across nodes through its networking model.
- **Pod-to-service communication:** Services abstract a set of Pods and provide stable endpoints for communication.
- **External-to-service communication:** This allows traffic from outside the cluster to access Services, such as web applications or APIs.

In a Kubernetes cluster, multiple applications run on the same set of machines. This creates challenges, such as preventing conflicts when different applications use the same network ports.

IP address management in Kubernetes

Kubernetes clusters use non-overlapping IP address ranges for Pods, Services, and Nodes. The network model is implemented through the following configurations:

- **Pods:** A network plugin, often a **Container Network Interface** (CNI) plugin, assigns IP addresses to Pods.
- **Services:** The kube-apiserver handles assigning IP addresses to Services.
- **Nodes:** IP addresses for Nodes are managed by either the kubelet or the cloud-controller-manager.

You may refer to the following website, `https://kubernetes.io/docs/concepts/cluster-administration/networking/`, to learn more about networking in Kubernetes.

Now, before exploring the Pod networking and Services, let us understand some of the basic technologies in the Kubernetes networking space.

Learning about network plugins

CNI is a specification and set of libraries developed by the **Cloud Native Computing Foundation** (CNCF). Its primary purpose is to standardize the configuration of network interfaces on Linux containers, enabling seamless communication between containers and the external environment.

These plugins are essential for implementing the Kubernetes network model, ensuring connectivity and communication within the cluster. It's crucial to choose a CNI plugin that aligns with the needs and compatibility requirements of your Kubernetes cluster. With various plugins available in the Kubernetes ecosystem, both open-source and closed-source, selecting the appropriate plugin is vital for smooth cluster operations.

Here's a concise list of CNI plugins with a brief description of each:

Open-source CNI plugins:

- **Calico:** Provides networking and network security with a focus on Layer 3 connectivity and fine-grained policy controls.
- **Flannel:** A simple CNI that offers basic networking for Kubernetes clusters, ideal for overlay networks.
- **Weave Net:** Focuses on easy setup and encryption, suitable for both cloud and on-premises environments.
- **Cilium:** Utilizes **eBPF** for advanced security features and network observability, perfect for microservice architectures.

Closed-source CNI plugins:

- **AWS VPC CNI:** Integrates Kubernetes Pods directly with AWS VPC for seamless IP management and connectivity.

- **Azure CNI**: Allows Pods to use IP addresses from the Azure VNet, ensuring integration with Azure's networking infrastructure.
- **VMware NSX-T**: Provides advanced networking capabilities like micro-segmentation and security for Kubernetes in VMware environments.

In some of the exercises in this book, we will use Calico. Now, let's find out what a service mesh is.

What is a service mesh?

A **service mesh** is an essential infrastructure layer integrated into microservices architecture, facilitating inter-service communication. It encompasses functionalities like service discovery, load balancing, encryption, authentication, and monitoring, all within the network infrastructure. Typically, service meshes are realized through lightweight proxies deployed alongside individual services, enabling precise control over traffic routing and enforcing policies such as rate limiting and circuit breaking. By abstracting complex networking tasks from developers, service meshes streamline the development, deployment, and management of microservice applications, while also enhancing reliability, security, and observability. Prominent service mesh implementations include **Istio**, **Linkerd**, and **Consul Connect**. However, it's worth noting that this topic is beyond the scope of this book.

Next, we'll explain what they're used for and how they can help you expose your Pod-launched microservices.

Understanding Pod IP assignment

To understand what Services are, we need to talk about Pods for a moment once again. On Kubernetes, everything is Pod management: Pods host your applications, and they have a special property. Kubernetes assigns them a private IP address as soon as they are created on your cluster. Keep that in mind because it is super important: each Pod created in your cluster has its unique IP address assigned by Kubernetes.

To illustrate this, we'll start by creating a nginx Pod. We are using an nginx container image to create a Pod here, but in fact, it would be the same outcome for any container image being used to create a Pod.

Let's do this using the declarative way with the following YAML definition:

```
# new-nginx-pod.yaml
apiVersion: v1
kind: Pod
metadata:
  name: new-nginx-pod
spec:
  containers:
  - image: nginx:1.17
    name: new-nginx-container
    ports:
    - containerPort: 80
```

As you can see from the previous YAML, this Pod called `new-nginx-pod` has nothing special and will just launch a container named `new-nginx-container` based on the `nginx:1.17` container image.

Once we have this YAML file, we can apply it using the following command to get the Pod running on our cluster:

```
$ kubectl apply -f new-nginx-pod.yaml
pod/new-nginx-pod.yaml created
```

As soon as this command is called, the Pod gets created on the cluster, and as soon as the Pod is created on the cluster, Kubernetes will assign it an IP address that will be unique.

Let's now retrieve the IP address Kubernetes assigned to our Pod. To do that, we can use the `kubectl get pods` command to list the Pods. In the following code snippet, please note the usage of the `-o wide` option, which will display the IP address as part of the output:

```
$ kubectl get po -o wide
NAME             READY    STATUS     RESTARTS   AGE    IP            NODE
NOMINATED NODE    READINESS GATES
new-nginx-pod    1/1      Running    0          99s    10.244.0.109   minikube
<none>           <none>
```

In our case, the IP address is `10.244.0.109`. This IP address will be unique on my Kubernetes cluster and is assigned to this unique Pod.

Of course, if you're following along on your cluster, you will have a different IP address. This IP address is a private IP version 4 (v4) address and only exists in the Kubernetes cluster. If you try to type this IP address into your web browser, you won't get anything because this address does not exist on the outside network or public internet; it only exists within your Kubernetes cluster.

Regardless of the cloud platform you use—whether it's **Amazon Web Services** (**AWS**), **Google Cloud Platform** (**GCP**), or Azure—the Kubernetes cluster leverages a network segment provided by the cloud provider. This segment, usually referred to as a **virtual private cloud** (**VPC**), defines a private and isolated network, similar to the private IP range from your on-premise LAN. In all cases, the CNI plugin used by Kubernetes ensures that each Pod is assigned a unique IP address, providing granular isolation at the Pod level. This rule applies across all cloud and on-premises environments.

We will now discover that this IP address assignment is dynamic, as well as finding out about the issues it can cause at scale.

Understanding the dynamics of Pod IP assignment in Kubernetes

The IP addresses assigned to the Pods are not static, and if you delete and recreate a Pod, you're going to see that the Pod will get a new IP address that's totally different from the one used before, even if it's recreated with the exact same YAML configuration. To demonstrate that, let's delete the Pod and recreate it using the same YAML file, as follows:

```
$ kubectl delete -f new-nginx-pod.yaml
pod/new-nginx-pod.yaml deleted
$ kubectl apply -f new-nginx-pod.yaml
pod/new-nginx-pod.yaml created
```

We can now run the kubectl get pods -o wide command once more to figure out that the new IP address is not the same as before, as follows:

```
$ kubectl get pods -o wide

NAME                READY    STATUS    RESTARTS    AGE    IP              NODE
NOMINATED NODE      READINESS GATES
new-nginx-pod       1/1      Running   0           97s    10.244.0.110    minikube
<none>              <none>
```

Now, the IP address is 10.244.0.110. This IP address is different from the one we had before, 10.244.0.109.

As you can see, when a Pod is destroyed and then recreated, even if you recreate it with the same name and the same configuration, it's going to have a different IP address.

The reason is that technically, it is not the same Pod but two different Pods; that is why Kubernetes assigns two completely different IP addresses.

Now, imagine you have an application accessing that nginx Pod that's using its IP address to communicate with it. If the nginx Pod gets deleted and recreated for some reason, then your application will be broken because the IP address will not be valid anymore.

In the next section, we'll discuss why hardcoding Pod IP addresses in your application code is not recommended and explore the challenges it presents in a production environment. We'll also look at more reliable methods for ensuring stable communication between microservices in Kubernetes.

Not hardcoding the Pod's IP address in application development

In production environments, relying on Pod IP addresses for application communication poses a significant challenge. Microservices, designed to interact through HTTP and relying on TCP/IP, require a reliable method to identify and connect to each other.

Therefore, establishing a robust mechanism for retrieving Pod information, not just IP addresses, is crucial. This ensures consistent communication even when Pods are recreated or rescheduled across worker nodes.

Crucially, avoid hardcoding Pod IP addresses directly in applications because the Pod IP addresses are dynamic. The ephemeral nature of Pods, which means they can be deleted, recreated, or moved, renders the practice of hardcoding Pod IP addresses in an application's YAML unreliable. If a Pod with a hardcoded IP is recreated, applications dependent on it will lose connectivity due to the IP resolving to nothing.

There are very concrete cases that we can give where this problem can arise, as follows:

- A Pod running a microservice *A* has a dependency and calls a microservice *B* that is running as another Pod on the same Kubernetes cluster.
- An application running as a Pod needs to retrieve some data from a **MySQL** server also running as a Pod on the same Kubernetes cluster.
- An application uses a **Redis cluster** as a caching engine deployed in multiple Pods on the same cluster.

- Your end users access an application by calling an IP address, and that IP address changes because of a Pod failure.

Any time you have an interconnection between services or, indeed, any network communication, this problem will arise.

The solution to this problem is the usage of the Kubernetes **Service** resource.

The Service object will act as an intermediate object that will remain on your cluster. The Service is not meant to be destroyed, but even if it is destroyed, it can be recreated without any impact, as it is the Service name that it used, not the IP address. In fact, they can remain on your cluster in the long term without causing any issues. Service objects provide a layer of abstraction to expose your application running in Pod(s) at the network level without any code or configuration changes through its life cycle.

Understanding how Services route traffic to Pods

Kubernetes **Services** exist as resources within your cluster and act as an abstraction layer for network traffic management. Services utilize CNI plugins to facilitate communication between clients and the Pods behind the Services. Services achieve this by creating service endpoints, which represent groups of Pods, enabling load balancing and ensuring that traffic reaches healthy instances.

Kubernetes Services offer a static and reliable way to access Pods within a cluster. They provide a DNS name that remains constant even if the underlying Pods change due to deployments, scaling, or restarts. Services leverage **service discovery** mechanisms and internal **load balancing** to efficiently route traffic to healthy Pods behind the scenes.

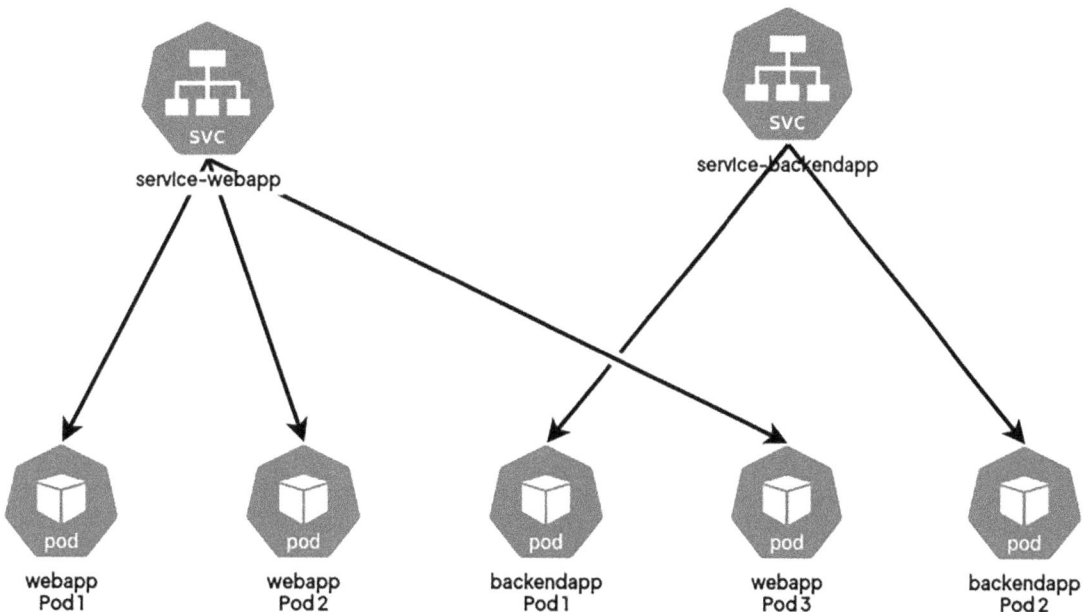

Figure 8.1: service-webapp is exposing webapp Pods 1, 2, and 3, whereas service-backendapp is exposing backendapp Pods 1 and 2

In fact, Services are deployed to Kubernetes as a resource type, and just as with most Kubernetes objects, you can deploy them to your cluster using interactive commands or declarative YAML files.

Like any other resources in Kubernetes, when you create a Service, you'll have to give it a name. This name will be used by Kubernetes to build a DNS name that all Pods on your cluster will be able to call. This DNS entry will resolve to your Service, which is supposed to remain on your cluster. The only part that is quite tricky at your end will be to give a list of Pods to expose to your Services: we will discover how to do that in this chapter. Don't worry—it's just a configuration based on **labels** and **selectors**.

Once everything is set up, you can just reach the Pods by calling the Service. This Service will receive the requests and forward them to the Pods. And that's pretty much it!

Understanding round-robin load balancing in Kubernetes

Kubernetes Services, once configured properly, can expose one or several Pods. When multiple Pods are exposed by the same Pod, the requests are evenly load-balanced to the Pods behind the Service using the round-robin algorithm, as illustrated in the following screenshot:

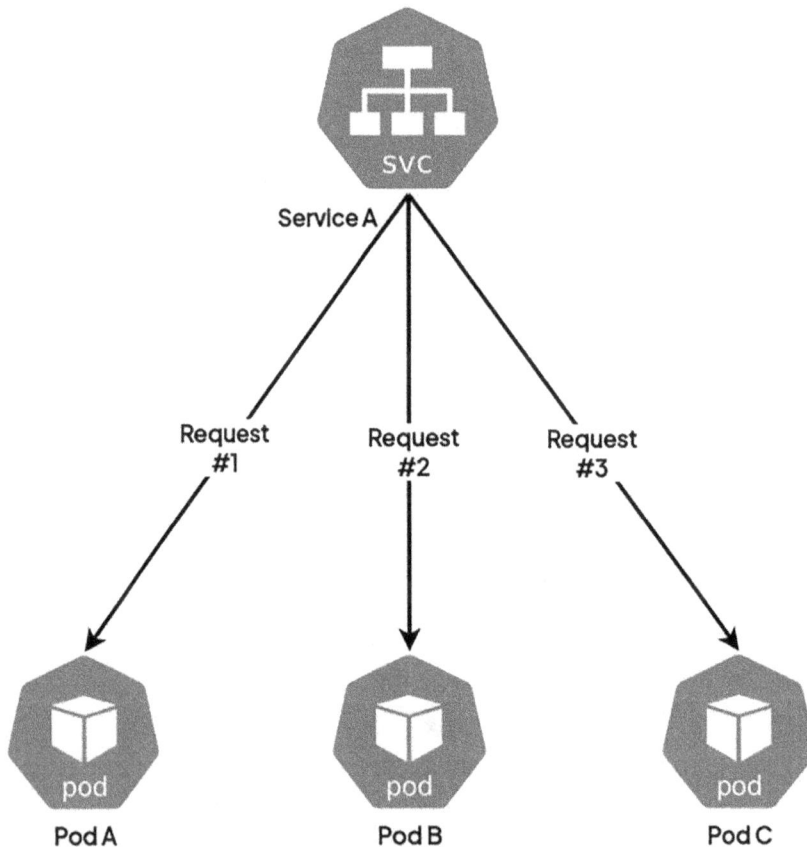

Figure 8.2: Service A proxies three requests to the three Pods behind it. At scale, each Service will receive 33% of the requests received by the Service

Scaling applications becomes easy. Adding more Pods behind the Service will be enough using Pod replicas. You will learn about Deployments and replicas in *Chapter 11, Using Kubernetes Deployments for Stateless Workloads*. As the Kubernetes Service has round-robin logic implemented, it can proxy requests evenly to the Pods behind it.

> Kubernetes Services offer more than just round-robin load balancing. While round-robin is commonly used in setups leveraging the IPVS mode of kube-proxy, it's important to note that iptables, the default mode in many distributions, often distributes traffic using random or hash-based methods instead.
>
> Kubernetes also supports additional load-balancing algorithms to meet various needs: the fewest connections for balanced workloads, source IP for consistent routing, and even custom logic for more intricate scenarios. Users should also be aware that IPVS offers more advanced traffic management features like session affinity and traffic shaping, which might not be available in iptables mode.
>
> Understanding the mode your cluster is using (either iptables or IPVS) can help in fine-tuning service behavior based on your scaling and traffic distribution requirements. Refer to the documentation to learn more (`https://kubernetes.io/docs/reference/networking/virtual-ips/`).

If the preceding Pod had four replicas, then each of them would receive roughly 25% of all the requests the service received. If 50 Pods were behind the Service, each of them would receive roughly 2% of all the requests received by the Service. All you need to understand is that Services behave like load balancers by following a specific load-balancing algorithm.

Let's now discover how you can call a Service in Kubernetes from another Pod.

Understanding how to call a Service in Kubernetes

When you create a Service in Kubernetes, it will be attached to two very important things, as follows:

- An IP address that will be unique and specific to it (just as Pods get their own IP)
- An automatically generated internal DNS name that won't change and is static

You'll be able to use any of the two in order to reach the Service, which will then forward your request to the Pod it is configured in the backend. Most of the time, though, you'll call the Service by its generated DNS name, which is easy to determine and predictable. Let's discover how Kubernetes assigns DNS names to services.

Understanding how DNS names are generated for Services

The DNS name generated for a Service is derived from its name. For example, if you create a Service named `my-app-service`, its DNS name will be `my-app-service.default.svc.cluster.local`.

This one is quite complicated, so let's break it into smaller parts, as follows:

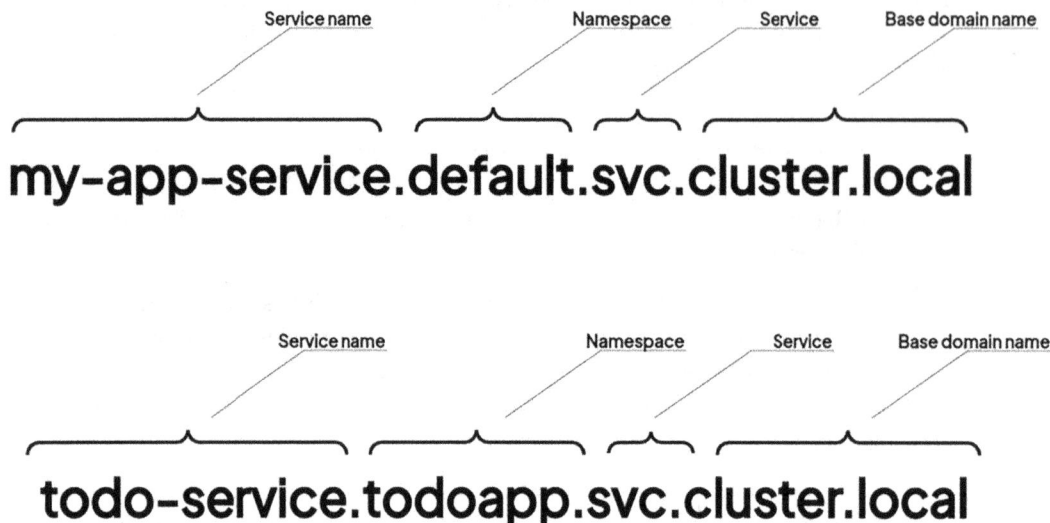

Figure 8.3: Service FQDN

The two moving parts are the first two, which are basically the Service name and the namespace where it lives. The DNS name will always end with the `.svc.cluster.local` string.

So, at any moment, from anywhere on your cluster, if you try to use `curl` or `wget` to call the `my-app-service.default.svc.cluster.local` address, you know that you'll reach your Service.

That name will resolve to the Service as soon as it's executed from a Pod within your cluster. But by default, Services won't proxy to anything if they are not configured to retrieve a list of the Pods to proxy. We will now discover how to do that!

How Services discover and route traffic to Pods in Kubernetes

When working with Services in Kubernetes, you will often come across the idea of *exposing* your Pods. Indeed, this is the terminology Kubernetes uses to tell that a Service is proxying network traffic to Pods. That terminology is everywhere: your colleagues might ask you one day, *"Which Service is exposing that Pod?"* The following screenshot shows Pods being exposed:

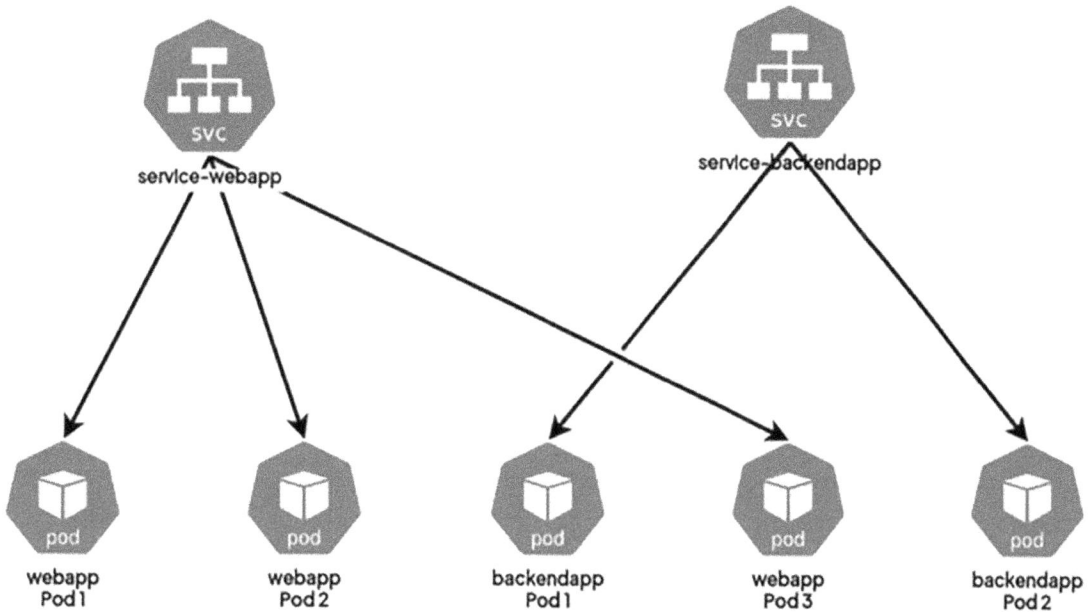

Figure 8.4: Webapp Pods 1, 2, and 3 are exposed by service-webapp, whereas backendapp Pods 1 and 2 are exposed by service-backendapp

You can successfully create a Pod and a Service to expose it using kubectl in literally one command, using the --expose parameter. For the sake of this example, let us create a nginx Pod with Service as follows.

We will also need to provide a port to the command to tell on which port the Service will be accessible:

```
$ kubectl run nginx --image nginx --expose=true --port=80
service/nginx created
pod/nginx created
```

Let's now list the Pods and the Service using kubectl to demonstrate that the following command created both objects:

```
$ kubectl get po,svc nginx
NAME         READY    STATUS     RESTARTS    AGE
pod/nginx    1/1      Running    0           24s


NAME            TYPE         CLUSTER-IP      EXTERNAL-IP    PORT(S)    AGE
service/nginx   ClusterIP    10.111.12.100   <none>         80/TCP     24s
```

As you can see based on the output of the command, both objects were created. We said earlier that Services can find the Pods they have to expose based on the Pods' labels. The nginx Pod we just created surely has some labels. To show them, let's run the kubectl get pods nginx --show-labels command.

In the following snippet, pay attention to the `--show-labels` parameter, which will display the labels as part of the output:

```
$  kubectl get po nginx --show-labels
NAME     READY    STATUS     RESTARTS    AGE    LABELS
nginx    1/1      Running    0           51s    run=nginx
```

As you can see, an `nginx` Pod was created with a label called `run` and a value of `nginx`. Let's now describe the `nginx` Service. It should have a selector that matches this label. The code is illustrated here:

```
$ kubectl describe svc nginx
Name:              nginx
Namespace:         default
Labels:            <none>
Annotations:       <none>
Selector:          run=nginx
Type:              ClusterIP
IP Family Policy:  SingleStack
IP Families:       IPv4
IP:                10.111.12.100
IPs:               10.111.12.100
Port:              <unset>  80/TCP
TargetPort:        80/TCP
Endpoints:         10.244.0.9:80
Session Affinity:  None
Events:            <none>
```

You can clearly see that the Service has a line called `Selector` that matches the label assigned to the `nginx` Pod. This way, the link between the two objects is made. We're now 100% sure that the Service can reach the `nginx` Pod and that everything should work normally.

Please note that if you are using the minikube in the lab, you will not be able to access the Service outside of your cluster as ClusterIP Services are only accessible from inside the cluster. You need to use debugging methods such as `kubectl port-forward` or `kubectl proxy` for such scenarios. You will learn how to test ClusterIP-type Services in the next section.

Also, to test the Service access, let us create a temporary port-forwarding, as follows:

```
$ kubectl port-forward pod/nginx 8080:80
Forwarding from 127.0.0.1:8080 -> 80
Forwarding from [::1]:8080 -> 80
```

Now, open another console and access URL as follows:

```
$ curl 127.0.0.1:8080
```

```
<!DOCTYPE html>
<html>
...<removed for brevity>...
```

In the preceding snippets, port 8080 is the localhost port we used for port-forwarding and 80 is the nginx port where the web Service is exposed.

Please note that the kubectl port-forward command will run until you break it using *Ctrl+C*.

Though it works, we strongly advise you to never do that in production. Services are very customizable objects, and the --expose parameter hides a lot of their features. Instead, you should really use declarative syntax and tweak the YAML to fit your exact needs.

Let's demonstrate that by using the dnsutils container image.

Using a utility Pod for debugging your Services

As your Services are created within your cluster, it is often hard to access them as we mentioned earlier, especially if our Pod is meant to remain accessible only within your cluster, or if your cluster has no internet connectivity, and so on.

In this case, it is good to deploy a debugging Pod in your cluster with just some binaries installed into it to run basic networking commands such as wget, nslookup, and so on. Let us use our custom utility container image quay.io/iamgini/k8sutils:debian12 for this purpose.

> You can add more tools or utilities inside the utility container image if required; refer to the Chapter-070/Containerfile for the source.

Here, we're going to curl the nginx Pod home page by calling the Service that is exposing the Pod. That Service's name is just nginx. Hence, we can forget the DNS name Kubernetes assigned to it: nginx.default.svc.cluster.local.

If you try to reach this **uniform resource locator** (URL) from a Pod within your cluster, you should successfully reach the nginx home page.

The following Pod definition will help us to create a debugging pod with the k8sutils image.

```
# k8sutils.yaml
apiVersion: v1
kind: Pod
metadata:
  name: k8sutils
  namespace: default
spec:
  containers:
    - name: k8sutils
```

```
    image: quay.io/iamgini/k8sutils:debian12
    command:
      - sleep
      - "infinity"
    # imagePullPolicy: IfNotPresent
  restartPolicy: Always
```

Let's run the following command to launch the k8sutils Pod on our cluster:

```
$ kubectl apply -f k8sutils.yaml
pod/k8sutils created
```

Now run the kubectl get pods command in order to verify that the Pod was launched successfully, as follows:

```
$ kubectl get po k8sutils
NAME        READY   STATUS    RESTARTS   AGE
k8sutils    1/1     Running   0          13m
```

That's perfect! Let's now run the nslookup command from the k8sutils Pod against the Service DNS name, as follows:

```
$ kubectl exec -it k8sutils -- nslookup nginx.default.svc.cluster.local
Server:        10.96.0.10
Address:       10.96.0.10#53
Name:   nginx.default.svc.cluster.local
Address: 10.106.124.200
```

In the previous snippet,

- Server: 10.96.0.10 - is the kube-dns Service IP (kubectl get svc kube-dns -n kube-system -o wide). If you are using a different DNS Service, check the Service details accordingly.
- nginx.default.svc.cluster.local is resolving to the IP address of nginx Service (kubectl get svc nginx -o wide), which is 10.106.124.

Everything looks good. Let's now run a curl command to check whether we can retrieve the nginx home page, as follows:

```
$ kubectl exec -it k8sutils -- curl nginx.default.svc.cluster.local
...<removed for brevity>....
<body>
<h1>Welcome to nginx!</h1>
<p>If you see this page, the nginx web server is successfully installed and
working. Further configuration is required.</p>

...<removed for brevity>...
```

Everything is perfect here! We successfully called the nginx Service by using the k8sutils debug Pod, as illustrated in the following screenshot:

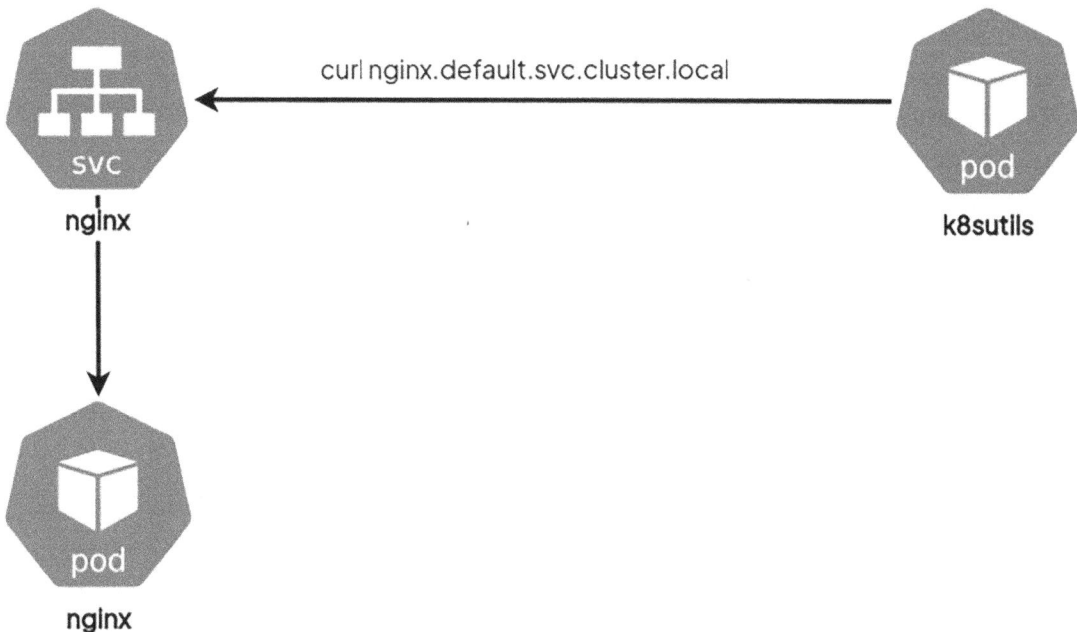

Figure 8.5: The k8sutils Pod is used to run curl against the nginx Service to communicate with the nginx Pod behind the Service

Keep in mind that you need to deploy a k8sutils Pod inside the cluster to be able to debug the Service. Indeed, the nginx.default.svc.cluster.local DNS name is not a public one and can only be accessible from within the cluster.

Let's explain why you should not use expose imperatively to expose your Pods in the next section.

Understanding the drawbacks of direct kubectl expose in Kubernetes

It is not recommended to use kubectl expose to create Services because you won't get much control over how the Service gets created. By default, kubectl expose will create a ClusterIP Service, but you might want to create a NodePort Service.

Defining the Service type is also possible using the imperative syntax, but in the end, the command you'll have to issue is going to be very long and complex to understand. That's why we encourage you to not use the expose option and stick with declarative syntax for complex objects such as Services.

Let's now discuss how DNS names are generated in Kubernetes when using Services.

Understanding how DNS names are generated for Services

You now know that Kubernetes Service-to-Pod communication relies entirely on labels on the Pod side and selectors on the Service side.

If you don't use both correctly, communication cannot be established between the two.

The workflow goes like this:

1. You create some Pods, and you set the labels arbitrarily.
2. You create a Service and configure its selector to match the Pods' labels.
3. The Service starts and looks for Pods that match its selector.
4. You call the Service through its DNS or its IP (DNS is way easier).
5. The Service forwards the traffic to one of the Pods that matches its labels.

If you look at the previous example achieved using the imperative style with the `kubectl expose` parameter, you'll notice that the Pod and the Services were respectively configured with proper labels (on the Pod side) and selector (on the Service side), which is why the Pod is successfully exposed. Please note that in your real-life cases, you will need to use the appropriate labels for your Pods instead of default labels.

Besides that, you must understand now that there are not one but several types of Services in Kubernetes—let us learn more about that.

Understanding the different types of Services

There are several types of Services in Kubernetes. Although there is only one kind called Service in Kubernetes, that kind can be configured differently to achieve different results.

Fortunately for us, no matter which type of Service you choose, the goal remains the same: to expose your Pods using a single static interface.

Each type of Service has its own function and its own use, so basically, there's one Service for each use case. A Service cannot be of multiple types at once, but you can still expose the same Pods using two Services' objects with different types as long as the Services' objects are named differently so that Kubernetes can assign different DNS names.

In this chapter, we will discover the three main types of Services, as follows:

- `NodePort`: This type binds a port from an ephemeral port range of the host machine (the worker node) to a port on the Pod, making it available publicly. By calling the port of the host machine, you'll reach the associated Kubernetes Pod. That's the way to reach your Pods for traffic coming from outside your cluster.
- `ClusterIP`: The `ClusterIP` Service is the one that should be used for private communication between Pods within the Kubernetes cluster. This is the one we experimented with in this chapter and is the one created by `kubectl expose` by default. This is certainly the most commonly used of them all because it allows inter-communication between Pods: as its name suggests, it has a static IP that is set cluster-wide. By reaching its IP address, you'll be redirected to the Pod behind it. If more than one Pod is behind it, the `ClusterIP` Service will provide a load-balancing mechanism following the round-robin or other algorithms.

- LoadBalancer: The LoadBalancer Service in Kubernetes streamlines the process of exposing your Pods to external traffic. It achieves this by automatically provisioning a cloud-specific load balancer, such as AWS ELB, when operating on supported cloud platforms. This removes the necessity for the manual setup of external load balancers within cloud environments. However, it's crucial to recognize that this Service is not compatible with bare-metal or non-cloud-managed clusters unless configured manually. While alternative solutions like Terraform exist for managing cloud infrastructure, the LoadBalancer Service provides a convenient option for seamlessly integrating cloud-native load balancers into your Kubernetes deployments, particularly in cloud-centric scenarios. Keep in mind that its suitability hinges on your specific requirements and infrastructure configuration.

Now, let's immediately dive into the first type of Service—the NodePort one.

As mentioned earlier, this one is going to be very useful to access our Pods from outside the cluster in our development environments, by attaching Pods to the Kubernetes node's port.

The NodePort Service

NodePort is a Kubernetes Service type designed to make Pods reachable from a port available on the host machine, the worker node. In this section, we're going to discover this type of port and be fully focused on NodePort Services!

Why do you need NodePort Services?

The first thing to understand is that NodePort Services allow us to access a Pod running on a Kubernetes node, on a port of the node itself. After you expose Pods using the NodePort type Service, you'll be able to reach the Pods by getting the IP address of the node and the port of the NodePort Service, such as <node_ip_address>:<node port>.

The port can be declared in your YAML declaration or can be randomly assigned by Kubernetes. Let's illustrate all of this by declaring some Kubernetes objects.

Most of the time, the NodePort Service is used as an entry point to your Kubernetes cluster. In the following example, we will create two Pods based on the containous/whoami container image available on Docker Hub, which is a very nice container image that will simply print the container hostname.

We will create two Pods so that we get two containers with different hostnames, and we will expose them using a NodePort Service.

Creating two containous/whoami Pods

Let's start by creating two Pods, without forgetting about adding one or more labels, because we will need labels to tell the Service which Pods it's going to expose.

We are also going to open the port on the Pod side. That won't make it exposed on its own, but it will open a port the Service will be able to reach. The code is illustrated here:

```
$ kubectl run whoami1 --image=containous/whoami --port 80 --labels="app=whoami"
pod/whoami1 created
$ kubectl run whoami2 --image=containous/whoami --port 80 --labels="app=whoami"
pod/whoami2 created
```

Now, we can run a `kubectl get pods` command in order to verify that our two Pods are running correctly. We can also add the `--show-labels` parameter in order to display the labels as part of the command output, as follows:

```
$ kubectl get pods --show-labels
NAME        READY    STATUS     RESTARTS    AGE      LABELS
whoami1     1/1      Running    0           3m5s     app=whoami
whoami2     1/1      Running    0           3m       app=whoami
```

Everything seems to be okay! Now that we have two Pods created with a label set for each of them, we will be able to expose them using a Service. We're now going to discover the YAML manifest file that will create the NodePort Service to expose these two Pods.

Understanding NodePort YAML definition

Since Services are quite complex resources, it is better to create Services using a YAML file rather than direct command input.

Here is the YAML file that will expose the whoami1 and whoamo2 Pods using a NodePort Service:

```
# ~/nodeport-whoami.yaml
apiVersion: v1
kind: Service
metadata:
  name: nodeport-whoami
spec:
  type: NodePort
  selector:
    app: whoami
  ports:
  - nodePort: 30001
    port: 80
    targetPort: 80
```

This YAML can be difficult to understand because it refers to three different ports as well as a `selector` block.

Before explaining the YAML file, let's apply it and check if the Service was correctly created afterwards, as follows:

```
$ kubectl apply -f nodeport-whoami.yaml
service/nodeport-whoami created

$ kubectl get service nodeport-whoami
NAME              TYPE       CLUSTER-IP     EXTERNAL-IP   PORT(S)        AGE
nodeport-whoami   NodePort   10.98.160.98   <none>        80:30001/TCP   14s
```

The previous `kubectl get services` command indicated that the Service was properly created!

The selector block is crucial for NodePort Services, acting as a label filter to determine which pods the Service exposes. Essentially, it tells the Service what pods to route traffic to. Without a selector, the Service remains inactive. In this example, the selector targets pods with the label key "app" and the value "whoami". This effectively exposes both the "whoami1" and "whoami2" pods through the Service.

Next, we have `type` as a child key under `spec,` where we specify the type of our Service. When we create `ClusterIP` or `LoadBalancer` Services, we will have to update this line. Here, we're creating a `NodePort` Service, so that's fine for us.

The last thing that is quite hard to understand is that `ports` block. Here, we define a map of multiple port combinations. We indicated three ports, as follows:

- `nodePort`: The port on the host machine/worker node you want this `NodePort` service to be accessible from. Here, we're specifying port `30001`, which makes this NodePort Service accessible from port `30001` on the IP address of the worker node. You'll be reaching this `NodePort` Service and the Pods it exposes by calling the following address: `<WORKER_NODE_IP_ADDRESS>:30001`. This `NodePort` setting cannot be set arbitrarily. Indeed, on a default Kubernetes installation, it can be a port from the `30000 - 32767` range.

- `port`: This setting indicates the port of the `NodePort` Service itself. It can be hard to understand, but `NodePort` Services do have a port of their own too, and this is where you specify it. You can put whatever you want here if it is a valid port.

- `targetPort`: As you might expect, `targetPort` is the port of the targeted Pods. It is where the application runs: the port where the NodePort will forward traffic to the Pod found by the selector mentioned previously.

Here is a quick diagram to sum all of this up:

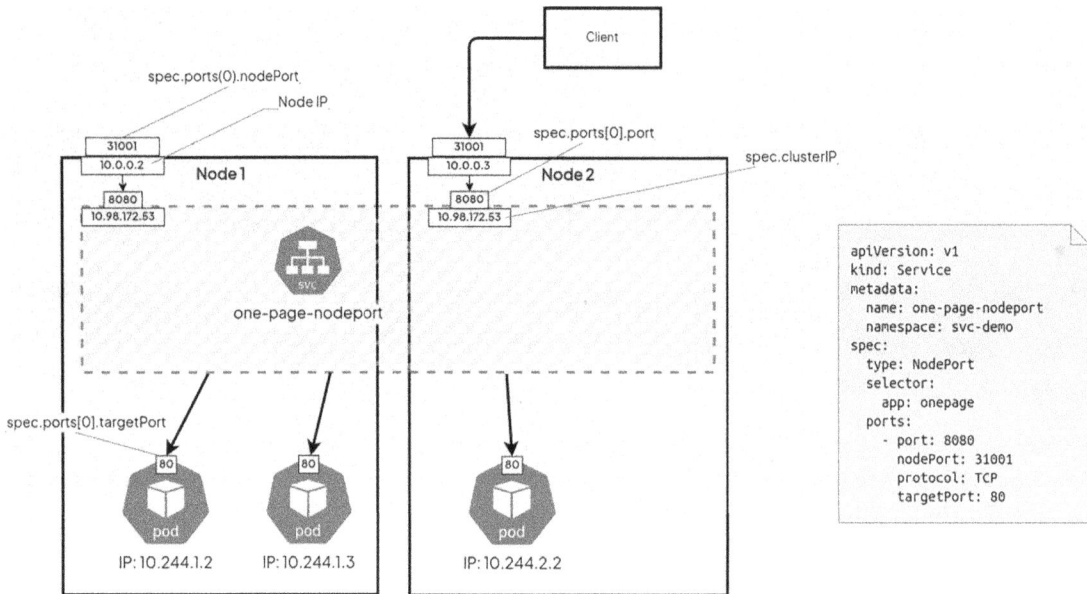

Figure 8.6: NodePort Service. There are three ports involved in the NodePort setup—nodePort is on the worker n006Fde, the port is on the Service itself, and targetPort is on the top

In this case, TCP port `31001` is used as the external port on each Node. If you do not specify `nodePort`, it will be allocated `dynamically` using the range. For internal communication, this Service still behaves like a simple `ClusterIP` Service, and you can use its `ClusterIP` address.

For convenience and to reduce complexity, the `NodePort` Service port and target port (the Pods' port) are often defined to the same value.

Making sure NodePort works as expected

To try out your `NodePort` setup, the first thing to do is to retrieve the public IP of your machine running it. In our example, we are running a single-machine Kubernetes setup with `minikube` locally. On AWS, GCP, or Azure, your node might have a public IP address or a private one if you access your node with a **virtual private network (VPN)**.

On `minikube`, the easiest way to retrieve the IP address is to issue the following command:

```
$ minikube ip
192.168.64.2
Or you can access the full URL as follows.
$ minikube service --url nodeport-whoami
http://192.168.49.2:30001
```

Now that we have all the information, we can open a web browser and enter the URL to access the `NodePort` Service and the Pods running. You should see the round-robin algorithm in place and reaching `whoami1` and then `whoami2`, and so on. The `NodePort` Service is doing its job!

Is this setup production-ready?

This question might not have a definitive answer as it depends on your configuration.

NodePort provides a way to expose Pods to the outside world by exposing them on a Node port. With the current setup, you have no HA: if your two Pods were to fail, you have no way to relaunch them automatically, so your Service wouldn't be able to forward traffic to anything, resulting in a poor experience for your end user.

> Please note that when we create the Pods using Deployments and replicasets, Kubernetes will create the new Pods in other available nodes. We will learn about Deployments and replicasets in *Chapter 11, Using Kubernetes Deployments for Stateless Workloads*.

Another problem is the fact that the choice of port is limited. Indeed, by default, you are just forced to use a port in the 30000-32767 range, and as it's forced, it will be inconvenient for a lot of people. Indeed, if you want to expose an HTTP application, you'll want to use port 80 or 443 of your frontal machine rather than a port in the 30000-32767 range, because all web browsers are configured with ports 80 and 443 as standard HTTP and **HTTP Secure (HTTPS)** ports.

The solution to this consists of using a tiered architecture. Indeed, a lot of Kubernetes architects tend to not expose a NodePort Service as the first layer in architecture but to put the Kubernetes cluster behind a reverse proxy, such as the AWS Application Load Balancer, and so on. Two other concepts of Kubernetes are the Ingress and IngressController objects: these two objects allow you to configure a reverse proxy such as nginx or HAProxy directly from Kubernetes objects and help you to make your application publicly accessible as the first layer of entry to Kubernetes. But this is way beyond the scope of Kubernetes Services.

Let us explore some more information about NodePort in the next sections, including how to list the Services and how to add Pods to the NodePort Services.

Listing NodePort Services

Listing NodePort Services is achieved through the usage of the kubectl command-line tool. You must simply issue a kubectl get services command to fetch the Services created within your cluster.

```
$ kubectl get service
NAME              TYPE        CLUSTER-IP       EXTERNAL-IP   PORT(S)        AGE
example-service   ClusterIP   10.106.224.122   <none>        80/TCP         26d
kubernetes        ClusterIP   10.96.0.1        <none>        443/TCP        26d
nodeport-whoami   NodePort    10.100.85.171    <none>        80:30001/TCP   21s
```

That being said, let's now discover how we can update NodePort Services to have them do what we want.

Adding more Pods to NodePort Services

If you want to add a Pod to the pool served by your Services, it's very easy. In fact, you just need to add a new Pod that matches the label selector defined on the Service—Kubernetes will take care of the rest. The Pod will be part of the pool served by the Service. If you delete a Pod, it will be deleted from the pool of Services as soon as it enters the `Terminating` state.

Kubernetes handles Service traffic based on Pod availability—for example, if you have three replicas of a web server and one goes down, creating an additional replica that matches the label selector on the Service will be enough. You'll discover later, in *Chapter 11, Using Kubernetes Deployments for Stateless Workloads*, that this behavior can be entirely automated.

Describing NodePort Services

Describing `NodePort` Services is super easy and is achieved with the help of the `kubectl describe` command, just as with any other Kubernetes object. Let us explore the details of the `nodeport-whoami` Service in the following command's output:

```
$ kubectl describe Service nodeport-whoami:
Name:                   nodeport-whoami
...<removed for brevity>...
Selector:               app=whoami
Type:                   NodePort
IP Family Policy:       SingleStack
IP Families:            IPv4
IP:                     10.98.160.98
IPs:                    10.98.160.98
Port:                   <unset>  80/TCP
TargetPort:             80/TCP
NodePort:               <unset>  30001/TCP
Endpoints:              10.244.0.16:80,10.244.0.17:80
...<removed for brevity>...
```

In the preceding output, we can see several details, including the following items:

- `IP:10.98.160.98`: The `ClusterIP` assigned to the Service. It's the internal IP address that other Services in the cluster can use to access this Service.

- `Port: <unset>  80/TCP`: The Service listens on port 80 with the TCP protocol. The `<unset>` means that no name was given for the port.

- `NodePort: <unset> 30001/TCP`: The `NodePort` is the port on each node in the cluster through which external traffic can access the Service. Here, it's set to port 30001, allowing external access to the Service via any node's IP at port 30001.

- `Endpoints: 10.244.0.16:80, 10.244.0.17:80`: The actual IP addresses and ports of the Pods behind the Service. In this case, two Pods are backing the Service, reachable at `10.244.0.16:80` and `10.244.0.17:80`.

In the next section, we will learn how to delete a Service using the `kubectl delete svc` command.

Deleting Services

Deleting a Service, whether it is a `NodePort` Service or not, should not be done often. Indeed, whereas Pods are supposed to be easy to delete and recreate, Services are supposed to be for the long term. They provide a consistent way to expose your Pod, and deleting them will impact how your applications can be reached.

Therefore, you should be careful when deleting Services: it won't delete the Pods behind it, but they won't be accessible anymore from outside of your cluster!

Here is the command to delete the Service created to expose the `whoami1` and `whoami2` Pods:

```
$ kubectl delete svc/nodeport-whoami
service "nodeport-whoami" deleted
```

You can run a `kubectl get svc` command now to check that the Service was properly destroyed, and then access it once more through the web browser by refreshing it. You'll notice that the application is not reachable anymore, but the Pods will remain on the cluster. Pods and Services have completely independent life cycles. If you want to delete Pods, then you'll need to delete them separately.

You probably remember the `kubectl port-forward` command we used when we created an nginx Pod and tested it to display the home page. You might think `NodePort` and `kubectl port-forward` are the same thing, but they are not. Let's explain quickly the difference between the two in the upcoming section.

NodePort or kubectl port-forward?

It might be tempting to compare `NodePort` Services with the `kubectl port-forward` command because, so far, we have used these two methods to access a running Pod in our cluster using a web browser.

The `kubectl port-forward` command is a testing tool, whereas `NodePort` Services are for real use cases and are a production-ready feature.

Keep in mind that `kubectl port-forward` must be kept open in your terminal session for it to work. As soon as the command is killed, the port forwarding is stopped too, and your application will become inaccessible from outside the cluster once more. It is only a testing tool meant to be used by the `kubectl` user and is just one of the useful tools bundled into the `kubectl` CLI.

`NodePort`, on the other hand, is really meant for production use and is a long-term production-ready solution. It doesn't require `kubectl` to work and makes your application accessible to anyone calling the Service, provided the Service is properly configured and the Pods are correctly labeled.

Simply put, if you just need to test your app, go for `kubectl port-forward`. If you need to expose your Pod to the outside world for real, go for `NodePort`. Don't create `NodePort` for testing, and don't try to use `kubectl port-forward` for production! Stick with one tool for each use case!

Now, we will discover another type of Kubernetes Service called `ClusterIP`. This one is probably the most widely used of them all, even more than the `NodePort` type!

The ClusterIP Service

We're now going to discover another type of Service called `ClusterIP`. Now, `ClusterIP` is, in fact, the simplest type of Service Kubernetes provides. With a `ClusterIP` Service, you can expose your Pod so that other Pods in Kubernetes can communicate with it via its IP address or DNS name.

Why do you need ClusterIP Services?

The `ClusterIP` Service type greatly resembles the `NodePort` Service type, but they have one big difference: `NodePort` Services are meant to expose Pods to the outside world, whereas `ClusterIP` Services are meant to expose Pods to other Pods inside the Kubernetes cluster.

Indeed, `ClusterIP` Services are the Services that allow different Pods in the same cluster to communicate with each other through a static interface: the `ClusterIP Service` object itself.

`ClusterIP` answers exactly the same need for a static DNS name or IP address we had with the `NodePort` Service: if a Pod fails, is recreated, deleted, relaunched, and so on, then Kubernetes will assign it another IP address. `ClusterIP` Services are here to remediate this issue, by providing an internal DNS name only accessible from within your cluster that will resolve to the Pods defined by the label selector.

As the name `ClusterIP` suggests, this Service grants a static IP within the cluster! Let's now discover how to expose our Pods using `ClusterIP`! Keep in mind that `ClusterIP` Services are not accessible from outside the cluster—they are only meant for inter-Pod communication.

How do I know if I need NodePort or ClusterIP Services to expose my Pods?

Choosing between the two types of Services is extremely simple, basically because they are not meant for the same thing.

If you need your app to be accessible from outside the cluster, then you'll need a `NodePort` Service (or other Services we will be exploring later in this chapter), but if your need is for the app to be accessible from inside the cluster, then you'll need a `ClusterIP` Service. `ClusterIP` Services are also good for stateless applications that can be scaled, destroyed, recreated, and so on. The reason is that the `ClusterIP` Service will maintain a static entry point to a whole pool of Pods without being constrained by a port on the worker node, unlike the `NodePort` Service.

The `ClusterIP` Service exposes Pods using internally visible virtual IP addresses managed by kube-proxy on each Node. This means that the Service will be reachable from within the cluster only. We have visualized the ClusterIP Service principles in the following diagram:

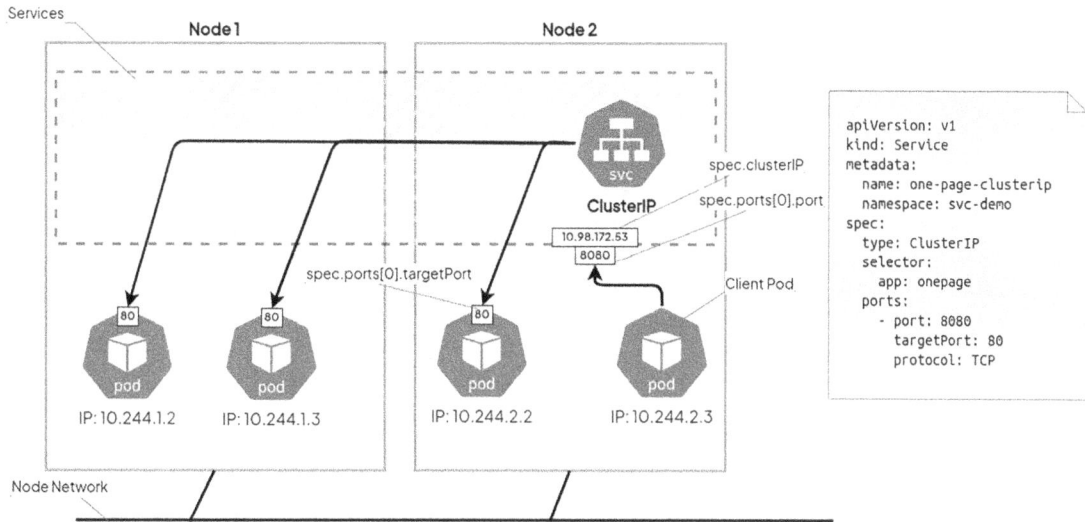

Figure 8.7: ClusterIP Service

In the preceding image, the ClusterIP Service is configured in such a way that it will map requests coming from its IP and TCP port 8080 to the container's TCP port 80. The actual ClusterIP address is assigned dynamically unless you specify one explicitly in the specifications. The internal DNS Service in a Kubernetes cluster is responsible for resolving the nginx-deployment-example name to the actual ClusterIP address as a part of Service discovery.

> kube-proxy handles the management of virtual IP addresses on the nodes and adjusts the forwarding rules accordingly. Services in Kubernetes are essentially logical constructs within the cluster. There isn't a separate physical process running inside the cluster for each Service to handle proxying. Instead, kube-proxy performs the necessary proxying and routing based on these logical Services.

Contrary to NodePort Services, ClusterIP Services will not take one port of the worker node, and thus it is impossible to reach it from outside the Kubernetes cluster.

Keep in mind that nothing prevents you from using both types of Services for the same pool of Pods. Indeed, if you have an app that should be publicly accessible but also privately exposed to other Pods, then you can simply create two Services, one NodePort Service and one ClusterIP Service.

In this specific use case, you'll simply have to name the two Services differently so that they won't conflict when creating them against kube-apiserver. Nothing else prevents you from doing so!

Listing ClusterIP Services

Listing ClusterIP Services is easy. It's basically the same command as the one used for NodePort Services. Here is the command to run:

```
$ kubectl get svc
```

As always, this command lists the Services with their type added to the output.

Creating ClusterIP Services using the imperative way

Creating ClusterIP Services can be achieved with a lot of different methods. Since it is an extremely used feature, there are lots of ways to create these, as follows:

- Using the --expose parameter or kubectl expose method (the imperative way)
- Using a YAML manifest file (the declarative way)

The imperative way consists of using the --expose method. This will create a ClusterIP Service directly from a kubectl run command. In the following example, we will create an nginx-clusterip Pod as well as a ClusterIP Service to expose them both at the same time. Using the --expose parameter will also require defining a ClusterIP port. ClusterIP will listen to make the Pod reachable. The code is illustrated here:

```
$ kubectl run nginx-clusterip --image nginx --expose=true --port=80
service/nginx-clusterip created
pod/nginx-clusterip created
```

As you can see, we get both a Pod and a Service to expose it. Let's describe the Service.

Describing ClusterIP Services

Describing ClusterIP Services is the same process as describing any type of object in Kubernetes and is achieved using the kubectl describe command. You just need to know the name of the Service in order to describe to achieve that.

Here, I'm going to the ClusterIP Service created previously:

```
$ kubectl describe svc/nginx-clusterip
Name:                nginx-clusterip
Namespace:           default
Labels:              <none>
Annotations:         <none>
Selector:            run=nginx-clusterip
Type:                ClusterIP
IP Family Policy:    SingleStack
IP Families:         IPv4
IP:                  10.101.229.225
IPs:                 10.101.229.225
Port:                <unset>  80/TCP
TargetPort:          80/TCP
Endpoints:           10.244.0.10:80
Session Affinity:    None
Events:              <none>
```

The output of this command shows us the `Selector` block, which shows that the `ClusterIP` Service was created by the `--expose` parameter with the proper label configured. This label matches the `nginx-clusterip` Pod we created at the same time. To be sure about that, let's display the labels of the said Pod, as follows:

```
$ kubectl get pods/nginx-clusterip --show-labels
NAME             READY   STATUS    RESTARTS   AGE   LABELS
nginx-clusterip  1/1     Running   0          76s   run=nginx-clusterip
```

As you can see, the selector on the Service matches the labels defined on the Pod. Communication is thus established between the two. We will now call the `ClusterIP` Service directly from another Pod on the cluster.

Since the `ClusterIP` Service is named `nginx-clusterip`, we know that it is reachable at this address: `nginx-clusterip.default.svc.cluster.local`.

Let's reuse the `k8sutils` container, as follows:

```
$ kubectl exec k8sutils -- curl nginx-clusterip.default.svc.cluster.local
  % Total    % Received % Xferd  Average Speed   Time    Time     Time  Current
                                 Dload  Upload   Total   Spent    Left  Speed
  0     0    0     0    0     0      0       0 --:--:-- --:--:-- --:--:--
0<!DOCTYPE ...<removed for brevity>...
<body>
<h1>Welcome to nginx!</h1>
<p>If you see this page, the nginx web server is successfully installed and
working. Further configuration is required.</p>
  ...<removed for brevity>...
```

The `ClusterIP` Service correctly forwarded the request to the nginx Pod, and we do have the nginx default home page. The Service is working!

We did not use `containous/whoami` as a web Service this time, but keep in mind that the `ClusterIP` Service is also doing load balancing internally following the round-robin algorithm. If you have 10 Pods behind a `ClusterIP` Service and your Service received 1,000 requests, then each Pod is going to receive 100 requests.

Let's now discover how to create a `ClusterIP` Service using YAML.

Creating ClusterIP Services using the declarative way

`ClusterIP` Services can also be created the declarative way by applying YAML configuration files against `kube-apiserver`.

Here's a YAML manifest file we can use to create the exact same `ClusterIP` Service we created before the imperative way:

```yaml
# clusterip-service.yaml
apiVersion: v1
kind: Service
metadata:
  name: nginx-clusterip
spec:
  type: ClusterIP # Indicates that the service is a ClusterIP
  ports:
    - port: 80 # The port exposed by the service
      protocol: TCP
      targetPort: 80 # The destination port on the pods
  selector:
    run: nginx-clusterip
```

Take some time to read the comments in the YAML, especially the `port` and `targetPort` ones.

Indeed, `ClusterIP` Services have their own port independent of the one exposed on the Pod side. You reach the `ClusterIP` Service by calling its DNS name and its port, and the traffic is going to be forwarded to the destination port on the Pods matching the labels and selectors.

Keep in mind that no worker node port is involved here. The ports we are mentioning when it comes to `ClusterIP` scenarios have absolutely nothing to do with the host machine!

Before we continue with the next section, you can clean up the environment by deleting the `nginx-clusterip` Service as follows:

```
$ kubectl delete  svc nginx-clusterip
service "nginx-clusterip" deleted
```

Keep in mind that deleting the cluster won't delete the Pods exposed by it. It is a different process; you'll need to delete Pods separately. We will now discover one additional resource related to `ClusterIP` Services, which are headless Services.

Understanding headless Services

Headless Services are derived from the `ClusterIP` Service. They are not technically a dedicated type of Service (such as `NodePort` or `ClusterIP`), but they are an option from `ClusterIP`.

Headless Services can be configured by setting the `.spec.clusterIP` option to `None` in a YAML configuration file for the `ClusterIP` Service. Here is an example derived from our previous YAML file:

```yaml
# clusterip-headless.yaml
apiVersion: v1
kind: Service
```

```
metadata:
  name: nginx-clusterip-headless
spec:
  clusterIP: None
  type: ClusterIP # Indicates that the service is a ClusterIP
  ports:
  - port: 80 # The port exposed by the service
    protocol: TCP
    targetPort: 80 # The destination port on the pods
  selector:
    run: nginx-clusterip
```

A headless Service roughly consists of a `ClusterIP` Service without load balancing and without a pre-allocated `ClusterIP` address. Thus, the load-balancing logic and the interfacing with the Pod are not defined by Kubernetes.

Since a headless Service has no IP address, you are going to reach the Pod behind it directly, without the proxying and the load-balancing logic. What the headless Service does is return you the DNS names of the Pods behind it so that you can reach them directly. There is still a little load-balancing logic here, but it is implemented at the DNS level, not as Kubernetes logic.

When you use a normal `ClusterIP` Service, you'll always reach one static IP address allocated to the Service and this is going to be your proxy to communicate with the Pod behind it. With a headless Service, the `ClusterIP` Service will just return the DNS names of the Pods behind it and the client will have the responsibility to establish a connection with the DNS name of its choosing.

Headless Services in Kubernetes are primarily used in scenarios where direct communication with individual Pods is required, rather than with a single endpoint or load-balanced set of Pods.

They are helpful when you want to build connectivity with clustered stateful Services such as **Lightweight Directory Access Protocol (LDAP)**. In that case, you may want to use an LDAP client that will have access to the different DNS names of the Pods hosting the LDAP server, and this can't be done with a normal `ClusterIP` Service since it will bring both a static IP and Kubernetes' implementation of load balancing. Let's now briefly introduce another type of Service called `LoadBalancer`.

The LoadBalancer Service

`LoadBalancer` Services are very interesting to explain because this Service relies on the cloud platform where the Kubernetes cluster is provisioned. For it to work, it is thus required to use Kubernetes on a cloud platform that supports the `LoadBalancer` Service type.

For cloud providers that offer external load balancers, specifying the `type` field as `LoadBalancer` configures a load balancer for your Service. The creation of the load balancer occurs asynchronously, and details about the provisioned balancer are made available in the `.status.loadBalancer` field of the Service.

Certain cloud providers offer the option to define the `loadBalancerIP`. In such instances, the load balancer is generated with the specified `loadBalancerIP`. If the `loadBalancerIP` is not provided, the load balancer is configured with an ephemeral IP address. However, if you specify a `loadBalancerIP` on a cloud provider that does not support this feature, the provided `loadBalancerIP` is disregarded.

For a Service with its type set to LoadBalancer, the `.spec.loadBalancerClass` field allows you to utilize a load balancer implementation other than the default provided by the cloud provider. When `.spec.loadBalancerClass` is not specified, the `LoadBalancer` type of Service will automatically use the default load balancer implementation provided by the cloud provider, assuming the cluster is configured with a cloud provider using the `--cloud-provider` component flag. However, if you specify `.spec.loadBalancerClass`, it indicates that a load balancer implementation matching the specified class is actively monitoring for Services. In such cases, any default load balancer implementation, such as the one provided by the cloud provider, will disregard Services with this field set. It's important to note that `spec.loadBalancerClass` can only be set on a Service of type `LoadBalancer` and, once set, it cannot be altered. Additionally, the value of `spec.loadBalancerClass` must adhere to a label-style identifier format, optionally with a prefix like `internal-vip` or `example.com/internal-vip`, with unprefixed names being reserved for end users.

The Kubernetes `Loadbalancer` type Service principles have been visualized in the following diagram:

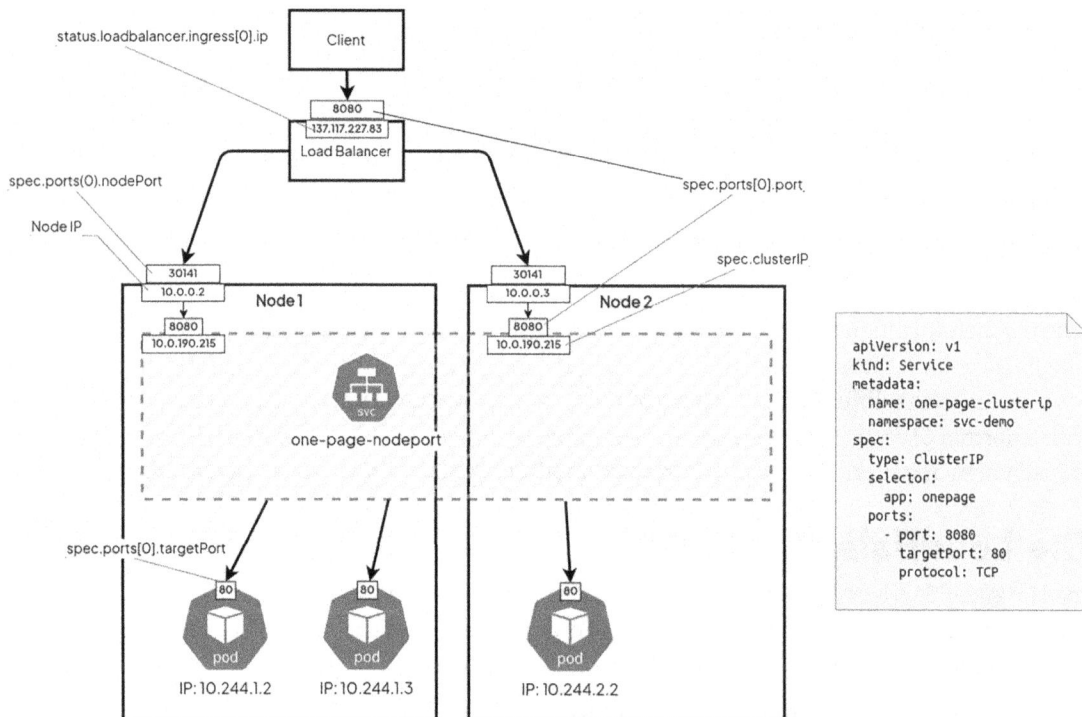

Figure 8.8: LoadBalancer Service

In the next section, we will learn about the supported cloud providers for `LoadBalancer` Service types.

Supported cloud providers for the LoadBalancer Service type

Not all cloud providers support the LoadBalancer Service type, but we can name a few that do support it. These are as follows:

- AWS
- GCP
- Azure
- OpenStack

The list is not exhaustive, but it's good to know that all three major public cloud providers are supported.

If your cloud provider is supported, keep in mind that the load-balancing logic will be the one implemented by the cloud provider: you have less control over how the traffic will be routed to your Pods from Kubernetes, and you will have to know how the load-balancer component of your cloud provider works. Consider it as a third-party component implemented as a Kubernetes resource.

Should the LoadBalancer Service type be used?

This question is difficult to answer but a lot of people tend to not use a LoadBalancer Service type for a few reasons.

The main reason is that LoadBalancer Services are nearly impossible to configure from Kubernetes. Indeed, if you must use a cloud provider, it is better to configure it from the tooling provided by the provider rather than from Kubernetes. The LoadBalancer Service type is a generic way to provision a LoadBalancer Service but does not expose all the advanced features that the cloud provider may provide.

Also, load balancers provided by cloud providers often come with additional costs, which can vary depending on the provider and the amount of traffic being handled.

In the next section, we will learn about another Service type called ExternalName Services.

The ExternalName Service type

ExternalName Services are a powerful way to connect your Kubernetes cluster to external resources like databases, APIs, or Services hosted outside the cluster. They work by mapping a Service in your cluster to a DNS name instead of Pods within the cluster. This allows your applications inside the cluster to seamlessly access the external resource without needing to know its IP address or internal details.

The following snippet shows a sample ExternalName type Service YAML definition:

```yaml
# externalname-service.yaml
apiVersion: v1
kind: Service
metadata:
  name: mysql-db
  namespace: prod
spec:
```

```
type: ExternalName
externalName: app-db.database.example.com
```

Here's how it works: instead of linking the external names or IP address to internal Pods, you simply define a DNS name like `app-db.database.example.com` in the Service configuration. Now, when your applications within the cluster try to access `mysql-db`, the magic happens—the cluster's DNS Service points them to your external database! They interact with it seamlessly, just like any other Service, but the redirection occurs at the DNS level, keeping things clean and transparent.

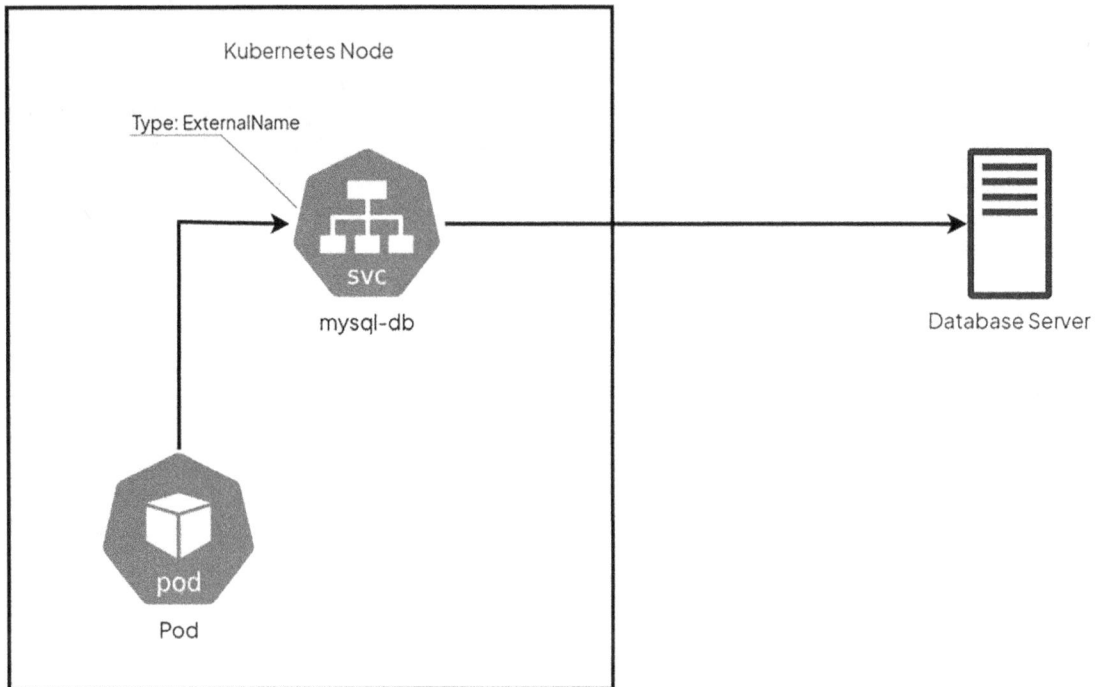

Figure 8.9: ExternalName Service type

The `ExternalName` Service can be a Service hosted in another Kubernetes namespace, a Service outside of the Kubernetes cluster, a Service hosted in another Kubernetes cluster, and so on.

This approach offers several benefits:

- **Simplified configuration:** Applications only need to know the Service name, not the external resource details, making things much easier.

- **Flexible resource management:** If you later move the database into your cluster, you can simply update the Service and manage it internally without affecting your applications.

- **Enhanced security:** Sensitive information like IP addresses stays hidden within the cluster, improving overall security.

Remember, `ExternalName` Services are all about connecting to external resources. For internal resource access within your cluster, stick to regular or headless Services.

Now that we have learned the different Service types in Kubernetes, let us explore how to use probes to ensure Service availability in the following sections.

Implementing Service readiness using probes

When you create a Service to expose an application running inside Pods, Kubernetes doesn't automatically verify the health of that application. The Pods may be up and running, but the application itself could still have issues, and the Service will continue to route traffic to it. This could result in users or other applications receiving errors or no response at all. To prevent this, Kubernetes provides health check mechanisms called probes. In the following section, we'll explore the different types of probes—liveness, readiness, and startup probes—and how they help ensure your application is functioning properly.

What is ReadinessProbe and why do you need it?

ReadinessProbe, along with LivenessProbe, is an important aspect to master if you want to provide the best possible experience to your end user. We will first discover how to implement ReadinessProbe and how it can help you ensure your containers are fully ready to serve traffic.

Readiness probes are technically not part of Services, but it is important to discover this feature alongside Kubernetes Services.

Just as with everything in Kubernetes, ReadinessProbe was implemented to bring a solution to a problem. This problem is this: how to ensure a Pod is fully ready before it can receive traffic, possibly from a Service.

Services obey a simple rule: they serve traffic to every Pod that matches their label selector. As soon as a Pod gets provisioned, if this Pod's labels match the selector of Service in your cluster, then this Service will immediately start forwarding traffic to it. This can lead to a simple problem: if the app is not fully launched, because it has a slow launch process or requires some configuration from a remote API, and so on, then it might receive traffic from Services before being ready for it. The result would be a poor **user experience (UX)**.

To make sure this scenario never happens, we can use the feature called ReadinessProbe, which is an additional configuration to add to a Pod's configuration.

When a Pod is configured with a readiness probe, it can send a signal to the control plane that it is not ready to receive traffic, and when a Pod is not ready, Services won't forward any traffic to it. Let's see how we can implement a readiness probe.

Implementing ReadinessProbe

ReadinessProbe implementation is achieved by adding some configuration data to a Pod YAML manifest. Please note that it has nothing to do with the Service object itself. By adding some configuration to the container spec in the Pod object, you can basically tell Kubernetes to wait for the Pod to be fully ready before it can receive traffic from Services.

ReadinessProbe can be of three different types, as outlined here:

- Command: Issue a command inside the pod that should complete with exit code 0, indicating the Pod is ready.
- HTTP: An HTTP request that should complete with a response code >= 200 and < 400, which indicates the Pod is ready.
- TCP: Issue a TCP connection attempt. If the connection is established, the Pod is ready.

Here is a YAML file configuring an nginx Pod with a readiness probe of type HTTP:

```
# nginx-pod-with-readiness-http.yaml
apiVersion: v1
kind: Pod
metadata:
  name: nginx-pod-with-readiness-http
spec:
  containers:
    - name: nginx-pod-with-readiness-http
      image: nginx
      readinessProbe:
        initialDelaySeconds: 5
        periodSeconds: 5
        httpGet:
          path: /ready
          port: 80
```

As you can see, we have two important inputs under the readinessProbe key, as follows:

- initialDelaySeconds, which indicates the number of seconds the probe will wait before running the first health check
- periodSeconds, which indicates the number of seconds the probe will wait between two consecutive health checks

The readiness probe will be replayed regularly, and the interval between two checks will be defined by the periodSeconds parameter.

In our case, our ReadinessProbe will run an HTTP call against the /ready path. If this request receives an HTTP response code >= 200 and < 400, then the probe will be a success, and the Pod will be considered healthy.

ReadinessProbe is important. In our example, the endpoint being called should test that the application is really in such a state that it can receive traffic. So, try to call an endpoint that is relevant to the state of the actual application. For example, you can try to call a page that will open a MySQL connection internally to make sure the application is capable of communicating with its database if it is using one, and so on. If you're a developer, do not hesitate to create a dedicated endpoint that will just open connections to the different backends to be fully sure that the application is definitely ready.

The Pod will then join the pool being served by the Service and will start receiving traffic. ReadinessProbe can also be configured as TCP and commands, but we will keep these examples for LivenessProbe. Let's discover them now!

What is LivenessProbe and why do you need it?

LivenessProbe resembles ReadinessProbe a lot. In fact, if you have used any cloud providers before, you might already have heard about something called health checks. LivenessProbe is basically a health check.

Liveness probes are used to determine whether a Pod is in a broken state or not, and the usage of LivenessProbe is especially suited for long-running processes such as web services. Imagine a situation where you have a Service forwarding traffic to three Pods and one of them is broken. Services cannot detect that on their own, and they will just continue to serve traffic to the three Pods, including the broken one. In such situations, 33% of your requests will inevitably lead to an error response, resulting in a poor UX, as illustrated in the following screenshot:

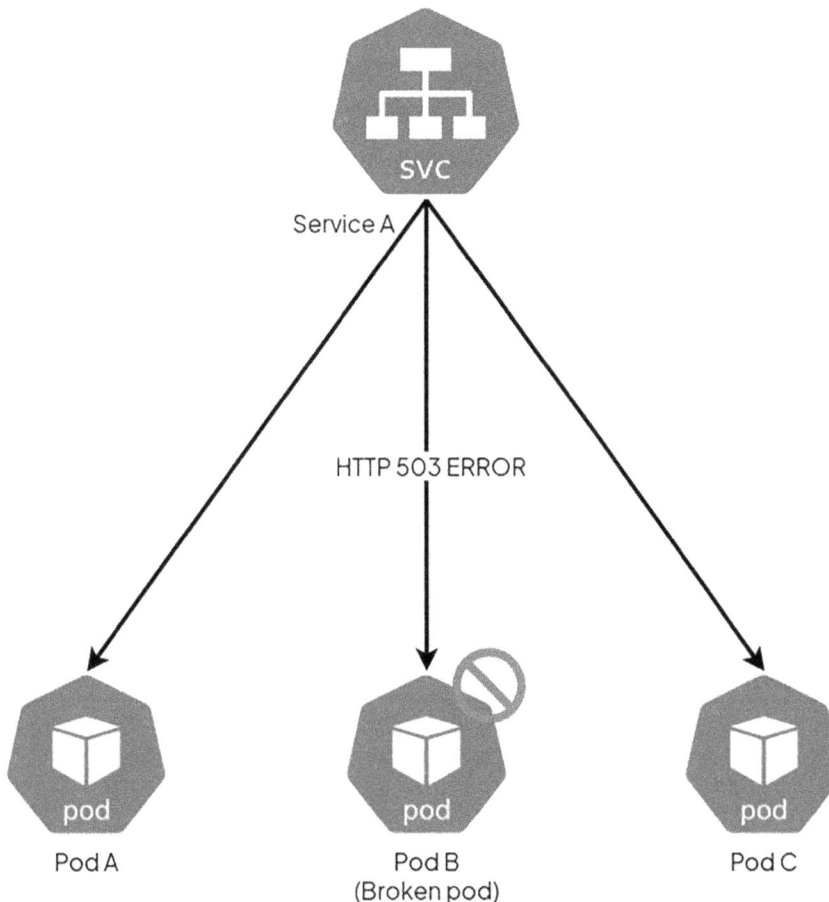

Figure 8.10: One of the Pods is broken but the Service will still forward traffic to it

You want to avoid such situations, and to do so, you need a way to detect situations where Pods are broken, plus a way to kill such a container so that it goes out of the pool of Pods being targeted by the Service.

LivenessProbe is the solution to this problem and is implemented at the Pod level. Be careful because LivenessProbe cannot repair a Pod: it can only detect that a Pod is not healthy and command its termination. Let's see how we can implement a Pod with LivenessProbe.

Implementing LivenessProbe

LivenessProbe is a health check that will be executed on a regular schedule to keep track of the application state in the long run. These health checks are executed by the kubelet component and can be of different types, as outlined here:

- **Command**, where you issue a command in the container and its result will tell whether the Pod is healthy or not (exit code = 0 means healthy)
- **HTTP**, where you run an HTTP request against the Pod, and its result tells whether the Pod is healthy or not (HTTP response code >= 200 and < 400 means the Pod is healthy)
- **TCP**, where you define a TCP call (a successful connection means the Pod is healthy)
- **GRPC**, if the application supports and implements the gRPC Health Checking Protocol

Each of these liveness probes will require you to input a parameter called periodSeconds, which must be an integer. This will tell the kubelet component the number of seconds to wait before performing a new health check. You can also use another parameter called initialDelaySeconds, which will indicate the number of seconds to wait before performing the very first health check. Indeed, in some common situations, a health check might lead to flagging an application as unhealthy just because it was performed too early. That's why it might be a good idea to wait a little bit before performing the first health check, and that parameter is here to help.

LivenessProbe configuration is achieved at the Pod YAML configuration manifest, not at the Service one. Each container in the Pod can have its own livenessProbe.

HTTP livenessProbe

HTTP probes in Kubernetes offer additional customizable fields, such as host, scheme, path, headers, and port, to fine-tune how the health check requests are made to the application. Here is a configuration file that checks if a Pod is healthy by running an HTTP call against a /healthcheck endpoint in a nginx container:

```
# nginx-pod-with-liveness-http.yaml
apiVersion: v1
kind: Pod
metadata:
  name: nginx-pod-with-liveness-http
spec:
  containers:
    - name: nginx-pod-with-liveness-http
```

```
      image: nginx
      livenessProbe:
        initialDelaySeconds: 5
        periodSeconds: 5
        httpGet:
          path: /healthcheck
          port: 80
          httpHeaders:
            - name: My-Custom-Header
              value: My-Custom-Header-Value
```

Please pay attention to all sections after the livenessProbe blocks. If you understand this well, you can see that we will wait 5 seconds before performing the first health check, and then, we will run one HTTP call against the /healthcheck path on port 80 every 5 seconds. One custom HTTP header was added. Adding such a header will be useful to identify our health checks in the access logs. Be careful because the /healthcheck path probably won't exist in our nginx container, and so this container will never be considered healthy because the liveness probe will result in a 404 HTTP response. Keep in mind that for an HTTP health check to succeed, it must answer an HTTP >= 200 and < 400. With 404 being out of this range, the answer Pod won't be healthy.

Command livenessProbe

You can also use a command to check if a Pod is healthy or not. Let's grab the same YAML configuration, but now, we will use a command instead of an HTTP call in the liveness probe, as follows:

```
# nginx-pod-with-liveness-command.yaml
apiVersion: v1
kind: Pod
metadata:
  name: nginx-pod-with-liveness-command
spec:
  containers:
    - name: nginx-pod-with-liveness-command
      image: nginx
      livenessProbe:
        initialDelaySeconds: 5
        periodSeconds: 5
        exec:
          command:
            - cat
            - /hello/world
```

If you check this example, you can see that it is much simpler than the HTTP one. Here, we are basically running a `cat /hello/world` command every 5 seconds. If the file exists and the `cat` command completes with an exit code equal to 0, then the health check will succeed. Otherwise, if the file is not present, the health check will fail, and the Pod will never be considered healthy and will be terminated.

TCP livenessProbe

In this situation, we will attempt a connection to a TCP socket on port 80. If the connection is successfully established, then the health check will pass, and the container will be considered ready. Otherwise, the health check will fail, and the Pod will be terminated eventually. The code is illustrated in the following snippet:

```yaml
# nginx-pod-with-liveness-tcp.yaml
apiVersion: v1
kind: Pod
metadata:
  name: nginx-pod-with-liveness-tcp
spec:
  containers:
    - name: nginx-pod-with-liveness-tcp
      image: nginx
      livenessProbe:
        initialDelaySeconds: 5
        periodSeconds: 5
        tcpSocket:
          port: 80
```

Using TCP health checks greatly resembles using HTTP ones since HTTP is based on TCP. But having TCP as a liveness probe is especially nice if you want to keep track of an application that is not based on using HTTP as protocol and if using that command is irrelevant to you, as when health-checking an LDAP connection, for example.

Using named Port with TCP and HTTP livenessProbe

You can use the named port to configure the `livenessProbe` for HTTP and TCP types (but not with gRPC) as follows:

```yaml
# nginx-pod-with-liveness-http-named-port.yaml
apiVersion: v1
kind: Pod
metadata:
  name: nginx-pod-with-liveness-http-named-port
spec:
  containers:
```

```
  - name: nginx-pod-with-liveness-http-named-port
    image: nginx
    ports:
      - name: liveness-port
        containerPort: 8080
        hostPort: 8080
    livenessProbe:
      initialDelaySeconds: 5
      periodSeconds: 5
      httpGet:
        path: /healthcheck
        port: liveness-port
```

In the preceding example, the `liveness-port` has been defined under the `ports` section and used under the `httpGet` for `livenessProbe`.

As we have explored multiple liveness probes, let us learn about startupProbe in the next section.

Using startupProbe

Legacy applications sometimes demand extra time on their first startup. This can create a dilemma when setting up liveness probes, as fast response times are crucial for detecting deadlocks.

The answer lies in using either initialDelaySeconds or a dedicated `startupProbe`. The `initialDelaySeconds` parameter allows you to postpone the first readiness probe, giving the application breathing room to initialize.

However, for more granular control, consider using a `startupProbe`. This probe mirrors your liveness probe (command, HTTP, or TCP check) but with a longer `failureThreshold` * `periodSeconds` duration. This extended waiting period ensures the application has ample time to initialize before being deemed ready for traffic, while still enabling the liveness probe to swiftly detect issues afterwards, as explained in the following YAML snippet:

```
# nginx-pod-with-startupprobe.yaml
apiVersion: v1
kind: Pod
metadata:
  name: nginx-pod-with-startupprobe
spec:
  containers:
    - name: nginx-pod-with-startupprobe
      image: nginx
      ports:
        - name: liveness-port
          containerPort: 8080
```

```
        hostPort: 8080
    livenessProbe:
      initialDelaySeconds: 5
      periodSeconds: 5
      httpGet:
        path: /healthcheck
        port: liveness-port
    startupProbe:
      httpGet:
        path: /healthz
        port: liveness-port
      failureThreshold: 30
      periodSeconds: 10
```

As you can see in the example code above, it is possible to combine multiple probes to ensure the application is ready to serve. In the following section, we will also learn how to use ReadinessProbe and LivenessProbe together.

Using ReadinessProbe and LivenessProbe together

You can use ReadinessProbe and LivenessProbe together in the same Pod.

They are configured almost the same way—they don't have exactly the same purpose, and it is fine to use them together. Please note that both the probes share these parameters:

- initialDelaySeconds: The number of seconds to wait before the first probe execution.
- periodSeocnds: The number of seconds between two probes.
- timeoutSeconds: The number of seconds to wait before timeout.
- successThreshold: The number of successful attempts to consider a Pod is ready (for ReadinessProbe) or healthy (for LivenessProbe).
- failureThreshold: The number of failed attempts to consider a Pod is not ready (for ReadinessProbe) or ready to be killed (for LivenessProbe).
- TerminationGracePeriodSeconds: Give containers a grace period to shut down gracefully before being forcefully stopped (default inherits Pod-level value).

We now have discovered ReadinessProbe and LivenessProbe, and we have reached the end of this chapter about Kubernetes Services and implementation methods.

Summary

This chapter was dense and contained a huge amount of information on networking in general when applied to Kubernetes. Services are just like Pods: they are the foundation of Kubernetes and mastering them is crucial to being successful with the orchestrator.

Overall, in this chapter, we discovered that Pods have dynamic IP assignments, and they get a unique IP address when they're created. To establish a reliable way to connect to your Pods, you need a proxy called a `Service` in Kubernetes. We've also discovered that Kubernetes Services can be of multiple types and that each type of Service is designed to address a specific need. We've also discovered what `ReadinessProbe` and `LivenessProbe` are and how they can help you in designing health checks to ensure your pods get traffic when they are ready and live.

In the next chapter, we'll continue to discover the basics of Kubernetes by discovering the concepts of `PersistentVolume` and `PersistentVolumeClaims`, which are the methods Kubernetes uses to deal with persistent data. It is going to be a very interesting chapter if you want to build and provision stateful applications on your Kubernetes clusters, such as database or file storage solutions.

Further reading

- Services, Load Balancing, and Networking: `https://kubernetes.io/docs/concepts/services-networking/`
- Kubernetes Headless Services: `https://kubernetes.io/docs/concepts/services-networking/service/#headless-services`
- Configure Liveness, Readiness and Startup Probes: `https://kubernetes.io/docs/tasks/configure-pod-container/configure-liveness-readiness-startup-probes/`
- Network Policies: `https://kubernetes.io/docs/concepts/services-networking/network-policies/`
- Declare Network Policy: `https://kubernetes.io/docs/tasks/administer-cluster/declare-network-policy/`
- Default NetworkPolicy: `https://kubernetes.io/docs/concepts/services-networking/network-policies/#default-policies`
- EndpointSlices: `https://kubernetes.io/docs/concepts/services-networking/endpoint-slices/`
- Network Plugins: `https://kubernetes.io/docs/concepts/extend-kubernetes/compute-storage-net/network-plugins/`
- Cluster Networking: `https://kubernetes.io/docs/concepts/cluster-administration/networking/`

Join our community on Discord

Join our community's Discord space for discussions with the authors and other readers:

`https://packt.link/cloudanddevops`

9

Persistent Storage in Kubernetes

In the previous chapters, we learned about Kubernetes' key concepts, and this chapter is going to be the last one about that. So far, we've discovered that Kubernetes is about representing a desired state for all the traditional IT layers by creating an object in its `etcd` datastore that will be converted into actual computing resources within your clusters.

This chapter will focus on persistent storage for stateful applications. As with any other resource abstraction, this will be another set of objects that we will master to get persistent storage on your clusters. Persistent storage is achieved in Kubernetes by using the `PersistentVolume` resource type, which has its own mechanics. Honestly, these can be relatively difficult to approach at first, but we are going to discover all of them and cover them in depth!

In this chapter, we're going to cover the following main topics:

- Why use persistent storage?
- Understanding how to mount `PersistentVolume` to your Pod
- Understanding the life cycle of the `PersistentVolume` object in Kubernetes
- Understanding static and dynamic `PersistentVolume` provisioning
- Advanced storage topics

Technical requirements

- A working Kubernetes cluster (either local or cloud-based)
- A working `kubectl` CLI configured to communicate with the cluster

If you do not meet these technical requirements, you can follow *Chapter 2, Kubernetes Architecture – from Container Images to Running Pods*, and *Chapter 3, Installing Your Kubernetes Cluster*, to get these two prerequisites.

You can download the latest code samples for this chapter from the official GitHub repository at `https://github.com/PacktPublishing/The-Kubernetes-Bible-Second-Edition/tree/main/Chapter09`.

Why use persistent storage?

Storage is an important resource within the IT world, as it provides a logical way to **create, read, update,** and **delete** (**CRUD**) information ranging from employee payslips in a PDF file format to petabytes of healthcare records. While storage is a key element in providing relevant information to the users, containers and microservices should be stateless. In other words, no information saved within a running container will be available when rescheduled or moved to a different cluster. The same goes for microservices; the data component should be decoupled, allowing the microservice to stay micro and not care about the data state and availability when being rescheduled.

So, where do we save the application data? In any sort of datastore, and from a business continuity perspective, if the related datastore runs on the same Kubernetes cluster as the microservices, it should have an application-aware replication mechanism. But remember, Kubernetes is a resource orchestrator that will act on the desired state you have defined for your application. When you're configuring your Pods, you have the opportunity to define the storage component to be used, providing your containers with a way to create, read, update, and delete data. Let's explore the different options Kubernetes has to offer to persist data.

Introducing Volumes

The first layer of storage abstraction is to access Kubernetes objects and mount them within a container like a data volume. This can be done for:

- A ConfigMap
- A Secret
- A ServiceAccount token (identical to a Secret)

This allows an application team to decouple the configuration of a microservice from the container or the deployment definition. If we consider the lifetime of an application, the credentials, certificates, or tokens to external services might need to be refreshed or a configuration parameter might need to be updated. We don't want these to be hardcoded in the deployment manifests or the container images for obvious security reasons.

Let's have a look at a configMap example with the manifest `nginx-configmap.yaml`:

```
# nginx-configmap.yaml
apiVersion: v1
kind: ConfigMap
metadata:
  name: nginx-hello
  labels:
    app: test
immutable: false
data:
  hello1.html: |
    <html>
```

```
      hello world 1
    </html>
  hello2.html: |
    <html>
      hello world 2
    </html>
```

This `ConfigMap` has two definitions for two different files, which we will mount within the NGINX Pod with the manifest `nginx-pod.yaml`:

```yaml
# nginx-pod.yaml
apiVersion: v1
kind: Pod
metadata:
  name: nginx-hello
  labels:
    app: test
spec:
  containers:
    - name: nginx
      image: nginx:1.14.2
      ports:
        - containerPort: 80
      volumeMounts:
        - name: nginx-hello
          mountPath: "/usr/share/nginx/html/hello"
  volumes:
    - name: nginx-hello
      configMap:
        name: nginx-hello
```

Let's apply these two manifests:

```
$ kubectl apply -f nginx-configmap.yaml
configmap/nginx-hello created
$ kubectl apply -f nginx-pod.yaml
pod/nginx-hello created
```

Let's verify the status of the two objects:

```
$ kubectl get pod,cm
NAME              READY    STATUS     RESTARTS    AGE
pod/nginx-hello   1/1      Running    0           7m26s
```

```
NAME                            DATA    AGE
configmap/kube-root-ca.crt      1       7d17h
configmap/nginx-hello           2       7m31s
```

Verify the files are available within the folder /usr/share/nginx/hello that we provided as a mount path:

```
$ kubectl exec -t pod/nginx-hello -- ls -al /usr/share/nginx/html/hello/
total 12
...<removed for brevity>...
lrwxrwxrwx 1 root root   18 Sep  7 21:19 hello1.html -> ..data/hello1.html
lrwxrwxrwx 1 root root   18 Sep  7 21:19 hello2.html -> ..data/hello2.html
```

Let's verify that the data is being served by NGINX via a port-forward to avoid setting up a service:

```
$ kubectl port-forward nginx-hello 8080:80
Forwarding from 127.0.0.1:8080 -> 80
Forwarding from [::1]:8080 -> 80
```

In a second terminal, you can then curl the two URLs:

```
$ curl 127.0.0.1:8080/hello/hello1.html
<html>
  hello world 1
</html>

$ curl 127.0.0.1:8080/hello/hello2.html
<html>
  hello world 2
</html>
```

While this is a great start, one limitation of these objects is the amount of data you can store. Since it depends on the etcd datastore, to avoid performance issues, the limitation is 1.5 MB (refer to https://etcd.io/docs/v3.5/dev-guide/limit). So, the next set of objects will allow your application to store much more data, in fact, as much data as the system hosting those volume objects can.

Let's consider a Kubernetes cluster with two worker nodes on which Pods can be scheduled, and explore the following five types of volumes:

- An emptyDir
- A hostPath
- A local volume
- A **Fiber Channel** (**FC**) block disk
- A **Network File System** (**NFS**) volume export

The first three types, `emptyDir`, `hostPath`, and local volumes, have two major limitations:

- They are limited to the disk space available on the worker node they are provisioned on.
- They are bound to the node on which the Pod will be deployed. If your Pod is provisioned on worker node 1, the data will only be stored on worker node 1.

These volume types could potentially lead to a degradation of service or worse, like a split-brain scenario. If worker node 1 becomes unhealthy, triggering a rescheduling of the Pod to worker node 2, the application will start without its data and could lead to a major outage.

Note that some applications have a native replication engine. A typical deployment for such an application would have two replicas running and creating a `hostPath` volume on each node. In this scenario, if one worker node becomes unhealthy, then the application becomes degraded but only from a high availability and performance perspective.

Being external to any of the compute resources of your Kubernetes cluster, the last two types, FC block disk and an NFS volume, address the above weaknesses but introduce a bit more complexity. While the first three types of volumes do not require you to interact with your storage administrators, the last two do. Without getting into too many details, your storage administrators will have:

- To provision a **Logical Unit Number** (**LUN** – the FC block disk) on their **Storage Area Network** (**SAN**) connected via an FC fabric to your Kubernetes worker nodes and allow access via a zoning configuration.
- To provision a data space on **Network Attached Storage** (**NAS**) connected to the corporate network and reachable by your Kubernetes worker nodes and allow access via an export policy.

Note that testing these two types of volumes requires specialized equipment with a nontrivial setup, although NAS is more and more popular within home labs. However, from a Kubernetes standpoint, these volumes can be configured as easily as with the configMap example. Here are the modified versions of the NGINX Pod definition:

- For an FC volume (`nginx-pod-fiberchannel.yaml`):

```
...
    - name: nginx
      image: nginx:1.14.2
      ports:
        - containerPort: 80
      volumeMounts:
        - name: fc-vol
          mountPath: "/usr/share/nginx/html/hello"
  volumes:
    - name: fc-vol
      fc:
        targetWWNs:
          - 500a0982991b8dc5
```

```
          - 500a0982891b8dc5
        lun: 2
        fsType: ext4
        readOnly: true
```

The fc part is where your SAN and LUN must be configured.

- For an NFS volume (nginx-pod-nfs-volume.yaml):

```
    ...
      containers:
        - name: nginx
          image: nginx:1.14.2
          ports:
            - containerPort: 80
          volumeMounts:
            - name: nfs-volume
              mountPath: "/usr/share/nginx/html/hello"
      volumes:
        - name: nfs-volume
          nfs:
            server: nfs.corp.mycompany.org
            path: /k8s-nginx-hello
            readOnly: true
```

The nfs part is where your NAS and exported volume must be configured.

Please note the following points:

- These two types of volumes, FC block disk and NFS, will be attached to the nodes as required by the Pod presence.
- While these two types of volumes can solve a series of challenges, they represent an anti-pattern to the decoupling of configurations and resources.
- While the configMap is mounted as a volume with the two HTML files on the container, the other types of volumes will require a different approach to have the data injected.
- There are other volume types available: https://kubernetes.io/docs/concepts/storage/volumes/.

The concept of volume within Kubernetes is an amazing starting point for deploying stateful applications. However, with the limitation of some, the complexity of others, and the storage knowledge it requires, it seems to be rather difficult to scale hundreds or thousands of microservices with such object definition. Thanks to an additional layer of abstraction, Kubernetes provides an agnostic approach to consume storage at scale with the usage of the PersistentVolume object, which we'll cover in the next section.

Introducing PersistentVolumes

Just like the Pod or ConfigMap, PersistentVolume is a resource type that is exposed through kube-apiserver; you can create, update, and delete **persistent volumes (PVs)** using YAML and kubectl just like any other Kubernetes objects.

The following command will demonstrate how to list the PersistentVolume resource type currently provisioned within your Kubernetes cluster:

```
$ kubectl get persistentvolume
No resource found
```

The persistentvolume object is also accessible with the plural form of persistentvolumes along with the alias of pv. The following three commands are essentially the same:

```
$ kubectl get persistentvolume
No resource found
$ kubectl get persistentvolumes
No resource found
$ kubectl get pv
No resource found
```

You'll find that the pv alias is very commonly used in the Kubernetes world, and a lot of people refer to PVs as simply pv, so be aware of that. As of now, no PersistentVolume object has been created within our Kubernetes cluster, and that is why we don't see any resource listed in the output of the preceding command.

PersistentVolume is the object and, essentially, represents a piece of storage that you can attach to your Pod. That piece of storage is referred to as a *persistent* one because it is not supposed to be tied to the lifetime of a Pod.

Indeed, as mentioned in *Chapter 5, Using Multi-Container Pods and Design Patterns*, Kubernetes Pods use the notion of volumes. Additionally, we discovered the emptyDir volumes, which initiate an empty directory that your Pods can share. It also defines a path within the worker node filesystem that will be exposed to your Pods. Both volumes were supposed to be attached to the life cycle of the Pod. This means that once the Pod is destroyed, the data stored within the volumes will be destroyed as well.

However, sometimes, you don't want the volume to be destroyed. You just want it to have its life cycle to keep both the volume and its data alive even if the Pod fails. That's where PersistentVolumes come into play: essentially, they are volumes that are not tied to the life cycle of a Pod. Since they are a resource type just like the Pods themselves, they can live on their own! In essence, PVs ensure that your storage remains available beyond the Pod's existence, which is crucial for maintaining data integrity in stateful applications. Now, let's break down PersistentVolumes objects: they consist of two key elements – a backend technology (the PersistentVolume type) and an access mode (like **ReadWriteOnce (RWO)**). Understanding these concepts is essential for effectively utilizing PVs within your Kubernetes environment.

Bear in mind that `PersistentVolumes` objects are just entries within the `etcd` datastore, and they are not actual disks on their own.

`PersistentVolume` is just a kind of pointer within Kubernetes to a piece of storage, such as an NFS, a disk, an Amazon **Elastic Block Store** (**EBS**) volume, and more. This is so that you can access these technologies from within Kubernetes and in a Kubernetes way.

In the next section, we'll begin by explaining what `PersistentVolume` types are.

Introducing PersistentVolume types

As you already know, the simplest Kubernetes setup consists of a simple `minikube` installation, whereas the most complex Kubernetes setup can be made of dozens of servers on a massively scalable infrastructure. All of these different setups will forcibly have different ways in which to manage persistent storage. For example, the three well-known public cloud providers have a lot of different solutions. Let's name a few, as follows:

- Amazon EBS volumes
- Amazon **Elastic File System** (**EFS**) filesystems
- Google GCE **Persistent Disk** (**PD**)
- Microsoft Azure disks

These solutions have their own design and set of principles, along with their own logic and mechanics. Kubernetes was built with the principle that all of these setups should be abstracted using just one object to abstract all of the different technologies; that single object is the `PersistentVolume` resource type. The `PersistentVolume` resource type is the object that is going to be attached to a running Pod. Indeed, a Pod is a Kubernetes resource and does not know what an EBS or a PD is; Kubernetes Pods only play well with `PersistentVolumes`, which is also a Kubernetes resource.

Whether your Kubernetes cluster is running on Google GKE or Amazon EKS, or whether it is a single minikube cluster on your local machine has no importance. When you wish to manage persistent storage, you are going to create, use, and deploy `PersistentVolumes` objects, and then bind them to your Pods!

Here are some of the backend technologies supported by Kubernetes out of the box:

- `csi`: **Container Storage Interface** (**CSI**)
- `fc`: FC storage
- `iscsi`: SCSI over IP
- `local`: Using local storage
- `hostPath`: HostPath volumes
- `nfs`: Regular network file storage

The preceding list is not exhaustive: Kubernetes is extremely versatile and can be used with many storage solutions that can be abstracted as `PersistentVolume` objects in your cluster.

Please note that in recent versions of Kubernetes, several `PersistentVolume` types have been deprecated or removed, indicating a shift in how storage is managed within Kubernetes environments. This change is part of the ongoing evolution of Kubernetes to streamline its APIs and improve compatibility with modern storage solutions.

For example, the following `PersistentVolume` types are either removed or deprecated in Kubernetes 1.29 onwards:

- `awsElasticBlockStore` – Amazon EBS
- `azureDisk` – Azure Disk
- `azureFile` – Azure File
- `portworxVolume` – Portworx volume
- `flexVolume` – FlexVolume
- `vsphereVolume` – vSphere VMDK volume
- `cephfs` – CephFS volume
- `cinder`

These changes reflect a broader trend toward standardized storage interfaces and an emphasis on more portable, cloud-agnostic solutions. For detailed guidance and updated information on PVs and supported types, you can refer to the official Kubernetes documentation at `https://kubernetes.io/docs/concepts/storage/persistent-volumes/#types-of-persistent-volumes`.

Benefits brought by PersistentVolume

PVs are an essential component in Kubernetes when managing stateful applications. Unlike ephemeral storage, PVs ensure data persists beyond the life cycle of individual Pods, making them ideal for applications requiring data retention and consistency. These storage resources bring flexibility and reliability to the Kubernetes ecosystem, enhancing both performance and resilience.

There are three major benefits of `PersistentVolume`:

- A PV in Kubernetes continues to exist independently of the Pod that uses it. This means that if you delete or recreate a Pod attached to a `PersistentVolume`, the data stored on that volume remains intact. The data's persistence depends on the reclaim policy of the `PersistentVolume`: with a retain policy, the data stays available for future use, while a delete policy removes both the volume and its data when the Pod is deleted. Thus, you can manage your Pods without worrying about losing data stored on `PersistentVolume`s.
- When a Pod crashes, the `PersistentVolume` object will survive the fault and not be removed from the cluster.
- `PersistentVolume` is cluster-wide; this means that it can be attached to any Pod running on any node. (You will learn about restrictions and methods later in this chapter.)

Bear in mind that these three statements are not always 100% valid. Indeed, sometimes, a `PersistentVolume` object can be affected by its underlying technology.

To demonstrate this, let's consider a `PersistentVolume` object that is, for example, a pointer to a `hostPath` storage on the compute node. In such a setup, `PersistentVolume` won't be available to any other nodes.

However, if you take another example, such as an NFS setup, it wouldn't be the same. Indeed, you can access an NFS from multiple machines at once. Therefore, a `PersistentVolume` object that is backed by an NFS would be accessible from several different Pods running on different nodes without much problem. To understand how to make a `PersistentVolume` object on several different nodes at a time, we need to consider the concept of access modes, which we'll be diving into in the next section.

Introducing PersistentVolume access modes

As the name suggests, access modes are an option you can set when you create a `PersistentVolume` type that will tell Kubernetes how the volume should be mounted.

`PersistentVolumes` supports four access modes, as follows:

- **ReadWriteOnce (RWO)**: This volume allows read/write by only one node at the same time.
- **ReadOnlyMany (ROX)**: This volume allows read-only mode by many nodes at the same time.
- **ReadWriteMany (RWX)**: This volume allows read/write by multiple nodes at the same time.
- **ReadWriteOncePod**: This is a new mode introduced recently and is already stable in the Kubernetes 1.29 version. In this access mode, the volume is mountable as read-write by a single Pod. Employ the `ReadWriteOncePod` access mode when you want only one Pod throughout the entire cluster to have the capability to read from or write to the **Persistent Volume Claim** (**PVC**).

It is necessary to set at least one access mode to a `PersistentVolume` type, even if said volume supports multiple access modes. Indeed, not all `PersistentVolume` types will support all access modes, as shown in the below table.

Volume Plugin	ReadWriteOnce	ReadOnlyMany	ReadWriteMany	ReadWriteOncePod
AzureFile	✓	✓	✓	-
CephFS	✓	✓	✓	-
CSI	depends on the driver	depends on the driver	depends on the driver	depends on the driver
FC	✓	✓	-	-
FlexVolume	✓	✓	depends on the driver	-
HostPath	✓	-	-	-
iSCSI	✓	✓	-	-
NFS	✓	✓	✓	-
RBD	✓	✓	-	-
VsphereVolume	✓	-	- (works when Pods are collocated)	-
PortworxVolume	✓	-	✓	-

Table 9.1: Access modes supported by different PersistentVolume types (Image source: kubernetes.io/docs/concepts/storage/persistent-volumes)

In Kubernetes, the access modes of a `PersistentVolume` type are closely tied to the underlying storage technology and how it handles data. Here's why different PV types support specific modes:

File vs. block storage:

- File storage (like **Network File System (NFS)** or **Common Internet File System (CIFS)**) allows multiple clients to access the same files concurrently. This is why file storage systems can support a variety of access modes, such as RWO, ROX, and RWX. They are built to handle multi-client access over a network, enabling several nodes to read and write from the same volume without data corruption.

- Block storage (like local storage or hostPath) is fundamentally different. Block storage is designed for one client to access at a time because it deals with raw disk sectors rather than files. Concurrent access by multiple clients would lead to data inconsistency or corruption. For this reason, block storage supports only the RWO mode, where a single node can both read and write to the volume.

Internal vs. external storage:

- hostPath volumes, which refer to storage on the same node as the workload, are inherently restricted to that node. Since this storage is tied to the physical node, it cannot be accessed by other nodes in the cluster. This makes it only compatible with the RWO mode.

- NFS or other external storage solutions, on the other hand, are designed to allow access over a network, enabling multiple nodes to share the same storage. This flexibility allows them to support additional modes like RWX.

Understanding this distinction helps to clarify why some `PersistentVolume` types support more flexible access modes, while others are constrained.

Now, let's create our first `PersistentVolume` object.

Creating our first PersistentVolume object

Let's create a `PersistentVolume` on the Kubernetes cluster using the declarative approach. Since `PersistentVolumes` are more complex resources, it's highly recommended to avoid using the imperative method. The declarative approach allows you to define and manage resources consistently in YAML files, making it easier to track changes, version control your configurations, and ensure repeatability across different environments. This approach also makes it simpler to manage large or complex resources like `PersistentVolumes`, where precise configurations and careful planning are essential.

See the example YAML definition below for creating a `PersistentVolume` object:

```
---
apiVersion: v1
kind: PersistentVolume
metadata:
  name: pv-hostpath
  labels:
    type: local
spec:
  storageClassName: manual
```

```
capacity:
  storage: 1Gi
accessModes:
  - ReadWriteOnce
hostPath:
  path: "/mnt/data"
```

This is the simplest form of PersistentVolume. Essentially, this YAML file creates a PersistentVolume entry within the Kubernetes cluster. So, this PersistentVolume will be a hostPath type.

> The hostPath type PersistentVolume is not recommended for production or critical workloads. We are using it here for demonstration purposes only.

Let's apply the PV configuration to the cluster as follows:

```
$ kubectl apply -f pv-hostpath.yaml
persistentvolume/pv-hostpath created
```

It could be a more complex volume, such as a cloud-based disk or an NFS, but in its simplest form, a PersistentVolume can simply be a hostPath type on the node running your Pod.

How does Kubernetes PersistentVolumes handle storage?

As we learned earlier, the PersistentVolume resource type is a pointer to a storage location and that can be, for example, a disk, an NFS drive, or a disk volume controlled by a storage operator. All of these different technologies are managed differently. However, fortunately for us, in Kubernetes, they are all represented by the PersistentVolume object.

Simply put, the YAML file to create a PersistentVolume will be a little bit different depending on the backend technology that the PersistentVolume is backed by. For example, if you want your PersistentVolume to be a pointer to an NFS share, you have to meet the following two conditions:

- The NFS share is already configured and reachable from the Kubernetes nodes.
- The YAML file for your PersistentVolume must include the NFS server details and NFS share information.

The following YAML definition is a sample for creating a PersistentVolume using NFS as the backend:

```
# pv-nfs.yaml
apiVersion: v1
kind: PersistentVolume
metadata:
  name: pv-nfs
spec:
  capacity:
```

```
      storage: 5Gi
    volumeMode: Filesystem
    accessModes:
      - ReadWriteOnce
    persistentVolumeReclaimPolicy: Recycle
    storageClassName: slow
    mountOptions:
      - hard
      - nfsvers=4.1
    nfs:
      path: /appshare
      server: nfs.example.com
```

For a `PersistentVolume` to work properly, it needs to be able to link Kubernetes and the actual storage. So, you need to create a piece of storage or provision it outside of Kubernetes and then create the `PersistentVolume` entry by including the unique ID of the disk, or the volume, that is backed by a storage technology that is external to Kubernetes. Next, let's take a closer look at some examples of `PersistentVolume` YAML files in the next section.

Creating PersistentVolume with raw block volume

This example displays a `PersistentVolume` object that is pointing to raw block volume:

```
# pv-block.yaml
apiVersion: v1
kind: PersistentVolume
metadata:
  name: pv-block
spec:
  capacity:
    storage: 100Gi
  accessModes:
    - ReadWriteOnce
  volumeMode: Block
  persistentVolumeReclaimPolicy: Retain
  fc:
    targetWWNs: ["50060e801049cfd1"]
    lun: 0
    readOnly: false
```

As you can see, in this YAML file, the fc section contains the FC volume details that this `PersistentVolume` object is pointing to. The exact raw volume is identified by the `targetWWNs` key. That's pretty much it. With this YAML file, Kubernetes is capable of finding the proper **World Wide Name** (WWN) and maintaining a pointer to it.

Now, let's discuss a little bit about the provisioning of storage resources.

Can Kubernetes handle the provisioning or creation of the resource itself?

The fact that you need to create the actual storage resource separately and then create a PersistentVolume in Kubernetes might be tedious.

Fortunately for us, Kubernetes is also capable of communicating with the **APIs** of your cloud provider or other storage backends in order to create volumes or disks on the fly. There is something called **dynamic provisioning** that you can use when it comes to managing PersistentVolume. It makes things a lot simpler when dealing with PersistentVolume provisioning, but it only works on supported storage backends or cloud providers.

However, this is an advanced topic, so we will discuss it in more detail later in this chapter.

Now that we know how to provision PersistentVolume objects inside our cluster, we can try to mount them. Indeed, in Kubernetes, once you create a PersistentVolume, you need to mount it to a Pod so that it can be used. Things will get slightly more advanced and conceptual here; Kubernetes uses an intermediate object in order to mount a PersistentVolume to Pods. This intermediate object is called PersistentVolumeClaim. Let's focus on it in the upcoming section.

Understanding how to mount a PersistentVolume to your Pod

We can now try to mount a PersistentVolume object to a Pod. To do that, we will need to use another object, which is the second object we need to explore in this chapter, called PersistentVolumeClaim.

Introducing PersistentVolumeClaim

Just like PersistentVolume or ConfigMap, PersistentVolumeClaim is another independent resource type living within your Kubernetes cluster.

First, bear in mind that even if both names are almost the same, PersistentVolume and PersistentVolumeClaim are two distinct resources that represent two different things.

You can list the PersistentVolumeClaim resource type created within your cluster using kubectl, as follows:

```
$ kubectl get persistentvolumeclaims
No resources found in default namespace.
```

The following output tells us that we don't have any PersistentVolumeClaim resources created within my cluster. Please note that the pvc alias works, too:

```
$ kubectl get pvc
No resources found in default namespace.
```

You'll quickly find that a lot of people working with Kubernetes refer to the `PersistentVolumeClaim` resources simply with pvc. So, don't be surprised if you see the term pvc here and there while working with Kubernetes. That being said, let's explain what `PersistentVolumeClaim` resources are in Kubernetes.

Splitting storage creation and storage consumption

The key to understanding the difference between `PersistentVolume` and `PersistentVolumeClaim` is to understand that one is meant to represent the storage itself, whereas the other one represents the request for storage that a Pod makes to get the actual storage.

The reason is that Kubernetes is typically supposed to be used by two types of people:

- **Kubernetes administrator:** This person is supposed to maintain the cluster, operate it, and also add computation resources and persistent storage.
- **Kubernetes application developer:** This person is supposed to develop and deploy an application, so, put simply, consume the computation resource and storage offered by the administrator.

In fact, there is no problem if you handle both roles in your organization; however, this information is crucial to understand the workflow to mount `PersistentVolume` to Pods.

Kubernetes was built with the idea that a `PersistentVolume` object should belong to the cluster administrator scope, whereas `PersistentVolumeClaim` objects belong to the application developer scope. It is up to the cluster administrator to add `PersistentVolumes` (or dynamic volume operators) since they might be hardware resources, whereas developers have a better understanding of what amount of storage and what kind of storage is needed, and that's why the `PersistentVolumeClaim` object was built.

Essentially, a Pod cannot mount a `PersistentVolume` object directly. It needs to explicitly ask for it. This *asking* action is achieved by creating a `PersistentVolumeClaim` object and attaching it to the Pod that needs a `PersistentVolume` object.

This is the only reason why this additional layer of abstraction exists. Now, let's understand the `PersistentVolume` workflow summarized in the next section.

Understanding the PersistentVolume workflow

Once the developer has built the application, it is their responsibility to ask for a `PersistentVolume` object if needed. To do that, the developer will write two YAML manifests:

- One manifest will be written for the Pod or deployment.
- The other manifest will be written for `PersistentVolumeClaim`.

The Pod must be written so that the `PersistentVolumeClaim` object is mounted as a `volumeMount` configuration key in the YAML file. Please note that for it to work, the `PersistentVolumeClaim` object needs to be in the same namespace as the application Pod that is mounting it. When both YAML files are applied and both resources are created in the cluster, the `PersistentVolumeClaim` object will look for a `PersistentVolume` object that matches the criteria required in the claim. Supposing that a `PersistentVolume` object capable of fulfilling the claim is created and ready in the Kubernetes cluster, the `PersistentVolume` object will be attached to the `PersistentVolumeClaim` object.

If everything is okay, the claim is considered fulfilled, and the volume is correctly mounted to the Pod: if you understand this workflow, essentially, you understand everything related to PersistentVolume usage.

The following diagram illustrates the workflow in static storage provisioning in Kubernetes.

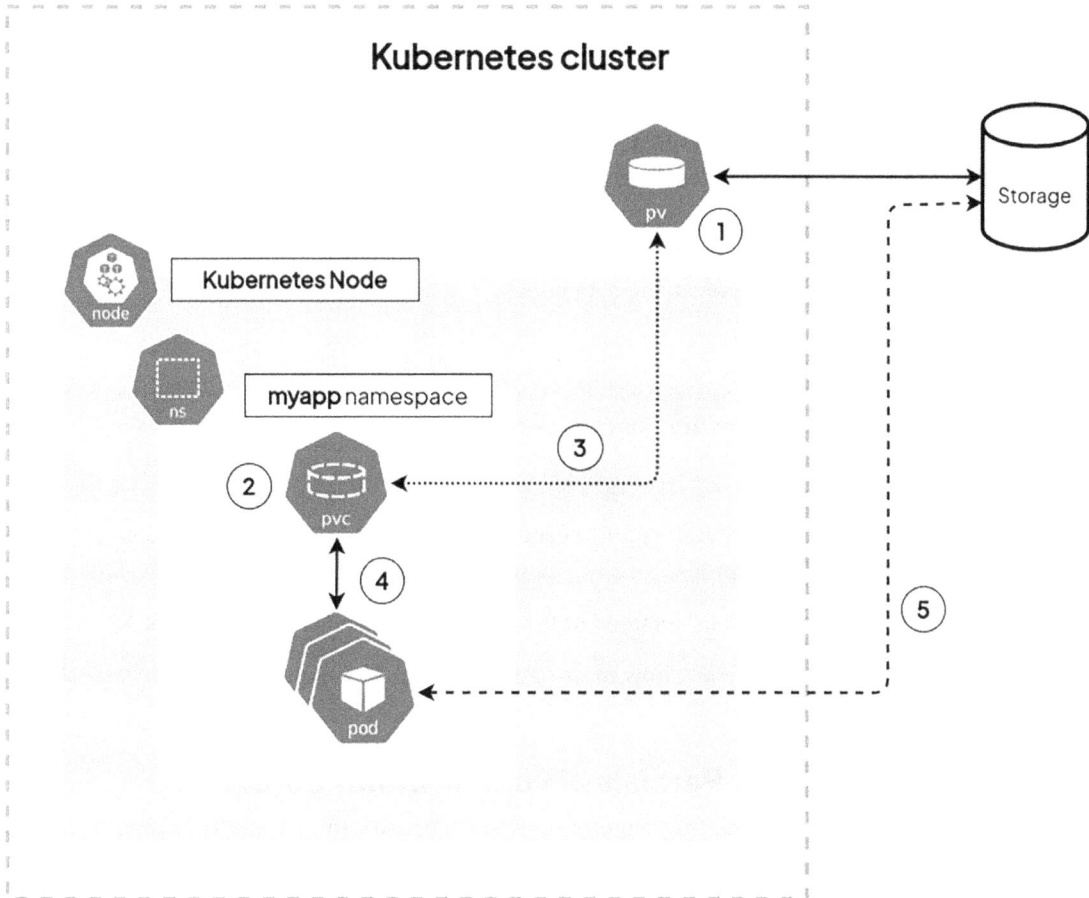

Figure 9.1: Static storage provisioning in Kubernetes

> You will learn about dynamic storage provisioning in a later section of this chapter, *Introducing dynamic provisioning*.

Imagine a developer needs persistent storage for their application running in Kubernetes. Here's the choreography that ensues:

1. The administrator prepares PersistentVolume: The Kubernetes administrator prepares the backend storage and creates a PersistentVolume object. This PV acts like a storage declaration, specifying details like capacity, access mode (read-write, read-only), and the underlying storage system (e.g., hostPath, NFS).

2. The developer makes a claim using `PersistentVolumeClaim`: The developer creates a `PersistentVolumeClaim` object. This PVC acts like a storage request, outlining the developer's needs. It specifies the size, access mode, and any storage class preferences (think of it as a wishlist for storage). The developer also defines a volume mount in the Pod's YAML file, specifying how the Pod should access the persistent storage volume.

3. Kubernetes fulfills the request: After the Pod and PVC are created, Kubernetes searches for a suitable PV that matches the requirements listed in the PVC. It's like a match-making service, ensuring the requested storage aligns with what's available.

4. The Pod leverages the storage using `volumeMount`: Once Kubernetes finds a matching PV, it binds it to the PVC. This makes the storage accessible to the Pod.

5. Data flow begins (**read/write operations**): Now, the Pod can interact with the persistent storage based on the access mode defined in the PV. It can perform read or write operations on the data stored in the volume, ensuring data persistence even if the Pod restarts.

> Please note that `PersistentVolume` is cluster-scoped, while `PersistentVolumeClaim`, Pod, and `volumeMount` are namespace-scoped objects.

This collaboration between PVs, PVCs, and Kubernetes ensures that the developers have access to persistent storage for their applications, enabling them to store and retrieve data across Pod life cycles.

This setup might seem complex to understand at first, but you will quickly get used to it.

In the following section, we will learn how to use the storage in Pods using `PersistentVolume` and `PersistentVolumeClaim`.

Creating a Pod with a PersistentVolumeClaim object

In this section, we will create a Pod that mounts `PersisentVolume` within a `minikube` cluster. This is going to be a kind of `PersisentVolume` object, but this time, it will not be bound to the life cycle of the Pod. Indeed, since it will be managed as a real `PersisentVolume` object, the `hostPath` type will get its life cycle independent of the Pod.

The very first thing to do is create the `PersisentVolume` object that will be a `hostPath` type. Here is the YAML file to do that. Please note that we created this `PersisentVolume` object with some arbitrary labels in the `metadata` section. This is so that it will be easier to fetch it from the `PersistentVolumeClaim` object later:

```
# pv.yaml
apiVersion: v1
kind: PersistentVolume
metadata:
  name: my-hostpath-pv
  labels:
    type: hostpath
```

```
      env: prod
  spec:
    capacity:
      storage: 1Gi
    accessModes:
      - ReadWriteOnce
    hostPath:
      path: "/tmp/test"
    storageClassName: slow
```

Please note the following items in the YAML, which we will use for matching the PVC later:

- labels
- capacity
- accessModes
- StorageClassName

We can now create and list the PersisentVolume entries available in our cluster, and we should observe that this one exists. Please note that the pv alias works, too:

```
$ kubectl apply -f pv.yaml
persistentvolume/my-hostpath-pv created

$ kubectl get pv
NAME              CAPACITY   ACCESS MODES   RECLAIM POLICY   STATUS      CLAIM
STORAGECLASS   VOLUMEATTRIBUTESCLASS   REASON   AGE
my-hostpath-pv    1Gi        RWO            Retain           Available
slow           <unset>                          3s
```

We can see that the PersisentVolume was successfully created, and the status is Available.

Now, we need to create two things to mount the PersisentVolume object:

- A PersistentVolumeClaim object that targets this specific PersisentVolume object
- A Pod to use the PersistentVolumeClaim object

To demonstrate the namespace scoped items and cluster scoped items, let us create a namespace for the PVC and Pod (refer to the pv-ns.yaml file):

```
$ kubectl apply -f pv-ns.yaml
namespace/pv-ns created
```

Let's proceed, in order, with the creation of the PersistentVolumeClaim object:

```
# pvc.yaml
apiVersion: v1
kind: PersistentVolumeClaim
metadata:
```

```
  name: my-hostpath-pvc
  namespace: pv-ns
spec:
  resources:
    requests:
      storage: 1Gi
  accessModes:
    - ReadWriteOnce
  selector:
    matchLabels:
      type: hostpath
      env: prod
  storageClassName: slow
```

Let's create the PVC and check that it was successfully created in the cluster. Please note that the pvc alias also works here:

```
$ kubectl apply -f pvc.yaml
persistentvolumeclaim/my-hostpath-pvc created


$ kubectl get pvc -n pv-ns
NAME                    STATUS    VOLUME          CAPACITY    ACCESS MODES
STORAGECLASS     VOLUMEATTRIBUTESCLASS    AGE
my-hostpath-pvc    Bound     my-hostpath-pv    1Gi         RWO              slow
<unset>                          2m29s
```

Please note the PVC status now – Bound – which means the PVC is already matched with a PV and ready to consume the storage.

Now that the `PersisentVolume` object and the `PersistentVolumeClaim` object exist, we can create a Pod that will mount the PV using the PVC.

Let's create an NGINX Pod that will do the job:

```
# pod.yaml
apiVersion: v1
kind: Pod
metadata:
  name: nginx
  namespace: pv-ns
spec:
  containers:
    - image: nginx
      name: nginx
```

```
    volumeMounts:
      - mountPath: "/var/www/html"
        name: mypersistentvolume
  volumes:
    - name: mypersistentvolume
      persistentVolumeClaim:
        claimName: my-hostpath-pvc
```

As you can see, in the volumeMounts section, the PersistentVolumeClaim object is referenced as a volume, and we reference the PVC by its name. Note that the PVC must live in the same namespace as the Pod that mounts it. This is because PVCs are **namespace-scoped** resources, whereas PVs are not. There are no labels and selectors for this one; to bind a PVC to a Pod, you simply need to use the PVC name.

That way, the Pod will become attached to the PersistentVolumeClaim object, which will find the corresponding PersisentVolume object. This, in the end, will make the host path available and mounted on my NGINX Pod.

Create the Pod and test the status:

```
$ kubectl apply -f pod.yaml
pod/nginx created

$ kubectl get pvc,pod -n pv-ns
NAME                                        STATUS    VOLUME           CAPACITY
ACCESS MODES    STORAGECLASS    VOLUMEATTRIBUTESCLASS    AGE
persistentvolumeclaim/my-hostpath-pvc   Bound     my-hostpath-pv   1Gi
RWO             slow            <unset>                  4m32s

NAME          READY    STATUS    RESTARTS    AGE
pod/nginx     1/1      Running   0           13s
```

The Pod is up and running with the hostPath /tmp/test mounted inside via the PV and PVC. So far, we have learned what PersistentVolume and PersistentVolumeClaim objects are and how to use them to mount persistent storage on your Pods.

Next, we must continue our exploration of the PersistentVolume and PersistentVolumeClaim mechanics by explaining the life cycle of these two objects. Because they are independent of the Pods, their life cycles have some dedicated behaviors that you need to be aware of.

Understanding the life cycle of a PersistentVolume object in Kubernetes

PersistentVolume objects are good if you want to maintain the state of your app without being constrained by the life cycle of the Pods or containers that are running them.

However, since `PersistentVolume` objects get their very own life cycle, they have some very specific mechanics that you need to be aware of when you're using them. We'll take a closer look at them next.

Understanding why PersistentVolume objects are not bound to namespaces

The first thing to be aware of when you're using `PersistentVolume` objects is that they are not namespaced resources, but `PersistentVolumeClaim` objects are.

```
$ kubectl api-resources --namespaced=false |grep -i volume
persistentvolumes                    pv          v1
false           PersistentVolume
volumeattachments                                storage.k8s.io/v1
false           VolumeAttachment
```

So, if the Pod wants to use the `PersistentVolume`, then the `PersistentVolumeClaim` must be created in the same namespace as the Pod.

The `PersistentVolume` will have the following life cycle stages typically:

* **Provisioning:** Admin creates the PV, defining capacity, access modes, and optional details like storage class and reclaim policy.
* **Unbound state:** Initially, the PV is available but not attached to any Pod (unbound).
* **Claiming:** The developer creates a PVC, specifying size, access mode, and storage class preference (request for storage).
* **Matching and binding:** Kubernetes finds an unbound PV that matches the PVC requirements and binds them together.
* **Using:** Pod accesses the bound PV through a volume mount defined in its YAML file.
* **Releasing:** When the Pod using the PVC is deleted, the PVC becomes unbound (the PV state depends on the reclaim policy).
* **Deletion:** The PV object itself can be deleted by the administrator, following the configured reclaim policy for the storage resource.

Now, let's examine another important aspect of `PersistentVolume`, known as reclaiming a policy. This is something that is going to be important when you want to unmount a PVC from a running Pod.

Reclaiming a PersistentVolume object

When it comes to `PersistentVolume`, there is a very important option that you need to understand, which is the reclaim policy. But what does this option do?

This option will tell Kubernetes what treatment it should give to your `PersistentVolume` object when you delete the corresponding `PersistentVolumeClaim` object that was attaching it to the Pods.

Indeed, deleting a `PersistentVolumeClaim` object consists of deleting the link between the Pods and your `PersistentVolume` object, so it's like you unmount the volume and then the volume becomes available again for another application to use.

However, in some cases, you don't want that behavior; instead, you want your PersistentVolume object to be automatically removed when its corresponding PersistentVolumeClaim object has been deleted. That's why the reclaim policy option exists, and it is what you should configure.

Let's explain these three reclaim policies:

- **Delete:** This is the simplest of the three. When you set your reclaim policy to delete, the PersistentVolume object will be wiped out and the PersistentVolume entry will be removed from the Kubernetes cluster when the corresponding PersistentVolumeClaim object is deleted. You can use this when you want your data to be deleted and not used by any other application. Bear in mind that this is a permanent option, so you might want to build a backup strategy with your underlying storage provider if you need to recover anything.

> In our example, the PV was created manually with hostPath and the path is /tmp/. The deletion operation will work without any issues here. However, the delete operation may not work for all PV types when you create it manually. It is highly recommended to use dynamic PV provisioning, which you will learn about later in this chapter.

- **Retain:** This is the second policy and is contrary to the delete policy. If you set this reclaim policy, the PersistentVolume object won't be deleted if you delete its corresponding PersistentVolumeClaim object. Instead, the PersistentVolume object will enter the released status, which means it is still available in the cluster, and all of its data can be manually retrieved by the cluster administrator.
- **Recycle:** This is a kind of combination of the previous two policies. First, the volume is wiped of all its data, such as a basic rm -rf volume/* volume. However, the volume itself will remain accessible in the cluster, so you can mount it again on your application.

> The recycle reclaim policy has been deprecated. It is now advised to utilize dynamic provisioning as the preferred approach.

The reclaim policy can be set in your cluster directly in the YAML definition file at the PersistentVolume level.

Updating a reclaim policy

The good news with a reclaim policy is that you can change it after the PersistentVolume object has been created; it is a mutable setting.

To demonstrate the reclaim policy differences, let us use the previously created Pod, PV, and PVC as follows:

```
$ kubectl get pod,pvc -n pv-ns
NAME          READY   STATUS    RESTARTS    AGE
```

```
pod/nginx   1/1      Running   0            30m

NAME                                           STATUS   VOLUME           CAPACITY
ACCESS MODES    STORAGECLASS    VOLUMEATTRIBUTESCLASS    AGE
persistentvolumeclaim/my-hostpath-pvc    Bound    my-hostpath-pv   1Gi
RWO             slow            <unset>                  34m
```

Delete the Pod first as it is using the PVC:

```
$ kubectl delete pod nginx -n pv-ns
pod "nginx" deleted
$ kubectl delete pvc my-hostpath-pvc -n pv-ns
persistentvolumeclaim "my-hostpath-pvc" deleted
```

Now, check the status of the PV:

```
$ kubectl get pv
NAME              CAPACITY    ACCESS MODES    RECLAIM POLICY    STATUS     CLAIM
STORAGECLASS    VOLUMEATTRIBUTESCLASS    REASON    AGE
my-hostpath-pv    1Gi         RWO             Retain            Released   pv-ns/
my-hostpath-pvc   slow                 <unset>                  129m
```

You can see from the output that the PV is in a `Released` state but not yet in an `Available` state for the next PVC to use.

Let us update the reclaim policy to `Delete` using the `kubectl patch` command as follows:

```
$ kubectl patch pv/my-hostpath-pv -p
'{"spec":{"persistentVolumeReclaimPolicy":"Delete"}}'
persistentvolume/my-hostpath-pv patched
Since the PV is not bound to any PVC, the PV will be instantly deleted due to
the Delete reclaim policy:
$ kubectl get pv
No resources found
```

As you can see in the preceding output, we have updated the reclaim policy of the PV and then the PV has been deleted from the Kubernetes cluster.

Now, let's discuss the different statuses that PVs and PVCs can have.

Understanding PersistentVolume and PersistentVolumeClaim statuses

Just like Pods can be in a different state, such as `Pending`, `ContainerCreating`, `Running`, and more, `PersistentVolume` and `PersistentVolumeClaim` can also hold different states. You can identify their state by issuing the `kubectl get pv` and `kubectl get pvc` commands.

PersistentVolume has the following different states that you need to be aware of:

- Available: This is the initial state for a newly created PV. It indicates the PV is ready to be bound to a PVC.
- Bound: This status signifies that the PV is currently claimed by a specific PVC and is in use by a Pod. Essentially, it indicates that the volume is currently in use. When this status is applied to a PersistentVolumeClaim object, this indicates that the PVC is currently in use: that is, a Pod is using it and has access to a PV through it.
- Terminating: The Terminating status applies to a PersistentVolumeClaim object. This is the status the PVC enters after you issue a kubectl delete pvc command.
- Released: If a PVC using the PV is deleted (and the reclaim policy for the PV is set to "Retain"), the PV will transition to this state. It's essentially unbound but still available for future PVCs to claim.
- Failed: This status indicates an issue with the PV, preventing it from being used. Reasons could be storage provider errors, access issues, or problems with the provisioner (if applicable).
- Unknown: In rare cases, the PV status might be unknown due to communication failures with the underlying storage system.

We now have all the basics relating to PersistentVolume and PersistentVolumeClaim, which should be enough to start using persistent storage in Kubernetes. However, there's still something important to know about this topic, and it is called dynamic provisioning. This is a very impressive aspect of Kubernetes that makes it able to communicate with cloud provider APIs to create persistent storage on the cloud. Additionally, it can make this storage available on the cluster by dynamically creating PV objects. In the next section, we will compare static and dynamic provisioning.

Understanding static and dynamic PersistentVolume provisioning

So far, we've only provisioned PersistentVolume by doing static provisioning. Now, we're going to discover dynamic PersistentVolume provisioning, which enables PersistentVolume provisioning directly from the Kubernetes cluster.

Static versus dynamic provisioning

So far, when using static provisioning, you have learned that you must follow this workflow:

1. You create the piece of storage against the cloud provider or the backend technology.
2. Then, you create the PersistentVolume object to serve as a Kubernetes pointer to this actual storage.
3. Following this, you create a Pod and a PVC to bind the PV to the Pod.

That is called static provisioning. It is static because you have to create the piece of storage before creating the PV and the PVC in Kubernetes. It works well; however, at scale, it can become more and more difficult to manage, especially if you are managing hundreds of PVs and PVCs. Let's say you are creating an Amazon EBS volume to mount it as a PersistentVolume object, and you would do it like this with static provisioning:

1. Authenticate against the AWS console.

2. Create an EBS volume.

3. Copy/paste its unique ID to a `PersistentVolume` YAML definition file.

4. Create the PV using your YAML file.

5. Create a PVC to fetch this PV.

6. Mount the PVC to the Pod object.

Again, it should work in a manual or automated way, but it would become complex and extremely time-consuming to do at scale, with possibly dozens and dozens of PVs and PVCs.

That's why Kubernetes developers decided that it would be better if Kubernetes was capable of provisioning the piece of actual storage on your behalf along with the `PersistentVolume` object to serve as a pointer to it. This is known as dynamic provisioning.

Introducing dynamic provisioning

When using dynamic provisioning, you configure your Kubernetes cluster so that it authenticates against the backend storage provider (such as AWS, Azure, or other storage devices). Then, you issue a command to provision a storage disk or volume and automatically create a `PersistentVolume` so that the PVC can use it. That way, you can save a huge amount of time by getting things automated. Dynamic provisioning is so useful because Kubernetes supports a wide range of storage technologies. We already introduced a few of them earlier in this chapter, when we mentioned NFS and other types of storage.

But how does Kubernetes achieve this versatility? Well, the answer is that it makes use of a third resource type, the `StorageClass` object, which we're going to learn about in this chapter.

Introduction to CSI

Before we talk about `StorageClass`, let us learn about CSI, which acts as a bridge between Kubernetes and various storage solutions. It defines a standard interface for exposing the storage to container workloads. CSI provides an abstraction layer to interact with Kubernetes primitives like `PersistentVolume`, enabling the integration of diverse storage solutions into Kubernetes, while maintaining a vendor-neutral approach.

Kubernetes dynamic storage provisioning involves the following steps typically:

1. Install and configure `StorageClass` and provisioner: The administrator installs a CSI driver (or in-tree provisioner) and configures a `StorageClass`, which defines the storage type, parameters, and reclaim policy.

2. Developer creates PVC with `StorageClass` information: The developer creates a `PersistentVolumeClaim`, specifying the desired size and access mode and referencing the `StorageClass` to request dynamic provisioning.

3. The `StorageClass`/CSI driver triggers a request to the backend provisioner: Kubernetes automatically triggers the CSI driver (or provisioner) when it detects the PVC, sending the request to provision storage from the backend storage system.

4. Provisioner communicates with backend storage and creates the volume: The provisioner communicates with the backend storage system, creates the volume, and generates a `PersistentVolume` in Kubernetes that binds to the PVC.

5. The PVC is mounted to the Pod, allowing storage access: The PVC is mounted to the requesting Pod, allowing the Pod to access the storage as specified by the `volumeMount` in the Pod's configuration.

The following diagram illustrates the dynamic PV provisioning workflow.

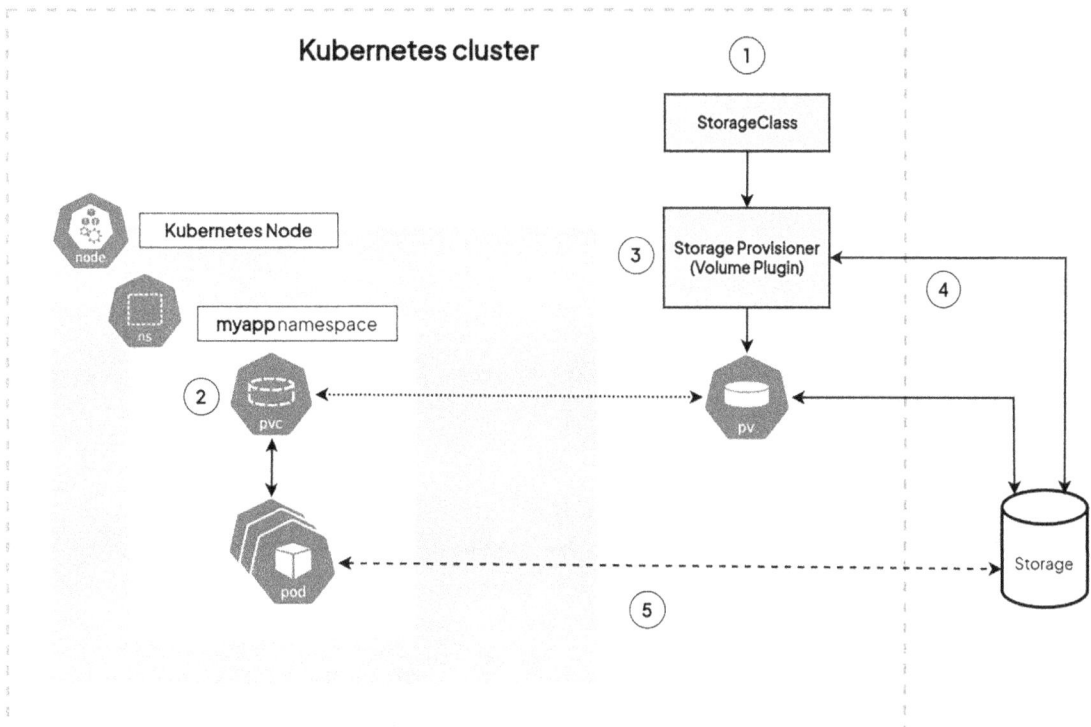

Figure 9.2: Dynamic PV provisioning in Kubernetes

CSI drivers are containerized implementations by storage vendors that adhere to the CSI specification and provide functionalities for provisioning, attaching, detaching, and managing storage volumes.

CSI node and controller services are Kubernetes services that run the CSI driver logic on worker nodes and a control plane respectively, facilitating communication between Pods and the storage system.

Once a CSI-compatible volume driver is deployed on a Kubernetes cluster, users can leverage the `csi` volume type. (Refer to the documentation at `https://kubernetes-csi.github.io/docs/drivers.html` to see the set of CSI drivers that can be used with Kubernetes). This allows them to attach or mount volumes exposed by the CSI driver. There are three ways to utilize a `csi` volume within a Pod:

- Referencing a `PersistentVolumeClaim`: This approach links the Pod to persistent storage managed by Kubernetes.

- Utilizing a generic ephemeral volume: This method provides temporary storage that doesn't persist across Pod restarts.
- Leveraging a CSI ephemeral volume (if supported by the driver): This offers driver-specific ephemeral storage options beyond the generic version.

Remember, you don't directly interact with CSI. StorageClasses can reference CSI drivers by name in the provisioner field, leveraging CSI for volume provisioning.

Introducing StorageClasses

StorageClass is another resource type exposed by kube-apiserver. You might already have noticed this field earlier in the kubectl get pv command output. This resource type is the one that grants Kubernetes the ability to deal with several underlying technologies transparently.

StorageClasses act as a user-facing interface for defining storage requirements. **CSI drivers**, referenced by StorageClasses, provide the actual implementation details for provisioning and managing storage based on the specific storage system. StorageClasses essentially bridge the gap between your storage needs and the capabilities exposed by CSI drivers.

You can access and list the storageclasses resources created within your Kubernetes cluster by using kubectl. Here is the command to list the storage classes:

```
$ kubectl get sc
NAME                   PROVISIONER                RECLAIMPOLICY
VOLUMEBINDINGMODE      ALLOWVOLUMEEXPANSION    AGE
standard (default)     k8s.io/minikube-hostpath
```

We can also check the details about the StorageClass using the -o yaml option:

```
$ kubectl get storageclasses standard  -o yaml
apiVersion: storage.k8s.io/v1
kind: StorageClass
...<removed for brevity>...
  name: standard
  resourceVersion: "290"
  uid: f41b765f-301f-4781-b9d0-46aec694336b
provisioner: k8s.io/minikube-hostpath
reclaimPolicy: Delete
volumeBindingMode: Immediate
```

Additionally, you can use the plural form of storageclasses along with the sc alias. The following three commands are essentially the same:

```
$ kubectl get storageclass
$ kubectl get storageclasses
$ kubectl get sc
```

Note that we haven't included the output of the command for simplicity, but it is essentially the same for the three commands. There are two fields within the command output that are important to us:

- NAME: This is the name and the unique identifier of the storageclass object.
- PROVISIONER: This is the name of the underlying storage technology: this is basically a piece of code the Kubernetes cluster uses to interact with the underlying technology.

> Note that you can create multiple StorageClass objects that use the same provisioner.

As we are currently using a minikube cluster in our lab environment, we have a storageclass resource called standard that is using the k8s.io/minikube-hostpath provisioner.

This provider deals with my host filesystem to automatically create provisioned host path volumes for my Pods, but it could be the same for Amazon EBS volumes or Google PDs.

In GKE, Google built a storage class with a provisioner that was capable of interacting with the Google PD's API, which is a pure Google Cloud feature, and you can implement it with StorageClass as follows:

```
# gce-pd-sc.yaml
apiVersion: storage.k8s.io/v1
kind: StorageClass
metadata:
  name: slow
provisioner: kubernetes.io/gce-pd
parameters:
  type: pd-standard
```

In contrast, in AWS, we have a storageclass object with a provisioner that is capable of dealing with EBS volume APIs. These provisioners are just libraries that interact with the APIs of these different cloud providers.

The storageclass objects are the reason why Kubernetes can deal with so many different storage technologies. From a Pod perspective, no matter if it is an EBS volume, NFS drive, or GKE volume, the Pod will only see a PersistentVolume object. All the underlying logic dealing with the actual storage technology is implemented by the provisioner the storageclass object uses.

The good news is that you can add as many storageclass objects with their provisioner as you want to your Kubernetes cluster in a plugin-like fashion.

By the way, nothing is preventing you from expanding your cluster by adding storageclasses to your cluster. You'll simply add the ability to deal with different storage technologies from your cluster. For example, we can add an Amazon EBS storageclass object to our minikube cluster. However, while it is possible, it's going to be completely useless. Indeed, if your minikube setup is not running on an EC2 instance but on your local machine, it won't be able to attach an EBS.

That said, for a more practical approach, you can consider using CSI drivers from providers that support local deployment, such as OpenEBS, TopoLVM, or Portworx. These allow you to work with persistent storage locally, even on minikube. Additionally, most cloud providers offer free tiers for small Kubernetes deployments, which could be useful for testing out storage solutions in a cloud environment without incurring significant costs.

In the next section, we will learn about the difference in dynamic storage provisioning with PVC.

Understanding the role of PersistentVolumeClaim for dynamic storage provisioning

When using dynamic storage provisioning, the PersistentVolumeClaim object will get an entirely new role. Since PersistentVolume is gone in this use case, the only object that will be left for you to manage is the PersistentVolumeClaim one because the PersistentVolume object will be managed by the StorageClass.

Let's demonstrate this by creating an NGINX Pod that will mount a hostPath type dynamically. In this example, the administrator won't have to provision a PersistentVolume object at all. This is because the PersistentVolumeClaim object and the StorageClass object will be able to create and provision the PersistentVolume together.

Let's start by creating a new namespace called dynamicstorage, where we will run our examples:

```
$ kubectl create ns dynamicstorage
namespace/dynamicstorage created
```

Now, let's run a kubectl get sc command to check that we have a storage class that is capable of dealing with the hostPath that is provisioned in our cluster.

For this specific storageclass object in this specific Kubernetes setup (minikube), we don't have to do anything to get the storageclass object, as it is created by default at cluster installation. However, this might not be the case depending on your Kubernetes distribution.

Bear this in mind because it is very important: clusters that have been set up on GKE might have default storage classes that are capable of dealing with Google's storage offerings, whereas an AWS-based cluster might have storageclass to communicate with Amazon's storage offerings and more. With minikube, we have at least one default storageclass object that is capable of dealing with a hostPath-based PersistentVolume object. If you understand that, you should understand that the output of the kubectl get sc command will be different depending on where your cluster has been set up:

```
$ kubectl get sc
NAME                    PROVISIONER                 RECLAIMPOLICY
VOLUMEBINDINGMODE       ALLOWVOLUMEEXPANSION    AGE
standard (default)      k8s.io/minikube-hostpath    Delete          Immediate
false                   21d
```

As you can see, we do have a storage class called standard on our cluster that is capable of dealing with hostPath.

Some complex clusters spanning across multiple clouds and or on-premises might be provisioned with a lot of different `storageclass` objects to be able to communicate with a lot of different storage technologies. Bear in mind that Kubernetes is not tied to any cloud provider and, therefore, does not force or limit you in your usage of backing storage solutions.

Now, we will create a `PersistentVolumeClaim` object that will dynamically create a `hostPath` type. Here is the YAML file to create the PVC. Please note that `storageClassName` is set to `standard`:

```yaml
# pvc-dynamic.yaml
apiVersion: v1
kind: PersistentVolumeClaim
metadata:
  name: my-dynamic-hostpath-pvc
spec:
  accessModes:
    - ReadWriteOnce
  storageClassName: standard # VERY IMPORTANT !
  resources:
    requests:
      storage: 1Gi
  selector:
    matchLabels:
      type: hostpath
      env: prod
```

Following this, we can create it in the proper namespace:

```
$ kubectl apply -f pvc-dynamic.yaml -n dynamicstorage
persistentvolumeclaim/my-dynamic-hostpath-pvc created
```

Let us check the status of the PV and PVC now:

```
$ kubectl get pod,pvc,pv -n dynamicstorage
NAME                                                STATUS   VOLUME
CAPACITY   ACCESS MODES   STORAGECLASS   VOLUMEATTRIBUTESCLASS    AGE
persistentvolumeclaim/my-dynamic-hostpath-pvc       Bound    pvc-4597ab27-c894-
40de-a7ac-1b6ca961bcdc   1Gi          RWO                  standard        <unset>
7s

NAME                                                                  CAPACITY
ACCESS MODES   RECLAIM POLICY   STATUS   CLAIM
STORAGECLASS   VOLUMEATTRIBUTESCLASS    REASON   AGE
```

```
persistentvolume/pvc-4597ab27-c894-40de-a7ac-1b6ca961bcdc     1Gi          RWO
Delete              Bound      dynamicstorage/my-dynamic-hostpath-pvc    standard
<unset>                                  7s
```

We can see that the PV has been created by the `StorageClass` and bound to the PVC as per the request.

Now that this PVC has been created, we can add a new Pod that will mount this `PersistentVolumeClaim` object. Here is a YAML definition file of a Pod that will mount the `PersistentVolumeClaim` object that was created earlier:

```yaml
# pod-with-dynamic-pvc.yaml
apiVersion: v1
kind: Pod
metadata:
  name: nginx-dynamic-storage
spec:
  containers:
    - image: nginx
      name: nginx
      volumeMounts:
        - mountPath: "/var/www/html"
          name: mypersistentvolume
  volumes:
    - name: mypersistentvolume
      persistentVolumeClaim:
        claimName: my-dynamic-hostpath-pvc
```

Now, let's create it in the correct namespace:

```
$ kubectl apply -f pod-with-dynamic-pvc.yaml -n dynamicstorage
pod/nginx-dynamic-storage created


$ kubectl get pod,pvc,pv -n dynamicstorage
NAME                          READY   STATUS    RESTARTS   AGE
pod/nginx-dynamic-storage     1/1     Running   0          45s

NAME                                               STATUS   VOLUME
CAPACITY    ACCESS MODES   STORAGECLASS   VOLUMEATTRIBUTESCLASS     AGE
persistentvolumeclaim/my-dynamic-hostpath-pvc      Bound    pvc-4597ab27-c894-
40de-a7ac-1b6ca961bcdc    1Gi        RWO              standard         <unset>
7m39s


NAME                                                          CAPACITY
ACCESS MODES   RECLAIM POLICY   STATUS   CLAIM
```

```
STORAGECLASS     VOLUMEATTRIBUTESCLASS     REASON     AGE
persistentvolume/pvc-4597ab27-c894-40de-a7ac-1b6ca961bcdc     1Gi          RWO
Delete           Bound     dynamicstorage/my-dynamic-hostpath-pvc     standard
<unset>                           7m39s
```

Everything is OK! We're finally done with dynamic provisioning! Please note that, by default, the re-claim policy will be set to `delete` so that the PV is removed when the PVC that created it is removed, too. Don't hesitate to change the reclaim policy if you need to retain sensitive data.

You can test it by deleting the Pod and PVC; the PV will be removed automatically by the `StorageClass`:

```
$ kubectl delete po nginx-dynamic-storage -n dynamicstorage
pod "nginx-dynamic-storage" deleted

$ kubectl delete pvc my-dynamic-hostpath-pvc -n dynamicstorage
persistentvolumeclaim "my-dynamic-hostpath-pvc" deleted

$ kubectl get pod,pvc,pv -n dynamicstorage
No resources found
```

We can see from the above snippet that the PV is also deleted automatically when the PVC gets deleted.

We've covered the basics of PVs, PVCs, `StorageClasses`, and the differences between static and dy-namic provisioning. In the next section, we'll dive into some advanced storage topics in Kubernetes, exploring how to optimize and extend your storage strategies.

Advanced storage topics

In addition to understanding the basics of PVs, PVCs, and `StorageClasses`, it's beneficial to delve into some advanced storage topics in Kubernetes. While not mandatory, having knowledge of these con-cepts can significantly enhance your expertise as a Kubernetes practitioner. In the following sections, we will introduce advanced topics such as ephemeral volumes for temporary storage, CSI Volume Cloning for flexible volume management, and expanding `PersistentVolumeClaims` to accommodate increased storage needs. These topics will provide you with a broader perspective on Kubernetes storage capabilities and practical applications.

Ephemeral volumes for temporary storage in Kubernetes

Ephemeral volumes offer a convenient way to provide temporary storage for Pods in Kubernetes. They're perfect for applications that need scratch space for caching or require read-only data, like configuration files or Secrets. Unlike PVs, ephemeral volumes are automatically deleted when the Pod terminates, simplifying deployment and management.

Here are a few key benefits of ephemeral volumes for temporary storage:

- Temporary storage for Pods
- Automatic deletion with Pod termination
- Simplified deployment and management

There are multiple types of ephemeral storage available in Kubernetes, as follows:

- emptyDir: This creates an empty directory on the node's local storage
- ConfigMap, downwardAPI, Secret: This injects data from Kubernetes objects into the Pod
- CSI ephemeral volumes: These are provided by external CSI drivers (requires specific driver support)
- Generic ephemeral volumes: These are offered by storage drivers supporting PVs

Now that we have learned some details with regard to ephemeral volumes, let's move on to gain some understanding of CSI volume cloning and volume snapshots.

CSI volume cloning and volume snapshots

CSI introduces a powerful feature: volume cloning. This functionality allows you to create an exact copy of an existing `PersistentVolumeClaim` as a new PVC.

The following YAML snippet illustrates a typical PVC cloning declaration:

```yaml
# pv-cloning.yaml
apiVersion: v1
kind: PersistentVolumeClaim
metadata:
  name: cloned-pvc
  namespace: mynamespace
spec:
  accessModes:
    - ReadWriteOnce
  storageClassName: custom-storage-class
  resources:
    requests:
      storage: 5Gi
  dataSource:
    kind: PersistentVolumeClaim
    name: original-pvc
```

Here are a few key benefits of CSI Volume Cloning:

- **Simplified workflows:** CSI Volume Cloning automates data replication, eliminating the need for manual copying and streamlining storage management.
- **Enhanced efficiency:** Easily create replicas of existing volumes, optimizing deployments and resource utilization.
- **Troubleshooting live data:** Instead of touching the production data, you can take a copy and use it for QA, troubleshooting, etc.

Refer to the documentation to learn more about CSI volume cloning: `https://kubernetes.io/docs/concepts/storage/volume-pvc-datasource`.

Like volume cloning, Kubernetes also offers another mechanism via CSI drivers to take data backups called volume snapshots. `VolumeSnapshot` provides a standardized way to create a point-in-time copy of a volume's data. Similar to `PersistentVolume` and `PersistentVolumeClaim` resources, Kubernetes uses VolumeSnapshot, `VolumeSnapshotContent`, and `VolumeSnapshotClass` resources to manage volume snapshots. VolumeSnapshots are user requests for snapshots, while VolumeSnapshotContent represents the actual snapshots on the storage system. These resources enable users to capture the state of their volumes without provisioning an entirely new volume, making it useful for scenarios such as database backups before performing critical updates or deletions. Unlike regular PVs, these snapshot resources are **Custom Resource Definitions (CRDs)** and require a CSI driver that supports snapshot functionality. The CSI driver uses a sidecar container called `csi-snapshotter` to handle `CreateSnapshot` and `DeleteSnapshot` operations.

When a user creates a snapshot, it can be either pre-provisioned by an administrator or dynamically provisioned from an existing PVC. The snapshot controller binds the VolumeSnapshot and VolumeSnapshotContent in both scenarios, ensuring that the snapshot content matches the user request. Snapshots can be easily deleted or retained based on the set `DeletionPolicy`, allowing flexibility in how data is managed. Furthermore, Kubernetes provides the option to convert a snapshot's volume mode (e.g., from filesystem to block) and restore data from a snapshot to a new PVC. This capability makes VolumeSnapshot a powerful tool in data protection, which can be complemented by CSI volume cloning to create efficient backups or test environments, adding another layer of flexibility to storage management in Kubernetes.

> Volume cloning in Kubernetes is ideal for creating identical copies of `PersistentVolumes`, often used for development and testing environments. Snapshots, on the other hand, capture the point-in-time state of a volume, making them useful for backup and restore purposes.
>
> Refer to the documentation (`https://kubernetes.io/docs/concepts/storage/volume-snapshots/`) to learn more about volume snapshots.

In the following section, we will learn how to expand a PVC.

Learning how to expand PersistentVolumeClaim

Kubernetes offers built-in support for expanding PVCs, allowing you to seamlessly increase storage capacity for your applications. This functionality is currently limited to volumes provisioned by CSI drivers (as of version 1.29, other volume types are deprecated).

To enable PVC expansion for a specific `StorageClass`, you need to set the `allowVolumeExpansion` field to `true` within the `StorageClass` definition. This flag controls whether PVCs referencing this `StorageClass` can request more storage space:

Example StorageClass Configuration:

```
# storageclass-expandable.yaml
apiVersion: storage.k8s.io/v1
```

```
kind: StorageClass
metadata:
  name: expadable-sc
provisioner: vendor-name.example/magicstorage
...<removed for brevity>...
allowVolumeExpansion: true
```

When your application requires additional storage, simply edit the PVC object and specify a larger size in the `resources.requests.storage` field. Kubernetes will then initiate the expansion process, resizing the underlying volume managed by the CSI driver. This eliminates the need to create a new volume and migrate data, streamlining storage management.

Refer to the documentation (`https://kubernetes.io/docs/concepts/storage/persistent-volumes/#expanding-persistent-volumes-claims`) to learn more.

Summary

We have arrived at the end of this chapter, which taught you how to manage persistent storage on Kubernetes. You discovered that `PersistentVolume` is a resource type that acts as a point to an underlying resource technology, such as `hostPath` and NFS, along with cloud-based solutions such as Amazon EBS and Google PDs.

Additionally, you discovered the relationship between `PersistentVolume`, `PersistentVolumeClaim`, and `storageClass`. You learned that `PersistentVolume` can hold different reclaim policies, which makes it possible to remove, recycle, or retain them when their corresponding `PersistentVolumeClaim` object gets removed.

Finally, we discovered what dynamic provisioning is and how it can help us. Bear in mind that you need to be aware of this feature because if you create and retain too many volumes, it can have a negative impact on your cloud bill at the end of the month, even though you can restrict storage usage using resource quotas for the namespaces.

We're now done with the basics of Kubernetes, and this chapter is also the end of this section. In the next section, you're going to discover Kubernetes controllers, which are objects designed to automate certain tasks in Kubernetes, such as maintaining a number of replicas of your Pods, either using the Deployment resource type or the StatefulSet resource type. There are still a lot of things to learn!

Further reading

- Persistent Volumes: `https://kubernetes.io/docs/concepts/storage/persistent-volumes/`
- Types of Persistent Volumes: `https://kubernetes.io/docs/concepts/storage/persistent-volumes/#types-of-persistent-volumes`
- Dynamic Volume Provisioning: `https://kubernetes.io/docs/concepts/storage/dynamic-provisioning/`

- Expanding Persistent Volumes Claims: `https://kubernetes.io/docs/concepts/storage/persistent-volumes/#expanding-persistent-volumes-claims`

- CSI Volume Cloning: `https://kubernetes.io/docs/concepts/storage/volume-pvc-datasource/`

Join our community on Discord

Join our community's Discord space for discussions with the authors and other readers:

`https://packt.link/cloudanddevops`

10

Running Production-Grade Kubernetes Workloads

In the previous chapters, we focused on containerization concepts and the fundamental Kubernetes building blocks, such as Pods, Jobs, and ConfigMaps. Our journey so far has covered mostly single-machine scenarios, where the application requires only one container host or Kubernetes node. For **production-grade** Kubernetes, you have to consider different aspects, such as **scalability**, **high availability** (HA), and **load balancing**, and this always requires **orchestrating** containers running on multiple hosts.

Briefly, **container orchestration** is a way of managing multiple containers' life cycles in large, dynamic environments—this can include deploying and maintaining the desired states for container networks, providing redundancy and HA of containers (using external components), scaling up and down the cluster and container replicas, automated health checks, and telemetry (log and metrics) gathering. Solving the problem of efficient container orchestration at cloud scale is not straightforward—this is why Kubernetes exists!

In this chapter, we will cover the following topics:

- Ensuring High Availability and Fault Tolerance on Kubernetes
- What is ReplicationController?
- What is ReplicaSet?

Technical requirements

For this chapter, you will need the following:

- A Kubernetes cluster deployed. You can use either a local or a cloud-based cluster, but in order to fully understand the concepts, we recommend using a multi-node, cloud-based Kubernetes cluster.
- The Kubernetes **command-line interface** (CLI) (kubectl) installed on your local machine and configured to manage your Kubernetes cluster.

Kubernetes cluster deployment (local and cloud-based) and `kubectl` installation were covered in *Chapter 3, Installing Your First Kubernetes Cluster*.

You can download the latest code samples for this chapter from the official GitHub repository at `https://github.com/PacktPublishing/The-Kubernetes-Bible-Second-Edition/tree/main/Chapter10`.

Ensuring High Availability and Fault Tolerance on Kubernetes

First, let's quickly recap how we define HA and **fault tolerance (FT)** and how they differ. These are key concepts in cloud applications that describe the ability of a system or a solution to be continuously operational for a desirably long length of time. From a system end user perspective, the aspect of availability, alongside data consistency, is usually the most important requirement.

High availability

In short, the term *availability* in systems engineering describes the percentage of time when the system is fully functional and operational for the end user. In other words, it is a measure of system uptime divided by the sum of uptime and downtime (which is basically the total time). For example, if, in the last 30 days (720 hours), your cloud application had 1 hour of unplanned maintenance time and was not available to the end user, it means that the availability measure of your application is $\frac{719h}{720h}$ = 99.861%. Usually, to simplify this notation when designing systems, the availability will be expressed in so-called nines: for example, if we say that a system has an availability of five nines, it means it is available at least 99.999% of the total time. To put this into perspective, such a system can have up to only 26 seconds per month of downtime! These measures are often the base indicators for defining **service-level agreements (SLAs)** for billed cloud services.

The definition of HA, based on that, is relatively straightforward, although not precise—a system is highly available if it is operational (available) without interruption for long periods of time. Usually, we can say that five nines of availability is considered the gold standard of HA.

Achieving HA in your system usually involves one or a combination of the following techniques:

- **Eliminating single points of failure (SPOFs) in the system**. This is usually achieved by components' redundancy.
- **Failover setup**, which is a mechanism that can automatically switch from the currently active (possibly unhealthy) component to a redundant one.
- **Load balancing**, which means managing traffic coming into the system and routing it to redundant components that can serve the traffic. This will, in most cases, involve proper failover setup, component monitoring, and telemetry.

Let's introduce the related concept of FT, which is also important in distributed systems such as applications running on Kubernetes.

Fault tolerance

Now, FT can be presented as a complement to the HA concept: a system is fault-tolerant if it can continue to be functional and operating in the event of the failure of one or more of its components. For example, FT mechanisms like RAID for data storage, which distributes data across multiple disks, or load balancers that redirect traffic to healthy nodes, are commonly used to ensure system resilience and minimize disruptions. Achieving full FT means achieving 100% HA, which, in many cases, requires complex solutions actively detecting faults and remediating the issues in the components without interruptions. Depending on the implementation, the fault may result in a graceful degradation of performance that is proportional to the severity of the fault. This means that a small fault in the system will have a small impact on the overall performance of the system while serving requests from the end user.

HA and FT for Kubernetes applications

In previous chapters, you learned about Pods and how Services expose them to external traffic (*Chapter 8*, *Exposing Your Pods with Services*). Services are Kubernetes objects that provide a stable network address for a set of healthy Pods. Internally, inside the Kubernetes cluster, the Service makes Pods addressable using virtual IP addresses managed by the kube-proxy component on each node. Externally, cloud environments typically use a cloud load balancer to expose the Service. This load balancer integrates with the Kubernetes cluster through a cloud-specific plugin within the cloud-controller-manager component. With an external load balancer in place, microservices or workloads running on Kubernetes can achieve load balancing across healthy Pods on the same or different nodes, which is a crucial building block for HA.

Services are required for load balancing requests to Pods, but we haven't yet covered how to maintain multiple replicas of the same Pod object definition that are possibly redundant and allocated on different nodes. Kubernetes offers multiple building blocks to achieve this goal, outlined as follows:

- A **ReplicationController object**—the original form of defining Pod replication in Kubernetes.
- A **ReplicaSet object**—the successor to ReplicationController. The main difference is that ReplicaSet has support for set-based requirement selectors for Pods.

> The preferred way to manage ReplicaSets is through a Deployment object, which simplifies updates and rollbacks.

- A **Deployment object**—another level of abstraction on top of ReplicaSet. This provides *declarative* updates for Pods and ReplicaSets, including rollouts and rollbacks. It is used for managing *stateless* microservices and workloads.
- A **StatefulSet object**—similar to Deployment but used to manage *stateful* microservices and workloads in the cluster. Managing the state inside a cluster is usually the toughest challenge to solve in distributed systems design.

- **A DaemonSet object**—used for running a singleton copy of a Pod on all (or some) of the nodes in the cluster. These objects are usually used for managing internal Services for log aggregation or node monitoring.

In the next sections, we will cover the basics of ReplicationController and ReplicaSets. The more advanced objects, such as Deployment, StatefulSet, and DaemonSet, will be covered in the next chapters.

> This chapter covers HA and FT for Kubernetes workloads and applications. If you are interested in how to ensure HA and FT for Kubernetes itself, please refer to the official documentation at `https://kubernetes.io/docs/setup/production-environment/tools/kubeadm/high-availability/`. Please note that in managed Kubernetes offerings in the cloud, such as **Azure Kubernetes Service (AKS)**, Amazon **Elastic Kubernetes Service (EKS)**, or **Google Kubernetes Engine (GKE)**, you are provided with highly available clusters, and you do not need to manage the master nodes yourself.

What is ReplicationController?

Achieving HA and FT requires providing redundancy of components and proper load balancing of incoming traffic between the replicas of components. Let's take a look at the first Kubernetes object that allows you to create and maintain multiple replicas of the Pods in your cluster: ReplicationController. Please note that we are discussing ReplicationController mainly for historical reasons as it was the initial way of creating multiple Pod replicas in Kubernetes. We advise you to use ReplicaSet whenever possible, which is basically the next generation of ReplicationController with an extended specification API.

> The Controller objects in Kubernetes have one main goal: to observe the current and the desired cluster state that is exposed by the Kubernetes API server and command changes that attempt to change the current state to the desired one. They serve as continuous feedback loops, doing all they can to bring clusters to the desired state described by your object templates.

> ReplicationController has a straightforward task—it needs to ensure that a specified number of Pod replicas (defined by a template) are running and healthy in a cluster at any time. This means that if ReplicationController is configured to maintain three replicas of a given Pod, it will try to keep exactly three Pods by creating and terminating Pods when needed. For example, right after you create a ReplicationController object, it will create three new Pods based on its template definition. If, for some reason, there are four such Pods in the cluster, ReplicationController will terminate one Pod, and if by any chance a Pod gets deleted or becomes unhealthy, it will be replaced by a new, hopefully healthy, one.

Since a Deployment, which configures a ReplicaSet, is now the recommended method for managing replication, we will not cover ReplicationController here. In the next section, we will focus on understanding and practicing the ReplicaSet concept. A detailed exploration of Deployments will follow in *Chapter 11, Using Kubernetes Deployments for Stateless Workloads*.

What is ReplicaSet?

Let's introduce another Kubernetes object: ReplicaSet. This is very closely related to ReplicationController, which we have just discussed. In fact, this is a **successor** to ReplicationController, which has a very similar specification API and capabilities. The purpose of ReplicaSet is also the same—it aims to maintain a fixed number of healthy, identical Pods (replicas) that fulfill certain conditions. So, again, you just specify a template for your Pod, along with appropriate label selectors and the desired number of replicas, and Kubernetes ReplicaSetController (this is the actual name of the controller responsible for maintaining ReplicaSet objects) will carry out the necessary actions to keep the Pods running.

Before we learn more about ReplicaSet, let us learn the major differences between ReplicationController and ReplicaSet in the next section.

How does ReplicaSet differ from ReplicationController?

The differences between ReplicaSet and ReplicationController are summarized in the following table:

Feature	ReplicaSet	ReplicationController
Label selectors	Supports set-based selectors (e.g., inclusion/exclusion of labels). Allows more complex logic, such as including `environment=test` or `environment=dev`, while excluding `environment=prod`.	Only supports equality-based selectors (e.g., `key=value`). No advanced label matching.
Integration with other Kubernetes objects	Acts as a foundation for more advanced objects like **Deployment** and **HorizontalPodAutoscaler** (**HPA**).	Primarily manages Pod replication directly, without such integrations.
Pod update rollout	Managed declaratively through Deployment objects, allowing for **staged rollouts** and **rollbacks.**	Managed manually with the now-deprecated `kubectl rolling-update` imperative command.
Future support	A more modern and flexible resource with future-proof features.	Expected to be deprecated in the future.

Table 10.1: Differences between ReplicaSet and ReplicationController

The bottom line—always choose ReplicaSet over ReplicationController. However, you should also remember that using bare ReplicaSets is generally not useful in production clusters, and you should use higher-level abstractions such as Deployment objects for managing ReplicaSets. We will introduce this concept in the next chapter.

In the next section, let us learn about creating and managing ReplicaSet objects.

Creating a ReplicaSet object

For the following demonstration, we are using a multi-node cluster using `kind`, which you have already learned about in *Chapter 3, Installing Your First Kubernetes Cluster*:

```
$ kind create cluster --config Chapter03/kind_cluster --image kindest/
node:v1.31.0

$ kubectl get nodes
NAME                  STATUS   ROLES           AGE   VERSION
kind-control-plane    Ready    control-plane   60s   v1.31.0
kind-worker           Ready    <none>          47s   v1.31.0
kind-worker2          Ready    <none>          47s   v1.31.0
kind-worker3          Ready    <none>          47s   v1.31.0
```

First of all, let us create a namespace to park our ReplicaSet resources.

```
$ kubectl create -f ns-rs.yaml
namespace/rs-ns created
```

Now, let's take a look at the structure of an `nginx-replicaset.yaml` example YAML manifest file that maintains three replicas of an `nginx` Pod, as follows:

```yaml
# nginx-replicaset-example.yaml
apiVersion: apps/v1
kind: ReplicaSet
metadata:
  name: nginx-replicaset-example
  namespace: rs-ns
spec:
  replicas: 4
  selector:
    matchLabels:
      app: nginx
      environment: test
  template:
    metadata:
      labels:
        app: nginx
        environment: test
    spec:
```

```
containers:
  - name: nginx
    image: nginx:1.17
    ports:
      - containerPort: 80
```

There are three main components of the ReplicaSet specification, as follows:

*   `replicas`: Defines the number of Pod replicas that should run using the given `template` and matching label `selector`. Pods may be created or deleted in order to maintain the required number.
*   `selector`: A label selector that defines how to identify Pods that the ReplicaSet object owns or acquires. Again, similar to the case of ReplicationController, please take note that this may have a consequence of existing bare Pods being acquired by ReplicaSet if they match the selector!
*   `template`: Defines a template for Pod creation. Labels used in `metadata` must match the `selector` label query.

These concepts have been visualized in the following diagram:

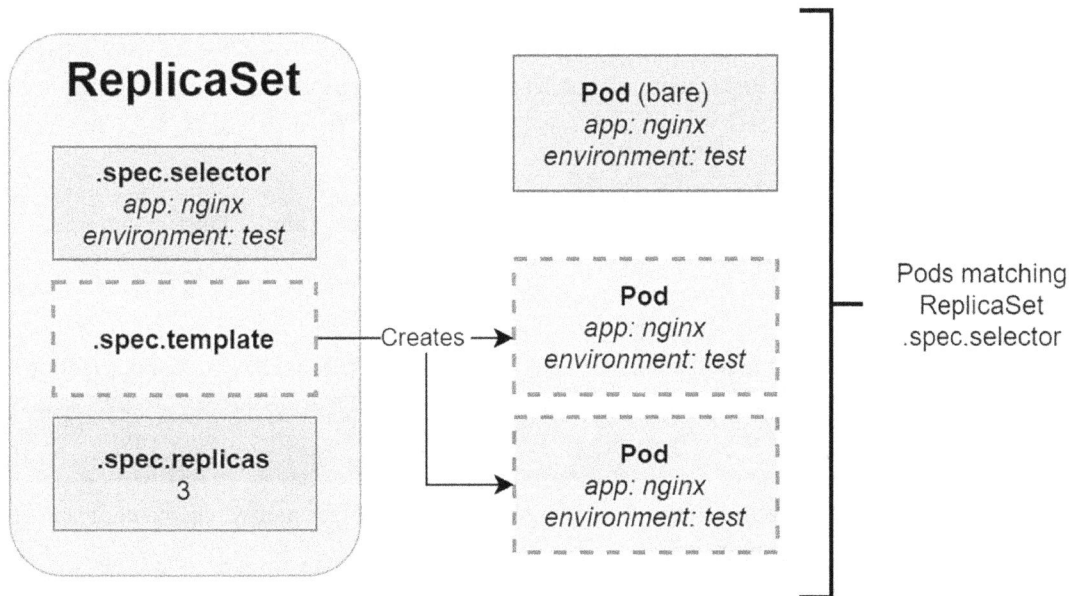

Figure 10.1: Kubernetes ReplicaSet

As you can see, the ReplicaSet object uses `.spec.template` in order to create Pods. These Pods must match the label selector configured in `.spec.selector`. Please note that it is also possible to acquire existing bare Pods that have labels matching the ReplicaSet object. In the case shown in *Figure 10.1*, the ReplicaSet object only creates two new Pods, whereas the third Pod is a bare Pod that was acquired.

In the preceding example, we have used a simple, **equality-based** selector specified by `spec.selector.matchLabels`. A more advanced, **set-based** selector can be defined using `spec.selector.matchExpressions`—for example, like this:

```yaml
# nginx-replicaset-expressions.yaml
...<removed for brevity>...
spec:
  replicas: 4
  selector:
    matchLabels:
      app: nginx
    matchExpressions:
      - key: environment
        operator: In
        values:
          - test
          - dev
  template:
...<removed for brevity>...
```

This specification would make ReplicaSet still match only Pods with `app=nginx`, and `environment=test` or `environment=dev`.

> When defining ReplicaSet, `.spec.template.metadata.labels` must match `spec.selector`, or it will be rejected by the API.

Now, let's apply the ReplicaSet manifest to the cluster using the `kubectl apply` command, as follows:

```
$ kubectl apply -f nginx-replicaset-example.yaml
replicaset.apps/nginx-replicaset-example created
```

You can immediately observe the status of your new ReplicaSet object named `nginx-replicaset-example` using the following command:

```
$ kubectl describe replicaset/nginx-replicaset-example -n rs-ns
...
Replicas:      4 current / 4 desired
Pods Status:   4 Running / 0 Waiting / 0 Succeeded / 0 Failed
...
```

You can use the `kubectl get pods -n rs-ns` command to observe the Pods that are managed by the ReplicaSet object. If you are interested, you can use the `kubectl describe pod <podId>` command in order to inspect the labels of the Pods and also see that it contains a `Controlled By: ReplicaSet/nginx-replicaset-example` property that identifies our example ReplicaSet object.

When using `kubectl` commands, you can use an `rs` abbreviation instead of typing `replicaset`.

Now let's move on to testing the behavior of ReplicaSet in the next section.

Testing the behavior of ReplicaSet

To demonstrate the agility of our ReplicaSet object, let's now delete one of the Pods that are owned by the `nginx-replicaset-example` ReplicaSet object using the following `kubectl delete` command:

```
$ kubectl delete po nginx-replicaset-example-6qc9p -n rs-ns
```

Now, if you are quick enough, you will be able to see from using the `kubectl get pods` command that one of the Pods is being terminated and ReplicaSet is immediately creating a new one in order to match the target number of replicas!

If you want to see more details about events that happened in relation to our example ReplicationController object, you can use the `kubectl describe` command, as illustrated in the following code snippet:

```
$ kubectl describe rs/nginx-replicaset-example -n rs-ns
...
Events:
  Type     Reason           Age    From                    Message
  ----     ------           ----   ----                    -------
...<removed for brevity>...
  Normal   SuccessfulCreate  9m9s   replicaset-controller   Created pod: nginx-
replicaset-example-r2qfn
  Normal   SuccessfulCreate  3m1s   replicaset-controller   Created pod: nginx-
replicaset-example-krdrs
```

In this case, the `nginx-replicaset-example-krdrs` Pod is a new Pod that was created by the ReplicaSet.

Now, let's try something different and create a bare Pod that matches the label selector of our ReplicaSet object. You can expect that the number of Pods that match the ReplicaSet will be four, so ReplicaSet is going to terminate one of the Pods to bring the replica count back to three.

Be careful with labels on bare Pods (Pods without a ReplicaSet manager). ReplicaSets can take control of any Pod with matching labels, potentially causing them to manage your bare Pod unintentionally. Use unique labels for bare Pods to avoid conflicts.

First, let's create a simple bare Pod manifest file named `nginx-pod-bare.yaml`, as follows:

```yaml
# nginx-pod-bare.yaml
apiVersion: v1
kind: Pod
metadata:
  name: nginx-pod-bare-example
  namespace: rs-ns
  labels:
    app: nginx
    environment: test
spec:
  containers:
    - name: nginx
      image: nginx:1.17
      ports:
        - containerPort: 80
```

The metadata of `Pod` must have labels matching the ReplicaSet selector. Now, apply the manifest to your cluster using the following command:

```
$ kubectl apply -f nginx-pod-bare.yaml
pod/nginx-pod-bare-example created
$ kubectl  get pods
NAME                            READY   STATUS        RESTARTS   AGE
nginx-pod-bare-example          0/1     Terminating   0          1s
nginx-replicaset-example-74kq9  1/1     Running       0          23h
nginx-replicaset-example-qfvx6  1/1     Running       0          23h
nginx-replicaset-example-s5cwc  1/1     Running       0          23h
```

Immediately after that, check the events for our example ReplicaSet object using the `kubectl describe` command, as follows:

```
$ kubectl describe rs/nginx-replicaset-example
...
Events:
  Type    Reason          Age   From                    Message
  ----    ------          ----  ----                    -------
...
  Normal  SuccessfulDelete  29s   replicaset-controller   Deleted pod: nginx-
pod-bare-example
```

As you can see, the ReplicaSet object has immediately detected that there is a new Pod created matching its label selector and has terminated the Pod.

Similarly, it is possible to remove Pods from a ReplicaSet object by modifying their labels so that they no longer match the selector. This is useful in various debugging or incident investigation scenarios.

In the following section, we will learn how ReplicaSet helps with the HA and FT of an application.

Testing HA and FT with a ReplicaSet

To demonstrate the scenario, let us use the previously deployed ReplicaSet `nginx-replicaset-example`. However, we will create a Service to expose this application, as follows.

```
$ kubectl apply -f nginx-service.yaml
service/nginx-service created
```

Let's forward the Service as follows for testing purposes:

```
$ kubectl port-forward svc/nginx-service 8080:80 -n rs-ns
Forwarding from 127.0.0.1:8080 -> 80
Forwarding from [::1]:8080 -> 80
```

Open another console and verify the application access.

```
$ curl localhost:8080
```

You should get the response with a default NGINX page output.

Previously we tested deleting the Pod and verified that the ReplicaSet will recreate the Pod based on the replica count. In this case, let us remove one of the Kubernetes nodes and see the behavior. Before we delete the node, let us check the current Pod placement as follows:

```
$ kubectl get po -n rs-ns -o wide
NAME                               READY    STATUS     RESTARTS    AGE     IP
NODE             NOMINATED NODE    READINESS GATES
nginx-replicaset-example-cfcfs     1/1      Running    0           24m     10.244.1.3
kind-worker2     <none>            <none>
nginx-replicaset-example-krdrs     1/1      Running    0           17m     10.244.3.5
kind-worker      <none>            <none>
nginx-replicaset-example-kw7cl     1/1      Running    0           24m     10.244.3.4
kind-worker      <none>            <none>
nginx-replicaset-example-r2qfn     1/1      Running    0           24m     10.244.2.4
kind-worker3     <none>            <none>
```

As per the preceding snippet, we have two Pods running on the `kind-worker` node; let us remove this node from the Kubernetes cluster as follows:

```
$ kubectl cordon kind-worker
node/kind-worker cordoned
```

```
$ kubectl drain kind-worker --ignore-daemonsets
...<removed for brevity>...
pod/nginx-replicaset-example-kw7cl evicted
node/kind-worker drained

$ kubectl delete node kind-worker
node "kind-worker" deleted
```

Now, let us check the Pod placement and number of replicas.

```
$ kubectl get po -n rs-ns -o wide
NAME                          READY   STATUS    RESTARTS   AGE     IP
NODE            NOMINATED NODE   READINESS GATES
nginx-replicaset-example-8lz9x   1/1     Running   0          2m33s
10.244.1.4   kind-worker2   <none>          <none>
nginx-replicaset-example-cfcfs   1/1     Running   0          30m
10.244.1.3   kind-worker2   <none>          <none>
nginx-replicaset-example-d8rz5   1/1     Running   0          2m33s
10.244.2.5   kind-worker3   <none>          <none>
nginx-replicaset-example-r2qfn   1/1     Running   0          30m
10.244.2.4   kind-worker3   <none>          <none>
```

As per the output, the ReplicaSet has already created the desired number of Pods on the available nodes.

If your kubectl port-forward is still running, you can again verify the application access (curl localhost:8080) and confirm the availability. Please note that for production environments, it is a best practice to integrate monitoring tools like **Prometheus** or **Grafana** for real-time health and resource visualization, and use **Fluentd** for logging to capture Pod logs to diagnose failures.

Now that we have learned the behavior of ReplicaSet with multiple examples, let us learn how to scale ReplicaSet.

Scaling ReplicaSet

For ReplicaSet, we can do a similar scaling operation as for ReplicationController in the previous section. In general, you will not perform manual scaling of ReplicaSets in usual scenarios. Instead, the size of the ReplicaSet object will be managed by another, higher-level object, such as Deployment.

Let's first scale up our example ReplicaSet object. Open the nginx-replicaset.yaml file and modify the replicas property to 5, as follows:

```
...
spec:
  replicas: 5
...
```

Now, we need to declaratively apply the changes to the cluster state. Use the following `kubectl apply` command to do this:

```
$ kubectl apply -f ./nginx-replicaset.yaml
```

To see that the number of Pods controlled by the ReplicaSet object has changed, you can use the `kubectl get pods` or `kubectl describe rs/nginx-replicaset-example` command.

> You can achieve similar results to the case of ReplicationController using the `kubectl scale rs/nginx-replicaset-example --replicas=5` imperative command. In general, such imperative commands are recommended only for development or learning scenarios.

Similarly, if you would like to scale down, you need to open the `nginx-replicaset.yaml` file and modify the `replicas` property to 2, as follows:

```
...
spec:
  replicas: 2
...
```

Again, declaratively apply the changes to the cluster state. Use the following `kubectl apply` command to do this:

```
$ kubectl apply -f ./nginx-replicaset.yaml
```

At this point, you can use the `kubectl get pods` or `kubectl describe rs/nginx-replicaset-example` command to verify that the number of Pods has been reduced to just 2.

Using Pod liveness probes together with ReplicaSet

Sometimes, you may want to consider a Pod unhealthy and require a container restart, even if the main process in the container has not crashed. You already learned about probes in *Chapter 8, Exposing Your Pods with Services*. We will quickly demonstrate how you can use **liveness probes** together with ReplicaSet to achieve even greater resilience to failures of containerized components.

In our example, we will create a ReplicaSet object that runs `nginx` Pods with an additional liveness probe on the main container, which checks whether an `HTTP GET` request to the path: / responds with a *successful* HTTP status code. You can imagine that, in general, your `nginx` process running in the container will always be healthy (until it crashes), but that doesn't mean that the Pod can be considered healthy. If the web server is not able to successfully provide content, it means that the web server process is running but something else might have gone wrong, and this Pod should no longer be used. We will simulate this situation by simply deleting the `/index.html` file in the container, which will cause the liveness probe to fail.

First, let's create a YAML manifest file named `nginx-replicaset-livenessprobe.yaml` for our new `nginx-replicaset-livenessprobe-example` ReplicaSet object with the following content:

```yaml
# nginx-replicaset-livenessprobe.yaml
apiVersion: apps/v1
kind: ReplicaSet
metadata:
  name: nginx-replicaset-livenessprobe-example
spec:
  replicas: 3
  selector:
    matchLabels:
      app: nginx
      environment: test
  template:
    metadata:
      labels:
        app: nginx
        environment: test
    spec:
      containers:
        - name: nginx
          image: nginx:1.17
          ports:
            - containerPort: 80
          livenessProbe:
            httpGet:
              path: /
              port: 80
            initialDelaySeconds: 2
            periodSeconds: 2
```

The highlighted part of the preceding code block contains the liveness probe definition and is the only difference between our earlier ReplicaSet examples. The liveness probe is configured to execute an HTTP GET request to the / path at port 80 for the container every 2 seconds (`periodSeconds`). The first probe will start after 2 seconds (`initialDelaySeconds`) from the container start.

> If you are modifying an existing ReplicaSet object, you need to first delete it and recreate it in order to apply changes to the Pod template.

Now, apply the manifest file to the cluster using the following command:

```
$ kubectl apply -f ./nginx-replicaset-livenessprobe.yaml
```

Verify that the Pods have been successfully started using the following command:

```
$ kubectl get pods
```

Now, you need to choose one of the ReplicaSet Pods in order to simulate the failure inside the container that will cause the liveness probe to fail. In the case of our example, we will be using the first Pod in the list and removing the index.html file inside the Pod. To simulate the failure, run the following command. This command will remove the index.html file served by the nginx web server and will cause the HTTP GET request to fail with a non-successful HTTP status code:

```
$ kubectl exec -it nginx-replicaset-livenessprobe-example-lgvqv -- rm /usr/
share/nginx/html/index.html
```

Inspect the events for this Pod using the kubectl describe command, as follows:

```
$ kubectl describe pod/nginx-replicaset-livenessprobe-example-lgvqv
Name:               nginx-replicaset-livenessprobe-example-lgvqv
...<removed for brevitt>...
Events:
  Type      Reason      Age                    From                Message
  ----      ------      ----                   ----                -------
  Normal    Scheduled   2m9s                   default-scheduler   Successfully
assigned default/nginx-replicaset-livenessprobe-example-lgvqv to kind-worker2
  Normal    Pulled      60s (x2 over 2m8s)     kubelet             Container image
"nginx:1.17" already present on machine
  Normal    Created     60s (x2 over 2m8s)     kubelet             Created container
nginx
  Normal    Started     60s (x2 over 2m8s)     kubelet             Started container
nginx
  Warning   Unhealthy   60s (x3 over 64s)      kubelet             Liveness probe
failed: HTTP probe failed with statuscode: 403
  Normal    Killing     60s                    kubelet             Container nginx
failed liveness probe, will be restarted
```

As you can see, the liveness probe has correctly detected that the web server became unhealthy and restarted the container inside the Pod.

However, please note that the ReplicaSet object itself did not take part in the restart in any way—the action was performed at the Pod level. This demonstrates how individual Kubernetes objects provide different features that can work together to achieve improved FT. Without the liveness probe, the end user could be served by a replica that is not able to provide content, and this would go undetected!

Deleting a ReplicaSet object

Lastly, let's take a look at how you can delete a ReplicaSet object. There are two possibilities, outlined as follows:

1. Delete the ReplicaSet object together with the Pods that it owns—this is performed by first scaling down automatically.
2. Delete the ReplicaSet object and leave the Pods unaffected.

To delete the ReplicaSet object together with the Pods, you can use the regular `kubectl delete` command, as follows:

```
$ kubectl delete rs/nginx-replicaset-livenessprobe-example
```

You will see that the Pods will first get terminated and then the ReplicaSet object is deleted.

Now, if you would like to delete just the ReplicaSet object, you need to use the `--cascade=orphan` option for `kubectl delete`, as follows:

```
$ kubectl delete rs/nginx-replicaset-livenessprobe-example --cascade=orphan
```

After this command, if you inspect which Pods are in the cluster, you will still see all the Pods that were owned by the `nginx-replicaset-livenessprobe-example` ReplicaSet object. These Pods can now, for example, be acquired by another ReplicaSet object that has a matching label selector.

Summary

In this chapter, you learned about the key building blocks for providing HA and FT for applications running in Kubernetes clusters. First, we explained why HA and FT are important. Next, you learned more details about providing component replication and failover using ReplicaSet, which is used in Kubernetes in order to provide multiple copies (replicas) of identical Pods. We demonstrated the differences between ReplicationController and ReplicaSet and explained why using ReplicaSet is currently the recommended way to provide multiple replicas of Pods.

The next chapters in this part of the book will give you an overview of how to use Kubernetes to orchestrate your container applications and workloads. You will familiarize yourself with concepts relating to the most important Kubernetes objects, such as Deployment, StatefulSet, and DaemonSet. Also, in the next chapter, we will focus on the next level of abstraction over ReplicaSets: Deployment objects. You will learn how to deploy and easily manage rollouts and rollbacks of new versions of your application.

Further reading

- ReplicaSet: https://kubernetes.io/docs/concepts/workloads/controllers/replicaset
- ReplicationController: https://kubernetes.io/docs/concepts/workloads/controllers/replicationcontroller

Join our community on Discord

Join our community's Discord space for discussions with the authors and other readers:

`https://packt.link/cloudanddevops`

11

Using Kubernetes Deployments for Stateless Workloads

The previous chapter introduced two important Kubernetes objects: **ReplicationController** and **Replica-Set**. At this point, you already know that they serve similar purposes in terms of maintaining identical, healthy replicas (copies) of Pods. In fact, ReplicaSet is a successor of ReplicationController and, in the most recent versions of Kubernetes, ReplicaSet should be used in favor of ReplicationController.

Now, it is time to introduce the **Deployment** object, which provides easy scalability, rolling updates, and versioned rollbacks for your stateless Kubernetes applications and services. Deployment objects are built on top of ReplicaSets and they provide a declarative way of managing them – just describe the desired state in the Deployment manifest and Kubernetes will take care of orchestrating the underlying ReplicaSets in a controlled, predictable manner. Alongside StatefulSet, which will be covered in the next chapter, it is the most important workload management object in Kubernetes. This will be the bread and butter of your development and operations on Kubernetes! The goal of this chapter is to make sure that you have all the tools and knowledge you need to deploy your stateless application components using Deployment objects, as well as to safely release new versions of your components using rolling updates of Deployments.

This chapter will cover the following topics:

- Introducing the Deployment object
- How Kubernetes Deployments seamlessly handle revisions and version rollouts
- Deployment object best practices

Technical requirements

For this chapter, you will need the following:

- A Kubernetes cluster that has been deployed. You can use either a local or cloud-based cluster, but to fully understand the concepts shown in this chapter, we recommend using a multi-node, cloud-based Kubernetes cluster if available.

- The Kubernetes CLI (kubectl) must be installed on your local machine and configured to manage your Kubernetes cluster.

Kubernetes cluster deployment (local and cloud-based) and kubectl installation were covered in *Chapter 3, Installing Your First Kubernetes Cluster*.

You can download the latest code samples for this chapter from the official GitHub repository: https://github.com/PacktPublishing/The-Kubernetes-Bible-Second-Edition/tree/main/Chapter11.

Introducing the Deployment object

Kubernetes gives you out-of-the-box flexibility when it comes to running different types of workloads, depending on your use cases. By knowing which workload type fits your application needs, you can make more informed decisions, optimize resource usage, and ensure better performance and reliability in your cloud-based applications. This foundational knowledge helps you unlock the full potential of Kubernetes' flexibility, allowing you to deploy and scale applications with confidence. Let's have a brief look at the supported workloads to understand where the Deployment object fits, as well as its purpose.

The following diagram demonstrates the different types of application workloads in Kubernetes, which we will explain in the following sections.

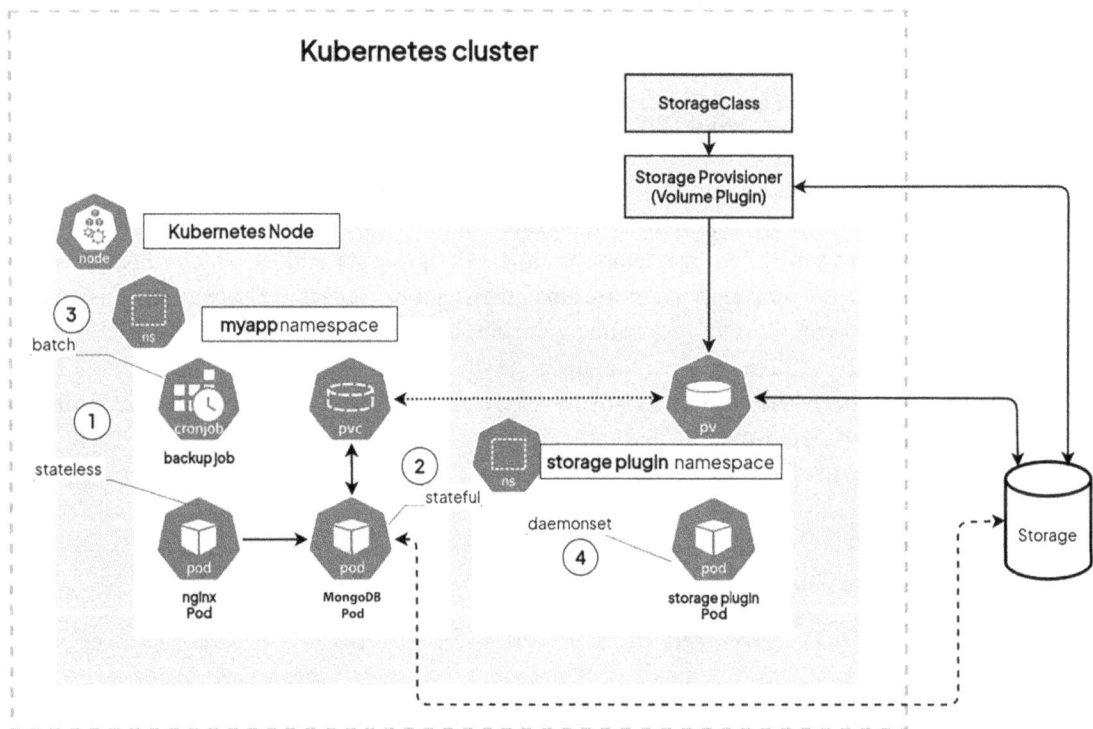

Figure 11.1: Application workload types in Kubernetes

When implementing cloud-based applications, you will generally need the following types of workloads:

- **Stateless:** In the world of containers, stateless applications are those that don't hold onto data (state) within the container itself. Imagine two Nginx containers serving the same purpose: one stores user data in a file inside the container, while the other uses a separate container like MongoDB for data persistence. Although they both achieve the same goal, the first Nginx container becomes stateful because it relies on internal storage. The second Nginx container, utilizing an external database, remains stateless. This stateless approach makes applications simpler to manage and scale in Kubernetes, as they can be easily restarted or replaced without worrying about data loss. Typically, Deployment objects in Kubernetes are used to manage these stateless workloads.

- **Stateful:** In the case of containers and Pods, we call them stateful if they store any modifiable data inside themselves. A good example of such a Pod is a MySQL or MongoDB Pod that reads and writes the data to a PersistentVolume. Stateful workloads are much harder to manage – you need to carefully manage sticky sessions or data partitions during rollouts, rollbacks, and when scaling. As a rule of thumb, try to keep stateful workloads outside your Kubernetes cluster if possible, such as by using cloud-based **Software-as-a-Service** (SaaS) database offerings. In Kubernetes, StatefulSet objects are used to manage stateful workloads. *Chapter 12, StatefulSet – Deploying Stateful Applications,* provides more details about these types of objects.

- **Job or CronJob:** This type of workload performs job or task processing, either scheduled or on demand. Depending on the type of application, batch workloads may require thousands of containers and a lot of nodes – this can be anything that happens *in the background*. Containers that are used for batch processing should also be stateless to make it easier to resume interrupted jobs. In Kubernetes, Job and CronJob objects are used to manage batch workloads. *Chapter 4, Running Your Containers in Kubernetes,* provides more details about these types of objects.

- **DaemonSet:** There are cases where we want to run workloads on every Kubernetes node to support the Kubernetes functionality. It can be a monitoring application, logging application, storage management agent (for PersistentVolumes), etc. For such workloads, we can use a special deployment type called a DaemonSet, which will guarantee that a copy of the workload will be running on every node in the cluster. *Chapter 13, DaemonSet – Maintaining Pod Singletons on Nodes,* provides more details about these types of objects.

With this concept regarding the different types of workloads in Kubernetes, we can dive deeper into managing stateless workloads using Deployment objects. In short, they provide declarative and controlled updates for Pods and ReplicaSets.

You can declaratively perform operations such as the following by using Deployment objects:

- **Rollout of a New ReplicaSet:** Deployments excel at managing controlled rollouts. You can define a new ReplicaSet with the desired Pod template within a Deployment object. Kubernetes then orchestrates a gradual rollout by scaling up the new ReplicaSet while scaling down the old one, minimizing downtime and ensuring a smooth transition.

- **Controlled Rollout with Pod Template Change:** Deployments allow you to update the Pod template within the Deployment definition. When you deploy the updated Deployment, Kubernetes performs a rolling update, scaling up the new ReplicaSet with the modified template and scaling down the old one. This enables you to introduce changes to your application's Pods in a controlled manner.

- **Rollback to a Previous Version:** Deployments keep track of their revision history. If you encounter any issues with the new version, you can easily roll back to a previous stable Deployment version. This allows you to revert changes quickly and minimize disruption.

- **Scaling ReplicaSets:** Scaling a Deployment directly scales the associated ReplicaSet. You can specify the desired number of replicas for the Deployment, and Kubernetes automatically scales the underlying ReplicaSet to meet that requirement.

- **Pausing and Resuming Rollouts:** Deployments offer the ability to pause the rollout of a new ReplicaSet if you need to address any issues or perform additional configuration. Similarly, you can resume the rollout once the issue is resolved. This provides flexibility during the deployment process.

In this way, Deployment objects provide an end-to-end pipeline for managing your stateless components running in Kubernetes clusters. Usually, you will combine them with Service objects, as presented in *Chapter 8, Exposing Your Pods with Services*, to achieve high fault tolerance, health monitoring, and intelligent load balancing for traffic coming into your application.

Now, let's have a closer look at the anatomy of the Deployment object specification and how to create a simple example deployment in our Kubernetes cluster.

Creating a Deployment object

Do not worry about the Deployment skeleton as you can use any supported tools for creating the Deployment YAMLs. It is possible to create a Deployment skeleton with `kubectl create deployment` commands as follows, which will basically display the YAML with the values:

```
$ kubectl create deployment my-deployment --replicas=1 --image=my-image:latest
--dry-run=client --port=80 -o yaml
```

You can redirect the output to a file and use it as the base for your deployment configurations:

```
$ kubectl create deployment my-deployment --replicas=1 --image=my-image:latest
--dry-run=client --port=80 -o yaml >my-deployment.yaml
```

First, let's take a look at the structure of an example Deployment YAML manifest file, `nginx-deployment.yaml`, that maintains three replicas of an `nginx` Pod:

```yaml
# nginx-deployment.yaml
apiVersion: apps/v1
kind: Deployment
metadata:
  name: nginx-deployment-example
spec:
  replicas: 3
  selector:
    matchLabels:
      app: nginx
```

```
      environment: test
  minReadySeconds: 10
  strategy:
    type: RollingUpdate
    rollingUpdate:
      maxUnavailable: 1
      maxSurge: 1
  template:
    metadata:
      labels:
        app: nginx
        environment: test
    spec:
      containers:
        - name: nginx
          image: nginx:1.17
          ports:
            - containerPort: 80
```

As you can see, the structure of the Deployment spec is almost identical to ReplicaSet, although it has a few extra parameters for configuring the strategy for rolling out new versions. The preceding YAML specification has four main components:

- `replicas`: Defines the number of Pod replicas that should run using the given `template` and matching label `selector`. Pods may be created or deleted to maintain the required number. This property is used by the underlying ReplicaSet.

- `selector`: A label selector, which defines how to identify Pods that the underlying ReplicaSet owns. This can include set-based and equality-based selectors. In the case of Deployments, the underlying ReplicaSet will also use a generated `pod-template-hash` label to ensure that there are no conflicts between different child ReplicaSets when you're rolling out a new version. Additionally, this generally prevents accidental acquisitions of bare Pods, which could easily happen with simple ReplicaSets. Nevertheless, Kubernetes does not prevent you from defining overlapping Pod selectors between different Deployments or even other types of controllers. However, if this happens, they may conflict and behave unexpectedly.

- `template`: Defines the template for Pod creation. Labels used in `metadata` must match our `selector`.

- `strategy`: Defines the details of the strategy that will be used to replace existing Pods with new ones. You will learn more about such strategies in the following sections. In this example, we showed the default `RollingUpdate` strategy. In short, this strategy works by slowly replacing the Pods of the previous version, one by one, by using the Pods of the new version. This ensures zero downtime and, together with Service objects and readiness probes, provides traffic load balancing to Pods that can serve the incoming traffic.

The Deployment spec provides a high degree of reconfigurability to suit your needs. We recommend referring to the official documentation for all the details: `https://kubernetes.io/docs/reference/generated/kubernetes-api/v1.31/#deployment-v1-apps` (refer to the appropriate version of your Kubernetes cluster, e.g., v1.31)

To better understand the relationship of Deployment, its underlying child ReplicaSet, and Pods, look at the following diagram:

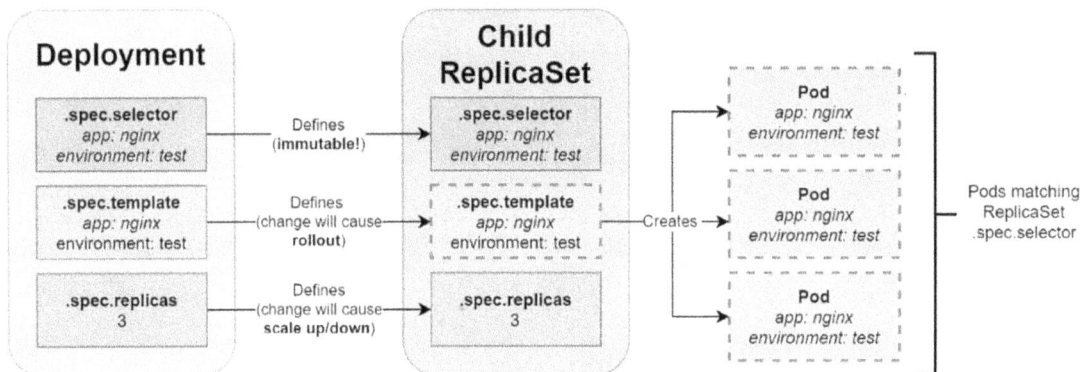

Figure 11.2: Kubernetes Deployment

Once you have defined and created a Deployment, it is not possible to change its `selector`. This is desired because, otherwise, you could easily end up with orphaned ReplicaSets. There are two important actions that you can perform on existing Deployment objects:

- **Modify template**: Usually, you would like to change the Pod definition to a new image version of your application. This will cause a rollout to begin, according to the rollout `strategy`.
- **Modify replica number**: Just changing the number will cause ReplicaSet to gracefully scale up or down.

Now, let's declaratively apply our example Deployment YAML manifest file, `nginx-deployment.yaml`, to the cluster using the `kubectl apply` command:

```
$ kubectl apply -f ./nginx-deployment.yaml
```

Using the `--record` flag is useful for tracking the changes that are made to the objects, as well as to inspect which commands caused these changes. You will then see an additional automatic annotation, `kubernetes.io/change-cause`, which contains information about the command. But the `--record` flag has been deprecated, and will be removed in the future. So, if you have a dependency on this annotation, it is a best practice to include the annotation manually as part of the Deployment update.

If you wish, add an annotation to the Deployment manually using the `kubectl annotate` command, as follows:

```
$ kubectl annotate deployment/ nginx-deployment-example kubernetes.io/change-
cause='Updated image to 1.2.3'
```

Immediately after the Deployment object has been created, use the `kubectl rollout` command to track the status of your Deployment in real time:

```
$ kubectl rollout status deployment nginx-deployment-example
Waiting for deployment "nginx-deployment-example" rollout to finish: 0 of 3
updated replicas are available...
deployment "nginx-deployment-example" successfully rolled out
```

This is a useful command that can give us a lot of insight into what is happening with an ongoing Deployment rollout. You can also use the usual `kubectl get` or `kubectl describe` commands:

```
$ kubectl get deploy nginx-deployment-example
NAME                        READY   UP-TO-DATE   AVAILABLE   AGE
nginx-deployment-example    3/3     3            3           6m21s
```

As you can see, the Deployment has been successfully created and all three Pods are now in the ready state.

> Instead of typing `deployment`, you can use the `deploy` abbreviation when using `kubectl` commands.

You may also be interested in seeing the underlying ReplicaSets:

```
$ kubectl get rs
NAME                                  DESIRED   CURRENT   READY   AGE
nginx-deployment-example-5b8dc6b8cd   3         3         3       2m17s
```

Please take note of the generated hash, `5b8dc6b8cd`, in the name of our ReplicaSet, which is also the value of the `pod-template-hash` label, which we mentioned earlier.

Lastly, you can see the Pods in the cluster that were created by the Deployment object using the following command:

```
$ kubectl get pods
NAME                                        READY   STATUS    RESTARTS   AGE
nginx-deployment-example-5b8dc6b8cd-lj2bz   1/1     Running   0          3m30s
nginx-deployment-example-5b8dc6b8cd-nxkbj   1/1     Running   0          3m30s
nginx-deployment-example-5b8dc6b8cd-shzmd   1/1     Running   0          3m30s
```

Congratulations – you have created and inspected your first Kubernetes Deployment! Next, we will take a look at how Service objects are used to expose your Deployment to external traffic coming into the cluster.

Exposing Deployment Pods using Service objects

Service objects were covered in detail in *Chapter 8*, *Exposing Your Pods with Services*, so, in this section, we will provide a brief recap about the role of Services and how they are usually used with Deployments.

The following diagram can be used as a base reference for the different networks in a Kubernetes cluster.

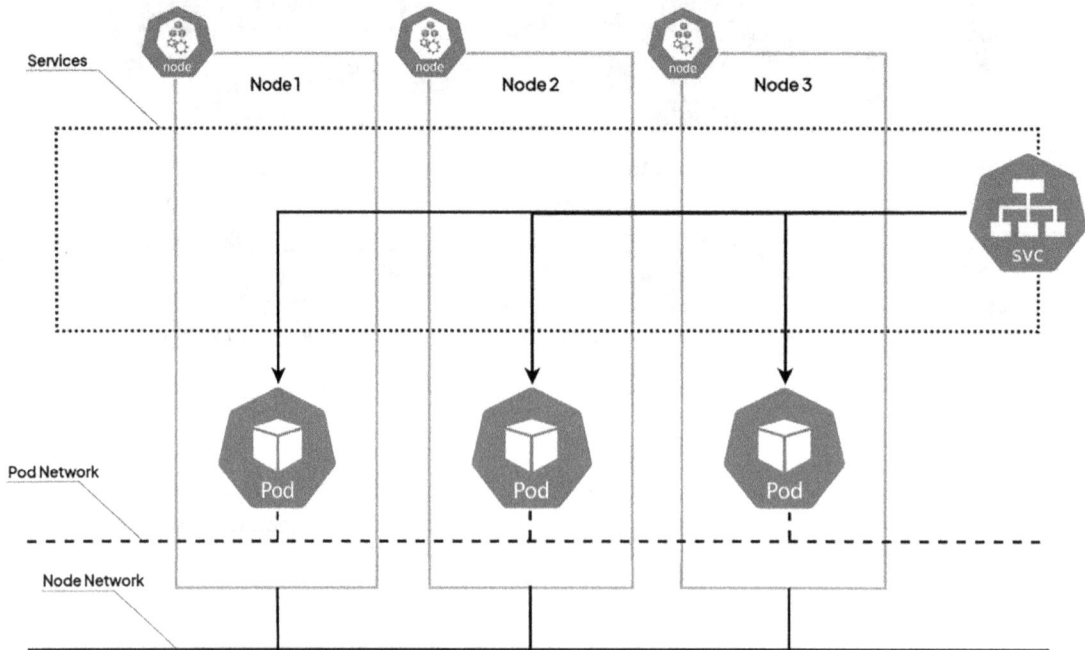

Figure 11.3: Different networks in Kubernetes

Services are Kubernetes objects that allow you to expose your Pods, either to other Pods in the cluster or to end users. They are the crucial building blocks for highly available and fault-tolerant Kubernetes applications since they provide a load balancing layer that actively routes incoming traffic to ready and healthy Pods.

Deployment objects, on the other hand, provide Pod replication, automatic restarts when failures occur, easy scaling, controlled version rollouts, and rollbacks. But there is a catch: Pods that are created by ReplicaSets or Deployments have a finite life cycle. At some point, you can expect them to be terminated; then, new Pod replicas with new IP addresses will be created in their place. So, what if you have a Deployment running web server Pods that need to communicate with Pods that have been created as a part of another Deployment such as backend Pods? Web server Pods cannot assume anything about the IP addresses or the DNS names of backend Pods, as they may change over time. This issue can be resolved with Service objects, which provide reliable networking for a set of Pods.

In short, Services target a set of Pods, and this is determined by label selectors. These label selectors work on the same principle that you have learned about for ReplicaSets and Deployments. The most common scenario is exposing a Service for an existing Deployment by using the same label selector.

The Service is responsible for providing a reliable DNS name and IP address, as well as for monitoring selector results and updating the associated endpoint object with the current IP addresses of the matching Pods. For internal cluster communication, this is usually achieved using simple `ClusterIP` Services, whereas to expose them to external traffic, you can use the `NodePort` Service or, more commonly in cloud deployments, the `LoadBalancer` Service.

To visualize how Service objects interact with Deployment objects in Kubernetes, look at the following diagram:

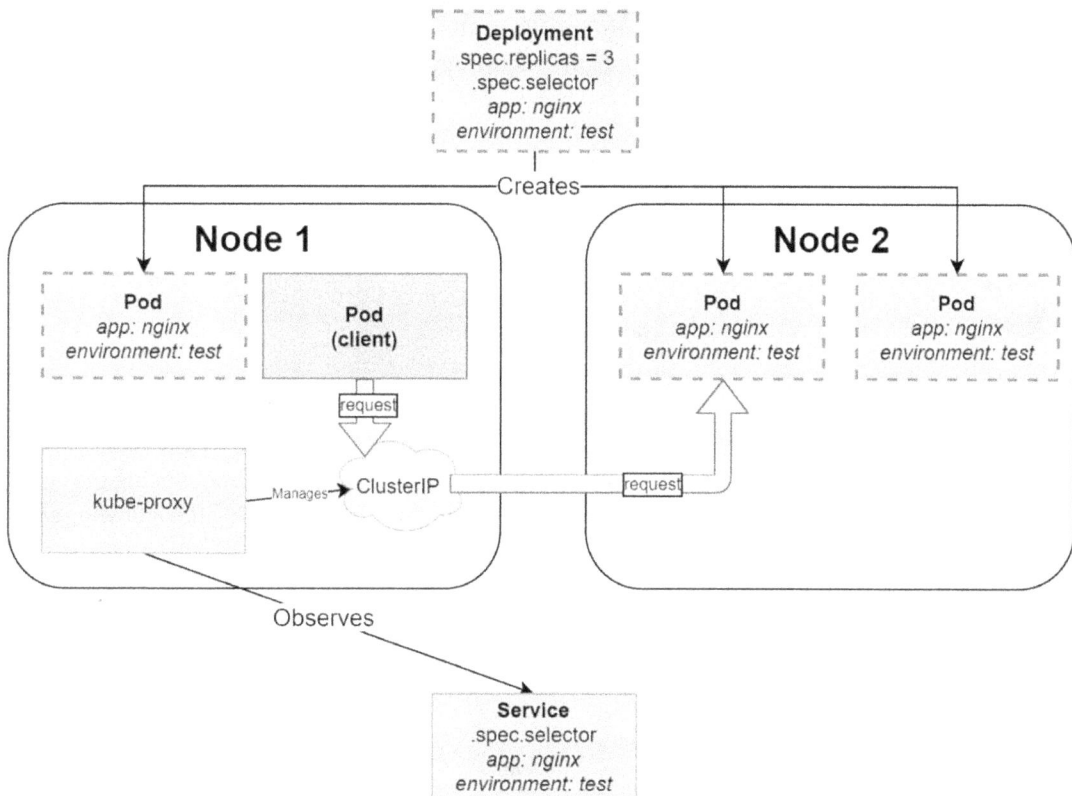

Figure 11.4: Client Pod performing requests to the Kubernetes Deployment, exposed by the ClusterIP Service

This diagram visualizes how any client Pod in the cluster can transparently communicate with the nginx Pods that are created by our Deployment object and exposed using the `ClusterIP` Service. ClusterIPs are essentially virtual IP addresses that are managed by the `kube-proxy` service that is running on each Node. `kube-proxy` is responsible for all the clever routing logic in the cluster and ensures that the routing is entirely transparent to the client Pods – they do not need to know if they are communicating with the same Node, a different Node, or even an external component. In the backend, `kube-proxy` watches the updates in the Service object and maintains all necessary routing rules required on each Node to ensure the proper traffic. `kube-proxy` generally uses `iptables` or **IP Virtual Server (IPVS)** to manage the traffic routing.

The role of the Service object is to define a set of ready Pods that should be *hidden* behind a stable ClusterIP. Usually, the internal clients will not be calling the Service pods using the ClusterIP, but they will use a DNS short name, which is the same as the Service name – for example, `nginx-service-example`. This will be resolved to the ClusterIP by the cluster's internal DNS service. Alternatively, they may use a DNS **Fully Qualified Domain Name (FQDN)** in the form of `<serviceName>.<namespaceName>.svc.<clusterDomain>`; for example, `nginx-service-example.default.svc.cluster.local`.

> For `LoadBalancer` or `NodePort` Services that expose Pods to external traffic, the principle is similar to internally; they also provide a ClusterIP for internal communication. The difference is that they also configure more components so that external traffic can be routed to the cluster.

Now that you're equipped with the necessary knowledge about Service objects and their interactions with Deployment objects, let's put what we've learned into practice!

> Refer to *Chapter 8, Exposing Your Pods with Services*, to learn more about Services and the different types of Services available in Kubernetes.

Creating a Service declaratively

In this section, we are going to expose our `nginx-deployment-example` Deployment using the `nginx-service-example` Service object, which is of the `LoadBalancer` type, by performing the following steps:

1. Create an `nginx-service.yaml` manifest file with the following content:

```yaml
# nginx-service.yaml
apiVersion: v1
kind: Service
metadata:
  name: nginx-service-example
spec:
  selector:
    app: nginx
    environment: test
  type: LoadBalancer
  ports:
    - port: 80
      protocol: TCP
      targetPort: 80
```

The label selector of the Service is the same as the one we used for our Deployment object. The specification of the Service instructs us to expose our Deployment on port 80 of the cloud load balancer, and then route the traffic from target port 80 to the underlying Pods.

Depending on how your Kubernetes cluster is deployed, you may not be able to use the `LoadBalancer` type. In that case, you may need to use the `NodePort` type for this exercise or stick to the simple `ClusterIP` type and skip the part about external access. For local development deployments such as `minikube`, you will need to use the `minikube service` command to access your Service. You can find more details in the documentation: `https://minikube.sigs.k8s.io/docs/commands/service/`.

2. Create the `nginx-service-example` Service and use the `kubectl get` or `kubectl describe` command to gather information about the status of our new Service and associated load balancer:

```
$ kubectl apply -f nginx-service.yaml
service/nginx-service-example created

$ kubectl describe service nginx-service-example
Name:                     nginx-service-example
Namespace:                default
Labels:                   <none>
Annotations:              <none>
Selector:                 app=nginx,environment=test
...<removed for brevity>...
Endpoints:                10.244.1.2:80,10.244.2.2:80,10.244.2.3:80
Session Affinity:         None
External Traffic Policy:  Cluster
Events:                   <none>
```

3. Now, let us try to access the Service from another Pod, as we did in *Chapter 8, Exposing Your Pods with Services*. Let us create a `k8sutils` Pod as follows and test the Service access:

```
$ kubectl apply -f ../Chapter07/k8sutils.yaml
pod/k8sutils created
$ kubectl exec -it k8sutils -- curl nginx-service-example.default.svc.
cluster.local |grep Welcome -A2
<title>Welcome to nginx!</title>
<style>
    body {
--
    <h1>Welcome to nginx!</h1>
<p>If you see this page, the nginx web server is successfully installed and
working. Further configuration is required.</p>
```

This shows how Services are used to expose Deployment Pods to external traffic. Now, we will quickly show you how to achieve a similar result using imperative commands to create a Service for our Deployment object.

Creating a Service imperatively

A similar effect can be achieved using the imperative kubectl expose command – a Service will be created for our Deployment object named nginx-deployment-example. Use the following command:

```
$ kubectl expose deployment --type=LoadBalancer nginx-deployment-example
service/nginx-deployment-example exposed
```

Let us explain the preceding code snippet:

- This will create a Service with the same name as the Deployment object – that is, nginx-deployment-example. If you would like to use a different name, as shown in the declarative example, you can use the --name=nginx-service-example parameter.
- Additionally, port 80, which will be used by the Service, will be the same as the one that was defined for the Pods. If you want to change this, you can use the --port=<number> and --target-port=<number> parameters.

> Check kubectl expose deployment --help to see the options available for exposing the Deployment.

Please note that this imperative command is recommended for use in development or debugging scenarios only. For production environments, you should leverage declarative *Infrastructure-as-Code* and *Configuration-as-Code* approaches as much as possible.

In the next section, let us learn how to use readiness, liveness, and startup probes with the Deployment.

Role of readiness, liveness, and startup probes

In *Chapter 8, Exposing Your Pods with Services*, we learned that there are three types of probes that you can configure for each container running in a Pod:

- Readiness probe
- Liveness probe
- Startup probe

All these probes are incredibly useful when you're configuring your Deployments – always try to predict possible life cycle scenarios for the processes running in your containers and configure the probes accordingly for your Deployments.

Please note that, by default, no probes are configured on containers running in Pods. Kubernetes will serve traffic to Pod containers behind the Service, but only if the containers have successfully started, and restart them if they have crashed using the default always-restart policy. This means that it is your responsibility to figure out what type of probes and what settings you need for your particular case. You will also need to understand the possible consequences and caveats of incorrectly configured probes – for example, if your liveness probe is too restrictive and has timeouts that are too small, it may wrongfully restart your containers and decrease the availability of your application.

Now, let's demonstrate how you can configure a **readiness probe** on your Deployment and how it works in real time.

> If you are interested in the configuration details for other types of probes, refer to *Chapter 8, Exposing Your Pods with Services,* and also to the official documentation: https://kubernetes.io/docs/tasks/configure-pod-container/configure-liveness-readiness-startup-probes/.

The nginx Deployment that we use is very simple and does not need any dedicated readiness probe. Instead, we will arrange the container's setup so that we can have the container's readiness probe fail or succeed on demand. The idea is to create an empty file called /usr/share/nginx/html/ready during container setup, which will be served on the /ready endpoint by nginx (just like any other file) and configure a readiness probe of the httpGet type to query the /ready endpoint for a successful HTTP status code. Now, by deleting or recreating the ready file using the kubectl exec command, we can easily simulate failures in our Pods that cause the readiness probe to fail or succeed.

Follow these steps to configure and test the readiness probe:

1. Delete the existing Deployment using the following command:

```
$ kubectl delete deployment nginx-deployment-example
```

2. Create a new Deployment YAML as follows:

```yaml
# nginx-deployment-readinessprobe.yaml
apiVersion: apps/v1
kind: Deployment
metadata:
  name: nginx-deployment-readiness
spec:
  replicas: 3
  selector:
    matchLabels:
      app: nginx
      environment: test
  minReadySeconds: 10
  strategy:
    type: RollingUpdate
    rollingUpdate:
      maxUnavailable: 1
      maxSurge: 1
...
```

Now we have the `spec.template` section as follows:

```yaml
# nginx-deployment-readinessprobe.yaml
...<continues>...
  template:
    metadata:
      labels:
        app: nginx
        environment: test
    spec:
      containers:
        - name: nginx
          image: nginx:1.25.4
          ports:
            - containerPort: 80
          command:
            - /bin/sh
            - -c
            - |
              touch /usr/share/nginx/html/ready
              echo "You have been served by Pod with IP address:
$(hostname -i)" > /usr/share/nginx/html/index.html
              nginx -g "daemon off;"
          readinessProbe:
            httpGet:
              path: /ready
              port: 80
            initialDelaySeconds: 5
            periodSeconds: 2
            timeoutSeconds: 10
            successThreshold: 1
            failureThreshold: 2
```

There are multiple parts changing in the Deployment manifest, all of which have been highlighted. First, we have overridden the default container entry point command using `command` and passed additional arguments. `command` is set to `/bin/sh` to execute a custom shell command. The additional arguments are constructed in the following way:

- `-c` is an argument for `/bin/sh` that instructs it that what follows is a command to be executed in the shell.
- `touch /usr/share/nginx/html/ready` is the first command that's used in the container shell. This will create an empty `ready` file that can be served by `nginx` on the `/ready` endpoint.

- echo "You have been served by Pod with IP address: $(hostname -i)" > /usr/share/ nginx/html/index.html is the second command that sets the content of index.html to information about the internal cluster Pod's IP address. hostname -i is the command that's used to get the container IP address. This value will be different for each Pod running in our Deployment.

- nginx -g "daemon off;": Finally, we execute the default entrypoint command for the nginx:1.25.4 image. This will start the nginx web server as the main process in the container.

> Usually, you would perform such customization using a new container image, which inherits from a generic container image (e.g., nginx image) as a base and dedicated application script. The method shown here is being used for demonstration purposes and shows how flexible the Kubernetes runtime is. Refer to the sample Chapter11/Containerfile in the GitHub repository for creating custom container images.

The second set of changes we made in the YAML manifest for the Deployment was for the definition of readinessProbe, which is configured as follows:

- The probe is of the httpGet type and executes an HTTP GET request to the /ready HTTP endpoint on port 80 of the container.

- initialDelaySeconds: This is set to 5 seconds and configures the probe to start querying after 5 seconds from container start.

- periodSeconds: This is set to 2 seconds and configures the probe to query in 2-second intervals.

- timeoutSeconds: This is set to 10 seconds and configures the number of seconds, after which the HTTP GET request times out.

- successThreshold: This is set to 1 and configures the minimum number of consecutive success queries of the probe before it is considered to be successful once it has failed.

- failureThreshold: This is set to 2 and configures the minimum number of consecutive failed queries of the probe before it is considered to have failed. Setting it to a value that's greater than 1 ensures that the probe is not providing false positives.

To create the Deployment, follow these steps:

1. Apply the new YAML manifest file to the cluster using the following command:

```
$ kubectl apply -f ./nginx-deployment-readinessprobe.yaml
```

2. Verify that the nginx-service-example Service is displaying with backend Pod IP addresses. You can see that the Service has three endpoints that map to our Deployment Pods, all of which are ready to serve traffic:

```
$  kubectl describe svc nginx-service-example
Name:                    nginx-service-example
Namespace:               default
Labels:                  <none>
Annotations:             <none>
Selector:                app=nginx,environment=test
Type:                    LoadBalancer
```

```
IP Family Policy:        SingleStack
IP Families:             IPv4
IP:                      10.96.231.126
IPs:                     10.96.231.126
Port:                    <unset>  80/TCP
TargetPort:              80/TCP
NodePort:                <unset>  32563/TCP
Endpoints:               10.244.1.6:80,10.244.1.7:80,10.244.2.6:80
Session Affinity:        None
External Traffic Policy: Cluster
Events:                  <none>
```

3. Test the nginx web server access using the k8sutils Pod that we created earlier in this chapter. You will notice that the responses iterate over different Pod IP addresses. This is because our Deployment has been configured to have three Pod replicas. Each time you perform a request, you may hit a different Pod:

```
$ kubectl exec -it k8sutils -- curl nginx-service-example.default.svc.
cluster.local
You have been served by Pod with IP address: 10.244.1.7
Chapter07 $ kubectl exec -it k8sutils -- curl nginx-service-example.
default.svc.cluster.local
You have been served by Pod with IP address: 10.244.2.6
```

4. Now, let's simulate a readiness failure for the first Pod. In our case, this is nginx-deployment-readiness-69dd4cfdd9-4pkwr, which has an IP address of 10.244.1.7. To do this, we need to simply delete the ready file inside the container using the kubectl exec command:

```
$ kubectl exec -it nginx-deployment-readiness-69dd4cfdd9-4pkwr -- rm /
usr/share/nginx/html/ready
```

5. The readiness probe will now start to fail, but not immediately! We have set it up so that it needs to fail at least two times, and each check is performed in 2-second intervals. Later, you will notice that you are only served by two other Pods that are still ready.

6. Now, if you describe the nginx-service-example Service, you will see that it only has two endpoints available, as expected:

```
$  kubectl describe svc nginx-service-example |grep Endpoint
Endpoints:               10.244.1.6:80,10.244.2.6:80
```

7. In the events for the Pod, you can also see that it is considered not ready:

```
$  kubectl describe pod nginx-deployment-readiness-69dd4cfdd9-4pkwr
Name:              nginx-deployment-readiness-69dd4cfdd9-4pkwr
...<removed for brevity>...
  Normal  Started    21m                    kubelet          Started
```

```
container nginx
  Warning  Unhealthy  72s (x25 over 118s)  kubelet              Readiness
probe failed: HTTP probe failed with statuscode: 404
```

We can push this even further. Delete the ready files in the other two Pods to make the whole Service fail:

```
$ kubectl exec -it nginx-deployment-readiness-69dd4cfdd9-7n2kz -- rm /
usr/share/nginx/html/ready
$ kubectl exec -it nginx-deployment-readiness-69dd4cfdd9-t7rp2 -- rm /
usr/share/nginx/html/ready
You can see, the Pods are Running but none of them are Ready to serve the
webservice due to readinessProbe failure.
$ kubectl get po -w
NAME                                            READY  STATUS    RESTARTS
AGE
k8sutils                                        1/1    Running   0
166m
nginx-deployment-readiness-69dd4cfdd9-4pkwr     0/1    Running   0
25m
nginx-deployment-readiness-69dd4cfdd9-7n2kz     0/1    Running   0
25m
nginx-deployment-readiness-69dd4cfdd9-t7rp2     0/1    Running   0
25m
```

Now, when you check the Service from the k8sutils Pod, you will see that the request is pending and that, eventually, it will fail with a timeout. We are now in a pretty bad state – we have a total readiness failure for all the Pod replicas in our Deployment!

```
$ kubectl exec -it k8sutils -- curl nginx-service-example.default.svc.
cluster.local
curl: (7) Failed to connect to nginx-service-example.default.svc.cluster.
local port 80 after 5 ms: Couldn't connect to server
command terminated with exit code 7
```

8. Finally, let's make one of our Pods ready again by recreating the file. You can refresh the web page so that the request is pending and, at the same time, execute the necessary command to create the ready file:

```
$ kubectl exec -it nginx-deployment-readiness-69dd4cfdd9-4pkwr -- touch /
usr/share/nginx/html/ready
After about 2 seconds (this is the probe interval), the pending request
in the web browser should succeed and you will be presented with a nice
response from nginx:
$ kubectl exec -it k8sutils -- curl nginx-service-example.default.svc.
cluster.local
You have been served by Pod with IP address: 10.244.1.7
```

Congratulations – you have successfully configured and tested the readiness probe for your Deployment Pods! This should give you a good insight into how the probes work and how you can use them with Services that expose your Deployments.

Next, we will take a brief look at how you can scale your Deployments.

Scaling a Deployment object

The beauty of Deployments is that you can almost instantly scale them up or down, depending on your needs. When the Deployment is exposed behind a Service, the new Pods will be automatically discovered as new endpoints when you scale up or automatically removed from the endpoints list when you scale down. The steps for this are as follows:

1. First, let's scale up our Deployment declaratively. Open the `nginx-deployment-readinessprobe.yaml` manifest file and modify the number of replicas:

   ```
   apiVersion: apps/v1
   kind: Deployment
   metadata:
     name: nginx-deployment-readiness
   spec:
     replicas: 10
   ...
   ```

2. Apply these changes to the cluster using the `kubectl apply` command:

   ```
   $ kubectl apply -f ./nginx-deployment-readinessprobe.yaml
   deployment.apps/nginx-deployment-readiness configured
   ```

3. Now, if you check the Pods using the `kubectl get pods` command, you will immediately see that new Pods are being created. Similarly, if you check the output of the `kubectl describe` command for the Deployment, you will see the following in the events:

   ```
   $ kubectl describe deployments.apps nginx-deployment-readiness
   Name:                   nginx-deployment-readiness
   ...<removed fro brevity>...
   Events:
     Type    Reason            Age   From                   Message
     ----    ------            ----  ----                   -------
     Normal  ScalingReplicaSet 32m   deployment-controller  Scaled up
   replica set nginx-deployment-readiness-69dd4cfdd9 to 3
     Normal  ScalingReplicaSet 9s    deployment-controller  Scaled up
   replica set nginx-deployment-readiness-69dd4cfdd9 to 10 from 3
   ```

You can achieve the same result using the imperative command, which is only recommended for development scenarios:

```
$ kubectl scale deploy nginx-deployment-readiness --replicas=10
deployment.apps/nginx-deployment-readiness scaled
```

4. To scale down our Deployment declaratively, simply modify the nginx-deployment-readinessprobe.yaml manifest file and change the number of replicas:

```
apiVersion: apps/v1
kind: Deployment
metadata:
  name: nginx-deployment-readiness
spec:
  replicas: 2
...
```

5. Apply the changes to the cluster using the kubectl apply command:

```
$ kubectl apply -f ./nginx-deployment-readinessprobe.yaml
```

6. You can achieve the same result using imperative commands. For example, you can execute the following command:

```
$ kubectl scale deploy nginx-deployment-readiness --replicas=2
```

If you describe the Deployment, you will see that this scaling down is reflected in the events:

```
$ kubectl describe deploy nginx-deployment-readiness
...
Events:
  Type     Reason            Age    From                    Message
  ----     ------            ----   ----                    -------

  Normal   ScalingReplicaSet  27s    deployment-controller   Scaled down
replica set nginx-deployment-readiness-69dd4cfdd9 to 2 from 10
```

Deployment events are very useful if you want to know the exact timeline of scaling and the other operations that can be performed with the Deployment object.

> It is possible to autoscale your deployments using HorizontalPodAutoscaler. This will be covered in *Chapter 20, Autoscaling Kubernetes Pods and Nodes*.

Next, you will learn how to delete a Deployment from your cluster.

Deleting a Deployment object

To delete a Deployment object, you can do two things:

- Delete the Deployment object along with the Pods that it owns. This can be done by first scaling down automatically.
- Delete the Deployment object and leave the other Pods unaffected.

To delete the Deployment object and its Pods, you can use the regular `kubectl delete` command:

```
$ kubectl delete deploy nginx-deployment-readiness
```

You will see that the Pods get terminated and that the Deployment object is then deleted.

Now, if you would like to delete just the Deployment object, you need to use the `--cascade=orphan` option for `kubectl delete`:

```
$ kubectl delete deploy nginx-deployment-readiness --cascade=orphan
```

After executing this command, if you inspect what Pods are in the cluster, you will still see all the Pods that were owned by the `nginx-deployment-example` Deployment.

In the following section, we will explore how to manage different revisions and roll out using the Deployment.

How Kubernetes Deployments seamlessly handle revisions and version rollouts

So far, we have only covered making one possible modification to a living Deployment – we have scaled up and down by changing the `replicas` parameter in the specification. However, this is not all we can do! It is possible to modify the Deployment's Pod template (`.spec.template`) in the specification and, in this way, trigger a rollout. This rollout may be caused by a simple change, such as changing the labels of the Pods, but it may be also a more complex operation when the container images in the Pod definition are changed to a different version. This is the most common scenario as it enables you, as a Kubernetes cluster operator, to perform a controlled, predictable rollout of a new version of your image and effectively create a new revision of your Deployment.

Your Deployment uses a rollout strategy, which can be specified in a YAML manifest using `.spec.strategy.type`. Kubernetes supports two strategies out of the box:

- `RollingUpdate`: This is the default strategy and allows you to roll out a new version of your application in a controlled way. This type of strategy uses two ReplicaSets internally. When you perform a change in the Deployment spec that causes a rollout, Kubernetes will create a new ReplicaSet with a new Pod template scaled to zero Pods initially. The old, existing ReplicaSet will remain unchanged at this point. Next, the old ReplicaSet will be scaled down gradually, whereas the new ReplicaSet will be scaled up gradually at the same time. The number of Pods that may be unavailable (readiness probe failing) is controlled using the `.spec.strategy.rollingUpdate.maxUnavailable` parameter.

The maximum number of extra Pods that can be scheduled above the desired number of Pods in the Deployment is controlled by the `.spec.strategy.rollingUpdate.maxSurge` parameter. Additionally, this type of strategy offers automatic revision history, which can be used for quick rollbacks in case of any failures.

- `Recreate`: This is a simple strategy that's useful for development scenarios where all the old Pods have been terminated and replaced with new ones. This instantly deletes any existing underlying ReplicaSet and replaces it with a new one. You should not use this strategy for production workloads unless you have a specific use case.

> Consider the Deployment strategies as basic building blocks for more advanced Deployment scenarios. For example, if you are interested in blue/green Deployments, you can easily achieve this in Kubernetes by using a combination of Deployments and Services while manipulating label selectors. You can find out more about this in the official Kubernetes blog post: `https://kubernetes.io/blog/2018/04/30/zero-downtime-deployment-kubernetes-jenkins/`.

Now, we will perform a rollout using the `RollingUpdate` strategy. The `Recreate` strategy, which is much simpler, can be exercised similarly.

Updating a Deployment object

We will explore the `RollingUpdate` strategy with a practical example in this section. First, let's recreate the Deployment that we used previously for our readiness probe demonstration:

1. Make a copy of the previous YAML manifest file:

```
$ cp nginx-deployment-readinessprobe.yaml nginx-deployment-rollingupdate.
yaml
```

2. Ensure that you have a strategy of the `RollingUpdate` type, called `readinessProbe`, set up and an image version of `nginx:1.17`. This should already be set up in the `nginx-deployment-readinessprobe.yaml` manifest file if you completed the previous sections:

```
# nginx-deployment-rollingupdate.yaml
apiVersion: apps/v1
kind: Deployment
metadata:
  name: nginx-deployment-rollingupdate
spec:
  replicas: 3
...
  minReadySeconds: 10
  strategy:
    type: RollingUpdate
    rollingUpdate:
```

```
        maxUnavailable: 1
        maxSurge: 1
  template:
...

    spec:
      containers:
        - name: nginx
          image: nginx:1.17

...

        readinessProbe:
          httpGet:

...
```

3. In this example, we are using maxUnavailable set to 1, which means that we allow only one Pod out of three, which is the target number, to be unavailable (not ready). This means that, at any time, there must be at least two Pods ready to serve traffic.

4. Similarly, setting maxSurge to 1 means that we allow one extra Pod to be created above the target number of three Pods during the rollout. This effectively means that we can have up to four Pods (ready or not) present in the cluster during the rollout. Please note that it is also possible to set up these parameters as percentage values (such as 25%), which is very useful in autoscaling scenarios.

5. Additionally, minReadySeconds (which is set to 10) provides an additional time span for which the Pod has to be ready before it can be *announced* as successful during the rollout.

6. Apply the manifest file to the cluster:

```
$ kubectl apply -f nginx-deployment-rollingupdate.yaml
deployment.apps/nginx-deployment-rollingupdate created

$ kubectl get pod
NAME                                                READY   STATUS
RESTARTS    AGE
nginx-deployment-rollingupdate-69d855cf4b-nshn2     1/1     Running    0
24s
nginx-deployment-rollingupdate-69d855cf4b-pqvjh     1/1     Running    0
24s
nginx-deployment-rollingupdate-69d855cf4b-tdxzl     1/1     Running    0
24s
```

With the Deployment ready in the cluster, we can start rolling out a new version of our application. We will change the image in the Pod template for our Deployment to a newer version and observe what happens during the rollout. To do this, follow the following steps:

7. Modify the container image that was used in the Deployment to `nginx:1.18`:

```
...
  template:
    metadata:
      labels:
        app: nginx
        environment: test
    spec:
      containers:
        - name: nginx
          image: nginx:1.18
...
```

8. Apply the changes to the cluster using the following command:

```
$ kubectl apply -f nginx-deployment-rollingupdate.yaml
deployment.apps/nginx-deployment-rollingupdate configured
```

9. Immediately after that, use the `kubectl rollout status` command to see the progress in real time:

```
$ kubectl rollout status deployment.apps/nginx-deployment-rollingupdate
deployment "nginx-deployment-rollingupdate" successfully rolled out
```

10. The rollout will take a bit of time because we have configured `minReadySeconds` on the Deployment specification and `initialDelaySeconds` on the Pod container readiness probe.

11. Similarly, using the `kubectl describe` command, you can see the events for the Deployment that inform us of how the ReplicaSets were scaled up and down:

```
$ kubectl describe deploy nginx-deployment-rollingupdate
Name:                    nginx-deployment-rollingupdate
...<removed for brevity>...
Events:
  Type    Reason            Age    From                  Message
  ----    ------            ----   ----                  -------
  Normal  ScalingReplicaSet 8m21s  deployment-controller  Scaled up
replica set nginx-deployment-rollingupdate-69d855cf4b to 3
  Normal  ScalingReplicaSet 2m51s  deployment-controller  Scaled up
replica set nginx-deployment-rollingupdate-5479f5d87f to 1
  Normal  ScalingReplicaSet 2m51s  deployment-controller  Scaled down
replica set nginx-deployment-rollingupdate-69d855cf4b to 2 from 3
  Normal  ScalingReplicaSet 2m51s  deployment-controller  Scaled up
replica set nginx-deployment-rollingupdate-5479f5d87f to 2 from 1
  Normal  ScalingReplicaSet 2m24s  deployment-controller  Scaled down
replica set nginx-deployment-rollingupdate-69d855cf4b to 1 from 2
```

```
  Normal  ScalingReplicaSet  2m24s  deployment-controller  Scaled up
replica set nginx-deployment-rollingupdate-5479f5d87f to 3 from 2
  Normal  ScalingReplicaSet  2m14s  deployment-controller  Scaled down
replica set nginx-deployment-rollingupdate-69d855cf4b to 0 from 1
```

12. Now, let's take a look at the ReplicaSets in the cluster:

```
$ kubectl get rs
NAME                                             DESIRED   CURRENT   READY
AGE
nginx-deployment-rollingupdate-5479f5d87f        3         3         3
4m22s
nginx-deployment-rollingupdate-69d855cf4b        0         0         0
9m52s
```

You will see something interesting here: the old ReplicaSet remains in the cluster but has been scaled down to zero Pods! The reason for this is that we're keeping the Deployment revision history – each revision has a matching ReplicaSet that can be used if we need to roll back. The number of revisions that are kept for each Deployment is controlled by the `.spec.revisionHistoryLimit` parameter – by default, it is set to `10`. Revision history is important, especially if you are making imperative changes to your Deployments. If you are using the declarative model and always committing your changes to a source code repository, then the revision history may be less relevant.

13. Lastly, we can check if the Pods were indeed updated to a new image version. Use the following command and verify one of the Pods in the output:

```
$ kubectl get po
NAME                                                READY   STATUS
RESTARTS     AGE
nginx-deployment-rollingupdate-5479f5d87f-2k7d6     1/1     Running    0
5m41s
nginx-deployment-rollingupdate-5479f5d87f-6gn9m     1/1     Running    0
5m14s
nginx-deployment-rollingupdate-5479f5d87f-mft6b     1/1     Running    0
5m41s

$ kubectl describe pod nginx-deployment-rollingupdate-5479f5d87f-
2k7d6|grep 'Image:'
    Image:          nginx:1.1
```

This shows that we have indeed performed a rollout of the new nginx container image version!

> You can change the Deployment container image *imperatively* using the kubectl set image deployment nginx-deployment-example nginx=nginx:1.18 command. This approach is only recommended for non-production scenarios, and it works well with *imperative* rollbacks.

Next, you will learn how to roll back a Deployment object.

Rolling back a Deployment object

If you are using a declarative model to introduce changes to your Kubernetes cluster and are committing each change to your source code repository, performing a rollback is very simple and involves just reverting the commit and applying the configuration again. Usually, the process of applying changes is performed as part of the CI/CD pipeline for the source code repository, instead of the changes being manually applied by an operator (such as an application team or administrators). This is the easiest way to manage Deployments, and this is generally recommended in the Infrastructure-as-Code and Configuration-as-Code paradigms.

> A powerful example of using a declarative model in practice is **Flux** (https://fluxcd.io/). While originally incubated by the **Cloud Native Computing Foundation** (CNCF), Flux has since graduated and become a full-fledged project within the CNCF landscape. Flux is the core of the GitOps approach, a methodology for implementing continuous deployment for cloud-native applications. It prioritizes a developer-centric experience by leveraging familiar tools like Git and continuous deployment pipelines, streamlining infrastructure management for developers.

However, Kubernetes still provides an imperative way to roll back a Deployment using revision history. Imperative rollbacks can also be performed on Deployments that have been updated declaratively. Now, we will demonstrate how to use kubectl for rollbacks. Follow the next steps:

1. First, let's imperatively roll out another version of our Deployment. This time, we will update the nginx image to version 1.19:

```
$ kubectl set image deployment nginx-deployment-rollingupdate
nginx=nginx:1.19
deployment.apps/nginx-deployment-rollingupdate image updated
```

2. Please note, nginx=nginx:1.19 means we are setting the nginx:1.19 image for the container called nginx in the nginx-deployment-rollingupdate Deployment.

3. Using kubectl rollout status, wait for the end of the Deployment:

```
$ kubectl rollout status deployment.apps/nginx-deployment-rollingupdate
deployment "nginx-deployment-rollingupdate" successfully rolled out
```

4. Now, let's suppose that the new version of the application image, 1.19, is causing problems and that your team decided to roll back to the previous version of the image, which was working fine.

5. Use the following `kubectl rollout history` command to see all the revisions that are available for the Deployment:

```
$ kubectl rollout history deployment.apps/nginx-deployment-rollingupdate
deployment.apps/nginx-deployment-rollingupdate
REVISION   CHANGE-CAUSE
1          <none>
2          <none>
3          <none>
```

6. As you can see, we have three revisions. The first revision is our initial creation of the Deployment. The second revision is the declarative update of the Deployment to the `nginx:1.18` image. Finally, the third revision is our last imperative update to the Deployment that caused the `nginx:1.19` image to be rolled out. `CHANGE-CAUSE` is empty here because we haven't used the `--record` flag, as the `--record` flag has been deprecated and will be removed in the future version. If you have a requirement to update `CHANGE-CAUSE`, you need to manually update the Deployment annotation, as we learned earlier in this chapter.

7. The revisions that were created as declarative changes do not contain too much information in `CHANGE-CAUSE`. To find out more about the second revision, you can use the following command:

```
$ kubectl rollout history deploy nginx-deployment-rollingupdate
--revision=2
deployment.apps/nginx-deployment-rollingupdate with revision #2
Pod Template:
...<removed for brevity>...
  Containers:
   nginx:
    Image:        nginx:1.18
...<removed for brevity>...
```

8. Now, let's perform a rollback to this revision. Because it is the previous revision, you can simply execute the following command:

```
$ kubectl rollout undo deploy nginx-deployment-rollingupdate
deployment.apps/nginx-deployment-rollingupdate rolled back
```

9. This would be equivalent to executing a rollback to a specific revision number:

```
$ kubectl rollout undo deploy nginx-deployment-rollingupdate --to-
revision=2
```

10. Again, as in the case of a normal rollout, you can use the following command to follow the rollback:

```
$ kubectl rollout status deploy nginx-deployment-rollingupdate
$ kubectl rollout history deployment.apps/nginx-deployment-rollingupdate
```

```
deployment.apps/nginx-deployment-rollingupdate
REVISION    CHANGE-CAUSE
1           <none>
3           <none>
4           <none>
```

11. You will notice that version 2 is missing, and a new version of the deployment has been created.

Please note that you can also perform rollbacks on currently ongoing rollouts. This can be done in both ways; that is, declaratively and imperatively.

> If you need to pause and resume the ongoing rollout of a Deployment, use the kubectl rollout pause deployment nginx-deployment-example and kubectl rollout resume deployment nginx-deployment-example commands.

Congratulations – you have successfully rolled back your Deployment. In the next section, we will provide you with a set of best practices for managing Deployment objects in Kubernetes.

Canary deployment strategy

Canary deployments offer a valuable strategy for minimizing risk during application rollouts. They take inspiration from the practice of sending canaries (birds) into coal mines to detect dangerous gases. Similarly, canary deployments introduce a new version of an application to a limited subset of users before exposing it to the entire production environment.

This controlled rollout allows for real-world testing of the new version while minimizing potential disruptions. Here's how it works: imagine you're deploying an update to your e-commerce website. Traditionally, the entire website would be switched to the new version simultaneously. With a canary deployment, however, it is possible to create two deployments: a stable deployment running the current version serving the majority of users, and a canary deployment with the new version serving a small percentage of users. Traffic routing mechanisms like ingress controllers or service annotations then direct a specific portion of traffic (e.g., 10%) to the canary deployment.

By closely monitoring the canary Deployment's performance through metrics like error rates, response times, and user feedback, you can assess the new version's stability. If everything runs smoothly, you can gradually increase the percentage of traffic directed to the canary deployment until it serves all users. Conversely, if issues arise, you can easily roll back the canary deployment and maintain the stable version, preventing a wider impact on your user base.

Consider this YAML example demonstrating a canary deployment for a frontend application: We have the stable app Deployment configuration with image: frontend-app:1.0 as follows.

```
# Stable app
...
name: frontend-stable
replicas: 3
```

```
...
labels:
  app: myapp
  tier: frontend
  version: stable
...
image: frontend-app:1.0
```

Now we have a canary deployment configuration for the same application with `image: frontend-app:2.0`, as follows.

```
# Canary version
...
name: frontend-canary
replicas: 1
...
labels:
  app: myapp
  tier: frontend
  version: canary
...
image: frontend-app:2.0
```

And we need a Service to route the traffic to both versions of the Deployment. Please note, we only used `app: myapp` and `tier: frontend` labels as selectors so that the Service will use both `stable` and Canary Pods to serve the traffic:

```
# Service (routes traffic to both deployments)
apiVersion: v1
kind: Service
metadata:
  name: frontend-service
spec:
  selector:
    app: myapp
    tier: frontend
  ports:
  - protocol: TCP
    port: 80
    targetPort: 8080
```

By adjusting the replicas in each Deployment, you control the percentage of users directed to each version.

Alternatively, it is also possible to create different Service objects for stable and canary versions and use a mechanism like ingress controllers or service annotations to route a specific percentage of traffic (e.g., 10%) to the canary deployment.

The important part is monitoring the canary deployment and ensuring the new version is working as per expectation. Based on the result, you can either promote the canary as the stable Deployment or delete the canary and update the new application image (e.g., `image: frontend-app:2.0`) in the stable Deployment.

Refer to the documentation (`https://kubernetes.io/docs/concepts/workloads/management/#canary-deployments`) to learn more about canary deployments.

Now let's move on to the next section on the best practices of working with Deployment objects.

Deployment object best practices

This section will summarize the best practices when working with Deployment objects in Kubernetes. This list is by no means complete, but it is a good starting point for your journey with Kubernetes.

Use declarative object management for Deployments

In the DevOps and containerized application world, it is a good practice to stick to declarative models when introducing updates to your infrastructure and applications. This is at the core of the *Infrastructure-as-Code* and *Configuration-as-Code* paradigms. In Kubernetes, you can easily perform declarative updates using the `kubectl apply` command, which can be used on a single file or even a whole directory of YAML manifest files.

> To delete objects, it is still better to use imperative commands. It is more predictable and less prone to errors. Declaratively deleting resources in your cluster is mostly useful in CI/CD and GitOps scenarios, where the whole process is entirely automated.

The same principle also applies to Deployment objects. Performing a rollout or rollback when your YAML manifest files are versioned and kept in a source control repository is easy and predictable. Using the `kubectl rollout undo` and `kubectl set image deployment` commands is generally not recommended in production environments. Using these commands gets much more complicated when more than one person is working on operations in the cluster.

Do not use the Recreate strategy for production workloads

Using the `Recreate` strategy may be tempting as it provides instantaneous updates for your Deployments. However, at the same time, this will mean downtime for your end users. This is because all the existing Pods for the old revision of the Deployment will be terminated at once and replaced with the new Pods. There may be a significant delay before the new Pods become ready, and this means downtime. This downtime can be easily avoided by using the `RollingUpdate` strategy in production scenarios.

> The Recreate deployment strategy in Kubernetes is best suited for specific scenarios where downtime is acceptable and offers some advantages over other strategies. For example, when deploying significant application changes or introducing entirely new versions, the Recreate strategy allows for a clean slate and ensures the new version runs independently from the previous one.

Do not create Pods that match an existing Deployment label selector

It is possible to create Pods with labels that match the label selector of some existing Deployment. This can be done using bare Pods or another Deployment or ReplicaSet. This leads to conflicts, which Kubernetes does not prevent, and makes the existing deployment *believe* that it has created the other Pods. The results may be unpredictable and, in general, you need to pay attention to how you label the resources in your cluster. We advise you to use semantic labeling here, which you can learn more about in the official documentation: https://kubernetes.io/docs/concepts/configuration/overview/#using-labels.

Carefully set up your container probes

The liveness, readiness, and startup probes of your Pod containers can provide a lot of benefits but, at the same time, if they have been misconfigured, they can cause outages, including cascading failures. You should always be sure that you understand the consequences of each probe going into a failed state and how it affects other Kubernetes resources, such as Service objects.

There are a few established best practices for readiness probes that you should consider:

- Use this probe whenever your containers may not be ready to serve traffic as soon as the container is started.
- Ensure that you check things such as cache warm-ups or your database migration status during readiness probe evaluation. You may also consider starting the actual process of a warm-up if it has not been started yet, but use this approach with caution – a readiness probe will be executed constantly throughout the life cycle of a Pod, which means you shouldn't perform any costly operations for every request. Alternatively, you may want to use a startup probe for this purpose.
- For microservice applications that expose HTTP endpoints, consider configuring the httpGet readiness probe. This will ensure that every basis is covered when a container is successfully running but the HTTP server has not been fully initialized.
- It is a good idea to use a separate, dedicated HTTP endpoint for readiness checks in your application. For example, a common convention is using /health.
- If you are checking the state of your dependencies (external database, logging services, etc.) in this type of probe, be careful with shared dependencies, such as databases. In this case, you should consider using a probe timeout, which is greater than the maximum allowed timeout for the external dependency – otherwise, you may get cascading failures and lower availability instead of occasionally increased latency.

Similar to readiness probes, there are a few guidelines on how and when you should use liveness probes:

- Liveness probes should be used with caution. Incorrectly configuring these probes can result in cascading failures in your services and container restart loops.

- Do not use liveness probes unless you have a good reason to do so. A good reason may be, for example, if there's a known issue with a deadlock in your application that has an unknown root cause.

- Execute simple and fast checks that determine the status of the process, not its dependencies. In other words, you do not want to check the status of your external dependencies in the liveness probe, since this can lead to cascading failures due to an avalanche of container restarts and overloading a small subset of Service Pods.

- If the process running in your container can crash or exit whenever it encounters an unrecoverable error, you probably do not need a liveness probe at all.

- Use conservative settings for `initialDelaySeconds` to avoid any premature container restarts and falling into a restart loop.

These are the most important points concerning probes for Pods. Now, let's discuss how you should tag your container images.

Use meaningful and semantic image tags

Managing Deployment rollbacks and inspecting the history of rollouts requires that we use good tagging for the container images. If you rely on the `latest` tag, performing a rollback will not be possible because this tag points to a different version of the image as time goes on. It is a good practice to use semantic versioning for your container images. Additionally, you may consider tagging the images with a source code hash, such as a Git commit hash, to ensure that you can easily track what is running in your Kubernetes cluster.

Migrate from older versions of Kubernetes

If you are working on workloads that were developed on older versions of Kubernetes, you may notice that, starting with Kubernetes 1.16, you can't apply the Deployment to the cluster because of the following error:

```
error: unable to recognize "deployment": no matches for kind "Deployment" in
version "extensions/v1beta1"
```

The reason for this is that in version 1.16, the Deployment object was removed from the `extensions/v1beta1` API group, according to the API versioning policy. You should use the `apps/v1` API group instead, which Deployment has been part of since 1.9. This also shows an important rule to follow when you work with Kubernetes: always follow the API versioning policy and try to upgrade your resources to the latest API groups when you migrate to a new version of Kubernetes. This will save you unpleasant surprises when the resource is eventually deprecated in older API groups.

Include resource management in the Deployment

Define resource requests (minimum guaranteed resources) and limits (maximum allowed resources) for containers within your Deployment. This helps the Kubernetes scheduler allocate resources efficiently and prevent resource starvation. Also, don't overestimate or underestimate resource needs. Analyze application behavior to determine appropriate resource requests and limits to avoid under-utilization or performance bottlenecks.

Scaling and replica management

Use **HorizontalPodAutoscaler (HPA)** to automatically scale your Deployment replicas up or down based on predefined metrics like CPU or memory usage.

The following image illustrates how HPA handles the desired count of replicas:

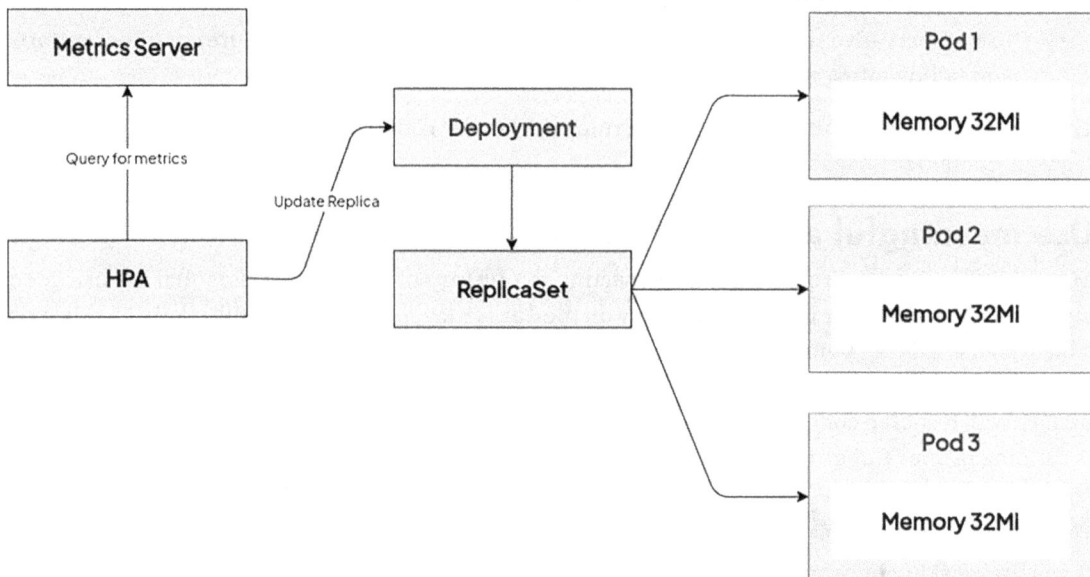

Figure 11.5: HPA handling the Deployment based on metrics information

Also, it is a best practice to set the initial replica count (number of Pod instances) for your Deployment considering factors like expected workload and resource availability.

Security considerations

Implement Deployment policies that will not allow running containers with unnecessary privileges (e.g., root user). This reduces the attack surface and potential security vulnerabilities. Also, remember to store sensitive information like passwords or configuration details in Secrets and ConfigMaps instead of embedding them directly in Deployments. Refer to *Chapter 18, Security in Kubernetes*, to explore different security aspects of Kubernetes.

Summary

In this chapter, you learned how to work with **stateless** workloads and applications on Kubernetes using Deployment objects. First, you created an example Deployment and exposed its Pods using a Service object of the LoadBalancer type for external traffic. Next, you learned how to scale and manage Deployment objects in the cluster. The management operations we covered included rolling out a new revision of a Deployment and rolling back to an earlier revision in case of a failure. Lastly, we equipped you with a set of known best practices when working with Deployment objects.

The next chapter will extend this knowledge with details about managing **stateful** workloads and applications. While doing so, we will introduce a new Kubernetes object: StatefulSet.

Further reading

- Kubernetes Deployments: `https://kubernetes.io/docs/concepts/workloads/controllers/deployment`
- StatefulSets: `https://kubernetes.io/docs/concepts/workloads/controllers/statefulset/`
- Canary deployments: `https://kubernetes.io/docs/concepts/workloads/management/#canary-deployments`

Join our community on Discord

Join our community's Discord space for discussions with the authors and other readers:

`https://packt.link/cloudanddevops`

12

StatefulSet – Deploying Stateful Applications

In the previous chapter, we explained how to use a Kubernetes cluster to run *stateless* workloads and applications and how to use Deployment objects for this purpose. Running stateless workloads in the cloud is generally easier to handle, as any container replica can handle the request without taking any dependencies on the results of previous operations by the end user. In other words, every container replica would handle the request in an identical way; all you need to care about is proper load balancing.

However, the main complexity is in managing the *state* of applications. By *state,* we mean any stored *data* that the application or component needs to serve the requests, and it can be modified by these requests. The most common example of a stateful component in applications is a database – for example, it can be a **relational MySQL database** or a **NoSQL MongoDB database**. In Kubernetes, you can use a dedicated object to run *stateful* workloads and applications: StatefulSet. When managing StatefulSet objects, you will usually need to work with Persistent Volumes (PVs), which have been covered in *Chapter 9, Persistent Storage in Kubernetes*. This chapter will provide you with knowledge about the role of StatefulSet in Kubernetes and how to create and manage StatefulSet objects to release new versions of your stateful applications.

In this chapter, we will cover the following topics:

- Introducing the StatefulSet object
- Managing StatefulSet
- Releasing a new version of an app deployed as a StatefulSet
- StatefulSet best practices

Technical requirements

For this chapter, you will need the following:

- A deployed Kubernetes cluster is needed. You can use either a local or a cloud-based cluster, but to fully understand the concepts, we recommend using a **multi-node**, cloud-based Kubernetes cluster. The cluster must support the creation of PersistentVolumeClaims. Any cloud-based cluster, or local cluster, for example, `minikube` with a `k8s.io/minikube-hostpath` provisioner, will be sufficient.

- A Kubernetes CLI (kubectl) must be installed on your local machine and configured to manage your Kubernetes cluster.

Kubernetes cluster deployment (local and cloud-based) and kubectl installation have been covered in *Chapter 3, Installing Your First Kubernetes Cluster*.

You can download the latest code samples for this chapter from the official GitHub repository at https://github.com/PacktPublishing/The-Kubernetes-Bible-Second-Edition/tree/main/Chapter12.

Introducing the StatefulSet object

You may wonder why running stateful workloads in the distributed cloud is generally considered **harder** than running stateless ones. In classic three-tier applications, all the states would be stored in a database (*data tier* or *persistence layer*) and there would be nothing special about it. For SQL servers, you would usually add a failover setup with data replication, and if you require superior performance, you would scale *vertically* by simply purchasing better hardware for hosting. Then, at some point, you might think about clustered SQL solutions, introducing *data sharding* (horizontal data partitions). But still, from the perspective of a web server running your application, the database would just be a single connection string to read and write the data. The database would be responsible for persisting a *mutable state*.

> Remember that every application *as a whole* is, in some way, stateful unless it only serves static content or just transforms user input. However, this does not mean that *every* component in the application is stateful. A web server that runs the application logic can be a *stateless* component, but the database where this application stores user input and sessions will be a *stateful* component.

We will first explain how you approach managing state in containers and what we consider an application or system state.

Managing state in containers

Now, imagine how this could work if you deployed your SQL server (single instance) in a container. The first thing you would notice is that after restarting the container, you would lose the data stored in the database – each time it is restarted, you get a fresh instance of the SQL server. Containers are *ephemeral*. This doesn't sound too useful for our use case. Fortunately, containers come with the option to mount data volumes. A volume can be, for example, a *host's directory or an external disk volume*, which will be *mounted* to a specific path in the container's filesystem. Whatever you store in this path will be kept in the volume even after the container is terminated or restarted. In a similar way, you can use NFS share or an external disk instance as a volume. Now, if you configure your SQL server to put its data files in the path where the volume is mounted, you achieve data persistence even if the container restarts. The container itself is still ephemeral, but the data (state) is *not*.

This is a high-level overview of how the state can be persisted for plain containers, without involving Kubernetes. But before we move on to Kubernetes, we need to clarify what we actually regard as a **state**.

Assume that you have a web server that serves just simple *static* content (which means it is always the same, as a simple HTML web page). There is still some data that has persisted, for example, the HTML files. However, this is *not* a state: user requests cannot modify this data, so *previous* requests from the user will not influence the result of the *current* request. In the same way, configuration files for your web server are not their state or log files written on the disk (well, that is arguable, but from the end user's perspective, it is not).

Now, if you have a web server that keeps user sessions and stores information about whether the user is logged in, then this is indeed the state. Depending on this information, the web server will return different web pages (responses) based on whether the user is logged in. Let's say that this web server runs in a container – there is a catch when it comes to whether it is *the* stateful component in your application. If the web server process stores user sessions as a file in the container (warning: this is probably quite a bad design), then the web server container is a *stateful* component. But if it stores user sessions in a database or a Redis cache running in separate containers, then the web server is *stateless*, and the database or Redis container becomes the stateful component.

This is briefly how it looks from a single container perspective. We need now to zoom out a bit and take a look at state management in **Kubernetes Pods**.

Managing state in Kubernetes Pods

In Kubernetes, the concept of container volumes is extended by **PersistentVolumes (PVs)**, **PersistentVolumeClaims (PVCs)**, and **StorageClasses (SCs)**, which are dedicated, storage-related objects. PVC aims to *decouple* Pods from the actual storage. PVC is a Kubernetes object that models a request for the storage of a specific type, class, or size – think of saying *I would like 10 GB of read/write-once SSD storage*. To fulfill such a request, a PV object is required, which is a piece of real storage that has been provisioned by the cluster's automation process – think of this as a directory on the host system or storage driver-managed disk. PV types are implemented as plugins, similar to volumes in Docker or Podman. Now, the whole process of provisioning PV can be *dynamic* – it requires the creation of an SC object and for this to be used when defining PVCs. When creating a new SC, you provide a *provisioner* (or plugin details) with specific parameters, and each PVC using the given SC will automatically create a PV using the selected provisioner. The provisioners may, for example, create cloud-managed disks to provide the backing storage. On top of that, containers of a given Pod can share data using the same PV and mount it to their filesystem.

This is just a brief overview of what Kubernetes provides for state storage. We have covered this in more detail in *Chapter 9, Persistent Storage in Kubernetes*.

On top of the management of state in a single Pod and its containers, there is the management of state in *multiple replicas* of a Pod. Let's think about what would happen if we used a Deployment object to run multiple Pods with MySQL Server. First, you would need to ensure that the state persisted on a volume in a container – for this, you can use PVs in Kubernetes. But then you actually get multiple, separate MySQL servers, which is not very useful if you would like to have high availability and fault tolerance. If you expose such a deployment using a service, it will also be useless because each time, you may hit a different MySQL Pod and get different data.

So, you arrive either at designing a *multi-node failover setup* with replication between the master and replicas, or a complex *cluster with data sharding*. In any case, your individual MySQL Server Pod replicas need to have a *unique identity* and, preferably, *predictable network names* so that the Nodes and clients can communicate.

> When designing your cloud-native application for the Kubernetes cluster, always analyze all the pros and cons of storing the state of the application as stateful components *running in Kubernetes*.

This is where StatefulSet comes in. Let's take a closer look at this Kubernetes object.

StatefulSet and how it differs from a Deployment object

Kubernetes StatefulSet is a similar concept to a Deployment object. It also provides a way of managing and scaling a set of Pods, but it provides guarantees about the *ordering and uniqueness* (unique identity) of the Pods. In the same way as Deployment, it uses a Pod template to define what each replica should look like. You can scale it up and down and perform rollouts of new versions. But now, in StatefulSet, the individual Pod replicas are *not interchangeable*. The unique, persistent identity for each Pod is maintained during any rescheduling or rollouts – this includes the **Pod name** and its **cluster DNS names**. This unique, persistent identity can be used to identify PVs assigned to each Pod, even if Pods are replaced following a failure. For this, StatefulSet provides another type of template in its specification named `volumeClaimTemplates`. This template can be used for the dynamic creation of the PVCs of a given SC. By doing this, the whole process of storage provisioning is fully dynamic – you just need to create a StatefulSet. The underlying storage objects are managed by the StatefulSet controller.

> Cluster DNS names of individual Pods in StatefulSet remain the same, but their cluster IP addresses are not guaranteed to stay the same. This means that if you need to connect to individual Pods in the StatefulSet, you should use cluster DNS names.

Basically, you can use StatefulSet for applications that require the following:

- Persistent storage managed by the Kubernetes cluster (this is the main use case, but not the only one)
- Stable and unique network identifiers (usually DNS names) for each Pod replica
- Ordered deployment and scaling
- Ordered, rolling updates

In the following diagram, you can see that StatefulSets can be seen as a more predictable version of a Deployment object, with the possibility to use persistent storage provided by PVCs.

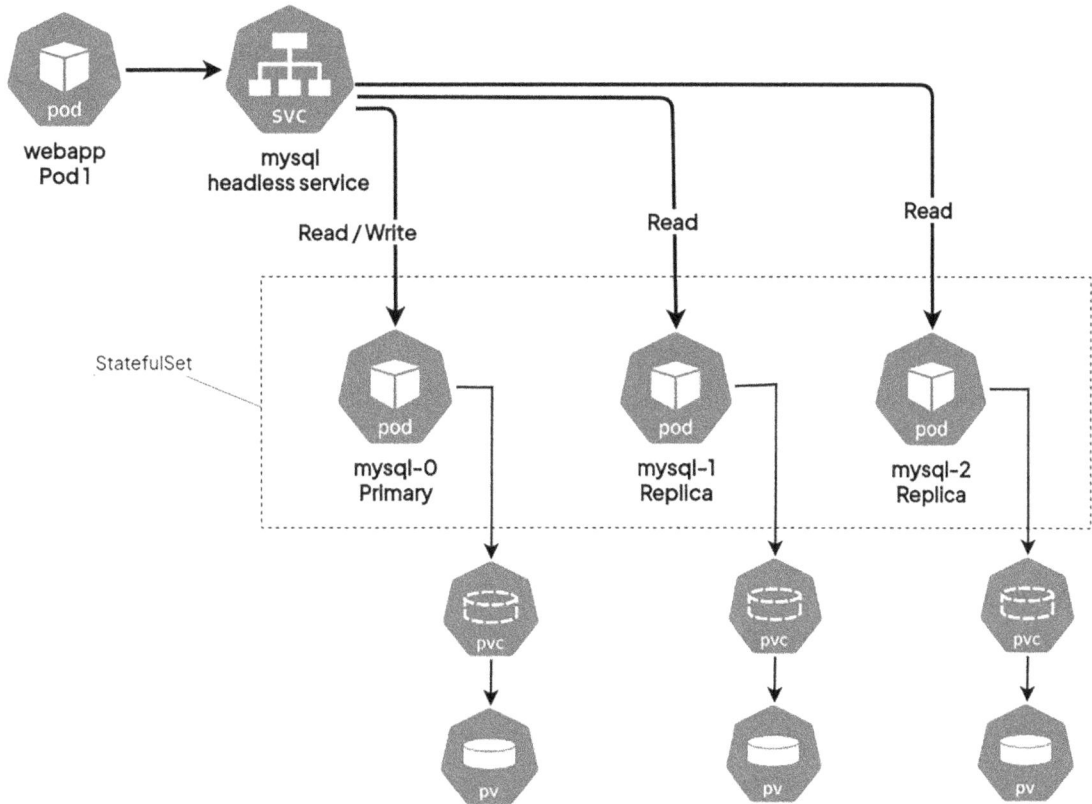

Figure 12.1: StatefulSet high-level view

To summarize, the key differences between StatefulSet and Deployment are as follows:

- StatefulSet ensures a *deterministic* (sticky) name for Pods, which consists of `<statefulSetName>-<ordinal>`. For Deployments, you would have a *random* name consisting of `<deploymentName>-<podTemplateHash>-<randomHash>`.

- For StatefulSet objects, the Pods are started and terminated in a *specific* and *predictable* order that ensures consistency, stability, and coordination while scaling the ReplicaSet. Let us take the preceding example diagram of MySQL StatefulSet; the Pods will be created in sequential order (mysql-0, mysql-1, and mysql-2). When you scale down the StatefulSet, the Pods will be terminated in the reverse order – mysql-2, mysql-1, and mysql-0 at the end.

- In terms of storage, Kubernetes creates PVCs based on `volumeClaimTemplates` of the StatefulSet specification for each Pod in the StatefulSet and always attaches this to the Pod with *the same* name. For Deployment, if you choose to use `persistentVolumeClaim` in the Pod template, Kubernetes will create a single PVC and attach the same to all the Pods in the deployment. This may be useful in certain scenarios but is not a common use case.

- You need to create a headless Service object that is responsible for managing the *deterministic network identity* (cluster DNS names) for Pods. The headless Service allows us to return *all* Pods' IP addresses behind the service as DNS A records instead of a single DNS A record with a ClusterIP Service. A headless Service is only required if you are not using a regular service. The specification of StatefulSet requires having the Service name provided in .spec.serviceName. Refer to *Understanding headless services* in *Chapter 8, Exposing Your Pods with Services*.

Before we explore the StatefulSet, we need to understand some of the limitations of the StatefulSet objects, as explained in the following section.

Exploring the limitations of StatefulSet

Here's a breakdown of some specific things to keep in mind when using StatefulSets:

- **Storage setup:** StatefulSets don't automatically create storage for your pods. You'll need to either use a built-in tool (like PersistentVolume Provisioner) or manually set up the storage beforehand.

- **Leftover storage:** When you scale down or remove a StatefulSet, the storage used by its pods sticks around. This is because your data is important and shouldn't accidentally be deleted. You'll need to clean up the storage yourself if needed. Otherwise, the leftover storage could become a concern because over time, all this unused storage can accumulate, leading to wasted resources and increased storage costs.

- **Pod address:** You'll need to set up a separate service (called a Headless Service) to give the Pod unique and stable network names.

- **Stopping StatefulSets:** There's no guarantee that pods will shut down in a specific order when you delete a StatefulSet. To ensure a clean shutdown, it's best to scale the StatefulSet down to zero Pods before removing it completely.

- **Updating StatefulSets:** Using the default update method with StatefulSets can sometimes lead to problems that require you to fix things manually. Be aware of this and consider alternative update strategies if needed.

Before you start exercises with StatefulSet, read important information about the StatefulSets in the next section.

Data management in Statefulset

Kubernetes offers StatefulSets as a powerful tool for managing stateful applications. However, successfully deploying and maintaining stateful applications requires user involvement beyond simply defining the StatefulSet itself. Here's a breakdown of key areas requiring your attention:

- **Data cloning and synchronization:** Unlike stateless applications, stateful applications rely on persistent data. While StatefulSets manage Pod ordering and identity, they don't handle data replication between Pods. You'll need to implement this functionality yourself. Common approaches include using init containers to copy data from a predefined source during Pod creation, leveraging built-in replication features within your application (like MySQL replication), or utilizing external scripts to manage data synchronization.

- **Remote storage accessibility:** StatefulSets ensure Pods can be rescheduled across available nodes in the cluster. To maintain data persistence during rescheduling, the storage provisioned by the PV needs to be accessible from all worker nodes. This means choosing a storage class that replicates data across nodes or using network-attached storage solutions accessible from all machines.

- **External backups:** StatefulSets are designed for managing Pod life cycles and data persistence within the cluster. However, they don't handle external backups. To ensure disaster recovery in case of catastrophic events, implementing a separate external backup solution is crucial. This could involve backing up your PVs to a cloud storage service or a dedicated backup server outside the Kubernetes cluster.

There are best practices and recommended approaches for handling the data in StatefulSets. The following section will explain some of the replication management techniques for StatefulSets.

Replication management

As the name suggests, Stateful applications handling data often require data initialization or synchronization. You might need to implement these using methods like the following:

- **init containers:** Copy data from a source (like a config map) to the persistent storage before starting the application.

- **Application-level replication:** Leverage built-in replication features within your application to handle data updates across Pods.

- **External scripts:** Use external scripts or tools to manage data migration during the update process.

Now, let's take a look at a concrete example of StatefulSet that deploys MySQL Pods with the backing of persistent storage.

Managing StatefulSet

To demonstrate how StatefulSet objects work, we will modify our MySQL deployment and adapt it to be a StatefulSet. A significant part of the StatefulSet specification is the same as for Deployments. As we would like to demonstrate how the automatic management of PVCs works in StatefulSet objects, we will use `volumeClaimTemplates` in the specification to create PVCs and PVs, which the Pods will consume. Each Pod will internally mount its assigned PV under the `/var/lib/mysql` path in the container filesystem, which is the default location of MySQL data files. In this way, we can demonstrate how the *state* persists, even if we forcefully restart Pods.

> The example that we are going to use in this chapter is for demonstration purposes only and is meant to be as simple as possible. If you are interested in *complex* examples, such as deploying and managing distributed databases in StatefulSets, please take a look at the official Kubernetes blog post about deploying the Cassandra database at `https://kubernetes.io/docs/tutorials/stateful-application/cassandra/`. Usually, the main source of complexity in such cases is handling the joining and removal of Pod replicas when scaling the StatefulSet.

We will now go through all the YAML manifests required to create our StatefulSet and apply them to the cluster.

Creating a StatefulSet

We have discussed the concepts of StatefulSet, and now it's time to learn how to manage them. First, let's take a look at the StatefulSet YAML manifest file named `mysql-statefulset.yaml`:

```yaml
# mysql-statefulset.yaml
apiVersion: apps/v1
kind: StatefulSet
metadata:
  name: mysql-stateful
  labels:
    app: mysql
  namespace: mysql
spec:
  serviceName: mysql-headless
  replicas: 3
  selector:
    matchLabels:
      app: mysql
      environment: test
# (to be continued in the next paragraph)
```

The first part of the preceding file is very similar to the Deployment object specification, where you need to provide the number of `replicas` and a `selector` for Pods. There is one new parameter, `serviceName`, which we will explain shortly.

The next part of the file concerns the specification of the Pod template that is used by the StatefulSet:

```yaml
# (continued)
  template:
    metadata:
      labels:
        app: mysql
        environment: test
    spec:
      containers:
        - name: mysql
          image: mysql:8.2.0
          ports:
            - containerPort: 3306
          volumeMounts:
```

```
          - name: mysql-data
            mountPath: /var/lib/mysql
        env:
          - name: MYSQL_ROOT_PASSWORD
            valueFrom:
              secretKeyRef:
                name: mysql-secret
                key: MYSQL_ROOT_PASSWORD
          - name: MYSQL_USER
            valueFrom:
              secretKeyRef:
                name: mysql-secret
                key: MYSQL_USER
          - name: MYSQL_PASSWORD
            valueFrom:
              secretKeyRef:
                name: mysql-secret
                key: MYSQL_PASSWORD
# (to be continued in the next paragraph)
```

If you look closely, you can observe that the structure is the same as for Deployments. Notice the environment variables we are providing via the Secret object, which needs to be created before creating the StatefulSet. Also, the last part of the file contains `volumeClaimTemplates`, which is used to define templates for PVC used by the Pod:

```
# (continued)
  volumeClaimTemplates:
    - metadata:
        name: mysql-data
      spec:
        accessModes:
          - "ReadWriteOnce"
        resources:
          requests:
            storage: 1Gi
```

As you can see, in general, the structure of the StatefulSet spec is similar to a Deployment, although it has a few extra parameters for configuring PVCs and associated Service objects. The specification has five main components:

- `replicas`: Defines the number of Pod replicas that should run using the given `template` and the matching label `selector`. Pods may be created or deleted to maintain the required number.

- serviceName: The name of the service that governs the StatefulSet and provides the network identity for the Pods. This Service must be created before the StatefulSet is created. We will create the mysql-headless Service in the next step.

- selector: A **label selector**, which defines how to identify Pods that the StatefulSet owns. This can include *set-based* and *equality-based* selectors.

- template: Defines the template for Pod creation. Labels used in metadata must match the selector. Pod names are not random and follow the <statefulSetName>-<ordinal> convention. You can optionally use .spec.ordinals to control the starting number for the unique identification number assigned to each pod in your StatefulSet.

- volumeClaimTemplates: Defines the template for PVC that will be created for each of the Pods. Each Pod in the StatefulSet object will get its own PVC that is assigned to a given Pod name persistently. In our case, it is a 1 GB volume with the ReadWriteOnce access mode. This access mode allows the volume to be mounted for reads and writes by a *single* Node only. We did not specify storageClassName, so the PVCs will be provisioned using the default SC in the cluster. PVC names are not random and follow the <volumeClaimTemplateName>-<statefulSetName>-<ordinal> convention.

> The default SC in your cluster is marked with the storageclass.kubernetes.io/is-default-class annotation. Whether you have a default SC, and how it is defined, depends on your cluster deployment. For example, in the Azure Kubernetes Service cluster, it will be an SC named default that uses the kubernetes.io/azure-disk provisioner. In minikube, it will be an SC named standard that uses the k8s.io/minikube-hostpath provisioner.

The specification also contains other fields that are related to rolling out new revisions of StatefulSet – we will explain these in more detail in the next section.

Next, let's have a look at our *headless* Service named mysql-headless. Create a mysql-headless-service.yaml file with the following content:

```yaml
# mysql-headless-service.yaml
apiVersion: v1
kind: Service
metadata:
  name: mysql-headless
  namespace: mysql
spec:
  selector:
    app: mysql
    environment: test
  clusterIP: None
  ports:
    - port: 3306
```

The specification is very similar to the normal Service that we created previously for the Deployment; the only difference is that it has the value None for the clusterIP field. This will result in the creation of a headless Service, mysql-headless. A headless Service allows us to return *all* Pods' IP addresses behind the Service as DNS A records instead of a single DNS A record with a clusterIP Service. We will demonstrate what this means in practice in the next steps.

With all the YAML manifest files, we can start deploying our example StatefulSet! Perform the following steps:

1. Create a namespace called mysql (using mysql-ns.yaml):

```
$ kubectl apply -f mysql-ns.yaml
namespace/mysql created
```

2. Create a Secret to store MySQL environment variables:

```
$ kubectl create secret generic mysql-secret \
  --from-literal=MYSQL_ROOT_PASSWORD='mysqlroot' \
  --from-literal=MYSQL_USER='mysqluser' \
  --from-literal=MYSQL_PASSWORD='mysqlpassword' \
  -n mysql
secret/mysql-secret created
```

> Please note that it is also possible to create the same secret using YAML and base64 encoded (e.g., echo -n 'mysqlroot' | base64) values inside (refer to mysql-secret.yaml in the repository to see the sample YAML file); we are using this imperative method to demonstrate the secret with actual values.

3. Create a headless Service, mysql-headless, using the following command:

```
$ kubectl apply -f mysql-headless-service.yaml
service/mysql-headless created
```

4. Create a StatefulSet object, mysql-stateful, using the following command:

```
$ kubectl apply -f mysql-statefulset.yaml
statefulset.apps/mysql-stateful created
```

5. Now, you can use the kubectl describe command to observe the creation of the StatefulSet object (alternatively, you can use sts as an abbreviation for StatefulSet when using kubectl commands):

```
$ kubectl describe statefulset mysql-stateful -n mysql

$ kubectl get sts -n mysql
NAME             READY   AGE
mysql-stateful   3/3     2m3s
```

6. Use the kubectl get pods command to see that the three desired Pod replicas have been created. Note that this can take a bit of time as the Pods have to get the PVs provisioned based on their PVCs:

```
$ kubectl get pod -n mysql
NAME                 READY    STATUS     RESTARTS    AGE
mysql-stateful-0     1/1      Running    0           2m32s
mysql-stateful-1     1/1      Running    0           2m29s
mysql-stateful-2     1/1      Running    0           2m25s
```

> Please note the ordered, deterministic Pod naming – this is the key to providing a unique identity to the Pods in the StatefulSet object.

7. If you describe one of the Pods, you will see more details about the associated PV and PVC:

```
$ kubectl -n mysql describe pod mysql-stateful-0
Name:             mysql-stateful-0
Namespace:        mysql
Priority:         0
Service Account:  default
...<removed for brevity>...
Volumes:
  mysql-data:
    Type:          PersistentVolumeClaim (a reference to a
PersistentVolumeClaim in the same namespace)
    ClaimName:     mysql-data-mysql-stateful-0
    ReadOnly:      false
...<removed for brevity>...
```

For the second Pod, you will see a similar output to the following, but with a different PVC:

```
$ kubectl -n mysql describe pod mysql-stateful-1
Name:             mysql-stateful-1
Namespace:        mysql
Priority:         0
...<removed for brevity>...
Volumes:
  mysql-data:
    Type:          PersistentVolumeClaim (a reference to a
PersistentVolumeClaim in the same namespace)
    ClaimName:     mysql-data-mysql-stateful-1
    ReadOnly:      false
...<removed for brevity>...
```

As you can see, the PVC used by this `mysql-stateful-0` Pod is named `mysql-data-mysql-stateful-0` and the PVC used by this `mysql-stateful-1` Pod is named `mysql-data-mysql-stateful-1`. Right after the Pod was scheduled on its target Node, the PVs are provisioned via the respective StorageClass and bound to the individual PVCs. After that, the actual container, which internally mounts this PV, has been created.

8. Using the `kubectl get` command, we can reveal more details about the PVC:

```
$  kubectl get pvc -n mysql
NAME                          STATUS   VOLUME
CAPACITY    ACCESS MODES    STORAGECLASS    VOLUMEATTRIBUTESCLASS    AGE
mysql-data-mysql-stateful-0   Bound    pvc-453dbfee-6076-48b9-8878-
e7ac6f79d271    1Gi         RWO             standard        <unset>
8m38s
mysql-data-mysql-stateful-1   Bound    pvc-36494153-3829-42aa-be6d-
4dc63163ea38    1Gi         RWO             standard        <unset>
8m35s
mysql-data-mysql-stateful-2   Bound    pvc-6730af33-f0b6-445d-841b-
4fbad5732cde    1Gi         RWO             standard        <unset>
8m31s
```

9. And finally, let's take a look at the PV that was provisioned:

```
$ kubectl get pv
NAME                                              CAPACITY
ACCESS MODES    RECLAIM POLICY    STATUS    CLAIM
STORAGECLASS    VOLUMEATTRIBUTESCLASS    REASON    AGE
pvc-36494153-3829-42aa-be6d-4dc63163ea38    1Gi          RWO
Delete            Bound    mysql/mysql-data-mysql-stateful-1    standard
<unset>                    11m
pvc-453dbfee-6076-48b9-8878-e7ac6f79d271    1Gi          RWO
Delete            Bound    mysql/mysql-data-mysql-stateful-0    standard
<unset>                    11m
pvc-6730af33-f0b6-445d-841b-4fbad5732cde    1Gi          RWO
Delete            Bound    mysql/mysql-data-mysql-stateful-2    standard
<unset>                    11m
```

Please note that, in our example, we are demonstrating this using the minikube `hostPath` type. If your Kubernetes cluster uses a different storage backend, you will see different outputs.

We have successfully created the StatefulSet object; now it is time to verify whether it works as expected in a basic scenario. To do this, let us use an updated k8sutils container image with the default MySQL client package installed. (Check Chapter12/Containerfile to see the details of the k8sutils image.) Create k8sutils.yaml as follows:

```
# k8sutils.yaml
apiVersion: v1
kind: Pod
metadata:
  name: k8sutils
  # namespace: default
spec:
  containers:
    - name: k8sutils
      image: quay.io/iamgini/k8sutils:debian12-1.1
      command:
        - sleep
        - "infinity"
      # imagePullPolicy: IfNotPresent
  restartPolicy: Always
```

Create the k8sutils Pod as follows:

```
$ kubectl apply -f k8sutils.yaml -n mysql
pod/k8sutils created
```

Please note that we used -n mysql while applying the YAML so that the resource will be created inside the mysql namespace.

Follow these steps to verify the content of different Pods in the StatefulSet:

1. Jump into the k8sutil Pod to execute our test commands:

```
$ kubectl exec -it -n mysql k8sutils -- /bin/bash
root@k8sutils:/#
```

2. Access the MySQL Stateful application using the default headless service we created earlier (remember the password you created using the Secret object earlier):

```
root@k8sutils:/# mysql -u root -p -h mysql-headless.mysql.svc.cluster.
local
Enter password: <mysqlroot>
Welcome to the MariaDB monitor.  Commands end with ; or \g.
Your MySQL connection id is 8
```

```
Server version: 8.2.0 MySQL Community Server - GPL

Copyright (c) 2000, 2018, Oracle, MariaDB Corporation Ab and others.

Type 'help;' or '\h' for help. Type '\c' to clear the current input
statement.

MySQL [(none)]> show databases;
+--------------------+
| Database           |
+--------------------+
| information_schema |
| mysql              |
| performance_schema |
| sys                |
+--------------------+
4 rows in set (0.003 sec)

MySQL [(none)]>
```

The basic MySQL connection is working, and we are able to access the MySQL server running as a StatefulSet application. We will now take a quick look at how the *headless* Service behaves.

Using the headless Service and stable network identities

Previously, we learned about headless service in Kubernetes and how we can use it for accessing StatefulSet applications. (Refer to *Understanding headless services* section in *Chapter 8, Exposing Your Pods with Services*). In this section, let us go deep and explore the headless service mechanism in the backend of StatefulSet.

Let's do an experiment that demonstrates how the `headless` Service is used to provide stable and predictable network identities for our Pods:

1. Log in to the same k8sutils Pod that we used in the previous test.
2. Perform DNS check for the headless Service, `mysql-headless`:

```
root@k8sutils:/# nslookup mysql-headless
Server:        10.96.0.10
Address:       10.96.0.10#53

Name:   mysql-headless.mysql.svc.cluster.local
Address: 10.244.0.14
Name:   mysql-headless.mysql.svc.cluster.local
Address: 10.244.0.15
Name:   mysql-headless.mysql.svc.cluster.local
Address: 10.244.0.16
```

We have received three `A` records that point directly to Pod IP addresses. Additionally, they have `CNAME` records in the form of `<podName>-<ordinal-number>.<headless-serviceName>.<namespace>.svc.cluster.local`. So, the difference with default Service is that a Service that has `ClusterIP` will get load balancing to a *virtual IP* level (which, on Linux, is usually handled at a kernel level by `iptables` rules configured by `kube-proxy`), whereas in the case of the headless Service, the responsibility for load balancing or choosing the target Pod is on the *client* making the request.

3. Having *predictable* FQDNs for Pods in the StatefulSet gives us the option to send the requests directly to individual Pods, without guessing their IP addresses or names. Let's try accessing the MySQL server served by the `mysql-stateful-0` using its short DNS name provided by the headless Service:

```
root@k8sutils:/# mysql -u root -p -h mysql-stateful-0.mysql-headless
Enter password: <mysqlroot>
Welcome to the MariaDB monitor.  Commands end with ; or \g.
Your MySQL connection id is 8
Server version: 8.2.0 MySQL Community Server - GPL

Copyright (c) 2000, 2018, Oracle, MariaDB Corporation Ab and others.

Type 'help;' or '\h' for help. Type '\c' to clear the current input
statement.

MySQL [(none)]>
```

As expected, you have connected directly to the Pod and have been served by the proper Pod.

4. Let us create a database inside the MySQL database server as follows:

```
MySQL [(none)]> create database ststest;
Query OK, 1 row affected (0.002 sec)
MySQL [(none)]> exit;
Bye
```

5. Now, we will show that this DNS name remains unchanged even if a Pod is restarted. The IP of the Pod will change, but the DNS name will not. What is more, the PV that is mounted will also stay the same, but we will investigate this in the next paragraphs. In another shell window, outside of the container, execute the following command to force a restart of the `mysql-stateful-0` Pod:

```
$ kubectl delete po -n mysql mysql-stateful-0
```

Check the Pods and you will see that `mysql-stateful-0` has been recreated and mounted with the same `mysql-data-mysql-stateful-0` PVC:

```
$ kubectl get po -n mysql
```

```
NAME                 READY   STATUS    RESTARTS   AGE
k8sutils             1/1     Running   0          35m
mysql-stateful-0     1/1     Running   0          6s
mysql-stateful-1     1/1     Running   0          52m
mysql-stateful-2     1/1     Running   0          51m
```

6. In the k8sutils shell, execute the MySQL client command to check the database server content:

```
root@k8sutils:/# mysql -u root -p -h mysql-stateful-0.mysql-headless
Enter password: <mysqlroot>
Welcome to the MariaDB monitor.  Commands end with ; or \g.
Your MySQL connection id is 8
Server version: 8.2.0 MySQL Community Server - GPL

Copyright (c) 2000, 2018, Oracle, MariaDB Corporation Ab and others.

Type 'help;' or '\h' for help. Type '\c' to clear the current input
statement.

MySQL [(none)]> show databases;
+--------------------+
| Database           |
+--------------------+
| information_schema |
| mysql              |
| performance_schema |
| ststest            |
| sys                |
+--------------------+
5 rows in set (0.003 sec)
```

You can see that the database ststest we created before the Pod deletion is still there, which means the data is persistent or stateful. Also, notice the IP address of the Pod changed but the DNS remained the same.

This explains how the headless Services can be leveraged to get a stable and predictable network identity that will not change when a Pod is restarted or recreated. You may wonder what the actual use of this is and why it is important for StatefulSet objects. There are a couple of possible use cases:

* Deploying clustered databases, such as etcd or MongoDB, requires specifying the network addresses of other Nodes in the database cluster. This is especially necessary if there are no *automatic discovery* capabilities provided by the database. In such cases, stable DNS names provided by headless Services help to run such clusters on Kubernetes as StatefulSets. There is still the problem of changing the configuration when Pod replicas are added or removed from the StatefulSet during scaling. In some cases, this is solved by the *sidecar container pattern*, which monitors the Kubernetes API to dynamically change the database configuration.

- If you decide to implement your own storage solution running as StatefulSet with advanced data sharding, you will most likely need mappings of logical shards to physical Pod replicas in the cluster. Then, the stable DNS names can be used as part of this mapping. They will guarantee that queries for each logical shard are performed against a proper Pod, irrespective of whether it was rescheduled to another Node or restarted.

Finally, let's take a look at the state persistence for Pods running in StatefulSet.

State persistence

As we demonstrated earlier, the data is persistent inside the PV and will bound back to the newly created Pod with the same ordinal number.

In the following example, we are deleting all of the Pods in the StatefulSet:

```
$ kubectl delete po -n mysql mysql-stateful-0 mysql-stateful-1 mysql-stateful-2
pod "mysql-stateful-0" deleted
pod "mysql-stateful-1" deleted
pod "mysql-stateful-2" deleted
```

Kubernetes will recreate all of the Pods in the same order as follows:

```
$ kubectl get pod -n mysql
NAME                  READY    STATUS     RESTARTS    AGE
k8sutils              1/1      Running    0           47m
mysql-stateful-0      1/1      Running    0           44s
mysql-stateful-1      1/1      Running    0           43s
mysql-stateful-2      1/1      Running    0           41s
```

We can also verify the PVs mounted to ensure the Pod to PVC binding was successful:

```
$ kubectl describe pod -n mysql -l app=mysql |egrep 'ClaimName|Name:'
Name:               mysql-stateful-0
    ClaimName:  mysql-data-mysql-stateful-0
    ConfigMapName:          kube-root-ca.crt
Name:               mysql-stateful-1
    ClaimName:  mysql-data-mysql-stateful-1
    ConfigMapName:          kube-root-ca.crt
Name:               mysql-stateful-2
    ClaimName:  mysql-data-mysql-stateful-2
    ConfigMapName:          kube-root-ca.crt
```

As you have learned, the PV will not be removed by the StatefulSet controller as part of Pod or StatefulSet deletion. It is your responsibility to clean up the data by removing the PVs manually if you are removing the StatefulSet objects completely.

Next, we will take a look at scaling the StatefulSet object.

Scaling StatefulSet

In the case of StatefulSets, you can do similar *scaling* operations as for Deployment objects by changing the number of replicas in the specification or using the kubectl scale imperative command. The new Pods will automatically be discovered as new Endpoints for the Service when you scale up, or automatically removed from the Endpoints list when you scale down.

However, there are a few differences when compared to Deployment objects:

- When you deploy a StatefulSet object of N replicas, the Pods during deployment are created sequentially, in order from 0 to N-1. In our example, during the creation of a StatefulSet object of three replicas, the first mysql-stateful-0 Pod is created, followed by n mysql-stateful-1, and finally mysql-stateful-2.

- When you scale *up* the StatefulSet, the new Pods are also created sequentially and in an ordered fashion.

- When you scale *down* the StatefulSet, the Pods are terminated sequentially, in *reverse order*, from N-1 to 0. In our example, while scaling down the StatefulSet object to zero replicas, the mysql-stateful-2 Pod is terminated, followed by mysql-stateful-1, and finally mysql-stateful-0.

- During the scaling up of the StatefulSet object, before the next Pod is created in the sequence, all its predecessors must be *running* and *ready*.

- During the scaling *down* of the StatefulSet object, before the next Pod is terminated in the reverse sequence, all its predecessors must be completely *terminated* and *deleted*.

- Also, in general, before *any* scaling operation is applied to a Pod in a StatefulSet object, all its predecessors must be running and ready. This means that if, during scaling down from four replicas to one replica, the mysql-stateful-0 Pod were to suddenly fail, then no further scaling operation would be performed on the mysql-stateful-1, mysql-stateful-2, and mysql-stateful-3 Pods. Scaling would resume when the mysql-stateful-0 Pod becomes ready again.

> This sequential behavior of scaling operations can be relaxed by changing the .spec.podManagementPolicy field in the specification. The default value is OrderedReady. If you change it to Parallel, the scaling operations will be performed on Pods in parallel, similar to what you know from Deployment objects. Note that this only affects scaling operations. The way of updating the StatefulSet object with updateStrategy of the RollingUpdate type does not change.

Equipped with this knowledge, let's *scale up* our StatefulSet imperatively to demonstrate it quickly:

1. Scale out the StatefulSet using the following command:

```
$ kubectl scale statefulset -n mysql mysql-stateful --replicas 4
statefulset.apps/mysql-stateful scaled
```

2. If you now check the Pods using the `kubectl get pods` command, you will see the sequential, ordered creation of new Pods:

```
$ kubectl get pod -n mysql
NAME                READY    STATUS     RESTARTS    AGE
k8sutils            1/1      Running    0           56m
mysql-stateful-0    1/1      Running    0           9m13s
mysql-stateful-1    1/1      Running    0           9m12s
mysql-stateful-2    1/1      Running    0           9m10s
mysql-stateful-3    1/1      Running    0           4s
```

Similarly, if you check the output of the `kubectl describe` command for the StatefulSet object, you will see the following in the events:

```
$ kubectl describe sts -n mysql mysql-stateful
Name:              mysql-stateful
Namespace:         mysql
...<removed for brevity>...
Events:
  Type      Reason                 Age                     From
Message
  ----      ------                 ----                    ----
-------
  Normal    SuccessfulCreate       23m (x2 over 75m)    statefulset-
controller   create Pod mysql-stateful-0 in StatefulSet mysql-stateful
successful
  Normal    RecreatingTerminatedPod  11m (x13 over 23m)    statefulset-
controller   StatefulSet mysql/mysql-stateful is recreating terminated Pod
mysql-stateful-0
  Normal    SuccessfulDelete       11m (x13 over 23m)    statefulset-
controller   delete Pod mysql-stateful-0 in StatefulSet mysql-stateful
successful
  Normal    SuccessfulCreate       2m28s                statefulset-
controller   create Claim mysql-data-mysql-stateful-3 Pod mysql-stateful-3
in StatefulSet mysql-stateful success
  Normal    SuccessfulCreate       2m28s                statefulset-
controller   create Pod mysql-stateful-3 in StatefulSet mysql-stateful
successful
Let us scale down our StatefulSet object imperatively and check the Pods
as follows:
$ kubectl scale statefulset -n mysql mysql-stateful --replicas 2
statefulset.apps/mysql-stateful scaled
```

You can see that the last two Pods – `mysql-stateful-3` and `mysql-stateful-2` – are deleted in an orderly way. Now, let us check the Pods in the `statefulset` as follows:

```
$ kubectl get pod -n mysql
NAME              READY   STATUS    RESTARTS   AGE
k8sutils          1/1     Running   0          61m
mysql-stateful-0  1/1     Running   0          15m
mysql-stateful-1  1/1     Running   0          15m
```

3. Check the PVCs now and you will see the PVCs are still there. This is an expected situation for StatefulSet, as we learned earlier:

```
$ kubectl get pvc -n mysql
NAME                               STATUS    VOLUME
CAPACITY   ACCESS MODES   STORAGECLASS   VOLUMEATTRIBUTESCLASS    AGE
mysql-data-mysql-stateful-0    Bound     pvc-453dbfee-6076-48b9-8878-
e7ac6f79d271   1Gi         RWO            standard       <unset>
79m
mysql-data-mysql-stateful-1    Bound     pvc-36494153-3829-42aa-be6d-
4dc63163ea38   1Gi         RWO            standard       <unset>
79m
mysql-data-mysql-stateful-2    Bound     pvc-6730af33-f0b6-445d-841b-
4fbad5732cde   1Gi         RWO            standard       <unset>
79m
mysql-data-mysql-stateful-3    Bound     pvc-6ec1ee2a-5be3-4bf9-84e5-
4f5aee566c11   1Gi         RWO            standard       <unset>
7m4s
```

> When managing StatefulSets with **Horizontal Pod Autoscaler (HPA)** or similar horizontal scaling tools, refrain from specifying a value for `.spec.replicas` in your manifest. Instead, leave it unset. The Kubernetes control plane will dynamically adjust the number of replicas as per resource requirements, facilitating efficient scaling of your application without the need for manual intervention.

Congratulations! We have learned how to deploy and scale StatefulSet objects. Next, we will demonstrate how you can delete a StatefulSet object.

Deleting a StatefulSet

To delete a StatefulSet object, there are two possibilities:

*   Delete the StatefulSet together with Pods that it owns
*   Delete the StatefulSet and leave the Pods unaffected

In both cases, the PVCs and PVs that were created for the Pods using `volumeClaimTemplates` will *not* be deleted by default. This ensures that state data is not lost accidentally unless you explicitly clean up the PVCs and PVs.

But with the latest Kubernetes versions (from v1.27 onwards), you can use the `.spec.persistentVo lumeClaimRetentionPolicy` field to control the deletion of PVCs as part of the StatefulSet life cycle:

```
apiVersion: apps/v1
kind: StatefulSet
...
spec:
  persistentVolumeClaimRetentionPolicy:
    whenDeleted: Retain
    whenScaled: Delete
...
```

Refer to the documentation to learn more about `persistentVolumeClaimRetentionPolicy` (https:// kubernetes.io/docs/concepts/workloads/controllers/statefulset/#persistentvolumeclaim- retention).

To delete the StatefulSet object together with Pods, you can use the regular `kubectl delete` command:

```
$ kubectl delete sts -n mysql mysql-stateful
statefulset.apps "mysql-stateful" deleted
```

You will see that the Pods will be terminated first, followed by the StatefulSet object. Please note that this operation is different from *scaling down* the StatefulSet object to zero replicas and then deleting it. If you delete the StatefulSet object with existing Pods, there are no guarantees regarding the order of termination of the individual Pods. In most cases, they will be terminated at once.

Optionally, if you would like to delete just the StatefulSet object, you need to use the `--cascade=orphan` option for `kubectl delete`:

```
$ kubectl delete sts -n mysql mysql-stateful --cascade=orphan
```

After this command, if you inspect what Pods are in the cluster, you will still see all the Pods that were owned by the `mysql-stateful` StatefulSet.

Lastly, if you would like to clean up PVCs and PVs after deleting the StatefulSet object, you need to perform this step manually. Use the following command to delete the PVC's created as part of the StatefulSet:

```
$ kubectl delete -n mysql pvc mysql-data-mysql-stateful-0 mysql-data-mysql-
stateful-1 mysql-data-mysql-stateful-2 mysql-data-mysql-stateful-3
persistentvolumeclaim "mysql-data-mysql-stateful-0" deleted
persistentvolumeclaim "mysql-data-mysql-stateful-1" deleted
persistentvolumeclaim "mysql-data-mysql-stateful-2" deleted
persistentvolumeclaim "mysql-data-mysql-stateful-3" deleted
```

This command will delete PVCs and associated PVs.

> IMPORTANT NOTE
>
> Please note that if you want to perform verifications of state persistence after exercising the new version rollout in the next section, you should not yet delete the PVCs. Otherwise, you will lose the MySQL files stored in the PVs.

With this section, we have completed our learning on basic operations with the StatefulSet objects in Kubernetes. Next, let's take a look at releasing new versions of apps deployed as StatefulSets and how StatefulSet revisions are managed.

Releasing a new version of an app deployed as a StatefulSet

We have just covered the *scaling* of StatefulSets in the previous section by the `kubectl scale` command (or by making changes to the `.spec.replicas` number in the specification). Everything you have learned about sequential and ordered changes to the Pods plays an important role in rolling out a new revision of a StatefulSet object when using the `RollingUpdate` strategy. There are many similarities between StatefulSets and Deployment objects. We covered the details of Deployment updates in *Chapter 11, Using Kubernetes Deployments for Stateless Workloads*. Making changes to the StatefulSet Pod *template* (`.spec.template`) in the specification will also cause the rollout of a new revision for StatefulSet.

StatefulSets support two types of *update strategies* that you define using the `.spec.updateStrategy.type` field in the specification:

- `RollingUpdate`: The default strategy, which allows you to roll out a new version of your application in a controlled way. This is slightly different from the `RollingUpdate` strategy known from Deployment objects. For StatefulSet, this strategy will terminate and recreate Pods in a sequential and ordered fashion and make sure that the Pod is recreated and in a ready state before proceeding to the next one.

- `OnDelete`: This strategy implements the legacy behavior of StatefulSet updates prior to Kubernetes 1.7. However, it is still useful! In this type of strategy, StatefulSet will *not* automatically update the Pod replicas by recreating them. You need to manually delete a Pod replica to get the new Pod template applied. This is useful in scenarios when you need to perform additional manual actions or verifications before proceeding to the next Pod replica. For example, if you are running a *Cassandra cluster* or an *etcd cluster* in a StatefulSet, you may want to verify whether the new Pod has correctly joined the existing cluster following the removal of the previous version of the Pod. Of course, it is possible to perform similar checks using the Pod template life cycle `postStart` and `preStop` hooks while using the `RollingUpdate` strategy, but this requires more sophisticated error handling in the hooks.

Let's now take a closer look at the `RollingUpdate` strategy, which is the most important and commonly used update strategy for StatefulSets. The key thing about this is that the strategy respects all the StatefulSet guarantees, which we explained in the previous section regarding scaling. The rollout is done in reverse order; for example, the first Pod, `mysql-stateful-2`, is recreated with the new Pod template, followed by `mysql-stateful-1`, and finally `mysql-stateful-0`.

If the process of rollout fails (not necessarily the Pod that was currently recreated), the StatefulSet controller is going to restore any failed Pod to its *current version*. This means that the Pods that have already received a *successful* update to the current version will remain at the current version, whereas the Pods that have not yet received the update will remain at the previous version. In this way, the StatefulSet attempts to always keep applications healthy and consistent. However, this can also lead to *broken* rollouts of StatefulSets. If one of the Pod replicas *never* becomes running and ready, then the StatefulSet will stop the rollout and wait for *manual* intervention. Applying the template again to the previous revision of StatefulSet is not enough – this operation will not proceed as the StatefulSet will wait for the failed Pod to become ready. The only resolution is manual deletion of the failed Pod and then having the StatefulSet apply the previous revision of the Pod template.

Lastly, the RollingUpdate strategy also provides the option to execute *staged* rollouts using the .spec.updateStrategy.rollingUpdate.partition field. This field defines a number for which all the Pod replicas that have a *lesser ordinal* number will not be updated, and, even if they are deleted, they will be recreated at the previous version. So, in our example, if the partition were to be set to 1, this means that during the rollout, only mysql-stateful-1 and mysql-stateful-2 would be updated, whereas mysql-stateful-0 would remain unchanged and run on the previous version. By controlling the partition field, you can easily roll out a single *canary* replica and perform *phased* rollouts. Please note that the default value is 0, which means that all Pod replicas will be updated.

Now, we will release a new version of our mysqlserver using the RollingUpdate strategy.

Updating StatefulSet

We will now demonstrate how to do a rollout of a new image version for a Pod container using the StatefulSet YAML manifest file that we created previously:

1. Make a copy of the previous YAML manifest file:

```
$ cp mysql-statefulset.yaml mysql-statefulset-rolling-update.yaml
```

2. Ensure that you have the RollingUpdate strategy type and partition set to 0. Also note that if you have attempted to create the StatefulSet object with a different strategy first, you will not be able to modify it without deleting the StatefulSet beforehand:

```
# mysql-statefulset-rolling-update.yaml
apiVersion: apps/v1
kind: StatefulSet
metadata:
  name: mysql-stateful
  labels:
    app: mysql
  namespace: mysql
spec:
  serviceName: mysql-headless
  podManagementPolicy: OrderedReady
```

```
  updateStrategy:
    type: RollingUpdate
    rollingUpdate:
      partition: 0
  replicas: 3
...<removed for brevity>...
(Refer to the GitHub repo for full YAML)
```

These values are the default ones, but it is worth specifying them explicitly to understand what is really happening.

3. Apply the manifest file to the cluster to create the `mysql-stateful` StatefulSet with new configurations:

```
$ kubectl apply -f mysql-statefulset-rolling-update.yaml
statefulset.apps/mysql-stateful created
```

Wait until the Pods are running before you continue the rolling update task. Let's verify the pods created by StatefulSet using the `kubectl get pods` command as follows:

```
$ kubectl get pods -n mysql
NAME                READY   STATUS    RESTARTS      AGE
k8sutils            1/1     Running   1 (21h ago)   23h
mysql-stateful-0    1/1     Running   0             65s
mysql-stateful-1    1/1     Running   0             62s
mysql-stateful-2    1/1     Running   0             58s
```

4. When the StatefulSet is ready in the cluster, let's create a new database inside the StatefulSet via the k8sutils Pod as follows:

```
$ kubectl exec -it -n mysql k8sutils -- /bin/bash
root@k8sutils:/# mysql -u root -p -h mysql-stateful-0.mysql-headless

Enter password: <mysqlroot>

Welcome to the MariaDB monitor.  Commands end with ; or \g.
...<removed for brevity>...

MySQL [(none)]> create database stsrolling;
Query OK, 1 row affected (0.027 sec)

MySQL [(none)]> exit;
```

Now, we have a new StatefulSet with `updateStrategy` and a new database created inside.

Next, we can roll out a new version of the MySQL container image for our StatefulSet object. To do that, perform the following steps:

1. Modify the container image used in the StatefulSet Pod template to mysql:8.3.0:

```
# mysql-statefulset-rolling-update.yaml
...<removed for brevity>...
    spec:
      containers:
        - name: mysql
          image: mysql:8.3.0
...<removed for brevity>...
```

2. Apply the changes to the cluster using the following command:

```
$ kubectl apply -f mysql-statefulset-rolling-update.yaml
statefulset.apps/mysql-stateful configured
```

3. Immediately after that, use the kubectl rollout status command to see the progress in real time. This process will be a bit longer than in the case of Deployment objects because the rollout is performed in a sequential and ordered fashion:

```
$ kubectl rollout status statefulset -n mysql
Waiting for 1 pods to be ready...
Waiting for 1 pods to be ready...
Waiting for 1 pods to be ready...
Waiting for 1 pods to be ready...
Waiting for partitioned roll out to finish: 1 out of 3 new pods have been
updated...
Waiting for 1 pods to be ready...
Waiting for 1 pods to be ready...
Waiting for 1 pods to be ready...
Waiting for 1 pods to be ready...
Waiting for partitioned roll out to finish: 2 out of 3 new pods have been
updated...
Waiting for 1 pods to be ready...
Waiting for 1 pods to be ready...
Waiting for 1 pods to be ready...
Waiting for 1 pods to be ready...
partitioned roll out complete: 3 new pods have been updated...
```

4. Similarly, using the kubectl describe command, you can see events for the StatefulSet that demonstrate precisely what the order of Pod replica recreation was:

```
$ kubectl describe sts -n mysql mysql-stateful
Name:              mysql-stateful
```

```
Namespace:            mysql
...<removed for brevity>...
Events:
  Type      Reason                        Age                   From
Message
  ----      ------                        ----                  ----
-------
...<removed for brevity>...
  Normal    SuccessfulDelete              72s (x7 over 73s)  statefulset-
controller  delete Pod mysql-stateful-2 in StatefulSet mysql-stateful
successful
  Normal    RecreatingTerminatedPod  72s (x7 over 72s)  statefulset-
controller  StatefulSet mysql/mysql-stateful is recreating terminated Pod
mysql-stateful-2
  Warning   FailedDelete                  72s                    statefulset-
controller  delete Pod mysql-stateful-2 in StatefulSet mysql-stateful
failed error: pods "mysql-stateful-2" not found
  Normal    SuccessfulDelete              70s (x2 over 71s)  statefulset-
controller  delete Pod mysql-stateful-1 in StatefulSet mysql-stateful
successful
  Normal    RecreatingTerminatedPod  70s                    statefulset-
controller  StatefulSet mysql/mysql-stateful is recreating terminated Pod
mysql-stateful-1
```

As expected, the rollout was done in *reverse* order. The first Pod to recreate was mysql-stateful-2 followed by mysql-stateful-1, and finally mysql-stateful-0. Also, because we have used the default partition value of 0, all the Pods were updated. This is because all ordinal numbers of Pod replicas are greater than or equal to 0.

5. Now, we can verify that the Pods were recreated with the new image. Execute the following command to verify the first Pod replica in the StatefulSet object:

```
$ kubectl describe pod -n mysql mysql-stateful-0|grep Image
    Image:          mysql:8.3.0
    Image ID:       docker.io/library/mysql@
sha256:9de9d54fecee6253130e65154b930978b1fcc336bcc86dfd06e89b72a2588ebe
```

6. And finally, you can verify that the *state* persisted because the existing PVCs were used for the new Pods. Please note that this will only work properly if you haven't deleted the PVCs for the StatefulSet manually in the previous section:

```
$ kubectl exec -it -n mysql k8sutils -- /bin/bash
root@k8sutils:/# mysql -u root -p -h mysql-stateful-0.mysql-headless
Enter password: <mysqlroot>
...<removed for brevity>...
Server version: 8.3.0 MySQL Community Server - GPL
```

```
Copyright (c) 2000, 2018, Oracle, MariaDB Corporation Ab and others.

Type 'help;' or '\h' for help. Type '\c' to clear the current input
statement.

MySQL [(none)]> show databases;
+--------------------+
| Database           |
+--------------------+
| information_schema |
| mysql              |
| performance_schema |
| stsrolling         |
| sys                |
+--------------------+
5 rows in set (0.004 sec)
MySQL [(none)]>
```

As you can see in the preceding output, the rollout of a new version of MySQL was completed successfully and the state has persisted even though the Pods were recreated; you can see the stsrolling database, which you created before the rolling update.

> You can change the StatefulSet container image *imperatively* using the kubectl -n mysql set image sts mysql-stateful mysql=mysql:8.3.0 command. This approach is only recommended for non-production and testing scenarios. In general, StatefulSets are much easier to manage declaratively than imperatively.

Now, let us learn how you can use the partition field to do a *phased* rollout with a *canary*. Assume that we would like to update the mysql image version to 8.4.0. You would like to make sure that the change is working properly in your environment using a canary deployment, which is a single (or some) Pod replica updated to the new image (or another image) version. Please refer to the following steps:

1. Modify the mysql-statefulset-rolling-update.yaml manifest file so that the partition number is equal to current replicas, in our case, 3:

```
...<removed for brevity>...
spec:
  serviceName: mysql-headless
  podManagementPolicy: OrderedReady
  updateStrategy:
    type: RollingUpdate
    rollingUpdate:
      partition: 3
```

```
...<removed for brevity>...
Also, update the image to 8.4.0 as follows:
...
    spec:
      containers:
        - name: mysql
          image: mysql:8.4.0
...<removed for brevity>...
```

When the `partition` number is the same as the number of `replicas`, we can apply the YAML manifest to the cluster and no changes to the Pods will be introduced yet. This is called **staging a rollout**.

2. Apply the manifest file to the cluster:

```
$ kubectl apply -f mysql-statefulset-rolling-update.yaml
statefulset.apps/mysql-stateful configured
```

3. Now, let's create a *canary* for our new version. Decrease the `partition` number by one to 2 in the manifest file. This means that all Pod replicas with an ordinal number of less than 2 will not be updated – in our case, that means updating the `mysql-stateful-2` Pod only. All others will remain unchanged:

```
...
spec:
  serviceName: mysql-headless
  podManagementPolicy: OrderedReady
  updateStrategy:
    type: RollingUpdate
    rollingUpdate:
      partition: 2
  replicas: 3
...
```

4. Apply the manifest file to the cluster again:

```
$ kubectl apply -f mysql-statefulset-rolling-update.yaml
statefulset.apps/mysql-stateful configured
```

5. Use the `kubectl rollout status` command to follow the process. As expected, only one Pod will be recreated:

```
$ kubectl rollout status statefulset -n mysql
Waiting for partitioned roll out to finish: 0 out of 1 new pods have been
updated...
Waiting for 1 pods to be ready...
```

```
Waiting for 1 pods to be ready...
Waiting for 1 pods to be ready...
partitioned roll out complete: 1 new pods have been updated...
```

6. If you describe the MySQL `mysql-stateful-0` and MySQL `mysql-stateful-2` Pods, you can see that the first one is using the old version of the image, whereas the second is using the new one:

```
$ kubectl describe pod -n mysql mysql-stateful-0|grep Image
    Image:          mysql:8.3.0
    Image ID:       docker-pullable://mysql@
sha256:9de9d54fecee6253130e65154b930978b1fcc336bcc86dfd06e89b72a2588ebe

$ kubectl describe pod -n mysql mysql-stateful-2|grep Image
    Image:          mysql:8.4.0
    Image ID:       docker-pullable://mysql@
sha256:4a4e5e2a19aab7a67870588952e8f401e17a330466ecfc55c9acf51196da5bd0
```

7. At this point, you can perform verifications and smoke tests on your canary. Log in to the k8sutils Pod and ensure the new Pod is running well with the new image (e.g., 8.4.0). The canary looks good, so we can continue with a *phased* rollout of our new version:

```
$ kubectl exec -it -n mysql k8sutils -- /bin/bash
root@k8sutils:/# mysql -u root -p -h mysql-stateful-2.mysql-headless
Enter password: <mysqlroot>
Welcome to the MariaDB monitor.  Commands end with ; or \g.
Your MySQL connection id is 11
Server version: 8.4.0 MySQL Community Server - GPL
...
```

8. For a phased rollout, you may use any *lower* `partition` number in the manifest. You can do a few small, phased rollouts or just proceed with a full rollout. Let's do a full rollout by decreasing `partition` to `0`:

```
...
  updateStrategy:
    type: RollingUpdate
    rollingUpdate:
      partition: 0
...
```

9. Apply the manifest file to the cluster again:

```
$ kubectl apply -f mysql-statefulset-rolling-update.yaml
statefulset.apps/mysql-stateful configured
```

10. Observe the next phase of the rollout using the kubectl rollout status command:

```
$ kubectl rollout status statefulset -n mysql
Waiting for partitioned roll out to finish: 1 out of 3 new pods have been
updated...
Waiting for 1 pods to be ready...
Waiting for 1 pods to be ready...
Waiting for partitioned roll out to finish: 2 out of 3 new pods have been
updated...
Waiting for 1 pods to be ready...
Waiting for 1 pods to be ready...
Waiting for 1 pods to be ready...
Waiting for 1 pods to be ready...
partitioned roll out complete: 3 new pods have been updated...
```

As you can see, the phased rollout to the mysql:8.4.0 image version was completed successfully.

> It is possible to do phased rollouts *imperatively*. To do that, you need to control the partition number using the kubectl patch command, for example, kubectl patch sts mysql-stateful -p '{"spec":{"updateStrategy":{"type":"RollingUpdat e","rollingUpdate":{"partition":3}}}}' -n mysql. However, this is much less readable and more error-prone than *declarative* changes.

We will now take a look at how you can do rollbacks of StatefulSets in the next section.

Rolling back StatefulSet

In the previous *Chapter 11, Using Kubernetes Deployments for Stateless Workloads*, we have described how you can do *imperative* rollbacks to Deployments. For StatefulSets, you can do exactly the same operations. To do that, you need to use the kubectl rollout undo commands. However, especially for StatefulSets, we recommend using a *declarative* model for introducing changes to your Kubernetes cluster. In this model, you usually commit each change to the source code repository. Performing rollback is very simple and involves just reverting the commit and applying the configuration again. Usually, the process of applying changes (both deployment and updates) can be performed as part of the CI/CD pipeline for the source code repository, instead of manually applying the changes by an operator. This is the easiest way to manage StatefulSets, and is generally recommended in Infrastructure-as-Code and Configuration-as-Code paradigms.

> When performing rollbacks to StatefulSets, you must be fully aware of the consequences of operations such as *downgrading* to an earlier version of the container image while persisting the state. For example, if your rollout to a new version has introduced *data schema changes* to the state, then you will not be able to safely roll back to an earlier version unless you ensure that the *downward migration* of state data is implemented!

In our example, if you would like to roll back to the mysql:8.3.0 image version for our StatefulSet, you would either modify the YAML manifest file manually or revert the commit in your source code repository if you use one. Then, all you would need to do is execute the kubectl apply command to the cluster.

Now, in the last section of this chapter, we will provide you with a set of best practices for managing StatefulSets in Kubernetes.

StatefulSet best practices

This section summarizes the known best practices when working with StatefulSet objects in Kubernetes. The list is by no means complete but is a good starting point for your journey with Kubernetes StatefulSet.

Use declarative object management for StatefulSets

It is a good practice in the DevOps world to stick to declarative models for introducing updates to your infrastructure and applications. Using the declarative way of updates is the core concept for paradigms such as Infrastructure-as-Code and Configuration-as-Code. In Kubernetes, you can easily perform declarative updates using the kubectl apply command, which can be used on a single file or even a whole directory of YAML manifest files.

> To delete objects, it is still better to use imperative commands. It is more predictable and less prone to errors. The declarative deletion of resources in the cluster is useful mostly in CI/CD scenarios, where the whole process is entirely automated.

The same principle also applies to StatefulSets. Performing a rollout or rollback when your YAML manifest files are versioned and kept in a source control repository is easy and predictable. Using the kubectl rollout undo method and kubectl set image deployment commands is generally not practiced in production environments. Using these commands gets much more complicated when more than one person is working on operations in the cluster.

Do not use the TerminationGracePeriodSeconds Pod with a 0 value for StatefulSets

The specification of Pod allows you to set TerminationGracePeriodSeconds, which informs kubelet how much time it should allow for a Pod to gracefully terminate when it attempts to terminate it. If you set TerminationGracePeriodSeconds to 0, this will effectively make Pods terminate *immediately*, which is strongly discouraged for StatefulSets. StatefulSets often need graceful cleanup or preStop life cycle hooks to run before the container is removed. Otherwise, there is a risk that the state of StatefulSet will become inconsistent. Refer to the Container hooks documentation (https://kubernetes.io/docs/concepts/containers/container-lifecycle-hooks/#container-hooks) to learn more.

Scale down StatefulSets before deleting

When you delete a StatefulSet and you intend to reuse the PVCs later, you need to ensure that the StatefulSet terminates gracefully, in an ordered manner, so that any subsequent redeployment will not fail because of an inconsistent state in PVCs. If you perform the kubectl delete operation on your StatefulSet, all the Pods will be terminated *at once*. This is often not desired, and you should first scale down the StatefulSet gracefully to zero replicas and then delete the StatefulSet itself.

Ensure state compatibility during StatefulSet rollbacks

If you ever intend to use StatefulSet rollbacks, you need to be aware of the consequences of operations such as downgrading to an earlier version of the container image while persisting the state. For example, if your rollout to a new version has introduced data schema changes in the state, then you will not be able to safely roll back to an earlier version unless you ensure that the downward migration of state data is implemented. Otherwise, your rollback will just recreate Pods with the older versions of the container image, and they will fail to start properly because of incompatible state data.

Do not create Pods that match an existing StatefulSet label selector

It is possible to create Pods with labels that match the label selector of some existing StatefulSet. This can be done using bare Pods or another Deployment or ReplicaSet. This leads to conflicts, which Kubernetes does not prevent, and makes the existing StatefulSet *believe* that it has created the other Pods. The results may be unpredictable and, in general, you need to pay attention to how you organize your labeling of resources in the cluster. It is advised to use semantic labeling. You can learn more about this approach in the official documentation: https://kubernetes.io/docs/concepts/configuration/overview/#using-labels.

Use Remote Storage for the PV

When using StatefulSets, it's important to ensure that you're utilizing remote storage. This means storing your application's data on a separate storage system, typically **Network-Attached Storage (NAS)**, **Storage Area Network (SAN)**, or a cloud storage service. By storing data remotely, you ensure that it's accessible from any instance of your application (or any nodes in the cluster), even if the instance is replaced or moved. This provides data persistence and resilience, helping to prevent data loss in case of failures or updates to your StatefulSet.

Define liveness and readiness probes

For stateful applications, a healthy Pod needs to not only be running but also be able to access and process its persistent state. Liveness probes help ensure this functionality. If a liveness probe fails consistently, it indicates a deeper issue with the Pod's ability to handle its state. Restarting the Pod in this case can potentially trigger recovery mechanisms or allow the StatefulSet controller to orchestrate a failover to another healthy Pod with the same state.

StatefulSets often manage services that rely on specific data or configurations to be available before serving traffic. Readiness probes can be tailored to check if the Pod's state is ready and operational. By preventing traffic from reaching unready Pods, you ensure a smooth user experience and avoid potential data inconsistencies.

Monitor your StatefulSets

Keeping an eye on your StatefulSets' health and performance is crucial. Utilize monitoring tools to track key metrics like Pod restarts, resource utilization, and application errors. This allows you to proactively identify and address potential issues before they impact your application's functionality.

Summary

This chapter demonstrated how to work with *stateful* workloads and applications on Kubernetes using StatefulSets. We first learned what the approaches to persisting states in containers and Kubernetes Pods are, and, based on that, we described how a StatefulSet object can be used to persist the state. Next, we created an example StatefulSet, together with a *headless* Service. Based on that, you learned how PVCs and PVs are used in StatefulSets to ensure that the state is persisted between Pod restarts. Next, we learned how you can scale the StatefulSet and how to introduce updates using *canary* and *phased* rollouts. And finally, we provided a set of known best practices when working with StatefulSets.

In the next chapter, you will learn more about managing special workloads where you need to maintain exactly one Pod per Node in Kubernetes. We will introduce a new Kubernetes object: DaemonSet.

Further reading

- StatefulSets: `https://kubernetes.io/docs/concepts/workloads/controllers/statefulset/`
- Headless Services: `https://kubernetes.io/docs/concepts/services-networking/service/#headless-services`
- Container hooks: `https://kubernetes.io/docs/concepts/containers/container-lifecycle-hooks/#container-hooks`

Join our community on Discord

Join our community's Discord space for discussions with the authors and other readers:

`https://packt.link/cloudanddevops`

13

DaemonSet — Maintaining Pod Singletons on Nodes

The previous chapters have explained and demonstrated how to use the most common Kubernetes controllers for managing Pods, such as ReplicaSet, Deployment, and StatefulSet. Generally, when running cloud application components that contain the actual *business logic*, you will need either Deployments or StatefulSets for controlling your Pods. In some cases, when you need to run batch workloads as part of your application, you will use Jobs and CronJobs.

However, in some cases, you will need to run components that have a supporting function and, for example, execute maintenance tasks or aggregate logs and metrics. More specifically, if you have any tasks that need to be executed for each Node in the cluster, they can be performed using a **DaemonSet**. The purpose of a DaemonSet is to ensure that *each* Node (unless specified otherwise) runs a *single* replica of a Pod. If you add a new Node to the cluster, it will automatically get a Pod replica scheduled. Similarly, if you remove a Node from the cluster, the Pod replica will be terminated – the DaemonSet will execute all the required actions.

In this chapter, we will cover the following topics:

- Introducing the DaemonSet object
- Creating and managing DaemonSets
- Common use cases for DaemonSets
- Alternatives to DaemonSets

Technical requirements

For this chapter, you will need the following:

- A Kubernetes cluster deployed. You can use either a local or cloud-based cluster, but in order to fully understand the concepts, we recommend using a *multi-node* Kubernetes cluster.
- The Kubernetes CLI (kubectl) installed on your local machine and configured to manage your Kubernetes cluster.

Kubernetes cluster deployment (local and cloud-based) and `kubectl` installation were covered in *Chapter 3, Installing Your First Kubernetes Cluster*.

You can download the latest code samples for this chapter from the official GitHub repository: `https://github.com/PacktPublishing/The-Kubernetes-Bible-Second-Edition/tree/main/Chapter13`.

Introducing the DaemonSet object

The term **daemon** in operating systems has a long history and, in short, is used to describe a program that runs as a background process, without interactive control from the user. In many cases, daemons are responsible for handling maintenance tasks, serving network requests, or monitoring hardware activities. These are often processes that you want to run reliably, all the time, in the background, from the time you boot the operating system to when you shut it down.

> Daemons are associated in most cases with Unix-like operating systems. In Windows, you will more commonly encounter the term *Windows service*.

Imagine needing a program to run on every computer in your office, making sure everything stays in order. In Kubernetes, that's where DaemonSets come in. They're like special managers for Pods, ensuring a single copy of a Pod runs on each machine (called a Node) in your cluster. Also, in use cases like gRPC, this is crucial as gRPC may require a dedicated socket to be created on the node's filesystem, which is easier to manage with one Pod per node.

These Pods handle crucial tasks for the entire cluster, like:

- **Monitoring:** Keeping an eye on the health of each Node
- **Logging:** Collecting information about what's happening on each Node and the Pods running on them
- **Storage management:** Handling requests for storage space for your applications
- **Network management:** Cluster components such as kube-proxy and **Container Network Interface (CNI)** (e.g., Calico) for connectivity

As your cluster grows (adding more Nodes), DaemonSets automatically add more Pods to manage the new machines. The opposite happens when Nodes are removed – the Pods on those Nodes are cleaned up automatically. Think of it as a self-adjusting team, always making sure every Node has the help it needs.

The following diagram shows the high-level details of a DaemonSet object.

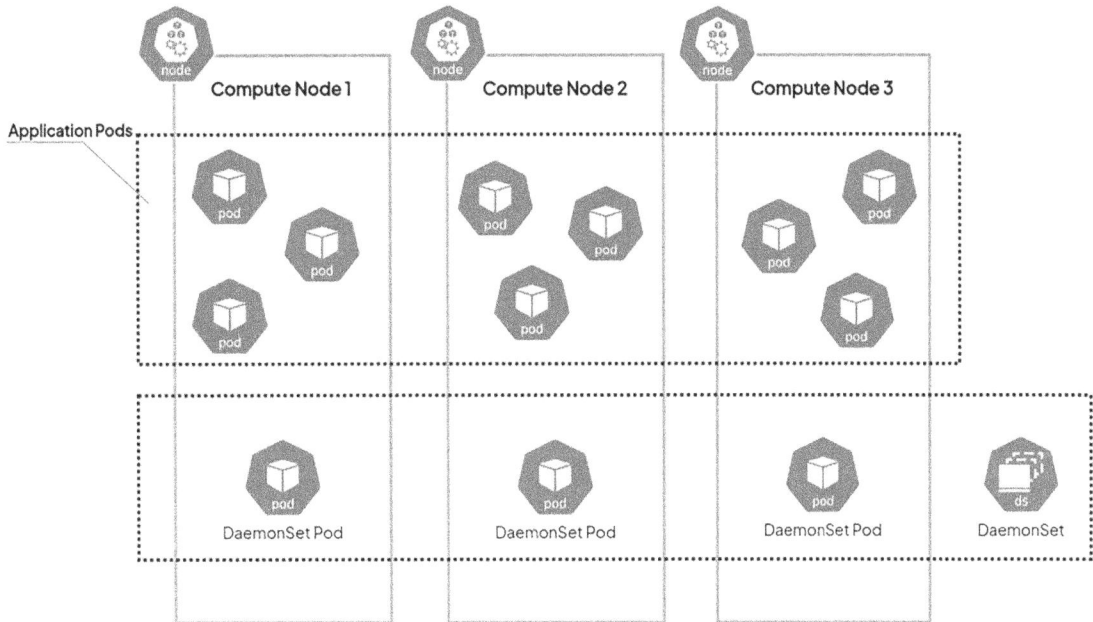

Figure 13.1: DaemonSet in Kubernetes

In simpler setups, one DaemonSet can handle everything for a particular task (like monitoring) across all Nodes. More complex situations might use multiple DaemonSets for the same task, but with different settings or resource needs depending on the type of Node (think high-powered machines vs. basic ones).

By using DaemonSets, you can ensure your Kubernetes cluster has the essential tools running on every Node, keeping things running smoothly and efficiently.

All you have learned in the previous chapters about ReplicaSets, Deployments, and StatefulSets applies more or less to the DaemonSet. Its specification requires you to provide a Pod template, Pod label selectors, and, optionally, Node selectors if you want to schedule the Pods only on a subset of Nodes.

Depending on the case, you may not need to communicate with the DaemonSet from other Pods or an external network. For example, if the job of your DaemonSet is just to perform a periodic cleanup of the filesystem on the Node, it is unlikely you would like to communicate with such Pods. If your use case requires any ingress or egress communication with the DaemonSet Pods, then you have the following common patterns:

- **Mapping container ports to host ports**: Since the DaemonSet Pods are guaranteed to be sin-gletons on cluster Nodes, it is possible to use mapped host ports. The clients must know the Node IP addresses.
- **Pushing data to a different service**: In some cases, it may be enough that the DaemonSet is responsible for sending updates to other services without needing to allow ingress traffic.
- **Headless service matching DaemonSet Pod label selectors**: This is a similar pattern to the case of StatefulSets, where you can use the cluster DNS to retrieve multiple A records for Pods using the DNS name of the headless service.

- **Normal service matching DaemonSet Pod label selectors:** Less commonly, you may need to reach *any* Pod in the DaemonSet. Using a normal Service object, for example, the `ClusterIP` type, will allow you to communicate with a random Pod in the DaemonSet.

As we discussed, DaemonSets ensures that the Pods for essential services will run on all or selected nodes. Let us explore in the next section how the scheduling works effectively for DaemonSets.

How DaemonSet Pods are scheduled

DaemonSets guarantee that a single Pod runs on every eligible node in your Kubernetes cluster. The DaemonSet controller creates Pods with node affinity rules targeting specific nodes. This ensures that Pods for the DaemonSet are only scheduled on specific nodes that meet the desired conditions, making it useful for targeting certain types of nodes in complex application setups. The default scheduler then binds the Pod to the intended node, potentially preempting existing Pods if resources are insufficient. While a custom scheduler can be specified, the DaemonSet controller ultimately ensures that the Pod placement aligns with the desired node affinity.

In the next section, we will learn how you can check the DaemonSet resources in the Kubernetes cluster.

Checking DaemonSets

Once you deploy a Kubernetes cluster, you might already be using some of the DaemonSets deployed as part of the Kubernetes or cluster support components, such as the DNS service or CNI.

For example, in the following snippet, we are creating a multi-node `minikube` Kubernetes cluster:

```
$ minikube start \
  --driver=virtualbox \
  --nodes 3 \
  --cni calico \
  --cpus=2 \
  --memory=2g \
  --kubernetes-version=v1.31.0 \
  --container-runtime=containerd
```

Once the cluster is created, verify the nodes in the cluster as follows:

```
$ kubectl get nodes
NAME            STATUS    ROLES           AGE       VERSION
minikube        Ready     control-plane   3m28s     v1.31.0
minikube-m02    Ready     <none>          2m29s     v1.31.0
minikube-m03    Ready     <none>          91s       v1.31.0
```

Now, let us check if any DaemonSet is available in the freshly installed system:

```
$ kubectl get daemonsets -A
NAMESPACE       NAME           DESIRED   CURRENT   READY   UP-TO-DATE   AVAILABLE
NODE SELECTOR            AGE
```

```
kube-system    calico-node    3         3         3         3         3
kubernetes.io/os=linux    63m
kube-system    kube-proxy    3         3         3         3         3
kubernetes.io/os=linux    63m
```

In the above code snippet, the `-A` or `--all` namespaces lists the requested objects across all namespaces.

If you are following the same method to create a `minikube` cluster, you will see a similar output with `calico-node` and `kube-proxy`, which are deployed as DaemonSets. (You can also install Calico in your other Kubernetes clusters and follow the remaining steps here.) You might have already noticed that we have enabled Calico as the CNI plugin in the `minikube` cluster earlier. Calico, when used for Kubernetes networking, is typically deployed as a DaemonSet.

> Ignore the `kube-proxy` DaemonSet for now as `minikube` runs kube-proxy as a DaemonSet. This guarantees that `kube-proxy`, which is responsible for managing network traffic within the cluster, is always up and running on every machine in your `minikube` environment.

Now, let us check the Pods deployed by the `calico-node` DaemonSet.

```
$ kubectl get po -n kube-system -o wide|grep calico
calico-kube-controllers-ddf655445-jx26x    1/1    Running    0    82m
10.244.120.65    minikube    <none>    <none>
calico-node-fkjxb    1/1    Running    0    82m
192.168.59.126    minikube    <none>    <none>
calico-node-nrzpb    1/1    Running    0    81m
192.168.59.128    minikube-m03    <none>    <none>
calico-node-sg66x    1/1    Running    0    82m
192.168.59.127    minikube-m02    <none>    <none>
```

From this output, we can see that:

- Calico Pods are deployed across all `minikube` nodes. The Pods reside on different nodes, which correspond to your `minikube` virtual machines (`minikube`, `minikube-m02`, and `minikube-m03`). This suggests Calico is using a DaemonSet to ensure a Pod is running on each node.
- The Pod `calico-kube-controllers-ddf655445-jx26x` is the controller of the Calico CNI.

Since the Calico DaemonSet is installed by `minikube` in this case, we will not explore much on that side. But in the next section, we will learn how to deploy a DaemonSet from scratch and explore it in detail.

Creating and managing DaemonSets

In order to demonstrate how DaemonSets work, we will use **Fluentd** Pods. Fluentd is a popular open-source log aggregator that centralizes log data from various sources. It efficiently collects, filters, and transforms log messages before forwarding them to different destinations for analysis and storage.

To access the DaemonSet endpoints, we will use a *headless* service, similar to what we did for Stateful-Set in *Chapter 12, StatefulSet – Deploy Stateful Applications*. Most of the real use cases of DaemonSets are rather complex and involve mounting various system resources to the Pods. We will keep our DaemonSet example as simple as possible to show the principles.

> If you would like to work on another example of a DaemonSet, we have provided a working version of Prometheus `node-exporter` deployed as a DaemonSet behind a headless Service: `node-exporter.yaml`. When following the guide in this section, the only difference is that you need to use `node-exporter` as the Service name, use port `9100`, and append the `/metrics` path for requests sent using `wget`. This DaemonSet exposes Node metrics in *Prometheus data model* format on port `9100` under the `/metrics` path.

We will now go through all the YAML manifests required to create our DaemonSet and apply them to the cluster.

Creating a DaemonSet

As a best practice, let us use the declarative way to create the DaemonSet for our hands-on practice. First, let's take a look at the DaemonSet YAML manifest file named `Fluentd-daemonset.yaml`.

The first part of the YAML is for creating a separate namespace for logging.

```
---
apiVersion: v1
kind: Namespace
metadata:
  name: logging
```

After that, we have our DaemonSet declaration details as follows.

```
apiVersion: apps/v1
kind: DaemonSet
metadata:
  name: fluentd-elasticsearch
  namespace: kube-system
  labels:
    k8s-app: fluentd-logging
spec:
  selector:
    matchLabels:
      name: fluentd-elasticsearch
# (to be continued in the next paragraph)
```

The first part of the preceding file contains the `metadata` and Pod label `selector` for the DaemonSet, quite similar to what you have seen in Deployments and StatefulSets. In the second part of the file, we present the Pod template that will be used by the DaemonSet:

```
# (continued)
  template:
    metadata:
      labels:
        name: fluentd-elasticsearch
    spec:
      containers:
        - name: fluentd-elasticsearch
          image: quay.io/fluentd_elasticsearch/fluentd:v4.7
          resources:
            limits:
              memory: 200Mi
            requests:
              cpu: 100m
              memory: 200Mi
          volumeMounts:
            - name: varlog
              mountPath: /var/log
      terminationGracePeriodSeconds: 30
      volumes:
        - name: varlog
          hostPath:
            path: /var/log
```

As you can see, the structure of the DaemonSet spec is similar to what you know from Deployments and StatefulSets. The general idea is the same; you need to configure the Pod template and use a proper label selector to match the Pod labels. Note that you do *not* see the replicas field here, as the number of Pods running in the cluster will be dependent on the number of Nodes in the cluster. The DaemonSet specification has two main components:

- spec.selector: A label selector, which defines how to identify Pods that the DaemonSet owns. This can include *set-based* and *equality-based* selectors.
- spec.template: This defines the template for Pod creation. Labels used in metadata must match the selector.

It is also common to specify .spec.template.spec.nodeSelector or .spec.template.spec. tolerations in order to control the Nodes where the DaemonSet Pods are deployed. We will cover Pod scheduling in detail in *Chapter 19, Advanced Techniques for Scheduling Pods*. Additionally, you can specify .spec.updateStrategy, .spec.revisionHistoryLimit, and .spec.minReadySeconds, which are similar to what you have learned about Deployment objects.

If you run hybrid Linux-Windows Kubernetes clusters, one of the common use cases for Node selectors or Node affinity for DaemonSets is ensuring that the Pods are scheduled only on Linux Nodes or only on Windows Nodes. This makes sense as the container runtime and operating system are very different between such Nodes.

Also, notice the volume mounting lines where the Fluentd pods will get access to the /var/log directory of the host (where the Pod is running) so that Fluentd can process the data and send it to the logging aggregator.

Please note that in actual Deployment, we need to provide the target Elasticsearch servers to the Fluentd Pods so that Fluentd can send the logs. In our demonstration, we are not covering the Elasticsearch setup and you may ignore this part for now.

It is possible to pass such parameters via environment variables to the containers as follows. (Refer to the fluentd-daemonset.yaml to learn more.)

```
env:
  - name:   FLUENT_ELASTICSEARCH_HOST
    value: "elasticsearch-logging"
  - name:   FLUENT_ELASTICSEARCH_PORT
    value: "9200"
```

We have all the required YAML manifest files for our demonstration and we can proceed with applying the manifests to the cluster. Please follow these steps:

1. Create the fluentd-elasticsearch DaemonSet using the following command:

    ```
    $ kubectl apply -f fluentd-daemonset.yaml
    namespace/logging created
    daemonset.apps/fluentd-elasticsearch created
    ```

2. Now, you can use the kubectl describe command to observe the creation of the DaemonSet:

    ```
    $ kubectl describe daemonset fluentd-elasticsearch -n logging
    ```

    Alternatively, you can use ds as an abbreviation for daemonset when using the kubectl commands.

3. Use the kubectl get pods command with the -w option and you can see that there will be one Pod scheduled for each of the Nodes in the cluster, as shown below:

    ```
    $ kubectl get po -n logging -o wide
    NAME                             READY   STATUS    RESTARTS   AGE      IP
    NODE              NOMINATED NODE    READINESS GATES
    fluentd-elasticsearch-cs4hm      1/1     Running   0          3m48s
    10.244.120.68     minikube         <none>            <none>
    fluentd-elasticsearch-stfqs      1/1     Running   0          3m48s
    10.244.205.194    minikube-m02     <none>            <none>
    ```

```
fluentd-elasticsearch-zk6pt    1/1      Running    0          3m48s
10.244.151.2    minikube-m03   <none>                 <none>
```

In our case, we have three Nodes in the cluster, so exactly three Pods have been created.

We have successfully deployed the DaemonSet and we can now verify that it works as expected. Ensure that the Fluentd Pods are able to access the Kubernetes node log files. To confirm that, log in to one of the Fluentd Pods and check the /var/log directory.

```
$ kubectl exec -n logging -it fluentd-elasticsearch-cs4hm -- /bin/bash
root@fluentd-elasticsearch-cs4hm:/# ls -l /var/log/
total 20
drwxr-xr-x   3 root root 4096 May 29 10:56 calico
drwxr-xr-x   2 root root 4096 May 29 12:40 containers
drwx------   3 root root 4096 May 29 10:55 crio
drwxr-xr-x   2 root root 4096 May 29 11:53 journal
drwxr-x--- 12 root root 4096 May 29 12:40 pods
root@fluentd-elasticsearch-cs4hm:/#
```

This demonstrates the most important principles underlying how DaemonSet Pods are scheduled in Kubernetes.

> It is best practice to use appropriate **taints** and **tolerations** for the nodes to implement DaemonSets. We will learn about taints and tolerations in *Chapter 19, Advanced Techniques for Scheduling Pods*.

Let us learn about some advanced configurations for the DaemonSet in the next section.

Prioritizing critical DaemonSets in Kubernetes

When managing critical system components using DaemonSets in Kubernetes, ensuring their uninterrupted operation is crucial. Here's how to leverage Pod priority and PriorityClasses to guarantee these essential Pods aren't disrupted by lower-priority tasks.

Kubernetes assigns a priority level to each Pod, determining its relative importance within the cluster. Higher-priority Pods are considered more significant compared to lower-priority ones.

By assigning a higher `PriorityClass` to your DaemonSet, you elevate the importance of its Pods. This ensures these critical Pods are not preempted by the scheduler to make way for lower-priority Pods when resource constraints arise.

A PriorityClass defines a specific priority level for Pods. Values in this class can range from negative integers to a maximum of 1 billion. Higher values represent higher priority.

A sample YAML definition for the **PriorityClass** is given below.

```yaml
# priorityclass.yaml
apiVersion: scheduling.k8s.io/v1
kind: PriorityClass
metadata:
  name: fluentd-priority
value: 100000  # A high value for criticality
globalDefault: false  # Not a default class for all Pods
description: "Fluentd Daemonset priority class"
```

Once you have created the PriorityClass, you can use the same in the DaemonSet configuration as follows.

```yaml
spec:
  template:
    spec:
      priorityClassName: fluentd-priority
```

For reference, system components like kube-proxy and cluster CNI (Calico) often utilize the built-in `system-node-critical` PriorityClass. This class possesses the highest priority, ensuring these vital Pods are never evicted under any circumstances.

We will now show how you can modify the DaemonSet to roll out a new version of a container image for the Pods.

Modifying a DaemonSet

Updating a DaemonSet can be done in a similar way as for Deployments. If you modify the *Pod template* of the DaemonSet, this will trigger a *rollout* of a new revision of DaemonSet according to its `updateStrategy`. There are two strategies available:

- `RollingUpdate`: The default strategy, which allows you to roll out a new version of your daemon in a controlled way. It is similar to rolling updates in Deployments in that you can define `.spec.updateStrategy.rollingUpdate.maxUnavailable` to control how many Pods in the clusters are unavailable at most during the rollout (defaults to 1) and `.spec.minReadySeconds` (defaults to 0). It is guaranteed that, *at most, one* Pod of DaemonSet will be in a running state on each node in the cluster during the rollout process.

- `OnDelete`: This strategy implements the legacy behavior of StatefulSet updates prior to Kubernetes 1.6. In this type of strategy, the DaemonSet will *not* automatically update the Pod by recreating them. You need to manually delete a Pod on a Node in order to get the new Pod template applied. This is useful in scenarios when you need to do additional manual actions or verifications before proceeding to the next Node.

The rollout of a new DaemonSet revision can be controlled in similar ways as for a Deployment object. You can use the kubectl rollout status command and perform *imperative* rollbacks using the kubectl rollout undo command. Let's demonstrate how you can *declaratively* update the container image in a DaemonSet Pod to a newer version:

1. Modify the fluentd-daemonset.yaml YAML manifest file so that it uses the quay.io/fluentd_elasticsearch/fluentd:v4.7.5 container image in the template:

```
...
      containers:
        - name: fluentd-elasticsearch
          image: quay.io/fluentd_elasticsearch/fluentd:v4.7.5
...
```

2. Apply the manifest file to the cluster:

```
$ kubectl apply -f fluentd-daemonset.yaml
namespace/logging unchanged
daemonset.apps/fluentd-elasticsearch configured
```

3. Immediately after that, use the kubectl rollout status command to see the progress in real time:

```
$ kubectl rollout status ds -n logging
Waiting for daemon set "fluentd-elasticsearch" rollout to finish: 2 out
of 3 new pods have been updated...
Waiting for daemon set "fluentd-elasticsearch" rollout to finish: 2 out
of 3 new pods have been updated...
Waiting for daemon set "fluentd-elasticsearch" rollout to finish: 2 of 3
updated pods are available...
daemon set "fluentd-elasticsearch" successfully rolled out
```

4. Similarly, using the kubectl describe command, you can see events for the DaemonSet that exactly show what the order was of the Pod recreation:

```
$ kubectl describe ds -n logging
Name:              fluentd-elasticsearch
Selector:          name=fluentd-elasticsearch
...<removed for brevity>...
Events:
  Type    Reason          Age     From                 Message
  ----    ------          ----    ----                 -------
...<removed for brevity>...
-elasticsearch-24v2z
  Normal  SuccessfulDelete  5m52s  daemonset-controller  Deleted pod:
fluentd-elasticsearch-zk6pt
```

```
Normal  SuccessfulCreate  5m51s  daemonset-controller  Created pod:
fluentd-elasticsearch-fxffp
```

You can see that the Pods were replaced one by one.

> You can change the DaemonSet container image *imperatively* using the `kubectl set image ds fluentd-elasticsearch fluentd-elasticsearch=quay.io/fluentd_ elasticsearch/fluentd:v4.7.5 -n logging` command. This approach is recommended only for non-production scenarios.

Additionally, the DaemonSet will automatically create Pods if a new Node joins the cluster (providing that it matches the selector and affinity parameters). If a Node is removed from the cluster, the Pod will also be terminated. The same will happen if you modify the labels or taints on a Node so that it matches the DaemonSet – a new Pod will be created for that Node. If you modify the labels or taints for a Node in a way that no longer matches the DaemonSet, the existing Pod will be terminated.

Next, we will learn how to roll back a DaemonSet update.

Rolling back the DaemonSet

As we learned in the previous chapters, it is also possible to roll back the DaemonSet using the `kubectl rollback` command as follows.

```
$ kubectl rollout undo daemonset fluentd-elasticsearch -n logging
```

However, it is highly recommended to update the YAML and apply configurations in a declarative method for the production environments.

Next, we will show how you can delete a DaemonSet.

Deleting a DaemonSet

In order to delete a DaemonSet object, there are two possibilities:

- Delete the DaemonSet together with the Pods that it owns.
- Delete the DaemonSet and leave the Pods unaffected.

To delete the DaemonSet together with Pods, you can use the regular `kubectl delete` command as follows:

```
$ kubectl delete ds fluentd-elasticsearch -n logging
```

You will see that the Pods will first get terminated and then the DaemonSet will be deleted.

Now, if you would like to delete just the DaemonSet, you need to use the `--cascade=orphan` option with `kubectl delete`:

```
$ kubectl delete ds fluentd-elasticsearch -n logging --cascade=orphan
```

After this command, if you inspect which Pods are in the cluster, you will still see all the Pods that were owned by the `fluentd-elasticsearch` DaemonSet.

> If you are draining a node using the kubectl drain command and this node is running Pods owned by a DaemonSet, you need to pass the --ignore-daemonsets flag to drain the node completely.

Let's now take a look at the most common use cases for DaemonSets in Kubernetes.

Common use cases for DaemonSets

At this point, you may wonder what is the actual use of the DaemonSet and what the real-life use cases are for this Kubernetes object. In general, DaemonSets are used either for very fundamental functions of the cluster, without which it is not usable, or for helper workloads performing maintenance or data collection. We have summarized the common and interesting use cases for DaemonSets in the following points:

- Depending on your cluster Deployment, the kube-proxy core service may be deployed as a DaemonSet instead of a regular operating system service. For example, in the case of **Azure Kubernetes Service (AKS)**, you can see the definition of this DaemonSet using the kubectl describe ds -n kube-system kube-proxy command. This is a perfect example of a backbone service that needs to run as a singleton on each Node in the cluster. You can also see an example YAML manifest for kube-proxy here: https://github.com/kubernetes/kubernetes/blob/master/cluster/addons/kube-proxy/kube-proxy-ds.yaml.

- Another example of fundamental services running as DaemonSets is running an installation of **CNI** plugins and agents for maintaining the network in a Kubernetes cluster. We have already tested this with the Calico CNI DaemonSet in our minikube cluster at the beginning of this chapter. Another good example of such a DaemonSet is the Flannel agent (https://github.com/flannel-io/flannel/blob/master/Documentation/kube-flannel.yml), which runs on each Node and is responsible for allocating a subnet lease to each host out of a larger, preconfigured address space. This, of course, depends on what type of networking is installed on the cluster.

- Cluster storage daemons will often be deployed as DaemonSets. A good example of a commonly used daemon is the **Object Storage Daemon (OSD)** for **Ceph**, which is a distributed object, block, and file storage platform. OSD is responsible for storing objects on the local filesystem of each Node and providing access to them over the network. You can find an example manifest file here (as part of a Helm chart template): https://github.com/ceph/ceph-container/blob/master/examples/helm/ceph/templates/osd/daemonset.yaml.

- Ingress controllers in Kubernetes are sometimes deployed as DaemonSets. We will take a closer look at Ingress in *Chapter 21, Advanced Kubernetes: Traffic Management, Multi-Cluster Strategies and More*. For example, when you deploy nginx as an Ingress controller in your cluster, you have an option to deploy it as a DaemonSet: https://github.com/nginxinc/kubernetes-ingress/blob/master/deployments/daemon-set/nginx-ingress.yaml. Deploying an Ingress controller as a DaemonSet is especially common if you do Kubernetes cluster deployments on bare-metal servers.

- Log gathering and aggregation agents are often deployed as DaemonSets. For example, `fluentd` can be deployed as a DaemonSet in a cluster. You can find multiple YAML manifest files with examples in the official repository: `https://github.com/fluent/fluentd-kubernetes-daemonset`.

- Agents for collecting Node metrics make a perfect use case for Deployment as DaemonSets. A well-known example of such an agent is Prometheus `node-exporter`: `https://github.com/prometheus-operator/kube-prometheus/blob/main/manifests/nodeExporter-daemonset.yaml`.

The list goes on – as you can see, a DaemonSet is another building block provided for engineers designing the workloads running on Kubernetes clusters. In many cases, DaemonSets are the hidden backbone of a cluster that makes it fully operational.

Let us now learn about the recommended best practices for DaemonSet implementations.

DaemonSet best practices

DaemonSets are powerful tools in Kubernetes for managing Pods that need to run on every node. But to ensure they work as expected, there are some key things to keep in mind:

- **Resource requests and limits:** Briefly mentioning the importance of setting appropriate resource requests and limits for DaemonSet Pods can help users manage resource allocation effectively. This can help prevent resource starvation for other Pods in the cluster.

- **Clean and separate:** Organize your DaemonSets by placing each one in its own separate namespace. This keeps things tidy and simplifies managing resources.

- **Scheduling smarts:** When creating a DaemonSet, it's recommended to use `preferredDuringSchedu2lingIgnoredDuringExecution` instead of `requiredDuringSchedulingIgnoredDuringExecution`. The first option allows for more flexibility if there aren't enough nodes available initially.

- **Wait for readiness** (optional): You can use the `minReadySeconds` setting in your Pod schema. This tells Kubernetes to wait a certain amount of time before creating new Pods during an update. This helps ensure all existing Pods are healthy before adding new ones.

- **Monitoring and logging:** A quick note about the importance of monitoring and logging for DaemonSet Pods can be helpful. This allows users to track the health and performance of their DaemonSets and identify any potential issues.

- **Always running:** Make sure your DaemonSet Pods have a **Restart Policy** set to `Always` (or leave it unspecified). This guarantees that the Pods automatically restart if they ever crash.

- **High priority:** Give your DaemonSet Pods a high priority (like `10000`) to ensure they get the resources they need and aren't evicted by other Pods.

- **Matching labels:** Define a pod selector that matches the labels of your DaemonSet template. This ensures the Pods deployed by the DaemonSet are the ones you intended.

By following these best practices, you can configure your DaemonSets to run smoothly and keep your Kubernetes cluster functioning optimally.

Next, let's discuss what possible alternatives there are to using DaemonSets.

Alternatives to DaemonSets

The reason for using DaemonSets is quite simple – you would like to have exactly one Pod with a particular function on each Node in the cluster. However, sometimes, you should consider different approaches that may fit your needs better:

- In log-gathering scenarios, you need to evaluate whether you want to design your log pipeline architecture based on DaemonSets or the *sidecar* container pattern. Both have their advantages and disadvantages, but in general, running sidecar containers may be easier to implement and more robust, even though it may require more system resources.

- If you just want to run periodic tasks, and you do not need to do it on each Node in the cluster, a better solution can be using **Kubernetes CronJobs**. Again, it is important to know what the actual use case is and whether running a separate Pod on each Node is a must-have requirement.

- Operating system daemons (for example, provided by `systemd` in Ubuntu) can be used to do similar tasks as DaemonSets. The drawback of this approach is that you cannot manage these native daemons using the same tools as you manage Kubernetes clusters with, for example, `kubectl`. But at the same time, you do not have the dependency on any Kubernetes service, which may be a good thing in some cases.

- Static Pods (`https://kubernetes.io/docs/tasks/configure-pod-container/static-pod/`) can be used to achieve a similar result. This type of Pod is created based on a specific directory watched by `kubelet` for static manifest files. Static Pods cannot be managed using `kubectl` and they are most useful for cluster bootstrapping functions.

Finally, we can now summarize our knowledge about DaemonSets.

Summary

In this chapter, you have learned how to work with DaemonSets in Kubernetes and how they are used to manage special types of workloads or processes that must run as a singleton on each Node in the cluster. You first created an example DaemonSet and learned what the most important parts of its specification are. Next, you practiced how to roll out a new revision of a DaemonSet to the cluster and saw how you can monitor the Deployment. Additionally, we discussed what the most common use cases are for this special type of Kubernetes object and what alternatives there are that you could consider.

This was the last type of Pod management controller that we discussed in this part of the book. In the next part of this book, we will examine some more advanced Kubernetes usage, starting with Helm charts and then Kubernetes operators.

Further reading

- DaemonSet: `https://kubernetes.io/docs/concepts/workloads/controllers/daemonset/`
- Network Policy: `https://minikube.sigs.k8s.io/docs/handbook/network_policy/`
- Fluentd Deployment on Kubernetes: `https://docs.fluentd.org/container-deployment/kubernetes`

- Perform a Rolling Update on a DaemonSet: `https://kubernetes.io/docs/tasks/manage-daemon/update-daemon-set/`

Join our community on Discord

Join our community's Discord space for discussions with the authors and other readers:

`https://packt.link/cloudanddevops`

14

Working with Helm Charts and Operators

In the Kubernetes ecosystem, it is very important to manage application redistribution and dependency management in cases when you want your applications to be easily downloadable and installable for customers, or when you want to share them between teams. One of the big differences between Linux package managers like APT or YUM and Kubernetes tools, such as Helm, is that they work universally and don't depend on a particular Kubernetes distribution. Helm enables the easily distributive creation of applications by means of packaging multiple resources into charts, supporting easy reuse and customization of applications across diverse environments. This relieves the struggle of managing a large number of YAML manifests in application deployments and therefore lightens the load on a developer or an operator.

Kubernetes Operators further complement Helm by adding the ability to manage an application's life cycle in automated ways such as upgrades, failovers, and backups, and by maintaining consistent configurations. Together, Helm and the Operators help with some of the challenges of scaling and managing applications within Kubernetes. This chapter explores these tools in depth: Helm chart development, how to install popular components, and successful redistribution of Kubernetes applications.

In this chapter, we will cover the following topics:

- Understanding Helm
- Releasing software to Kubernetes using Helm
- Installing Kubernetes Dashboard using Helm Charts
- Introducing Kubernetes Operators
- Enabling Kubernetes monitoring using Prometheus and Grafana

Technical requirements

For this chapter, you will need the following:

- A Kubernetes cluster deployed. We recommend using a *multi-node, or* cloud-based, Kubernetes cluster if possible. You need to ensure CPU, memory, and storage resources are allocated for the Kubernetes cluster to ensure multiple Pods can be scheduled (e.g., you can create a larger minikube cluster using the command `minikube start --kubernetes-version=1.31.0 --cpus=4 --memory=8g`).

- The Kubernetes **command-line interface (CLI)** (`kubectl`) installed on your local machine and configured to manage your Kubernetes cluster.

Basic Kubernetes cluster deployment (local and cloud-based) and `kubectl` installation have been covered in *Chapter 3, Installing Your First Kubernetes Cluster*. The upcoming chapters, *15*, *16*, and *17*, can give you an overview of how to deploy a fully functional Kubernetes cluster on different cloud platforms. You can download the latest code samples for this chapter from the official GitHub repository: `https://github.com/PacktPublishing/The-Kubernetes-Bible-Second-Edition/tree/main/Chapter14`

Understanding Helm

This approach is often used as a basic showcase of how you can run a given application as a container on Kubernetes. However, sharing raw YAML manifests has quite a few disadvantages:

- All values in YAML templates are *hardcoded*. This means that if you want to change the number of replicas of a Service object target or a value stored in the ConfigMap object, you need to go through the manifest files, find the values you want to configure, and then edit them. Similarly, if you want to deploy the manifests to a different namespace in the cluster than the creators intended, you need to edit all YAML files. On top of that, you do not really know which values in the YAML templates are intended to be configurable by the creator unless they document this.

- The deployment process can vary for each application. There is no standardized approach regarding which YAML manifests the creator provides and which components you are required to deploy manually.

- There is no *dependency management*. For example, if your application requires a **MySQL** server running as a StatefulSet in the cluster, you either need to deploy it yourself or rely on the creator of the application to provide YAML manifests for the MySQL server.

This is a bit similar to what you see with the other applications if you do not use **Application Store** or a package manager such as **Chocolatey**, **APT**, **YUM**, **DNF**, and so on. Some applications that you download will come with an installer as a `setup.sh` script file, some with a `.exe` file, some as a `.msi`, and others will just be `.zip` files that you need to extract and configure yourself.

In Kubernetes, you can use **Helm** (`https://helm.sh`), which is one of the most popular package managers for Kubernetes applications and services. If you are familiar with popular package managers such as APT, yum, npm, or Chocolatey, you will find many concepts in Helm similar and easy to understand. The following are the three most important concepts in Helm:

- A **chart** is a Helm *package*. This is what you *install* when you use the Helm CLI. A Helm chart contains all Kubernetes YAML manifests required to deploy the particular application on the cluster. Please note that these YAML manifests may be *parameterized*, so that you can easily inject configuration values provided by the user who installs the chart.

- A **repository** is a storage location for Helm charts, used to collect and share charts. They can be public or private – there are multiple public repositories that are available, which you can browse on Artifact Hub (`https://artifacthub.io`). Private repositories are usually used for distributing components running on Kubernetes between teams working on the same product.

- A **release** is an *instance* of a Helm chart that was installed and is running in a Kubernetes cluster. This is what you manage with the Helm CLI, for example, by upgrading or uninstalling it. You can install one chart many times on the same cluster and have multiple releases of it that are identified uniquely by release names.

> In short, Helm charts contain parameterizable YAML manifests that you store in a Helm repository for distribution. When you install a Helm chart, a Helm release is created in your cluster that you can further manage.

Let's quickly summarize some of the common use cases for Helm:

- Deploying popular software to your Kubernetes cluster. This makes *development* on Kubernetes much easier – you can deploy third-party components to the cluster in a matter of seconds. The same approach may be used in *production* clusters. You do not need to rely on writing your own YAML manifest for such third-party components.

- Helm charts provide *dependency management* capabilities. If chart A requires chart B to be installed first with specific parameters, Helm supports syntax for this out of the box.

- Sharing your own applications as Helm charts. This can include packaging a product for consumption by the end users or using Helm as an internal package and dependency manager for microservices in your product.

- Ensuring that the applications receive proper upgrades. Helm has its own process for upgrading Helm releases.

- Configuring software deployments for your needs. Helm charts are basically generic YAML templates for Kubernetes object manifests that can be *parameterized*. Helm uses **Go** templates (`https://godoc.org/text/template`) for parameterization.

Currently, Helm is distributed as a binary client (library) that has a CLI similar to kubectl. All operations that you perform using Helm do not require any additional components to be installed on the Kubernetes cluster.

Please note that the Helm architecture was changed with the release of version 3.0.0 of Helm. Previously, the architecture of Helm was different, and it required a special, dedicated service running on Kubernetes named Tiller. This was causing various problems, such as with security around **Role-Based Access Control** (**RBAC**) and elevated-privilege Pods running inside the cluster. You can read more about the differences between the latest major version of Helm and previous ones in the official FAQ:

https://helm.sh/docs/faq/#changes-since-helm-2

This is useful to know if you find any online guides that still mention Tiller – they are most likely intended for older versions of Helm.

Now that we have learned about Helm and its important concepts, we are going to install Helm and deploy a simple Helm chart from Artifact Hub to verify that it works correctly on your cluster.

Releasing software to Kubernetes using Helm

In this section, you will learn how to install Helm and how to test the installation by deploying an example Helm chart. Helm is provided as binary releases (https://github.com/helm/helm/releases) available for multiple platforms. You can use them or refer to the following guides for installation using a package manager on your desired operating system.

Installing Helm on Linux

To install Helm on Fedora, you need to ensure that the default Fedora repository is configured and working:

```
$ sudo dnf repolist | grep fedora
fedora                                    Fedora 39 – x86_64
```

Then install Helm as follows:

```
$ sudo dnf install helm
```

Once installed, you can verify the version of the installed Helm package:

```
$ helm version
version.BuildInfo{Version:"v3.11", GitCommit:"", GitTreeState:"",
GoVersion:"go1.21rc3"}
```

It is also possible to install Helm using the script (https://helm.sh/docs/intro/install/#from-script), which will automatically detect the platform, download the latest Helm, and install it on your machine.

Once installed, you can move on to *Deploying an example chart – WordPress* in this section.

Installing Helm on Windows

To install Helm on Windows, the easiest way is to use the Chocolatey package manager. If you have not used Chocolatey before, you can find more details and the installation guide in the official documentation at `https://chocolatey.org/install`.

Execute the following command in PowerShell or Command shell to install Helm:

```
PS C:\Windows\system32> choco install kubernetes-helm
PS C:\Windows\system32> helm version
version.BuildInfo{Version:"v3.15.0-rc.2",
GitCommit:"c4e37b39dbb341cb3f716220df9f9d306d123a58", GitTreeState:"clean",
GoVersion:"go1.22.3"}
```

Once installed, you can move on to *Deploying an example chart – WordPress*, which is later in this section.

Installing Helm on macOS

To install Helm on macOS, you can use the standard **Homebrew** package manager. Use the following command to install the Helm formula:

```
$ brew install helm
```

Verify that the installation was successful by trying to get the Helm version from the command line:

```
$ helm version
version.BuildInfo{Version:"v3.16.2",
GitCommit:"13654a52f7c70a143b1dd51416d633e1071faffb", GitTreeState:"dirty",
GoVersion:"go1.23.2"}
```

Once installed, we can deploy an example chart to verify that Helm works properly on your Kubernetes cluster.

Installing from the binary releases

It is also possible to install the latest Helm package from the binary itself. You need to find the latest or desired version of the binary from the release page (`https://github.com/helm/helm/releases`) and download it for your operating system. In the following example, we will see how to install the latest Helm from the binary on the Fedora workstation:

Download and install Helm:

```
$ wget https://get.helm.sh/helm-v3.15.1-linux-amd64.tar.gz
$ tar -zxvf helm-v3.15.1-linux-amd64.tar.gz
linux-amd64/
linux-amd64/README.md
linux-amd64/LICENSE
linux-amd64/helm
$ sudo mv linux-amd64/helm /usr/local/bin/helm
```

```
$ helm version
version.BuildInfo{Version:"v3.15.1",
GitCommit:"e211f2aa62992bd72586b395de50979e31231829", GitTreeState:"clean",
GoVersion:"go1.22.3"}
```

Now, we will test the Helm package by *Deploying an example chart - WordPress* in the next section.

Deploying an example chart — WordPress

By default, Helm comes with no repositories configured. One possibility, which is no longer recommended, is to add the stable repository so that you can browse the most popular Helm charts:

```
$ helm repo add stable https://charts.helm.sh/stable
"stable" has been added to your repositories
```

> Adding random Helm chart repositories for deployment can pose serious security risks. Security audits are a must to ensure that only trusted and secure payloads are deployed within your Kubernetes environment.

Please note that most charts are now in the process of deprecation as they are moved to different Helm repositories where they will be maintained by the original creators. You can see this if you try to search for available Helm charts using the helm search repo command:

```
$ helm search repo stable|grep -i deprecated|head
stable/acs-engine-autoscaler             2.2.2          2.1.1
DEPRECATED Scales worker nodes within agent pools
stable/aerospike                         0.3.5          v4.5.0.5
DEPRECATED A Helm chart for Aerospike in Kubern...
stable/airflow                           7.13.3         1.10.12
DEPRECATED ...<removed for brevity>...
```

Instead, the new recommended way is to use the helm search hub command, which allows you to browse the Artifact Hub directly from the CLI:

```
$ helm search hub|head
URL                                                CHART VERSION
APP VERSION                                        DESCRIPTION
https://artifacthub.io/packages/helm/mya/12factor  24.1.2
Easily deploy any application that conforms to ...
https://artifacthub.io/packages/helm/gabibbo97/... 0.1.0
fedora-32                                          389 Directory Server
...<removed for brevity>...
```

Now, let's try searching for some of the most popular Helm charts that we can use to test our installation. We would like to deploy **WordPress** on our Kubernetes cluster. We chose WordPress to demonstrate here because it is a typical three-tier application with a public access tier (The Service), a web tier (WordPress), and a database tier (MariaDB). First, let's check what the available charts are for WordPress on Artifact Hub:

```
$ helm search hub wordpress
URL                                             CHART VERSION   APP
VERSION         DESCRIPTION
https://artifacthub.io/packages/helm/kube-wordp...   0.1.0      1.1
this is my wordpress package
https://artifacthub.io/packages/helm/wordpress-...   1.0.2      1.0.0
A Helm chart for deploying Wordpress+Mariadb st...
https://artifacthub.io/packages/helm/bitnami-ak...   15.2.13    6.1.0
WordPress is the world's most popular blogging ...
...<removed for brevity>...
```

Similarly, you can directly use the Artifact Hub web UI and search for WordPress Helm charts as follows:

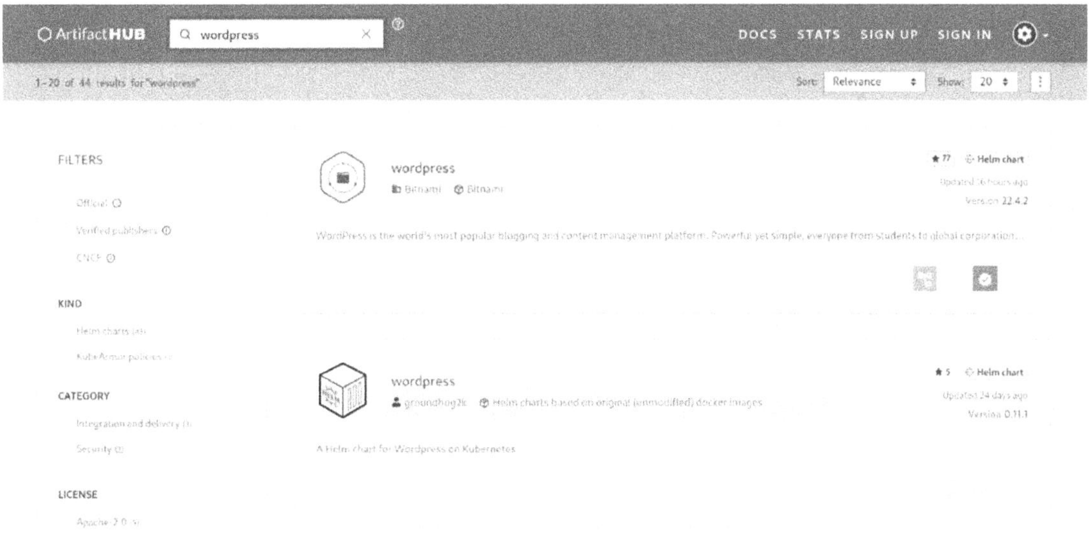

Figure 14.1: Artifact Hub search results for WordPress Helm charts

We will be using the Helm chart provided and maintained by **Bitnami,** a company specializing in distributing open-source software on various platforms, such as Kubernetes. If you navigate to the search result for WordPress charts by Bitnami, you will see the following:

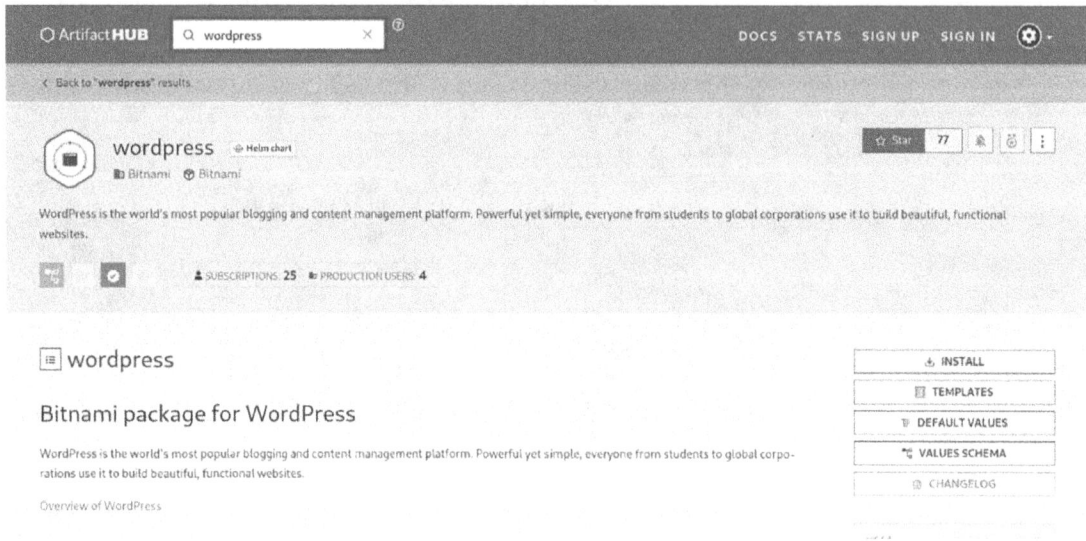

Figure 14.2: Bitnami WordPress Helm chart on Artifact Hub with install instructions

The page gives you detailed information about how you can add the `bitnami` repository and install the Helm chart for WordPress. Additionally, you will find a lot of details about available configurations, known limitations, and troubleshooting. You can also navigate to the home page of each of the charts in order to see the YAML templates that make up the chart (`https://github.com/bitnami/charts/tree/master/bitnami/wordpress`).

We can now do the installation by following the instructions on the web page. First, add the `bitnami` repository to your Helm installation:

```
$ helm repo add bitnami https://charts.bitnami.com/bitnami
"bitnami" has been added to your repositories
```

As a best practice, let us install WordPress inside a dedicated namespace called `wordpress`:

```
$ kubectl create ns wordpress
namespace/wordpress created
```

With the repository present, we can install the `bitnami/wordpress` Helm chart, but before that, we need to prepare some details for the deployment. Check the chart page in Artifact Hub (`https://artifacthub.io/packages/helm/bitnami/wordpress`). You will find a lot of parameters given here for you to configure and customize your WordPress deployment. If you do not provide any parameters, default values will be used for the Helm release deployment. For this demonstration, let us configure some of the parameters instead of using the default values.

You can pass the individual parameters using the `--set` argument as follows:

```
$ helm install wp-demo bitnami/wordpress -n wordpress --set
wordpressUsername=wpadmin
```

When you have multiple parameters to configure, you can pass multiple --set arguments but it is recommended to use variables in files; you can use one or more files to pass the variables.

Let us create a wp-values.yaml file to store the variables and values as follows:

```
# wp-values.yaml
wordpressUsername: wpadmin
wordpressPassword: wppassword
wordpressEmail: admin@example.com
wordpressFirstName: WP
wordpressLastName: Admin
service:
  type: NodePort
volumePermissions:
  enabled: true
```

As you can see, we are passing some of the WordPress configurations to the Helm chart. Please note that we are changing the default WordPress type to NodePort as we are using a minikube cluster here. If you are using a different Kubernetes cluster – for example, cloud-based Kubernetes – then you may leave it as the default, which is LoadBalancer.

Now that we have the Helm repo configured, a dedicated namespace created, and parameters configured in a variable file, let us deploy WordPress using Helm; we will use the name wp-demo for this release:

```
$ helm install wp-demo bitnami/wordpress -n wordpress --values wp-values.yaml
NAME: wp-demo
LAST DEPLOYED: Tue Jun  4 21:27:49 2024
NAMESPACE: wordpress
STATUS: deployed
REVISION: 1
TEST SUITE: None
NOTES:
CHART NAME: wordpress
CHART VERSION: 22.4.2
APP VERSION: 6.5.3
...<to be continued>...
```

Helm will show the basic release information, as shown here. After a while, you will also see the deployment details, including the service name, how to access the WordPress website, and more:

```
...
** Please be patient while the chart is being deployed **
```

```
Your WordPress site can be accessed through the following DNS name from within
your cluster:

    wp-demo-wordpress.wordpress.svc.cluster.local (port 80)

To access your WordPress site from outside the cluster follow the steps below:
Get the WordPress URL by running these commands:

    export NODE_PORT=$(kubectl get --namespace wordpress -o jsonpath="{.spec.
ports[0].nodePort}" services wp-demo-wordpress)
    export NODE_IP=$(kubectl get nodes --namespace wordpress -o jsonpath="{.
items[0].status.addresses[0].address}")
    echo "WordPress URL: http://$NODE_IP:$NODE_PORT/"
...<removed for brevity>...
```

This is the beauty of Helm – you have executed a single helm install command and you are presented with a detailed guide on how to use the deployed component on *your* cluster. Meanwhile, the WordPress instance deploys without any intervention from you!

> It is a good practice to first inspect what Kubernetes objects' YAML manifests were produced by Helm. You can do that by running the helm install command with additional flags: helm install wp-demo bitnami/wordpress --dry-run --debug. The output will contain the joint output of YAML manifests, and they will not be applied to the cluster.

You can also mention the specific version of the Helm chart using the --version argument as follows:

```
$ helm install my-wordpress bitnami/wordpress --version 22.4.2
```

Let's now follow the instructions from the Helm chart installation output:

1. Wait for a while as the database needs to initialize and deploy the Pods. Check the Pod status:

```
$ kubectl get po -n wordpress
NAME                                     READY   STATUS    RESTARTS
AGE
wp-demo-mariadb-0                        1/1     Running   9 (5m57s ago)
31m
wp-demo-wordpress-5d98c44785-9xd6h       1/1     Running   0
31m
```

2. Notice the database is deployed as a StatefulSet as follows:

```
$ kubectl get statefulsets.apps -n wordpress
NAME              READY   AGE
wp-demo-mariadb   1/1     99s
```

3. Wait for the `wp-demo` Service object (of the NodePort type) to acquire port details:

```
$ kubectl get svc -n wordpress
NAME                TYPE        CLUSTER-IP      EXTERNAL-IP   PORT(S)
AGE
wp-demo-mariadb     ClusterIP   10.100.118.79   <none>        3306/TCP
2m39s
wp-demo-wordpress   NodePort    10.100.149.20   <none>        80:30509/
TCP,443:32447/TCP   2m39s
```

In our case, the port will be `80:31979/TCP`.

4. Since we are using minikube in this case, let us find the IP address and port details (if you are using the `LoadBalancer` type, then you can directly access the IP address to see the WordPress site):

```
$ minikube service --url wp-demo-wordpress -n wordpress
http://192.168.59.150:30509
http://192.168.59.150:32447
```

5. Now open your web browser and navigate to the WordPress URL, `http://192.168.59.150:30509` (the other port is for the HTTPS URL):

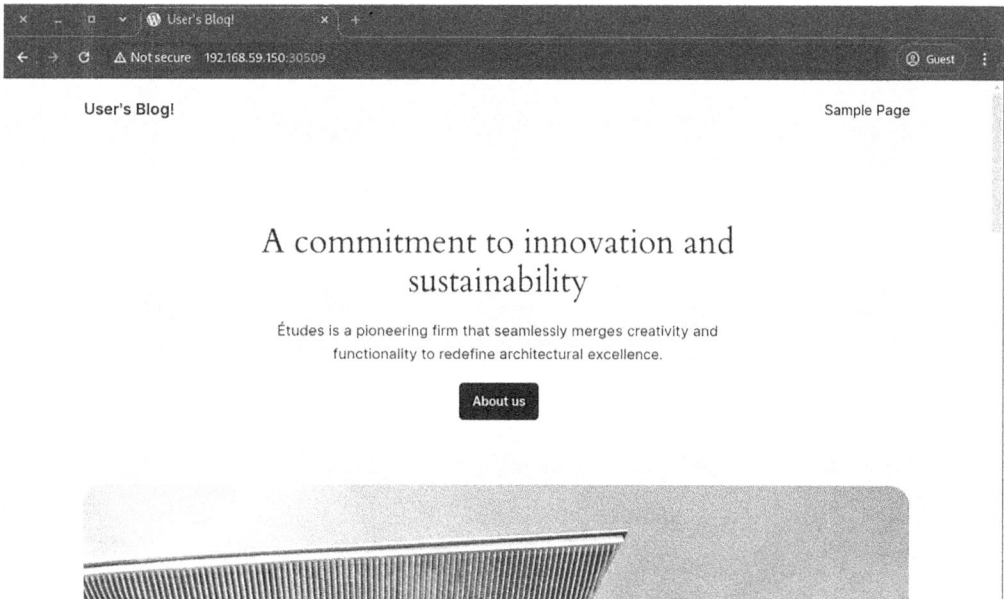

Figure 14.3: WordPress chart deployed on Kubernetes – main page

6. Now you can log in to the WordPress dashboard at `http://192.168.59.150:30509/wp-admin`. Please note that if you missed setting the WordPress parameters, including the password, you need to find the default values Helm has used. For example, to retrieve the WordPress login password, check the secret as follows. Use the following commands to obtain the credentials that are stored in a dedicated `wp-demo-wordpress` Secret object deployed as part of the chart:

```
$ kubectl get secret --namespace wordpress wp-demo-wordpress -o
jsonpath="{.data.wordp
ress-password}" | base64 --decode
wppassword
```

7. Use the credentials to log in as the WordPress admin:

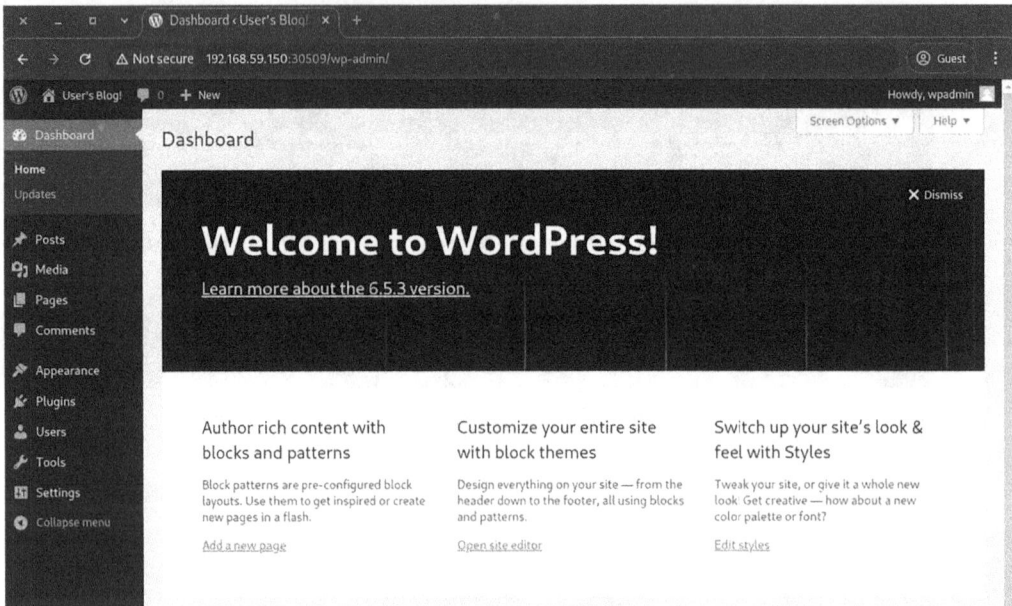

Figure 14.4: WordPress chart deployed on Kubernetes – admin dashboard

You can enjoy your WordPress now, congratulations! If you are interested, you can inspect the Pods, Services, Deployments, and StatefulSets that were deployed as part of this Helm chart. This will give you a lot of insight into what the components of the chart are and how they interact.

> The Helm CLI offers *autocompletion* for various shells. You can run the `helm completion` command to learn more.

If you want to get information about all Helm releases that are deployed in your Kubernetes cluster, use the following command:

```
$ helm list -n wordpress
```

```
NAME       NAMESPACE        REVISION        UPDATED
STATUS            CHART                 APP VERSION
wp-demo wordpress        1                2024-06-04 22:20:32.556683096 +0800 +08
deployed           wordpress-22.4.2    6.5.3
```

In the next section, we will learn how to delete a deployed release using Helm.

Deleting a Helm release

As we learned in the previous sections, the Helm chart has deployed several resources, including Deployment, PVC, Services, Secrets, and so on. It is not practical to find and delete these items one by one. But Helm provides an easy method to remove the deployment using the `helm uninstall` command. When you are ready, you can clean up the release by uninstalling the Helm release using the following command:

```
$ helm uninstall wp-demo -n wordpress
release "wp-demo" uninstalled
```

This will delete all Kubernetes objects that the release has created. Please note though that Persistent-Volumes and PersistentVolumeClaims, created by the Helm chart, will not be cleaned up – you need to clean them up manually. We will now take a closer look at how Helm charts are structured internally.

Helm chart anatomy

As an example, we will take the WordPress Helm chart by Bitnami (https://github.com/bitnami/charts/tree/master/bitnami/wordpress) that we have just used to perform a test Deployment in the cluster. Helm charts are simply directories with a specific structure (convention) that can live either in your local filesystem or in a Git repository. The directory name is at the same time the name of the chart – in this case, wordpress. The structure of files in the chart directory is as follows:

- `Chart.yaml`: YAML file that contains metadata about the chart such as version, keywords, and references to dependent charts that must be installed.
- `LICENSE`: Optional, plain-text file with license information.
- `README.md`: End user README file that will be visible on Artifact Hub.
- `values.yaml`: The default configuration values for the chart that will be used as YAML template parameters. These values can be overridden by the Helm user, either one by one in the CLI or as a separate YAML file with values. You have already used this method by passing the `--values wp-values.yaml` argument.
- `values.schema.json`: Optionally, you can provide a JSON schema that `values.yaml` must follow.
- `charts/`: Optional directory with additional, dependent charts.
- `crds/`: Optional Kubernetes custom resource definitions.
- `templates/`: The most important directory that contains all YAML *templates* for generating Kubernetes YAML manifest files. The YAML templates will be combined with the provided *values*. The resulting YAML manifest files will be applied to the cluster.
- `templates/NOTES.txt`: Optional file with short usage notes.

For example, if you inspect `Chart.yaml` in the WordPress Helm chart, you can see that it depends on the **MariaDB** chart by Bitnami, if an appropriate value of `mariadb.enabled` is set to `true` in the provided values:

```
...

appVersion: 6.5.3
dependencies:
- condition: memcached.enabled
  name: memcached
  repository: oci://registry-1.docker.io/bitnamicharts
  version: 7.x.x
- condition: mariadb.enabled
  name: mariadb
  repository: oci://registry-1.docker.io/bitnamicharts
  version: 18.x.x
...
```

Now, if you take a look at the `values.yaml` file with the default values, which is quite verbose, you can see that by default MariaDB is enabled:

```
...
## MariaDB chart configuration
## ref: https://github.com/bitnami/charts/blob/main/bitnami/mariadb/values.yaml
##
mariadb:
  ## @param mariadb.enabled Deploy a MariaDB server to satisfy the applications
database requirements
  ## To use an external database set this to false and configure the
`externalDatabase.*` parameters
  ##
  enabled: true
...
```

And lastly, let's check what one of the YAML templates looks like – open the `deployment.yaml` file (https://github.com/bitnami/charts/blob/master/bitnami/wordpress/templates/deployment.yaml), which is a template for the Kubernetes Deployment object for Pods with WordPress containers. For example, you can see how the number of `replicas` is referenced from the provided values:

```
kind: Deployment
...
spec:
...
  replicas: {{ .Values.replicaCount }}
...
```

This will be replaced by the `replicaCount` value (for which the default value of 1 is found in the `values.yaml` file). The details about how to use Go templates can be found at `https://pkg.go.dev/text/template`. You can also learn by example by analyzing the existing Helm charts – most of them use similar patterns for processing provided values.

> The detailed documentation on Helm chart structure can be found at `https://helm.sh/docs/topics/charts/`.

In most cases, you will need to override some of the default values from the `values.yaml` file during the installation of a chart, as we learned earlier.

Now, that you know the most important details about the Helm chart structure, in the next section, we can deploy Kubernetes Dashboard using Helm charts.

Installing Kubernetes Dashboard using Helm Charts

Kubernetes Dashboard is the official web UI for providing an overview of your cluster. The Helm chart for this component is officially maintained by the Kubernetes community (`https://artifacthub.io/packages/helm/k8s-dashboard/kubernetes-dashboard`). We are going to install it with the default parameters, as there is no need for any customizations at this point.

> For minikube clusters, you can enable the dashboard and access it using a single command: `minikube dashboard`. But our intention here is to learn how to deploy a dashboard for any type of Kubernetes cluster.

First, add the `kubernetes-dashboard` repository to Helm:

```
$ helm repo add kubernetes-dashboard https://kubernetes.github.io/dashboard/
"kubernetes-dashboard" has been added to your repositories
```

Now, we can install the Helm chart as a `kubernetes-dashboard` release in the cluster as follows:

```
$ helm upgrade --install kubernetes-dashboard kubernetes-dashboard/kubernetes-dashboard --create-namespace --namespace kubernetes-dashboard
```

Wait for the installation to finish and check the output messages. Notice the following message as we will use it later to access the dashboard WEBUI:

```
...
Congratulations! You have just installed Kubernetes Dashboard in your cluster.

To access the Dashboard, run the following:

  kubectl -n kubernetes-dashboard port-forward svc/kubernetes-dashboard-kong-
```

```
proxy 8443:443
...
```

Also, ensure Pods have a Running status:

```
$  kubectl get pod -n kubernetes-dashboard
NAME                                                                    READY    STATUS
RESTARTS     AGE
kubernetes-dashboard-api-86c68c7896-7jxwz                               1/1      Running    0
2m53s
kubernetes-dashboard-auth-59784dd8b-vsr99                               1/1      Running    0
2m53s
kubernetes-dashboard-kong-7696bb8c88-6q7zs                              1/1      Running    0
2m53s
kubernetes-dashboard-metrics-scraper-5485b64c47-9d69q                   1/1      Running    0
2m53s
kubernetes-dashboard-web-84f8d6fff4-nxt5f                               1/1      Running    0
2m53s
```

You may ignore other Pods deployed by the Helm chart for now. We will learn how to access the dashboard UI in the next section.

Secure access to the Kubernetes Dashboard

By default, the Kubernetes Dashboard prioritizes security by using a minimal RBAC configuration. This helps safeguard your cluster data. Currently, logging in to the dashboard requires a Bearer Token.

> This sample user creation guide will likely grant administrative privileges. Be sure to use it only for educational purposes and implement proper RBAC controls for production environments.

Follow these steps to create a token to access the Kubernetes Dashboard WEBUI:

1. Create a ServiceAccount; prepare the YAML as follows:

    ```yaml
    # dashboard-sa.yaml
    apiVersion: v1
    kind: ServiceAccount
    metadata:
      name: admin-user
      namespace: kubernetes-dashboard
    ```

2. Create the ServiceAccount as follows:

```
$ kubectl apply -f dashboard-sa.yaml
serviceaccount/admin-user created
```

3. Create `ClusterRoleBinding` to allow the access. Prepare the YAML as follows:

```yaml
# dashboard-rbac.yml
apiVersion: rbac.authorization.k8s.io/v1
kind: ClusterRoleBinding
metadata:
  name: admin-user
roleRef:
  apiGroup: rbac.authorization.k8s.io
  kind: ClusterRole
  name: cluster-admin
subjects:
  - kind: ServiceAccount
    name: admin-user
    namespace: kubernetes-dashboard
```

4. Create `ClusterRoleBinding` by applying the YAML definition:

```
$ kubectl apply -f dashboard-rbac.yml
clusterrolebinding.rbac.authorization.k8s.io/admin-user created
```

5. Create the token:

```
$ kubectl -n kubernetes-dashboard create token admin-user
```

Copy the long token string generated, and we will use it in the next section to log in to the cluster dashboard.

Accessing Dashboard WEBUI

The Kubernetes Dashboard offers various access methods. Here, we'll focus on the default approach. This method assumes you haven't altered the standard configuration during installation. If you've made modifications, the steps might differ.

Execute the following command (which you copied from the `helm install` output earlier) to get Dashboard access. The command will stay there with the status of `port-forward`; do not exit from the command:

```
$ kubectl -n kubernetes-dashboard port-forward svc/kubernetes-dashboard-kong-
proxy 8443:443
Forwarding from 127.0.0.1:8443 -> 8443
Forwarding from [::1]:8443 -> 8443
...
```

Now, access the URL `https://localhost:8443/` in a web browser. You can ignore the SSL certificate warning as the dashboard is using the self-signed SSL certificates. Enter the token you generated in *Step 3*, of *Secure Access to the Kubernetes Dashboard*, and log in to Dashboard as shown in the following figure.

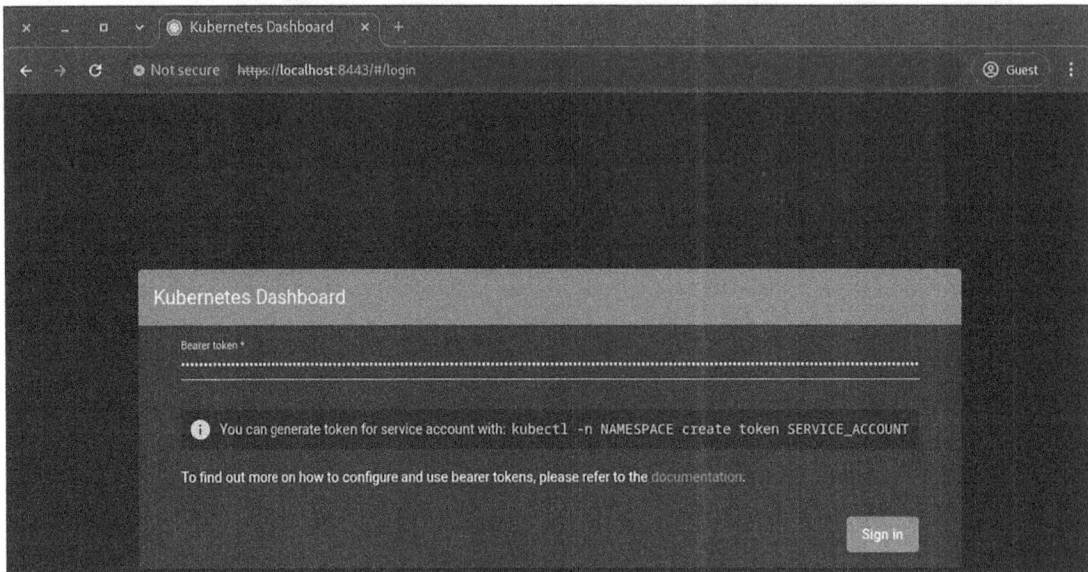

Figure 14.5: Kubernetes Dashboard chart – login page

At this point, you have access to the dashboard, and you can browse its functionalities:

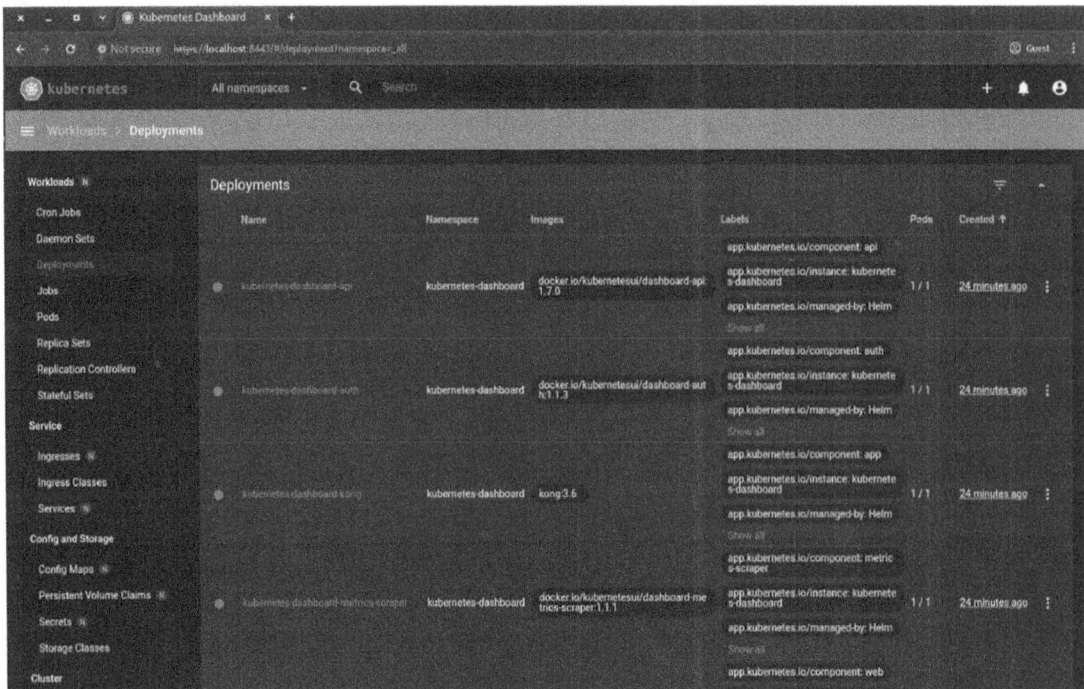

Figure 14.6: Kubernetes Dashboard chart – Deployments page

The bearer token is for a user with the `cluster-admin` role, so be careful, as you can perform any operations, including deleting resources.

Congratulations; you have successfully deployed the Kubernetes Dashboard using Helm charts, and verified the access using the token. You can explore more deployments using Helm charts as we explain in the next section.

Installing other popular solutions using Helm charts

To practice more with Helm charts, you can quickly install some of the other software for your Kubernetes cluster. It can be useful in your development scenarios or as building blocks of your cloud-native applications.

Elasticsearch with Kibana

Elasticsearch is a popular full-text search engine that is commonly used for log indexing and log analytics. Kibana, which is part of the Elasticsearch ecosystem, is a visualization UI for the Elasticsearch database. To install this stack, we will need to use two charts, both of which are maintained by Elasticsearch creators:

- **Elasticsearch chart** (https://artifacthub.io/packages/helm/elastic/elasticsearch)
- **Kibana chart** (https://artifacthub.io/packages/helm/elastic/kibana)

Prometheus with Grafana

Prometheus is a popular monitoring system with a time series database and Grafana is used as a visualization UI. Similar to the Elastic Stack, to install this Prometheus and Grafana stack, we will need to use two charts:

- **Prometheus** (https://artifacthub.io/packages/helm/prometheus-community/prometheus)
- **Grafana** (https://artifacthub.io/packages/helm/grafana/grafana)

Since we have already explained how to deploy Helm charts in the Kubernetes cluster, we will skip the step-by-step instructions for these. You may continue deploying the charts and exploring the functionalities.

In this section, we will explore some of the key security considerations for Helm Charts.

Security considerations for Helm Charts

Helm charts ease the deployment of applications into Kubernetes, but they can introduce several security vulnerabilities that need to be managed. Some of the most important things to bear in mind include the following:

- **Source Verification:** Always check the source for Helm charts before deploying and never install charts originating from non-trusted or less reputed repositories. Malicious/insecure applications may be contained. Verify the origin of a chart, and try using official ones or well-maintained community repositories.

- **Regular Audits:** Periodically run security audits of Helm charts and their dependencies. This process helps in identifying known vulnerabilities that, in turn, ensure applications deployed are secure enough to meet standards set by your organization. Perform vulnerability scanning within Helm charts using tools like Trivy or Anchore.

- **Chart Configuration:** Beware of the defaults that come in Helm charts. Most are shipped with configuration settings that are not appropriate to your production environment. Consider reviewing and adjusting the default settings as necessary, using your organizational security policy and best practices.

- **Role-Based Access Control (RBAC):** Implement this to restrict the deployment and management of Helm charts in your Kubernetes cluster. This will reduce unauthorized changes and also ensure that only trusted persons can deploy sensitive applications.

- **Dependency Management:** Monitor and manage the list of dependencies shown within your Helm charts. Regularly update these to avoid security gaps in applications and make sure they receive the latest security patches and improvements.

- **Namespace Isolation:** Consider using Helm charts, each in their own separate namespace, as this will increase security. If something bad happens, the blast radius will be limited, thus giving better isolation for applications and their resources.

By being proactive in these areas, you can go a long way toward improving the security posture of your Kubernetes deployments and ensuring that potential vulnerabilities do not bring down your applications and data.

Now, we have explored and practiced the Helm charts in the first half of this chapter. We also learned that the Helm chart is a great way to deploy and manage complex applications in Kubernetes.

As Kubernetes adoption increases, so does the complexity of the production environment for managing applications. While Helm and similar tools improve the process of deploying applications, they cannot independently address stateful applications' operational needs such as scaling, configuration management, and failure recovery at runtime. We need solutions that package application knowledge and best practices for operation to automate operational tasks at every stage of the applications' lives with health and high performance in mind. With these solutions, teams minimize human interaction as well as reducing human error, and focus on delivering value through the application and not managing the infrastructure. Kubernetes Operators were introduced to help with such requirements.

In the following section, we will learn what Kubernetes Operators are and how to install complex deployments with Kubernetes Operators.

Introducing Kubernetes Operators

We've explored the differences between Deployments and StatefulSets, with StatefulSets managing stateful applications that require persistent data storage. We also learned about the manual (and automated) operations needed for StatefulSets to function, such as data synchronization between Pod replicas and initialization tasks.

However, manual intervention goes against the core principles of Kubernetes, where automation and self-healing are paramount. This is where Kubernetes Operators step in.

From humans to software

Imagine replacing human Operators with software Operators. Kubernetes Operators are essentially software extensions that automate complex application management tasks, especially for stateful applications. Instead of relying on manual intervention to maintain application stacks, Operators leverage their built-in software components and intelligence to execute and manage these tasks effectively.

The following image shows the high-level relationship between the components in a Kubernetes cluster with Operators.

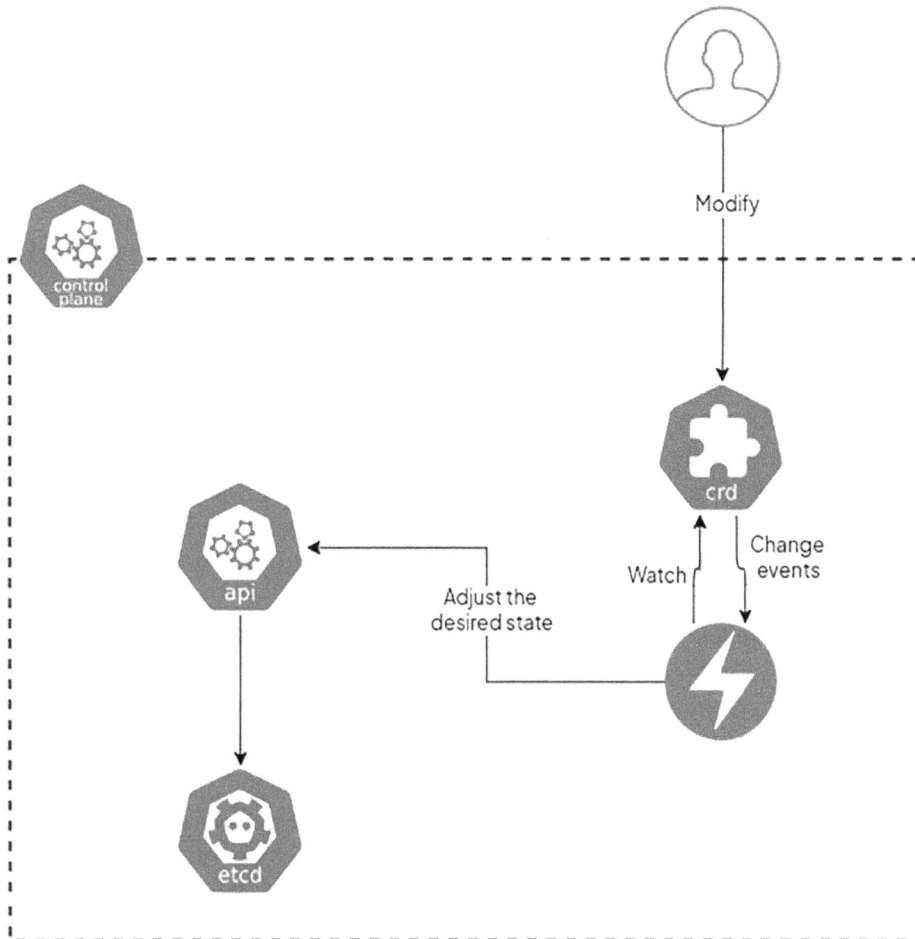

Figure 14.7: How Operators manage resources in Kubernetes

In the following sections, we will discuss the advantages and benefits of Kubernetes Operators.

Helm Charts versus Kubernetes Operators

While Helm charts are associated components related to Operators in application management on Kubernetes, they are different. With Helm, users can install applications much faster than with Operators because Helm is an effective package manager, using pre-designed charts that make the process easier.

It finds its best application in situations that demand speed and, therefore, it is ideal for a one-time installation or even for smaller applications. Yet Helm still focuses way more on the deployment phase rather than on the continued operational needs across an application's runtime.

Operators are software programs that intend to assume responsibility for the entire life cycle of complex and stateful applications. Using **Custom Resource Definitions** (**CRDs**), they encode knowledge about how an application operates and allow scaling, configuration management, automation of upgrades, and self-healing mechanisms. Operators are constantly observing the health of an application and taking corrective action to head it back toward the desired state. That's why they become so valuable for applications that are very management- and observation-intensive. To put it succinctly, Helm charts make the deployment easy, but Operators extend Kubernetes into production operability, letting teams automate and simplify how they manage their apps in production.

In the next section, let us explore some of the major features of Kubernetes Operators compared to Helm charts.

How can the Operators help in the application deployment?

Operators are created with the capabilities to manage and maintain the application with all possible operations, including the following:

- **Life Cycle Management:** Operators manage the life cycle of your application or application stack beyond mere initial deployment. They automate key operational tasks like upgrades, scaling, and recovery from failures to keep the application healthy and performing over time. Other approaches, like Helm charts, generally stop at deployment, whereas Operators constantly monitor the current application state and automatically reconcile it to the desired configuration in case changes or issues arise. This allows for automated upgrades, configuration changes, and status monitoring – all with no intervention required or even desired. For complex stateful applications with demanding life cycle management, Operators offer significant advantages over Helm and must be preferred for such applications that require continuous care and automated management.

- **Resource Orchestration:** Operators create essential resources like **ConfigMaps**, **Secrets**, and **PVCs** required by your application to function properly.

- **Automated Deployment:** Operators can deploy your application stack based on either user-provided configurations or default values, streamlining the deployment process.

- **Database Management:** Take PostgreSQL clusters, for example. Operators can leverage StatefulSets to deploy them and ensure data synchronization across replicas.

- **Self-Healing Capabilities:** Operators can detect and react to application failures, triggering recovery or failover mechanisms to maintain service continuity.

Reusability of the automation

Operators promote reusability. The same Operator can be utilized across different projects or in multiple Kubernetes clusters, ensuring consistent and efficient application management throughout your infrastructure.

How Operators ensure the application state

Kubernetes Operators function similarly to Kubernetes itself, utilizing a control loop to manage applications. This loop continuously monitors the desired state of your application, defined by a CRD, and compares it to the application's actual state within the cluster. Any discrepancies trigger corrective actions from the Operator. These actions can involve scaling the application, updating configurations, or restarting Pods. The control loop's continuous operation ensures your application remains aligned with the desired state, promoting self-healing and automated management.

Custom resource definitions – building blocks for Operators

Kubernetes Operators rely on CRDs. These are essentially extensions of the Kubernetes API that allow you to define custom resources specific to your application or its needs. Think of them as blueprints for your application's desired configuration within the Kubernetes ecosystem.

CRDs essentially extend the Kubernetes API, allowing you to define custom resources specific to your application. These resources represent the building blocks of your application within the Kubernetes cluster. They can specify details like the following:

- The desired number of application replicas (Pods).
- Resource requests and limits for memory and CPU.
- Storage configurations for persistent data.
- Environment variables and configuration settings.

Benefits of CRDs

There are multiple benefits of using CRDs to deploy applications, including the following:

- **Declarative Management:** Instead of manually configuring individual resources like Deployments or Services, CRDs let you define the desired state of your application in a declarative manner. The Operator then takes care of translating that desired state into actual running resources within the cluster.
- **Application-Specific Configuration:** CRDs cater to the unique needs of your application. You can define custom fields specific to your application logic or configuration requirements.
- **Simplified Management:** CRDs provide a central point for managing your application's configuration. This streamlines the process compared to scattered configurations across different resource types.

Operator distribution mechanism

Operators are primarily created and distributed by application vendors who possess the expertise in deploying those specific application stacks. However, a vibrant community actively develops, distributes, and maintains a vast array of Operators.

OperatorHub (`https://operatorhub.io/`) serves as a central repository where you can discover and install Operators for a wide range of applications and functionalities.

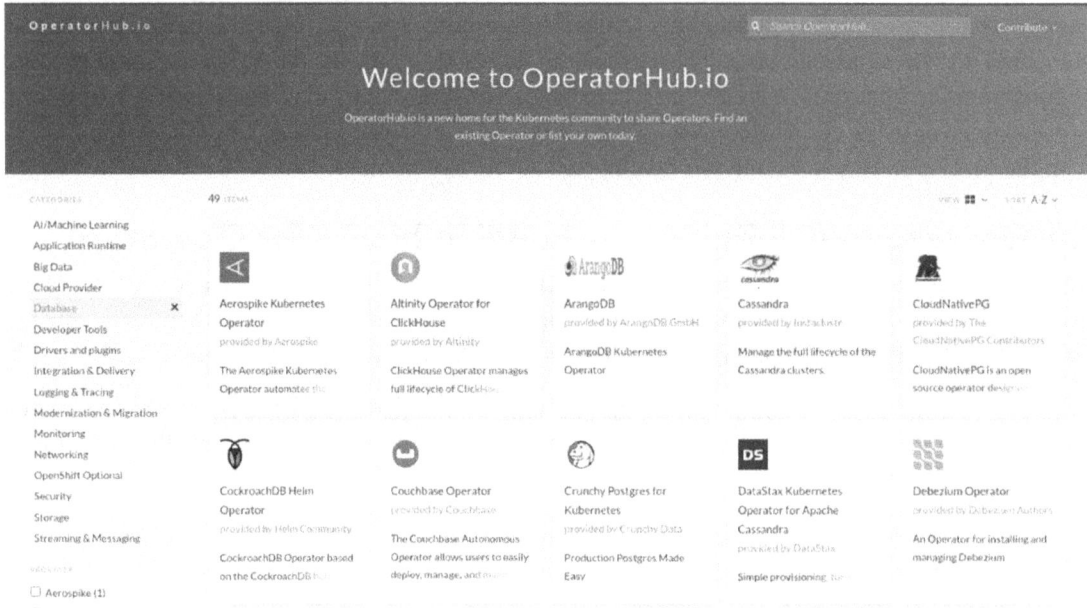

Figure 14.8: OperatorHub main page

Building your own Operator

The **Operator Framework** provides a powerful toolkit known as the **Operator SDK**. This open-source gem empowers you to develop and build custom Operators, taking control of your application management within Kubernetes. The Operator SDK streamlines the often-complex process of crafting Operators. It offers pre-built components and functionalities that handle common tasks like interacting with the Kubernetes API, managing custom resources, and implementing the control loop. This allows you to focus on the unique logic and configuration needs of your specific application.

Building your own Operators unlocks several advantages. Firstly, you gain fine-grained control over application management. The Operator can be tailored to your application's specific needs, handling configuration, deployment strategies, and scaling requirements perfectly. Secondly, Operators automate repetitive tasks associated with the application life cycle, leading to significant efficiency gains. Finally, custom Operators are reusable. Once built for a particular application, they can be applied across different deployments of the same application, saving you time and effort in managing your Kubernetes infrastructure.

You can learn more about the Operator SDK at `https://sdk.operatorframework.io/`.

Before we do some hands-on with the Operator, let us learn a few details about **Operator Lifecycle Manager (OLM)** in the next section.

Operator Lifecycle Manager (OLM)

Operator Lifecycle Manager in Kubernetes simplifies the process of deploying and managing applications packaged as Operators. OLM uses a declarative way of specifying the resources using YAML files. In this, there is no need for complex multi-file deployments that rely on specific ordering. Installation and automatic upgrades also keep your Operators up to date with OLM. It also exposes a package discovery capability called Catalog Sources, which enables the use of Operators from, for example, OperatorHub, or from sources of your choosing. With OLM, you will gain several advantages. First, it reduces deployment complexity by managing dependencies and order. OLM can manage thousands of Operators over large clusters. What's more, OLM enforces the desired configuration. This will ease rollouts and updates. Not to mention that OLM promotes standardization by providing a consistent way to package and deploy applications as Operators.

This is a question of your environment and personal preference. OLM is much more integrated and native to Kubernetes, while Helm, being familiar for other deployments, comes with a richer package ecosystem.

When we learn about Operators and OLM, it is also important to know about the **ClusterServiceVersion (CSV)**. The ClusterServiceVersion will be the central point of an important part of Kubernetes' Operator Lifecycle Manager, the core metadata processing and deploying information of the Operator. It goes on to define the name of the Operator and the version it holds, and gives a brief description. It also outlines the permissions or roles required by the Operator to operate correctly. It defines the CRDs that the Operator governs, installation strategy, and upgrade flows. Refer to the documentation (`https://olm.operatorframework.io/docs/concepts/crds/clusterserviceversion/`) to learn more about the CSV.

In the following section, we'll explore how to deploy Prometheus monitoring on Kubernetes using both OLM and the Prometheus Operator, showcasing the power of both approaches in application management.

Enabling Kubernetes monitoring using Prometheus and Grafana

Success in keeping your Kubernetes applications healthy and performant depends on different factors; one of those is having a robust, reliable environment. Here, monitoring tools like **Prometheus** and **Grafana** can help. Prometheus works behind the scenes; it gathers and stores valuable metrics about your Kubernetes cluster. Grafana visualizes this treasure trove of data, presenting it in an understandable format for you to gain deep insight into the health and behavior of your applications and infrastructure.

The following figure shows the high-level architecture of Prometheus components. (It's an official reference.)

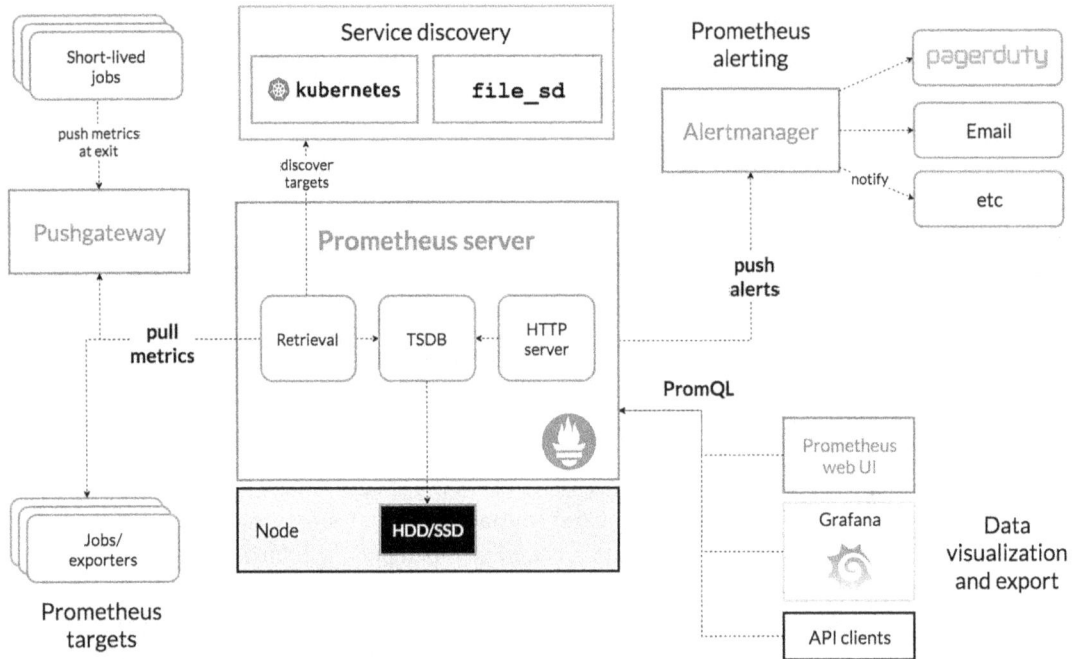

Figure 14.9: Architecture of Prometheus and some of its ecosystem components (Source: https://prometheus.io/docs/introduction/overview/)

Traditional deployments, say, of monitoring stacks comprising Prometheus and Grafana, are pretty cumbersome. Writing several YAML manifests by hand, with all the dependencies and the proper ordering of deployment, is rather laborious and prone to errors. The Prometheus and Grafana Operators offer an even more efficient and maintainable solution.

It is possible to utilize Helm charts to deploy this entire monitoring stack – including Operators and instances – using projects such as kube-prometheus-stack (https://artifacthub.io/packages/helm/prometheus-community/kube-prometheus-stack). But in the next section, our intention is to set up the same monitoring stack using OLM.

In the following section, we'll explore the process of deploying Prometheus and Grafana within your Kubernetes cluster, equipping you to effectively monitor your applications and ensure their smooth operation.

Installing Operator Lifecycle Manager

To utilize the OLM-based operator installation, we need to install OLM in the cluster. You can install OLM in the cluster using the operator-sdk utility, Helm charts, or even by applying the Kubernetes YAML manifests for OLM. For this exercise, let us use the install.sh script-based installation as follows:

```
$ curl -sL https://github.com/operator-framework/operator-lifecycle-manager/
releases/download/v0.28.0/install.sh | bash -s v0.28.0
```

Wait for the script to complete and configure the Kubernetes cluster with OLM. Once finished, verify the Pods are running in the olm namespace as follows:

```
$ kubectl get pods -n olm
NAME                                  READY   STATUS    RESTARTS   AGE
catalog-operator-9f6dc8c87-rt569      1/1     Running   0          36s
olm-operator-6bccddc987-nrlkm         1/1     Running   0          36s
operatorhubio-catalog-6l8pw           0/1     Running   0          21s
packageserver-6df47456b9-8fdt7        1/1     Running   0          24s
packageserver-6df47456b9-lrvzp        1/1     Running   0          24s
```

Also, check the ClusterServiceVersion details using the kubectl get csv command as follows:

```
$ kubectl get csv -n olm
NAME            DISPLAY         VERSION   REPLACES   PHASE
packageserver   Package Server  0.28.0               Succeeded
```

That has a success status – we can see the OLM is deployed and ready to manage the Operators. In the next section, we will deploy Prometheus and Grafana Operators using OLM.

Installing Prometheus and Grafana Operators using OLM

Once you configure OLM, deploying Operators is pretty straightforward. Most of the time, you will find the operator installation command and instructions on the Operators page at operatorhub.io.

To install the Prometheus operator (https://operatorhub.io/operator/prometheus) using OLM, follow these steps:

1. Search for the Prometheus Operator in OperatorHub:

    ```
    $ kubectl get packagemanifests | grep prometheus
    ack-prometheusservice-controller          Community Operators   12m
    prometheus                                Community Operators   12m
    prometheus-exporter-operator              Community Operators   12m
    ```

2. Install the Prometheus operator using OLM:

    ```
    $ kubectl create -f https://operatorhub.io/install/prometheus.yaml
    subscription.operators.coreos.com/my-prometheus created
    ```

> https://operatorhub.io/install/prometheus.yaml provides a basic YAML definition to create a subscription. You can create local YAML files with all the customization required.

3. Wait for a few minutes and ensure the Prometheus operator is deployed properly:

```
$ kubectl get csv -n operators
NAME                            DISPLAY              VERSION    REPLACES
PHASE
prometheusoperator.v0.70.0    Prometheus Operator   0.70.0
prometheusoperator.v0.65.1    Succeeded
```

4. Verify the Prometheus operator Pods are running:

```
$ kubectl get pods -n operators
NAME                                  READY   STATUS    RESTARTS   AGE
prometheus-operator-84f9b76686-2j27n  1/1     Running   0          87s
```

5. In the same way, find and install the Grafana operator using OLM (follow the previous steps for references):

```
$ kubectl create -f https://operatorhub.io/install/grafana-operator.yaml
```

6. Now, also verify the CRDs created in the backend as these entries are created as part of the Operator installation:

```
$ kubectl get crd
AME                                              CREATED AT
alertmanagerconfigs.monitoring.coreos.com
2024-10-18T09:14:03Z
alertmanagers.monitoring.coreos.com
2024-10-18T09:14:04Z
catalogsources.operators.coreos.com
2024-10-18T09:10:00Z
clusterserviceversions.operators.coreos.com
2024-10-18T09:10:00Z
grafanaalertrulegroups.grafana.integreatly.org
2024-10-18T09:25:19Z
...<removed for brevity>...
```

7. You will find multiple CRDs created in the Kubernetes cluster.

Now that we have deployed the Operators, it is time to create the Prometheus and Grafana instances and configure the stack to monitor the Kubernetes cluster. We will learn about these operations in the next section.

Configuring Prometheus and Grafana instances using Operators

To configure new instances, let us utilize standard YAML definitions with the CRD configuration. The instructions and the YAML definition files for this exercise are stored in the Chapter 14 directory of the GitHub repository.

Follow the steps to configure a monitoring stack in Kubernetes with Prometheus and Grafana:

1. As a best practice, let us create a namespace to deploy the monitoring solution (refer to monitoring-ns.yaml):

```
$ kubectl apply -f monitoring-ns.yaml
namespace/monitoring created
```

2. Configure a ServiceAccount with the appropriate role and RBAC (refer to monitoring-sa.yaml in the repo):

```
$ kubectl apply -f monitoring-sa.yaml
serviceaccount/prometheus created
role.rbac.authorization.k8s.io/prometheus-role created
rolebinding.rbac.authorization.k8s.io/prometheus-rolebinding created
```

3. Prepare YAML for the new Prometheus instance (refer to promethues-instance.yaml):

```
apiVersion: monitoring.coreos.com/v1
kind: Prometheus
metadata:
  name: example-prometheus
  namespace: monitoring
spec:
  replicas: 2
  serviceAccountName: prometheus
  serviceMonitorSelector:
    matchLabels:
      app.kubernetes.io/name: node-exporter
```

> Notice the kind: Prometheus in the preceding YAML definition, as we are using a CRD here; the Prometheus operator will understand this CRD and take necessary actions to create the deployment.

4. Create a new Prometheus instance by applying the configuration to the cluster:

```
$ kubectl apply -f promethues-instance.yaml
prometheus.monitoring.coreos.com/example-prometheus
```

5. In a similar way, deploy the Grafana instance using the operator:

```
# grafana-instnace.yaml
apiVersion: grafana.integreatly.org/v1beta1
kind: Grafana
metadata:
  labels:
    dashboards: grafana-a
    folders: grafana-a
```

```
  name: grafana-a
  namespace: monitoring
spec:
  config:
    auth:
      disable_login_form: 'false'
    log:
      mode: console
    security:
      admin_password: start
      admin_user: root
```

> We are using the configuration and passwords here in plain text directly inside the YAML definition. In a production environment, you should be using Kubernetes Secrets to store such sensitive data.

6. Apply the YAML definition to create a Grafana instance:

```
$ kubectl apply -f grafana-instance.yaml
grafana.grafana.integreatly.org/grafana-a created
```

7. Verify the objects created by Prometheus and Grafana Operators in the monitoring namespace:

```
$ kubectl get pod,svc,sts -n monitoring
NAME                                          READY   STATUS    RESTARTS   AGE
pod/grafana-a-deployment-69f8999f8-82zbv      1/1     Running   0          17m
pod/node-exporter-n7tlb                       1/1     Running   0          93s
pod/prometheus-example-prometheus-0           2/2     Running   0          20m
pod/prometheus-example-prometheus-1           2/2     Running   0          20m

NAME                          TYPE        CLUSTER-IP       EXTERNAL-IP
PORT(S)      AGE
service/grafana-a-service     ClusterIP   10.107.212.241   <none>
3000/TCP     17m
service/prometheus-operated   ClusterIP   None             <none>
9090/TCP     20m

NAME                                                   READY   AGE
statefulset.apps/prometheus-example-prometheus         2/2     20m
```

8. You can see, in the preceding output, that the Operators have created appropriate Kubernetes resources based on the CRD. You can even see the Prometheus and Grafana Service created and will be able to access it. We will demonstrate it at a later stage of this exercise.

For demonstration purposes, let us enable Node Exporter in the cluster and visualize it using Grafana. Node Exporter is one of the Prometheus exporters, which exposes detailed metrics of a host machine, including hardware and OS details, such as CPU usage, memory usage, and other system-level metrics. It runs on each node in a Kubernetes cluster – or physical or virtual servers – as a separate service and exposes these metrics through an HTTP endpoint. By scraping this data, Prometheus can know the health and performance of a node, thereby enabling an administrator to understand resource utilization and point out problems in the infrastructure.

Running Node Exporter as a DaemonSet would imply that each node in the Kubernetes cluster is running an instance of the exporter. In that way, Prometheus would be able to consistently scrape the system metrics across all nodes for effective observation of the overall health and performance of the cluster. Create the Node Exporter using `node-exporter-daemonset.yaml` as follows:

```
$ kubectl apply -f node-exporter-daemonset.yaml
daemonset.apps/node-exporter created
```

A Node Exporter Service (`svc`) should be created to expose the metrics that the Node Exporter is collecting to Prometheus. The service provides a way for Prometheus to discover and scrape metrics from the Node Exporter pods running on each node, thus providing a capability of centralized monitoring of node performance across the Kubernetes cluster. Create a Service for Node Exporter as follows:

```
$ kubectl apply -f node-exporter-svc.yaml
service/node-exporter created
```

A Node Exporter serviceMonitor conventionally enables Prometheus to discover and scrape the Node Exporter service for metrics. This described configuration simplifies the whole process of monitoring by defining how and where Prometheus should scrape for metrics, like specifying the target service, interval, labels, and others serving to ensure consistency in the collection without an administrator's intervention. Create a serviceMonitor CRD as follows:

```
$ kubectl apply -f servicemonitor-instance.yaml
servicemonitor.monitoring.coreos.com/node-exporter created
Now it is our time to verify the monitoring stack deployment in Kubernetes
cluster.
```

Let us verify the Prometheus portal to ensure the details are collected there. In one of your consoles, start a `kubectl port-forward` command to expose the Prometheus service as follows (you can end the `port-forward` using *Ctrl+C* later once you finish testing):

```
$ kubectl port-forward -n monitoring svc/prometheus-operated 9091:9090
Forwarding from 127.0.0.1:9091 -> 9090
Forwarding from [::1]:9091 -> 9090
```

Open a browser and access `http://localhost:9091/targets` to ensure `node-exporter` is visible to Prometheus.

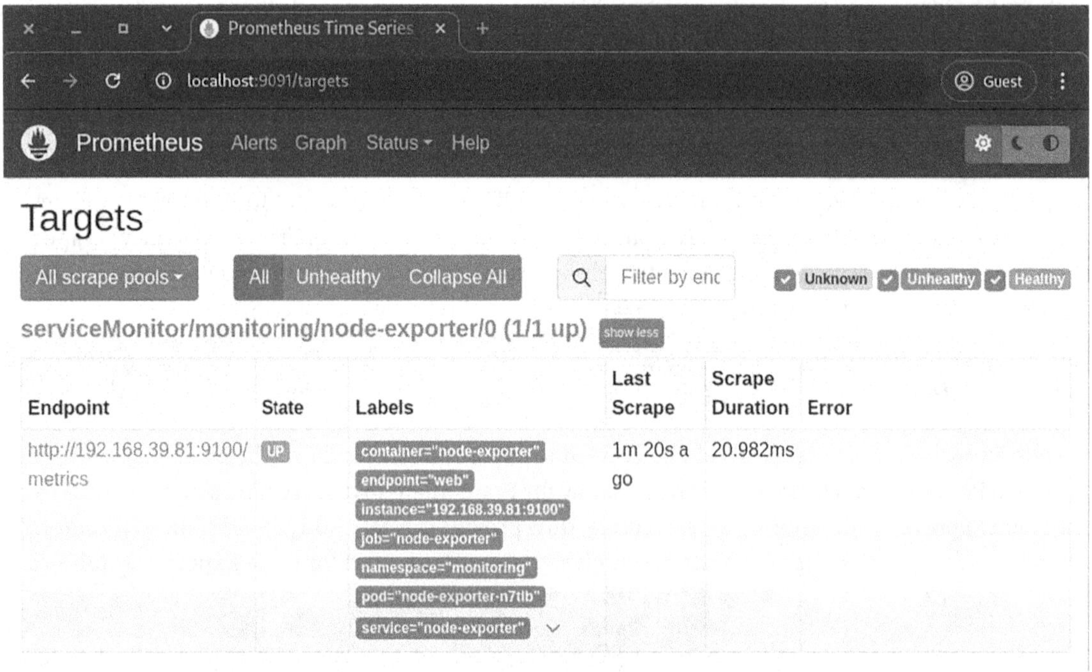

Figure 14.10: Prometheus portal with node-exporter visible for Kubernetes node

You can confirm from the preceding screenshot that Prometheus is getting the node metrics successfully.

Now, let us visualize the metrics and monitoring data using our visualization tool, Grafana. Open a console and use the `kubectl port-foward` command to expose the Grafana service as follows:

```
$ kubectl port-forward -n monitoring service/grafana-a-service 3000:3000
Forwarding from 127.0.0.1:3000 -> 3000
Forwarding from [::1]:3000 -> 3000
```

Open a browser and access the URL http://localhost:3000 to see the Grafana dashboard.

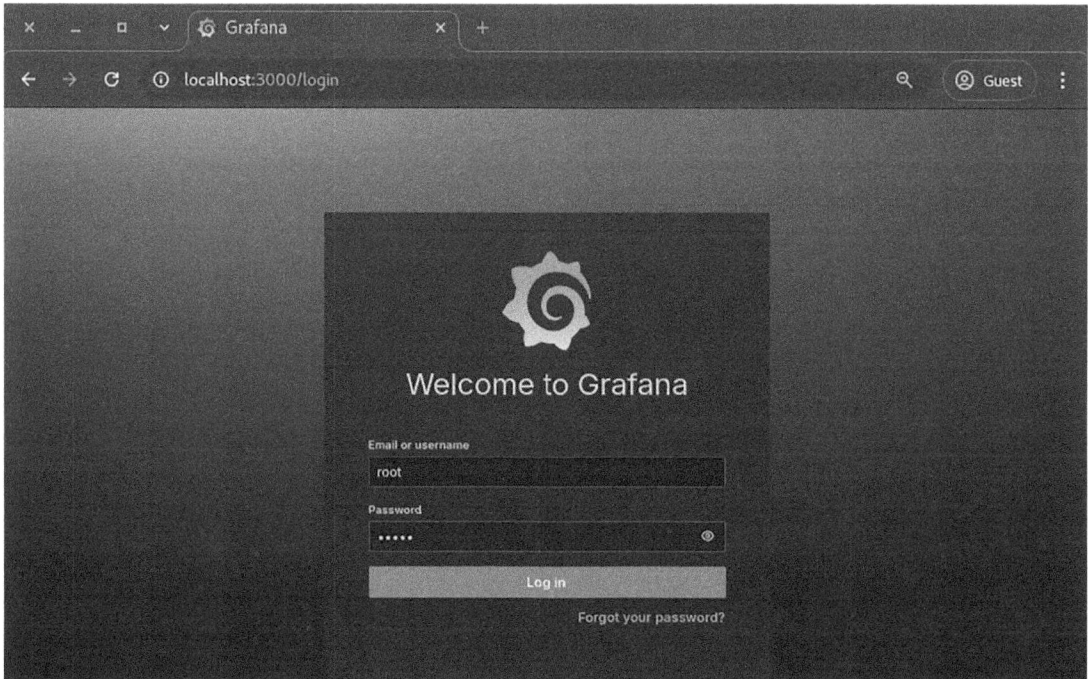

Figure 14.11: Grafana portal

Use the login credential you have configured in the `grafana-instance.yaml` (or the secret if you used one) and log in to the Grafana dashboard.

You will only find a default dashboard there as we need to configure a new dashboard for our own purpose.

From the left-side menu, go to **Connections | Data sources** as shown in the following figure.

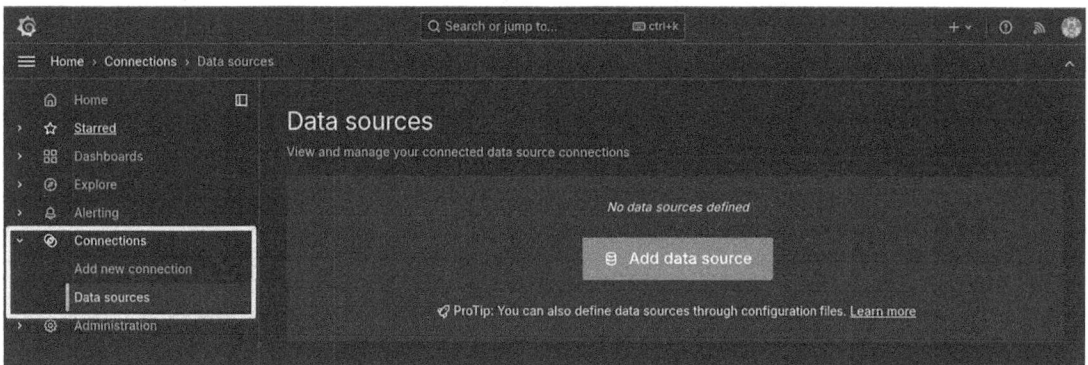

Figure 14.12: Adding a data source in Grafana

In the next window, select **Prometheus** as the data source and enter the Prometheus URL as shown in the following figure. Remember to enter the FQDN (e.g., `http://prometheus-operated.monitoring.svc.cluster.local:9090` – r to *Chapter 8, Exposing Your Pods with Services*, to learn more about Services and FQDNs).

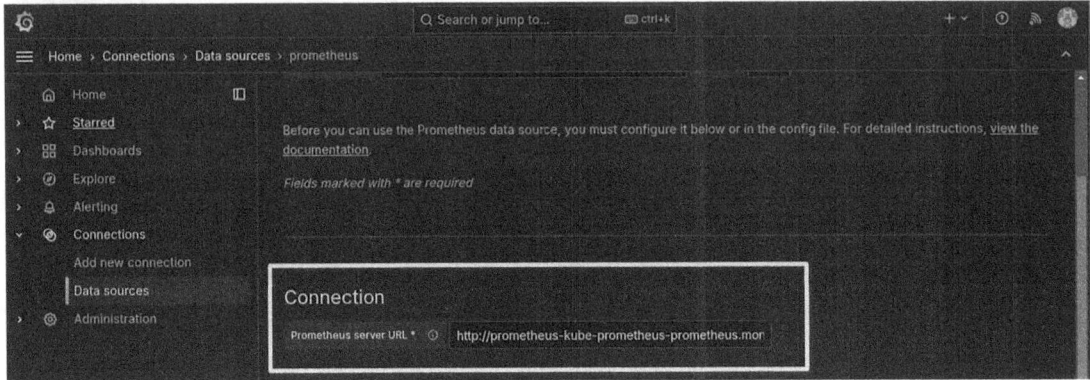

Figure 14.13: Configure Grafana data source

Click on the **Save & test** button at the bottom of the page and you will receive a success message as shown in the following figure. (If you get any error messages, then please check the Prometheus URL you have used, including the FQDN and the port number.)

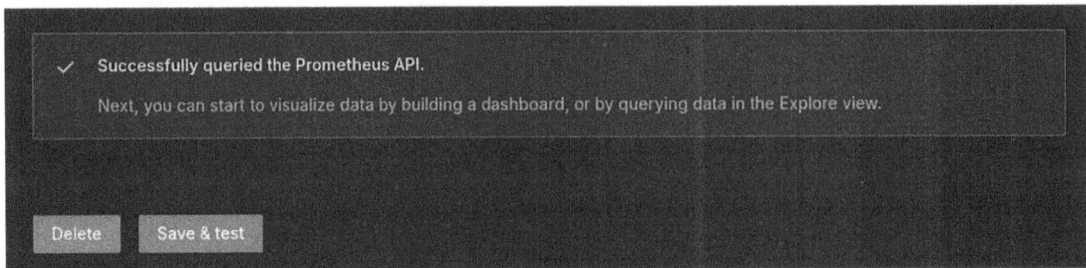

Figure 14.14: Grafana data source configured successfully

Now that we have the data source, we need to create a dashboard to visualize the data. You can either create a dashboard from scratch or import the dashboard with predefined configurations. For that, visit `https://grafana.com/grafana/dashboards/` and find the **Node Exporter Full** dashboard. Click on the **Copy ID to clipboard** button as follows.

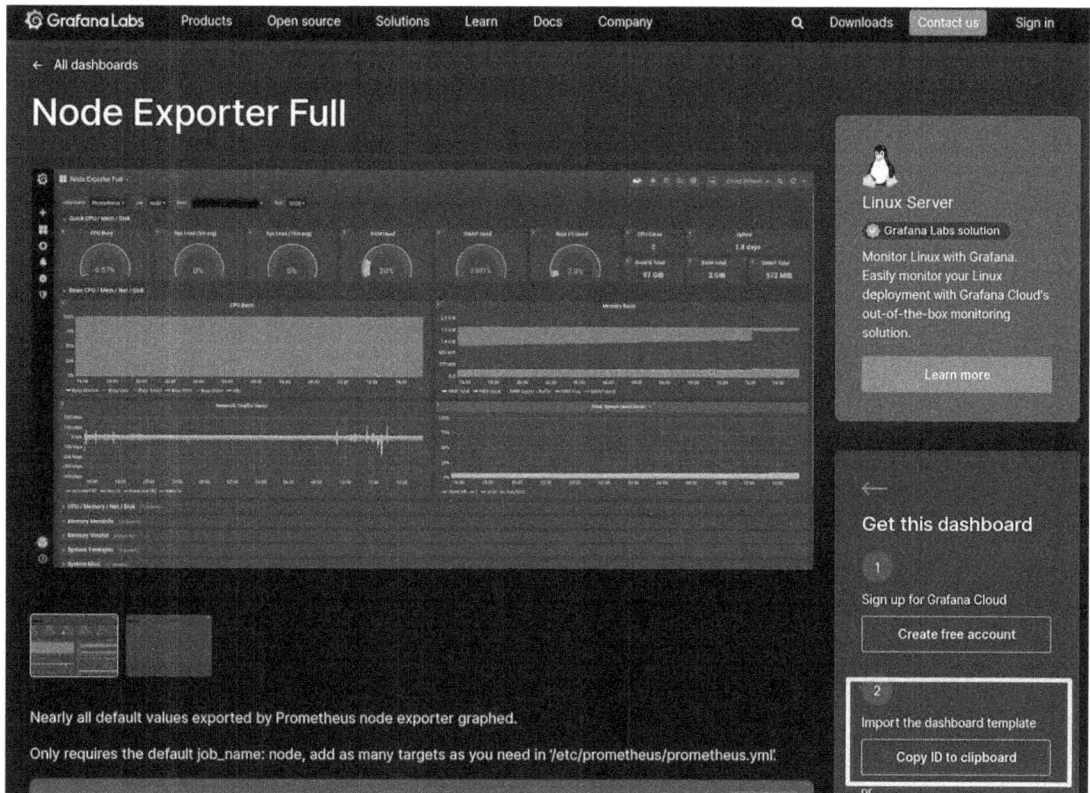

Figure 14.15: Copy the Grafana dashboard ID to import

Go back to the Grafana **WEBUI** | **Dashboards** | **New** | **Import**, as shown in the following figure.

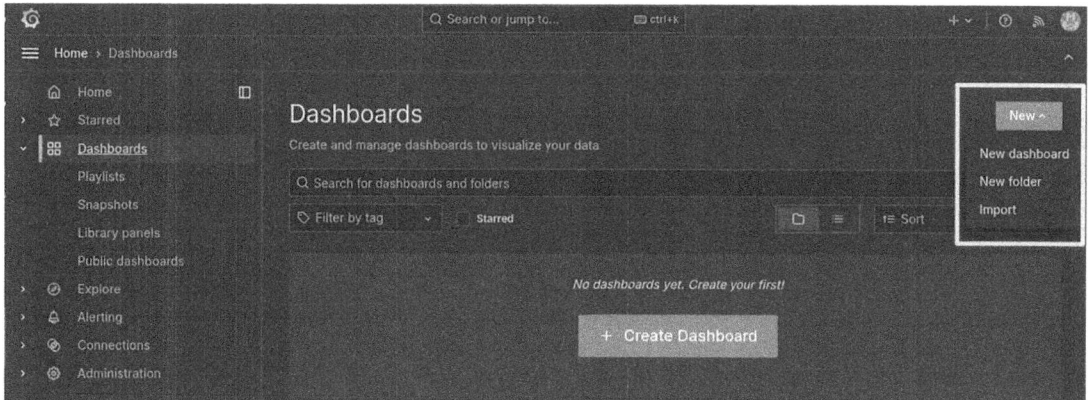

Figure 14.16: Importing a new dashboard to Grafana

Enter the **Node Exporter Full** dashboard ID that you already copied in the previous step, as shown here, and click on the **Load** button.

Figure 14.17: Provide the dashboard ID to import in Grafana

On the next screen, select **Prometheus** as the data source (which you configured earlier) and click the **Import** button as follows.

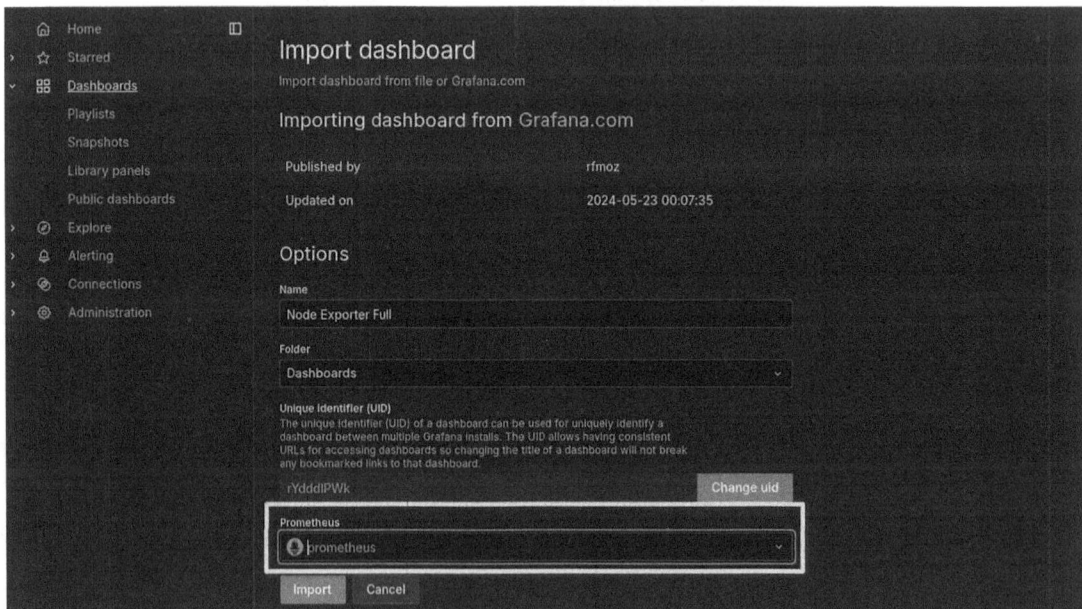

Figure 14.18: Complete dashboard import in Grafana

That's it; you will see the nice dashboard with preconfigured widgets and graphs, as shown in the following figure. You can explore the dashboard and find the details about your cluster, such as CPU, memory, network traffic, and so on.

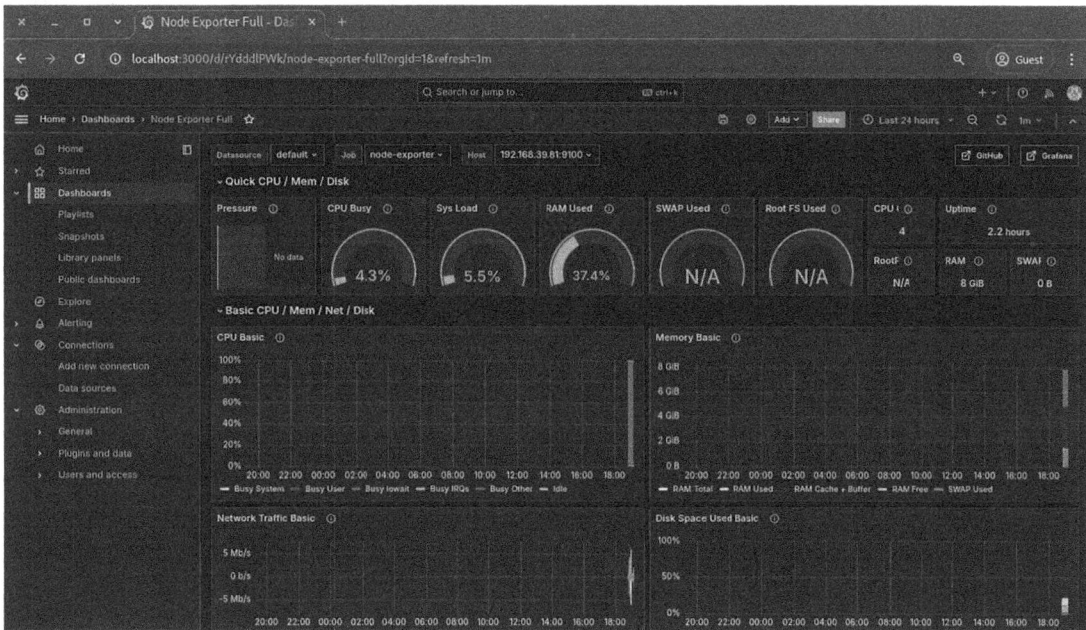

Figure 14.19: Node Exporter: full dashboard imported in Grafana

Congratulations, you have successfully deployed the Prometheus and Grafana stack on your Kubernetes cluster with Node Exporter enabled!

That is all for this chapter. As you can see, working with Helm charts and Operators, even for complex, multi-component solutions, is easy and can provide a lot of benefits for your development and production environments.

Summary

This chapter covered the details of working with Helm, Helm charts, and Kubernetes Operators. First, you learned what the purpose of package management is and how Helm works as a package manager for Kubernetes. We demonstrated how you can install Helm on your local machine, and how you can deploy a WordPress chart to test the installation. Then, we went through the structure of Helm charts, and we showed how the YAML templates in charts can be configured using user-provided values. Next, we showed the installation of popular solutions on a Kubernetes cluster using Helm. We installed Kubernetes Dashboard and explored the components. After that, we learned about Kubernetes Operators and other components, including customer resource definitions. We also deployed the Prometheus stack, including Grafana using Helm and Operators.

In the next part, you will get all the details required to effectively deploy Kubernetes clusters in different cloud environments. We will first take a look at working with clusters deployed on Google Kubernetes Engine.

Further reading

- **Helm website:** `https://helm.sh/`
- **Kubernetes Dashboard:** `https://kubernetes.io/docs/tasks/access-application-cluster/web-ui-dashboard/`
- **Accessing Dashboard:** `https://github.com/kubernetes/dashboard/blob/master/docs/user/accessing-dashboard/README.md`
- **Deploy and Access the Kubernetes Dashboard:** `https://kubernetes.io/docs/tasks/access-application-cluster/web-ui-dashboard/`
- **Creating a sample user for dashboard access:** `https://github.com/kubernetes/dashboard/blob/master/docs/user/access-control/creating-sample-user.md`
- **Build a Kubernetes Operator in six steps:** `https://developers.redhat.com/articles/2021/09/07/build-kubernetes-operator-six-steps`
- **Custom resources:** `https://kubernetes.io/docs/concepts/extend-kubernetes/api-extension/custom-resources/`

For more information regarding Helm and Helm charts, please refer to the following *Packt Publishing* book:

- *Learn Helm,* by *Andrew Block, Austin Dewey* (`https://www.packtpub.com/product/learn-helm/9781839214295`)

You can learn more about Elasticsearch and Prometheus in the following *Packt Publishing* books:

- *Learning Elasticsearch,* by *Abhishek Andhavarapu* (`https://www.packtpub.com/product/learning-elasticsearch/9781787128453`)
- *Hands-On Infrastructure Monitoring with Prometheus,* by *Joel Bastos, Pedro Araujo* (`https://www.packtpub.com/product/hands-on-infrastructure-monitoring-with-prometheus/9781789612349`)

Join our community on Discord

Join our community's Discord space for discussions with the authors and other readers:

`https://packt.link/cloudanddevops`

15

Kubernetes Clusters on Google Kubernetes Engine

In this chapter, we will launch a Kubernetes cluster on **Google Cloud Platform** (**GCP**), the first of the three public cloud providers we will cover over the next few chapters.

By the end of this chapter, you will have signed up for the Google Cloud Platform, launched a Kubernetes workload using **Google Kubernetes Engine** (**GKE**), and gained an understanding of GKE's features.

We will be covering the following topics in this chapter:

- What are Google Cloud Platform and Google Kubernetes Engine?
- Preparing your local environment
- Launching your first Google Kubernetes Engine cluster
- Deploying a workload and interacting with your cluster
- More about cluster nodes

Technical requirements

To follow along with this chapter, you will need a Google Cloud Platform account with a valid payment method attached.

> Following the instructions in this chapter will incur a financial cost. It is, therefore, important that you terminate any resources you launch once you have finished using them. All prices quoted in this chapter are correct at the time of writing, and we recommend that you review the current costs before you launch any resources.

What are Google Cloud Platform and Google Kubernetes Engine?

Before we roll up our sleeves and look at signing up for a Google Cloud Platform account and installing the tools, we will need to launch our GKE-powered cluster. We should also discuss Google Kubernetes Engine and how it came to be.

> Henceforth, I will refer to Google Cloud Platform as GCP and Google Kubernetes Engine as GKE in the body text of the chapter.

Google Cloud Platform

Of the **"big three"** public cloud providers, GCP is the newest. In the following two chapters, we will examine **Amazon Web Services (AWS)** and **Microsoft Azure**.

Google's foray into public cloud technology started differently from the other two providers. In April 2008, Google launched the public preview of its Google App Engine, which was the first component of its cloud offering. As a service, App Engine is still available at the time of writing, in mid-2024. The service allows developers to deploy their applications in Google-managed runtimes, including PHP, Java, Ruby, Python, Node.js, and C#, along with Google's programming language, Go, which was open sourced in 2009.

The following service under the GCP banner arrived in May 2010: **Google Cloud Storage**, followed by **Google BigQuery** and a preview version of its Prediction API. A year later, in October 2011, **Google Cloud SQL** was launched. Then, in June 2012, the **Google Compute Engine** preview was launched.

As you can see, four years had passed, and we then had what most would consider the core services that go into making a public cloud service.

However, most of the services were still in preview; it wouldn't be until 2013 that many of these core services would move out of preview and become **generally available (GA)**. This meant that it was possible to safely run production workloads at scale and, more importantly, with a **Service-Level Agreement (SLA)**, which is critical for both **small and medium-sized enterprises (SMEs)** and large enterprise companies in adopting the service. All of this was just a year before Google would launch Kubernetes.

Toward the end of 2014, Google brought out the first alpha version of GKE.

Google Kubernetes Engine

Given that Kubernetes was developed at Google and Google had vast experience running container workloads at scale with the Borg project, it came as no surprise that Google was one of the first public cloud providers to offer its own Kubernetes offering in GKE.

After Kubernetes v1 came out and was then handed over to the **Cloud Native Computing Foundation (CNCF)** to maintain in July 2015, it was only a month later that the GKE service became GA.

The GKE service lets you launch and manage a CNCF-certified Kubernetes cluster, powered by GCP's native compute, storage, and network services. It also allows deep integration with the platform's monitoring, identity, and access management functionality.

> We will look at the differences between the various Cloud-based Kubernetes offerings, and why you should use them, at the end of *Chapter 17*.

Now that we know a little bit of the history behind the service, we can sign up and install some of the management tools we will use to launch our cluster.

Preparing your local environment

The first thing we need to do is get you access to GCP. To do this, you must either sign up for an account or log in to your existing one. Let's learn how.

To sign up for a GCP account, you will need to visit `https://cloud.google.com/`. Here, you should be greeted by a page that looks something like the following:

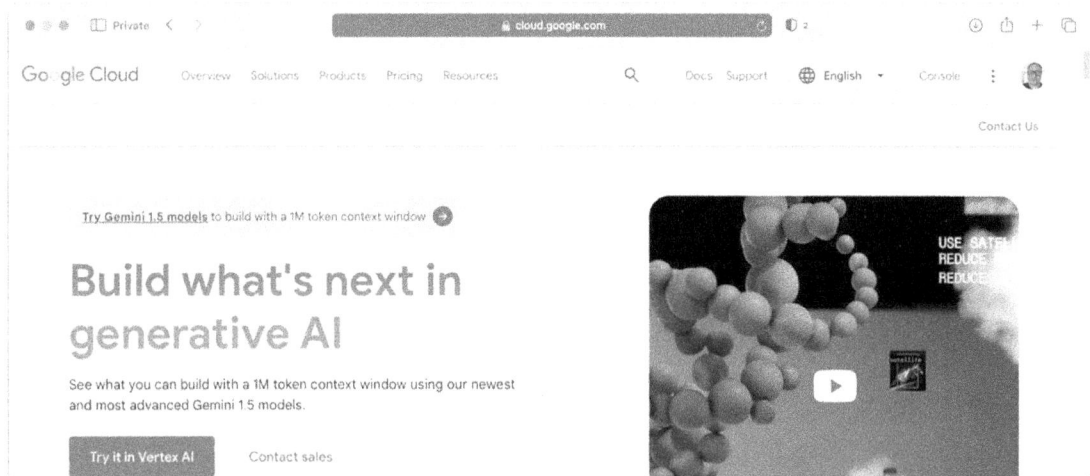

Figure 15.1: The Google Cloud home page

If you already use a Google service such as Gmail or YouTube or have an Android phone, you will possess a Google account. You can use this account to enroll, as shown in *Figure 15.1*. I am already logged in to my own Google account, as indicated by the avatar on the top-right-hand side of the screenshot.

> At the time of writing, Google offers $300 of credits for you to use over 90 days. You must still enter a valid payment method to take advantage of the free credits. Google does this to ensure that it is not an automated bot signing up for the account, abusing the credits they offer. Once the credits have been used or expired, you will be given the option to upgrade your account to a paid account. For more details on this offer, see the following link: `https://cloud.google.com/free/docs/free-cloud-features`.

If you want to take advantage of the free credits, follow this link: `https://cloud.google.com/free/` `docs/free-cloud-features#try-it-for-yourself`. Click on **Get started for free**; if you qualify for the offer, follow the onscreen prompts, ensuring that you read the terms and conditions. Once you have enrolled, you will be taken to the GCP console. Alternatively, if you already have a GCP account, you can log in to the GCP console directly at `https://console.cloud.google.com/`.

Now that we have an account, let's learn how to create a project where we will launch our resources.

Creating a project

GCP has a concept whereby resources are launched into projects. If you have just signed up for an account, then a "My First Project" project will have been created for you as part of the enrollment process.

If you are using an existing account, I recommend creating a new project to launch your GKE cluster. To do this, click on the **Select** menu, which is located in the top bar just to the right of the GCP logo at the top left of the page.

This will display all the projects you can access and allow you to create a new project. To do so, follow these steps:

1.   Click on **NEW PROJECT**.

2.   You will be asked to do the following:

     •   Give your new project a name.

     •   Select an organization to attach the project to.

     •   Select the location; this is either an organization or a folder to store the project in.

3.   Once details from the previous step have been entered, click the **CREATE** button.

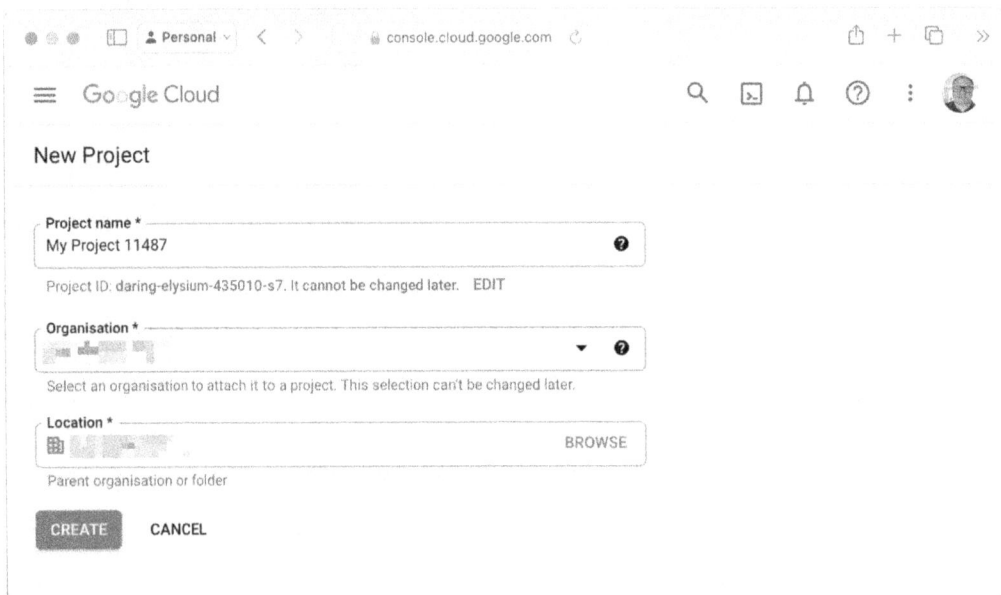

Figure 15.2: Creating a new project

Now that we have a place to launch our resources, we can look at installing the GCP command-line tool.

Installing the GCP command-line interface

Here, we are going to cover the GCP **Command-Line Interface** (CLI) on your local machine. Don't worry if you do not want to or are unable to install it, as there is a way to run the CLI in the cloud. But let's start with my operating system of choice, macOS.

Installing on macOS

If you are like me and you do a lot of work on macOS using Terminal, there is a high likelihood that you have installed and used Homebrew at some point.

> For the most up-to-date installation instructions for Homebrew, see the project's website at `https://brew.sh/`.

With Homebrew installed, you will be able to install the GCP CLI using the following command:

```
$ brew install --cask google-cloud-sdk
```

You can test the installation by running the following command:

```
$ gcloud version
```

If everything goes as planned, you should see something like the following output:

Figure 15.3: Checking the version number on macOS

If you are having problems, you can check the documentation linked above; for information on the GCP CLI installation, see:

```
$ brew info --cask google-cloud-sdk
```

Once you have the GCP CLI installed and working, you can move on to the *Initialization* section of this chapter.

Installing on Windows

There are a few options for installing the GCP CLI on Windows. The first is to open a PowerShell session and run the following command to install it using Chocolatey.

> Like Homebrew on macOS, Chocolatey is a package manager that lets you easily and consistently install a wide variety of packages on Windows via PowerShell, using the same command syntax rather than having to worry about the numerous installation methods that exist on Windows – see `https://chocolatey.org/` for more information.

If you don't have Chocolatey installed, then you can run the following command in a PowerShell session that has been launched with administrative privileges:

```
$ Set-ExecutionPolicy Bypass -Scope Process -Force; [System.
Net.ServicePointManager]::SecurityProtocol = [System.Net.
ServicePointManager]::SecurityProtocol -bor 3072; iex ((New-Object System.Net.
WebClient).DownloadString('https://chocolatey.org/install.ps1'))
```

Once Chocolatey is installed, or if you already have it installed, run the following:

```
$ choco install --ignore-checksum gcloudsdk
```

The other way to install it is to download the installer from the following URL:

`https://dl.google.com/dl/cloudsdk/channels/rapid/GoogleCloudSDKInstaller.exe`

Once downloaded, run the executable by double-clicking on it and following the onscreen prompts.

Once installed, open a new PowerShell window and run the following command:

```
$ gcloud version
```

If everything goes as planned, you should see something like the output shown in the screenshot from the macOS version in the previous section. Again, once installed and working, you can move on to the *Initialization* section of this chapter.

Installing on Linux

While the Google Cloud CLI packages are available for most distributions, we would need a lot more space to cover the various package managers here, so instead, we will use the global installation script provided by Google.

To run this, you need to use the following command:

```
$ curl https://sdk.cloud.google.com | bash
```

The script will ask you several questions during the installation. For most people, answering Yes will be fine. Once the GCP CLI is installed, run the command below to reload your session so that it picks up all of the changes made by the installer:

```
$ exec -l $SHELL
```

As per the macOS and Windows installations, which we have covered in the previous sections, you can run the following command to find out details on the GCP CLI version that has been installed:

```
$ gcloud version
```

You may have noticed from the three installations that, while the installation method differs, we use the same `gcloud` command once the package has been installed and get the same results. From here, it shouldn't matter which operating system you are running, as the commands will apply to all three.

Cloud Shell

Before we started installing the Google Cloud CLI, I did mention that there was a fourth option. That option is Google Cloud Shell, built into the Google Cloud console. To access this, click on the **Shell** icon, which can be found on the right of the top menu.

Once configured, you should see what looks to be a web-based terminal from which you can run the following:

```
$ gcloud version
```

The output differs slightly here, as Google has provided the full suite of supporting tools. However, you will notice from the following screen that the versions do match the ones we installed locally:

Figure 15.4: Using the GCP Cloud Shell

If you are using Google Cloud Shell, you can skip the initialization step below, as this has already been done for you.

Initialization

If you have chosen to install the client locally, you will need to take one final step to link it to your GCP account. To do this, run the following command:

```
$ gcloud init
```

This will immediately run a quick network diagnostic to ensure that the client has the connectivity it needs to run. Once the diagnostic has passed, you will be prompted with the following question:

```
You must log in to continue. Would you like to log in (Y/n)?
```

Answering Y will open a browser window. If it doesn't, then copy and paste the provided URL into your browser where, once you have selected the account you wish to use, you will be presented with an overview of the permissions that the client requests, as seen in the following screen:

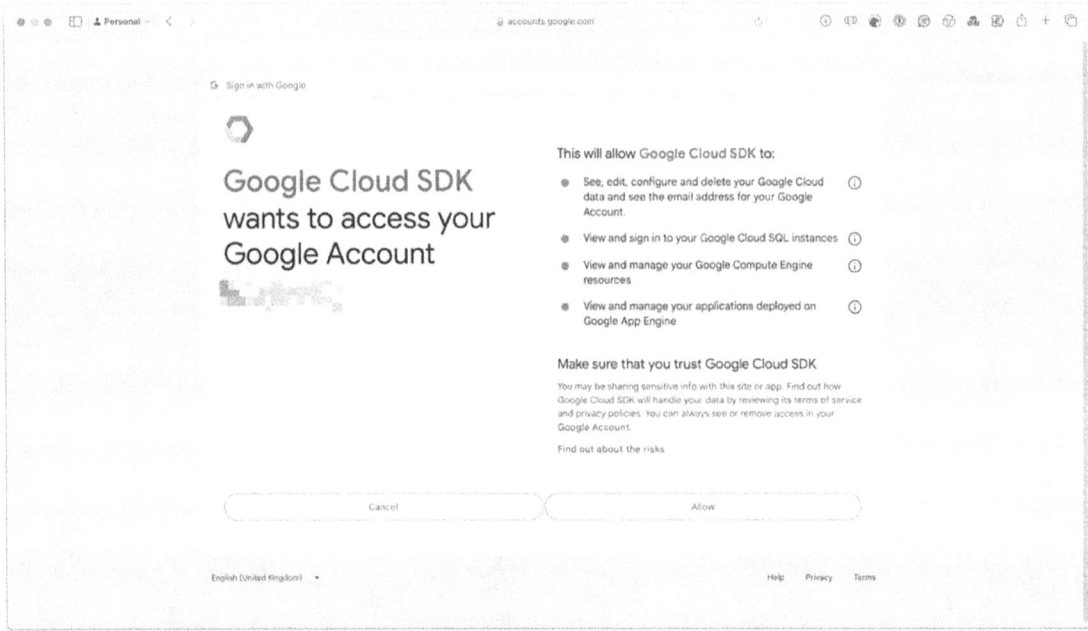

Figure 15.5: Reviewing the permissions

If you are happy to grant the permissions, click the **Allow** button. Back on your terminal, you will get confirmation of the user you are logged in as. Then, you will be asked to pick a cloud project to use. The list will only contain the unique ID of the project, not the friendly name that you saw or set up in Google Cloud Console earlier. If you have more than one project, please pick the correct one.

Should you need to update the project that the client uses at any point, you can run the following command:

```
$ gcloud config set project PROJECT_ID
```

Make sure you replace PROJECT_ID with the unique ID of the project you wish to switch to.

Now that you have installed the Google Cloud CLI and have configured your account, you are ready to launch your GKE cluster.

Launching your first Google Kubernetes Engine cluster

As the cluster will take a few minutes to deploy fully, let's run the command to initiate the process and then discuss in more detail what happens while it launches.

Before we launch our cluster, we must ensure that the container.googleapis.com service is enabled. To do this, run the following command:

```
$ gcloud services enable container.googleapis.com
```

We then need to install a plugin that will allow us to authenticate against the cluster using kubectl; to do this, run the following:

```
$ gcloud components install gke-gcloud-auth-plugin
```

Once the service and plugin have been enabled, the command to launch a two-node cluster called myfirstgkecluster, which will be hosted in a single zone in the Central US region, is as follows:

```
$ gcloud container clusters create myfirstgkecluster --num-nodes=2 --zone=us-
central1-a
```

After about five minutes, you should see something that looks like the following output:

```
) gcloud container clusters create myfirstgkecluster --num-nodes=2 --zone=us-central1-a
Default change: VPC-native is the default mode during cluster creation for versions greater than 1.2
1.0-gke.1500. To create advanced routes based clusters, please pass the `--no-enable-ip-alias` flag
Note: Your Pod address range (`--cluster-ipv4-cidr`) can accommodate at most 1008 node(s).
Creating cluster myfirstgkecluster in us-central1-a... Cluster is being health-checked (master is h
ealthy)...done.
Created [https://container.googleapis.com/v1/projects/smart-cove-423815-v4/zones/us-central1-a/clust
ers/myfirstgkecluster].
To inspect the contents of your cluster, go to: https://console.cloud.google.com/kubernetes/workload
_/gcloud/us-central1-a/myfirstgkecluster?project=smart-cove-423815-v4
kubeconfig entry generated for myfirstgkecluster.
NAME              LOCATION       MASTER_VERSION    MASTER_IP      MACHINE_TYPE  NODE_VERSION
   NUM_NODES  STATUS
myfirstgkecluster  us-central1-a  1.28.8-gke.1095000  34.67.204.195  e2-medium     1.28.8-gke.109500
0  2          RUNNING
```

Figure 15.6: Launching the cluster

Once the cluster has launched, you should be able to follow the URL in the output and view it in Google Cloud Console, as seen below:

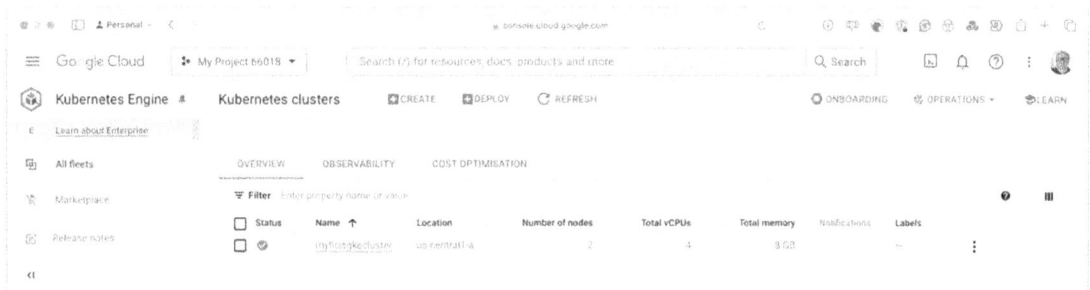

Figure 15.7: Viewing the cluster in Google Cloud Console

> You can access the Google Cloud Console at `https://console.cloud.google.com/`.

Now that our cluster is up and running, we can deploy a workload and then examine Google Cloud Console in more detail.

Deploying a workload and interacting with your cluster

One of the things to note from the feedback when we launched our cluster by running:

```
$ gcloud container clusters create myfirstgkecluster --num-nodes=2 --zone=us-
central1-a
```

was the following line in the output:

```
kubeconfig entry generated for myfirstgkecluster.
```

As you may have guessed, this has downloaded and configured all the necessary information to connect your local copy of kubectl to the newly deployed GKE cluster.

You can confirm this by running the following command:

```
$ kubectl get nodes
```

The output you get from the command should show two nodes with a prefix of `gke`, so it should appear something like the following Terminal output:

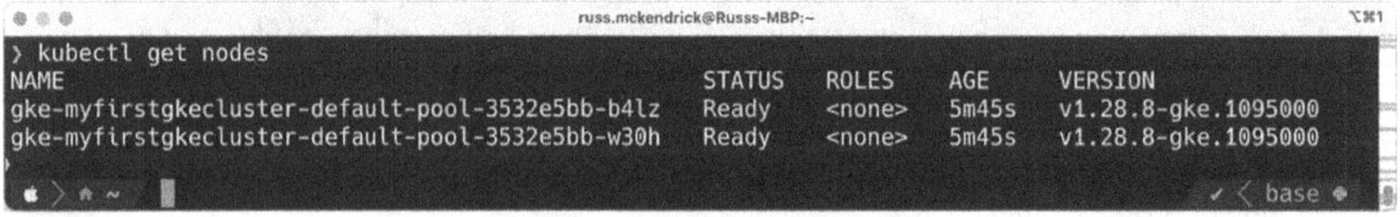

```
> kubectl get nodes
NAME                                                  STATUS   ROLES    AGE     VERSION
gke-myfirstgkecluster-default-pool-3532e5bb-b4lz      Ready    <none>   5m45s   v1.28.8-gke.1095000
gke-myfirstgkecluster-default-pool-3532e5bb-w30h      Ready    <none>   5m45s   v1.28.8-gke.1095000
```

Figure 15.8: Using kubectl to list the nodes

If your GKE cluster nodes are listed when running the preceding output and you are happy to proceed with the current kubectl configuration you are using, you can skip the next section of the chapter and move straight on to the *Launching an example workload* section.

The *Further reading* section at the end of this chapter also provides a link to the official GKE documentation.

Configuring your local client

If you need to configure another or your kubectl installation to connect to your cluster, the GCP CLI has a command.

Running the following command is based on the assumption that you have the GCP CLI installed and configured. If you don't, please follow the instructions in the *Installing the GCP command-line interface* section of this chapter.

The command you need to run to download the credentials and configure kubectl is as follows:

```
$ gcloud container clusters get-credentials myfirstgkecluster --zone=us-
central1-a
```

You can run the following commands if you need to switch to or from another configuration (or context as it is known). The first command lists the current context:

```
$ kubectl config current-context
```

The following command lists the names of all of the contexts that you have configured:

```
$ kubectl config get-contexts -o name
```

Once you know the name of the context that you need to use, you can run the following command, making sure to replace CONTEXT_NAME with the name of the context that you change it to, as in the following:

```
$ kubectl config use-context CONTEXT_NAME
```

Now that your kubectl client is configured to talk to and interact with your GKE cluster, we can launch an example workload.

Launching an example workload

The example workload we will launch is the PHP/Redis Guestbook example used throughout the official Kubernetes documentation. The first step in launching the workload is to create the Redis Leader deployment and service. To do this, we use the following commands:

```
$ kubectl apply -f https://raw.githubusercontent.com/GoogleCloudPlatform/
kubernetes-engine-samples/main/quickstarts/guestbook/redis-leader-deployment.
yaml
$ kubectl apply -f https://raw.githubusercontent.com/GoogleCloudPlatform/
kubernetes-engine-samples/main/quickstarts/guestbook/redis-leader-service.yaml
```

Next up, we need to repeat the process but, this time, to launch the Redis Follower deployment and service, as shown here:

```
$ kubectl apply -f https://raw.githubusercontent.com/GoogleCloudPlatform/
kubernetes-engine-samples/main/quickstarts/guestbook/redis-follower-deployment.
yaml
$ kubectl apply -f https://raw.githubusercontent.com/GoogleCloudPlatform/
kubernetes-engine-samples/main/quickstarts/guestbook/redis-follower-service.
yaml
```

Now that we have Redis up and running, it is time to launch the frontend deployment and service; this is the application itself:

```
$ kubectl apply -f https://raw.githubusercontent.com/GoogleCloudPlatform/
kubernetes-engine-samples/main/quickstarts/guestbook/frontend-deployment.yaml
$ kubectl apply -f https://raw.githubusercontent.com/GoogleCloudPlatform/
kubernetes-engine-samples/main/quickstarts/guestbook/frontend-service.yaml
```

After a few minutes, you should be able to run the following command to get information on the service you have just launched:

```
$ kubectl get service frontend
```

The output of the command should give you an external IP address that looks like the following Terminal output:

Figure 15.9: Getting information on the frontend service

Now that we have launched our application, copy the `EXTERNAL-IP` value and put the IP address into your browser; make sure you use `http://EXTERNAL-IP` rather than `https://EXTERNAL-IP`, as the Guestbook listens on port 80 and not 443.

You should be presented with the simple Guestbook application when you open the address in your browser. Try submitting a few messages, as shown in the following screen:

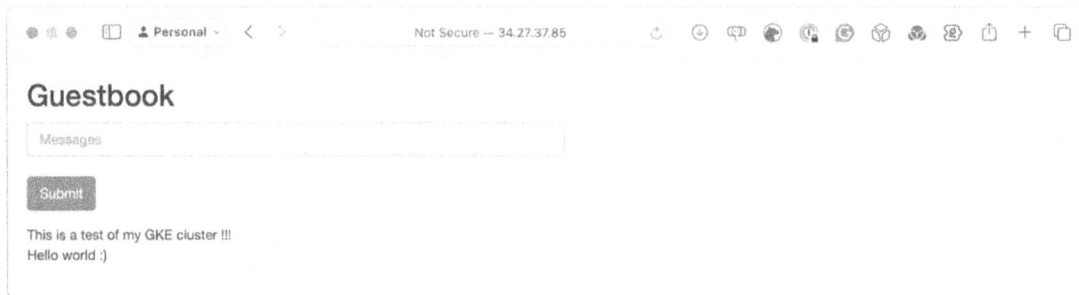

Figure 15.10: The Guestbook application with a few test messages

Now that we have launched our basic workload, let's return to and explore Google Cloud Console.

Exploring Google Cloud Console

We have already seen our cluster in Google Cloud Console, so next, click on the **Workloads** link, which is in the left-hand menu of the **Kubernetes Engine** section.

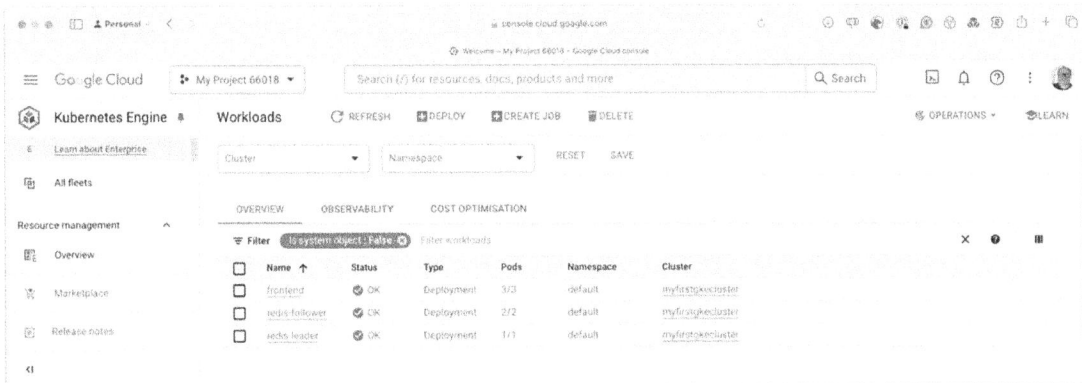

Figure 15.11: Viewing the workload in the console

As you can see from the preceding screen, the three deployments are listed, along with confirmation of the namespace they are in and the cluster that the workload belongs to.

If we had more than one cluster with multiple namespaces and deployments, we could use the filter to drill down into our GKE workloads.

Workloads

Clicking on the **frontend** deployment will let you view more information:

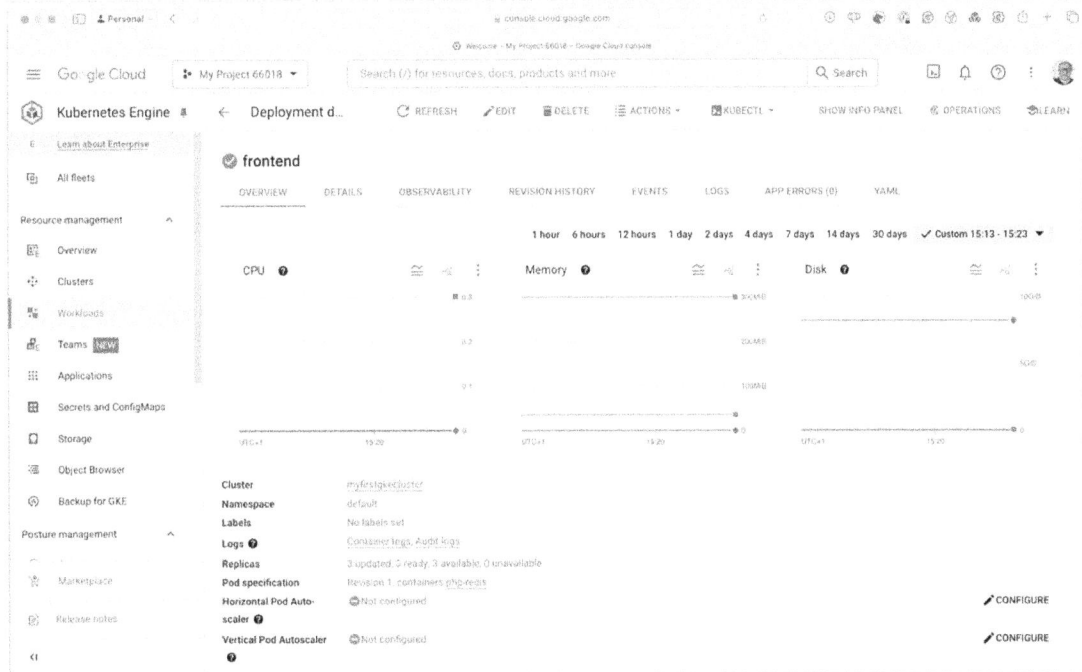

Figure 15.12: Getting an overview of the frontend deployment

On this page, you will be able to drill further down into your deployment using the tabs below the deployment name, as follows:

- **OVERVIEW:** This view gives you an overview of the selected workload. As you can see from the previous screenshot, you can see the CPU, memory, and disk utilization, along with other information.

- **DETAILS:** This lists more information about the environment and workload itself. Here, you can find out when it was created, annotations, labels (as well as details about the replicas), the update strategy, and Pod information.

- **OBSERVABILITY:** Here, you can view all metrics on your workload's infrastructure.

- **REVISION HISTORY:** Here, you will find a list of all the workload revisions. This is useful if your deployment is updated frequently and you need to keep track of when the deployment was updated.

- **EVENTS:** If you have any problems with your workload, this is the place to look. All events, such as scaling, pod availability, and other errors, will be listed here.

- **LOGS:** This is a searchable list of the logs from all containers running the pod.

- **APP ERRORS:** Here, you will find any application errors.

- **YAML:** This is an exportable YAML file containing the full deployment configuration.

This information is available for all the deployments across all the GKE clusters you have launched within the project.

Gateways, Services, and Ingress

In this section, we will look at the **Gateways, Services and Ingress** section. As you may have already guessed, this lists all the services launched across your GKE clusters. You can do this by clicking on **Gateways, Services and Ingress**.

When you first go to this section, it will default to the **GATEWAYS** tab, but we haven't launched a gateway, so you need to click on the **SERVICES** tab. When this loads, you will be presented with a screen that looks like the following:

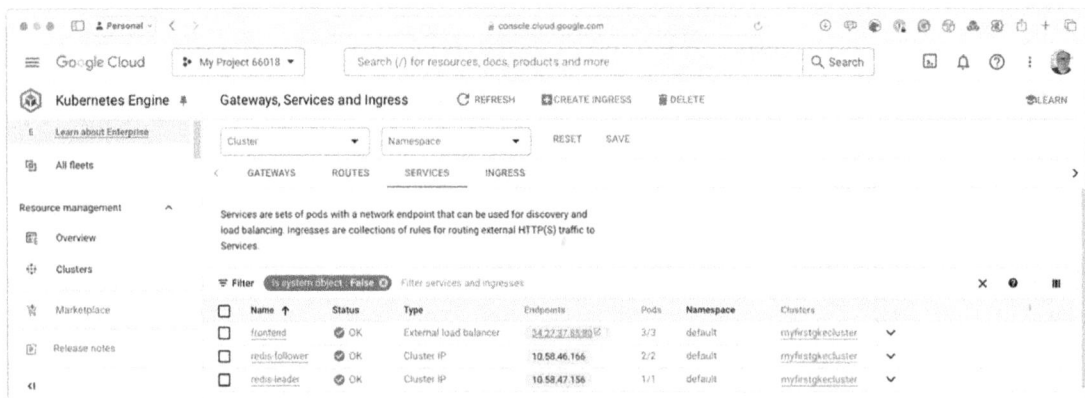

Figure 15.13: Viewing the services in the console

As you can see, the three services we launched are listed. However, the **frontend** service has an **External load balancer** type and a public IP address listed, as opposed to the **redis-leader** and **redis-follower** services, which only have a **Cluster IP**.

This is because, of the three services, we only want the frontend service to be accessible publicly, as only our frontend service uses the two Redis services.

When the external load balancer was launched, as we were running our cluster in a public cloud provider, the Kubernetes scheduler knew to contact the Google Cloud API and launch a load balancer for use. It then configured it to talk back to our cluster nodes, exposing the workload; this deployment runs on port **80** externally, and traffic is passed back to port **31740** on the nodes.

As before, clicking on one of the three running services will give you several pieces of information:

- **OVERVIEW:** This view gives you a summary of the service configuration and utilization.
- **DETAILS:** Here, you can find more details on the service and a link to view the load balancer resource that has been launched within our Google Cloud project.
- **EVENTS:** As before, you can find any events affecting the service here. This is useful for troubleshooting.
- **LOGS:** This is a repeat of the logs shown in the **Workloads** section.
- **APP ERRORS:** You will be able to find any application errors here.
- **YAML:** Again, this is a way to export the complete service configuration as a YAML file.

While this covers the basic workload we launched, GKE offers plenty of other options and features. Let's quickly work through the functionality hidden behind the other items in the left-hand menu.

Other GKE features

There are two types of features supported by GKE; the ones listed below are the standard ones. There are also Enterprise features that can be enabled when you run GKE Enterprise, which we will discuss at the end of this section:

- **Clusters:** Here, you can manage your clusters; you will see the cluster we launched using the GCP CLI and have the option of creating more clusters, using the guided web interface.
- **Workloads:** As already covered, you will find the workloads you have deployed to your clusters.
- **Applications:** This will show you any applications you have deployed from the Marketplace, which we will cover in a moment.
- **Secrets and ConfigMaps:** Here, you can securely manage your Secrets and ConfigMaps.
- **Storage:** You can view any persistent storage claims you have configured here and manage storage classes.
- **Object Browser:** Think of this as an explorer for the K8s and GCP API; clicking through it gives you an overview of all the possible APIs and endpoints you can interact with and what objects/resources are currently deployed.

The following services bring native Google services into GKE:

- **Backup for GKE:** Here, you can use GCP's backup service, which is outside the GKE offering, to back up your GKE-hosted workloads and applications. Personally, when I run a Kubernetes workload in the cloud, the data within it is ephemeral. For services such as databases, where I would want to have backups, they run in an SLA-backed, cloud-native service outside of my cluster – but if, for whatever reason, you can't do that, then this service has you covered.

- **Gateways, Services and Ingress:** We have already covered this option.

- **Network Function Optimizer:** You can link your GKE workloads to other GCP resources' private networking here. This allows you to extend your cluster network so that services like Google Cloud databases can be accessed privately by your cluster resources.

- **Marketplace:** Here, you will find pre-packaged applications to run in your GKE cluster. They are published by various software vendors and include both free and commercial offerings.

As mentioned, the features above are available in the standard edition of GKE; this edition is like rolling your own Kubernetes deployment, except Google takes care of the node deployment and management plan for you.

Then, there is the Enterprise edition; as you might have guessed, this introduces additional features that an enterprise deploying Kubernetes may require, such as managed services meshes and security, compliance, and role-based access control options. These features come at an extra cost but are essential for larger organizations that may work in more regulated industries.

Deleting your cluster

Once you finish with your cluster, you can remove it by running the following commands. The first removes the service we created, and the second terminates the cluster itself:

```
$ kubectl delete service frontend
$ gcloud container clusters delete myfirstgkecluster --zone=us-central1-a
```

It will take a few minutes to remove the cluster; please ensure that you wait until this process finishes.

> This should also delete any services launched by your workloads, such as load balancers. However, I recommend double-checking the Google Cloud Console for any orphaned services or resources to ensure that you do not incur unexpected costs.

We have been using the `--zone=us-central1-a` flag, which launches our cluster in a single availability zone in the US Central region. Let's discuss what other cluster options are available.

More about cluster nodes

At the end of the previous section, I mentioned availability zones and regions. Before we discuss some of the cluster deployment options, we should get a slightly better understanding of what we mean by availability zones and regions:

- **Region:** A region is made up of zones. Zones have great low-latency network connectivity to other zones within the same region. This gives you a way of deploying highly available, always-on, fault-tolerant workloads.

- **Availability zone:** Think of availability zones as separate data centers within a region. The zones have diverse networks and power, meaning that if a single zone has an issue and you run your workload across multiple zones, your workload shouldn't be impacted.

The one thing to note with zones is that you might find that not all machine types are available across all zones within a region. Therefore, please check before attempting to deploy your workload.

Google best practices recommend that you deploy your workload across the maximum number of zones within a single region for optimum performance and availability.

However, it is possible to split your workloads across multiple regions by using Multi Cluster Ingress – all you have to do is account for and allow the increased latency between regions for shared services, such as databases. For more information on Multi Cluster Ingress, see the *Further reading* section at the end of the chapter.

Let's get back to looking at launching a cluster in a single region across multiple zones, by examining the command we used to launch the test cluster at the start of the chapter:

```
$ gcloud container clusters create myfirstgkecluster --num-nodes=2 --zone=us-
central1-a
```

As we know, it will launch two nodes but only in a single zone. We have only passed one using the -zone flag, which, in our case, was the us-central1 region and zone a.

Let's run the following command but use the --region flag rather than the --zone one:

```
$ gcloud container clusters create myfirstgkecluster --num-nodes=2 --region=us-
central1
```

Once launched, run the following:

```
$ kubectl get nodes
```

This should output something that looks like the following:

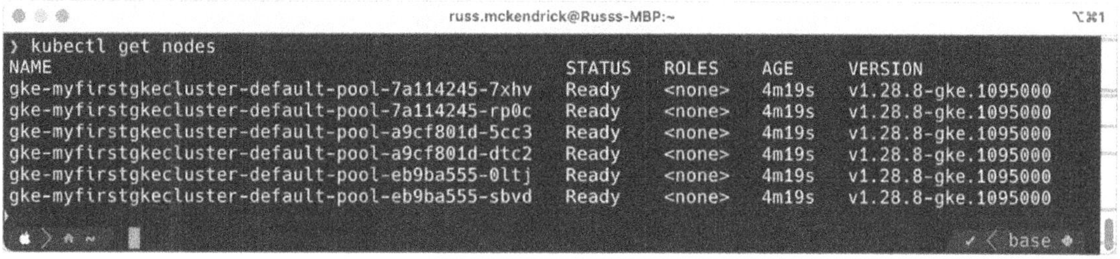

```
> kubectl get nodes
NAME                                              STATUS   ROLES    AGE     VERSION
gke-myfirstgkecluster-default-pool-7a114245-7xhv  Ready    <none>   4m19s   v1.28.8-gke.1095000
gke-myfirstgkecluster-default-pool-7a114245-rp0c  Ready    <none>   4m19s   v1.28.8-gke.1095000
gke-myfirstgkecluster-default-pool-a9cf801d-5cc3  Ready    <none>   4m19s   v1.28.8-gke.1095000
gke-myfirstgkecluster-default-pool-a9cf801d-dtc2  Ready    <none>   4m19s   v1.28.8-gke.1095000
gke-myfirstgkecluster-default-pool-eb9ba555-0ltj  Ready    <none>   4m19s   v1.28.8-gke.1095000
gke-myfirstgkecluster-default-pool-eb9ba555-sbvd  Ready    <none>   4m19s   v1.28.8-gke.1095000
```

Figure 15.14: Viewing the nodes running in a region

As you can see, we have two nodes in each zone, giving us a total cluster size of six nodes. This is because when you define a region by default, your cluster is spread across three zones. You can override this using the `--node-locations` flag.

This makes our command look like the following if we wanted to deploy a cluster from scratch; as you can see, we now have a comma-separated list of availability zones:

```
$ gcloud container clusters create myfirstgkecluster --num-nodes=2 --region=us-
central1 --node-locations us-central1-a,us-central1-b,us-central1-c,us-
central1-f
```

We are still using the `us-central1` region but deploying to the a, b, c, and f zones.

Given that we already have a cluster running, we can run the following command to update the cluster, adding the new availability zone and two new nodes:

```
$ gcloud container clusters update myfirstgkecluster --min-nodes=2 --region=us-
central1 --node-locations us-central1-a,us-central1-b,us-central1-c,us-
central1-f
```

Running the `kubectl get nodes` command now shows the following:

```
russ.mckendrick@Russs-MBP:~                                             ⌥⌘1
> kubectl get nodes
NAME                                            STATUS   ROLES    AGE   VERSION
gke-myfirstgkecluster-default-pool-7a114245-7xhv   Ready    <none>   8m5s  v1.28.8-gke.1095000
gke-myfirstgkecluster-default-pool-7a114245-rp0c   Ready    <none>   8m5s  v1.28.8-gke.1095000
gke-myfirstgkecluster-default-pool-7dea94ee-bgpj   Ready    <none>   11s   v1.28.8-gke.1095000
gke-myfirstgkecluster-default-pool-7dea94ee-r130   Ready    <none>   16s   v1.28.8-gke.1095000
gke-myfirstgkecluster-default-pool-a9cf801d-5cc3   Ready    <none>   8m5s  v1.28.8-gke.1095000
gke-myfirstgkecluster-default-pool-a9cf801d-dtc2   Ready    <none>   8m5s  v1.28.8-gke.1095000
gke-myfirstgkecluster-default-pool-eb9ba555-0ltj   Ready    <none>   8m5s  v1.28.8-gke.1095000
gke-myfirstgkecluster-default-pool-eb9ba555-sbvd   Ready    <none>   8m5s  v1.28.8-gke.1095000
>
                                                                ✓ < base ◆
```

Figure 15.15: Our cluster now running across four availability zones

As you can see, Google has made deploying your clusters straightforward across multiple zones. This means that you can deploy your workload across a fully redundant cluster with minimal effort; given the complexity involved in replicating this type of deployment manually using virtual machines and networks, the ease is very welcome.

To remove clusters deployed using the `--region` flag, you should use the following command:

```
$ gcloud container clusters delete myfirstgkecluster --region=us-central1
```

At the time of writing, the simple two-node cluster we launched at the start of the chapter costs around $55 per month, and the price increases to around $347 per month for the eight-node cluster we just launched. For more cost information, see the Google Cloud Pricing Calculator link in the *Further reading* section.

This concludes our look at GKE. Before we move on to the next public cloud provider, let's summarize what we have covered.

Summary

In this chapter, we discussed the origins of GCP and the GKE service, before walking through how to sign up for an account and how to install and configure the Google Cloud command-line tool.

We then launched a simple two-node cluster using a single command, and then we deployed and interacted with a workload, using the kubectl command and Google Cloud Console.

Finally, again using only a single command, we redeployed our cluster to take advantage of multiple availability zones, quickly scaling to a fully redundant and highly available eight-node cluster that runs across four availability zones.

I am sure you will agree that Google has done an excellent job of making deploying and maintaining a complex infrastructure configuration a relatively trivial and quick task.

Also, once your workloads are deployed, you can manage them the same way you would if your cluster were deployed elsewhere – we really haven't had to make any allowances for our cluster being run on GCP.

The next chapter will examine deploying a Kubernetes cluster in AWS using Amazon Elastic Kubernetes Service, Amazon's fully managed Kubernetes offering.

Further reading

Here are links to more information on some of the topics and tools that we have covered in this chapter:

- Google Kubernetes Engine: https://cloud.google.com/kubernetes-engine/
- Google Kubernetes Engine Documentation: https://cloud.google.com/kubernetes-engine/docs
- Google Cloud Pricing Calculator: https://cloud.google.com/products/calculator
- The Guestbook Sample Application: https://github.com/GoogleCloudPlatform/kubernetes-engine-samples/tree/main/quickstarts/guestbook
- The Google Cloud Kubernetes comic: https://cloud.google.com/kubernetes-engine/kubernetes-comic
- Regional Clusters: https://cloud.google.com/kubernetes-engine/docs/concepts/regional-clusters
- Multi Cluster Ingress: https://cloud.google.com/kubernetes-engine/docs/concepts/multi-cluster-ingress

Join our community on Discord

Join our community's Discord space for discussions with the authors and other readers:

`https://packt.link/cloudanddevops`

16

Launching a Kubernetes Cluster on Amazon Web Services with Amazon Elastic Kubernetes Service

Let's build on what we learned in the previous chapter. We launched a Kubernetes cluster in a public cloud, taking our first steps to run Kubernetes on one of the "big three" public cloud providers.

Now that we know what the **Google Cloud Platform (GCP)** Kubernetes offering looks like, we will move on to **Amazon Elastic Kubernetes Service (EKS)** by **Amazon Web Services (AWS)**.

In this chapter, you will learn how to set up an AWS account and install the supporting toolsets on macOS, Windows, and Linux, before finally launching and interacting with an Amazon EKS cluster.

We will be covering the following topics in this chapter:

- What are Amazon Web Services and Amazon Elastic Kubernetes Service?
- Preparing your local environment
- Launching your Amazon Elastic Kubernetes Service cluster
- Deploying a workload and interacting with your cluster
- Deleting your Amazon Elastic Kubernetes Service cluster

Technical requirements

If you plan to follow along with this chapter, you will need an AWS account with a valid payment attached.

Following the instructions in this chapter will incur a cost, and you must terminate any resources you launch once you have finished using them. All prices quoted in this chapter are correct as of when this book was written, and we recommend that you review the current costs before launching any resources.

What are Amazon Web Services and Amazon Elastic Kubernetes Service?

You may have already heard of Amazon Web Services, or AWS, as we will refer to it. It was one of the first public cloud providers and has the largest market share at the time of writing (June 2024): 31%, with Microsoft Azure second at 25% and GCP third at 11%.

Amazon Web Services

As you may have already guessed, Amazon owns and operates AWS. What began as Amazon experimenting with cloud services in 2000, by developing and deploying **application programming interfaces (APIs)** for its retail partners, has since evolved into AWS, a global leader in cloud computing, powering businesses of all sizes across various industries.

Based on this work, Amazon realized that they needed to build a better and more standardized infrastructure platform not only to host the services they had been developing but also to ensure they could quickly scale as more of their retail outlets consumed the software services and grew at an expedient rate.

Chris Pinkham and Benjamin Black wrote a white paper in 2003 that was approved by Jeff Bezos in 2004, which described an infrastructure platform where compute and storage resources can all be deployed programmatically.

The first public acknowledgment of AWS's existence was made in late 2004; however, the term was used to describe a collection of tools and APIs that would allow third parties to interact with Amazon's retail product catalog, rather than what we know today.

It wasn't until 2006 that a rebranded AWS was launched, starting in March with **Simple Storage Service (S3)**. This service allowed developers to write and serve individual files using a web API, rather than writing and reading from a traditional local or remote filesystem.

The next service to launch, **Amazon Simple Queue Service (SQS)**, had formed part of the original AWS collection of tools. This was a distributed message system that, again, could be controlled and consumed by developers using an API.

The final service, launched in 2006, was a beta of **Amazon Elastic Compute Cloud (Amazon EC2)**, limited to existing AWS customers. Again, you could use the APIs developed by Amazon to launch Amazon EC2 resources.

This was the final piece of the puzzle for Amazon. They now had the foundations of a public cloud platform on which they could not only use their own retail platform but also sell space to other companies and the public, such as you and me.

Let's fast-forward from 2006, when there were 3 services, to the time of writing, mid-2024, where there are over 200 services available, all of which run in over 125 physical data centers across 39 Regions, which comprise over 38 million square feet in total.

All 200+ services adhere to the core principles laid out in the 2003 white paper. Each service is software-defined, meaning that developers must make a simple API request to launch, configure, and, in some cases, consume the service before finally making another API request to terminate it.

> You may have already noticed this from the services that have already been mentioned up to this point, but services running in AWS are prefixed with either Amazon or AWS – why is this? Well, services that start with Amazon are standalone services, unlike those prefixed with AWS, which are services designed to be used alongside the services that are prefixed with Amazon.

Long gone are the days of having to order a service, have someone build and deploy it, and then hand it over to you; this reduces deployment times to seconds from what could sometimes take weeks or months.

Rather than discussing all 200+ services, which would be an entire series of books, we should discuss the service we will examine in this chapter.

Amazon Elastic Kubernetes Service

While AWS was the first of the major public cloud providers, it was one of the last to launch a standalone Kubernetes service. Amazon EKS was first announced in late 2017. It became generally available in the United States (US), starting with the East (N. Virginia) and West (Oregon) Regions in June 2018.

The service is built to work with and take advantage of other AWS services and features, such as the following:

- **AWS Identity and Access Management (IAM)** allows you to control and manage end-user and programmatic access to other AWS services.
- **Amazon Route 53** is Amazon's **Domain Name System (DNS)** service. Amazon EKS can use it as a source of DNS for clusters, meaning service discovery and routing can easily be managed within your cluster.
- **Amazon Elastic Block Store (EBS)**: If you need persistent block storage for containers running within your Amazon EKS cluster, **Amazon Elastic Block Store (EBS)** provides this storage, just as it does for your EC2 compute resources
- **EC2 Auto Scaling**: If your cluster needs to scale, the same technology is employed to scale your EC2 instances.
- **Multi-Availability Zones (AZs)** can be a useful feature. The Amazon EKS management layer and cluster nodes can be configured to be spread across multiple AZs within a given Region to bring **High Availability (HA)** and resilience to your deployment.

Before we launch our Amazon EKS cluster, we will need to download, install, and configure a few tools.

Preparing your local environment

We need to install two command-line tools, but before we do, we should quickly discuss the steps to sign up for a new AWS account. If you already have an AWS account, skip this task and move straight to the *Installing the AWS command-line interface* section.

Signing up for an AWS account

Signing up for an AWS account is a straightforward process, as detailed here:

1. Go to `https://aws.amazon.com/` and click the **Create an AWS account** button at the top right of the page.

   > Amazon offers a free tier for new users. It is limited to certain services and instance sizes and lasts 12 months. For information on the AWS Free Tier, see `https://aws.amazon.com/free/`.

2. Fill out the initial form that asks for an email address. This will be used for account recovery and some basic administrative functions. Also, provide your chosen AWS account name. Don't worry if you change your mind; you can change this in your account settings after you sign up. You will then need to verify your email address. Once the email address has been verified, you will be asked to set the password for your "root" account.

3. Once you enter your password, click **Continue** and follow the onscreen instructions. There are five steps; these will involve you confirming your payment details and identity via an automated phone call.

Once you have created and enabled your account, you will be notified when you can start using AWS services – most of the time, this will be straight away, but it can take up to 48 hours.

Now, we need to install the command-line tools that we will be using to launch our Amazon EKS cluster.

Installing the AWS command-line interface

Next on our list of tasks is to install the AWS **Command-Line Interface (CLI)**. As we did in the previous chapter, *Chapter 15, Kubernetes Clusters on Google Kubernetes Engine*, we will target Windows, Linux, and macOS, which we will look at first.

Installing on macOS

Installing the AWS CLI on macOS using Homebrew is as simple as running the following command:

```
$ brew install awscli
```

Once it's installed, run the following command, which should give you the version number:

```
$ aws -version
```

This will output the version of the AWS CLI, along with some of the support services it needs, as illustrated in the following screenshot:

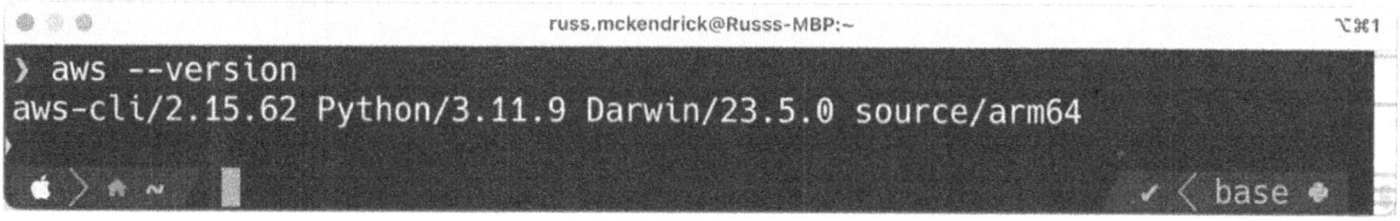

```
) aws --version
aws-cli/2.15.62 Python/3.11.9 Darwin/23.5.0 source/arm64
>
```

Figure 16.1: Checking the version of the AWS CLI

Once installed, you can move on to the AWS CLI configuration section.

Installing on Linux

While packages are available for each distribution, the easiest way to install the AWS CLI on Linux is to download and run the installer.

> These instructions assume that you have the curl and unzip packages installed. If you don't, please install them using your distribution's package manager. For example, on Ubuntu, you would need to run sudo apt-get install unzip curl to install both packages.

To download and install the AWS CLI, run the following commands:

```
$ curl "https://awscli.amazonaws.com/awscli-exe-linux-x86_64.zip" -o "awscliv2.
zip"
$ unzip awscliv2.zip
$ sudo ./aws/install
```

Once installed, you should be able to execute the aws --version command, and you will get something like the output shown in the *Installing on macOS* section, and the Windows version, which we will look at next.

Installing on Windows

As with macOS, you can install the AWS CLI with a package manager. As in *Chapter 15, Kubernetes Clusters on Google Kubernetes Engine*, we will use Chocolatey. The command you need to run is shown here:

```
$ choco install awscli
```

Once installed using Chocolatey, executing the command below will give you a similar output to what we saw on macOS, with changes to the OS and Python versions being used:

```
$ aws --version
```

Again, once installed, you can move on to the *AWS CLI configuration* section below.

AWS CLI configuration

Once you have installed the AWS CLI and checked that it is running properly by issuing the `aws --version` command, you must link your local CLI install to your AWS account. To do this, you must log in to the AWS console, which can be accessed at `http://console.aws.amazon.com/`.

Once logged in, type **IAM** into the search box, which is located at the top left of the page, next to the **Services** button. Then, click on the **IAM Identity Center** service link in the Services results to be taken to the **IAM Identity Center** page.

We need to create a user with programmatic access; to do this, follow these steps:

1. Depending on your AWS account's age or access level, you may need to enable **IAM Identity Center**. My AWS account is for my own projects, so after clicking the **Enable** button, I chose the **Enable in only this AWS account** option rather than the recommended **Enable with AWS Organizations**. I did this because I do not have or require managing multiple AWS accounts within a single organization. Follow the onscreen instructions to enable the service.

2. Once the **IAM Identity Center** service is enabled, we must create a user with programmatic-only access. To do this, return to the search box in the menu at the top of the screen and search for **IAM** again, but this time select **IAM**, which is listed as **Manage access to AWS resources**. Once the page loads, click on **Users**, which can be found in the **Access Management** section of the left-hand-side menu, and then click the **Create user** button.

3. Enter the username of **ekscluster** and ensure you do not select the **Provide user access to the AWS Management Console – optional** option, then click **Next**. We will discuss this option later in the chapter, once we have launched our cluster.

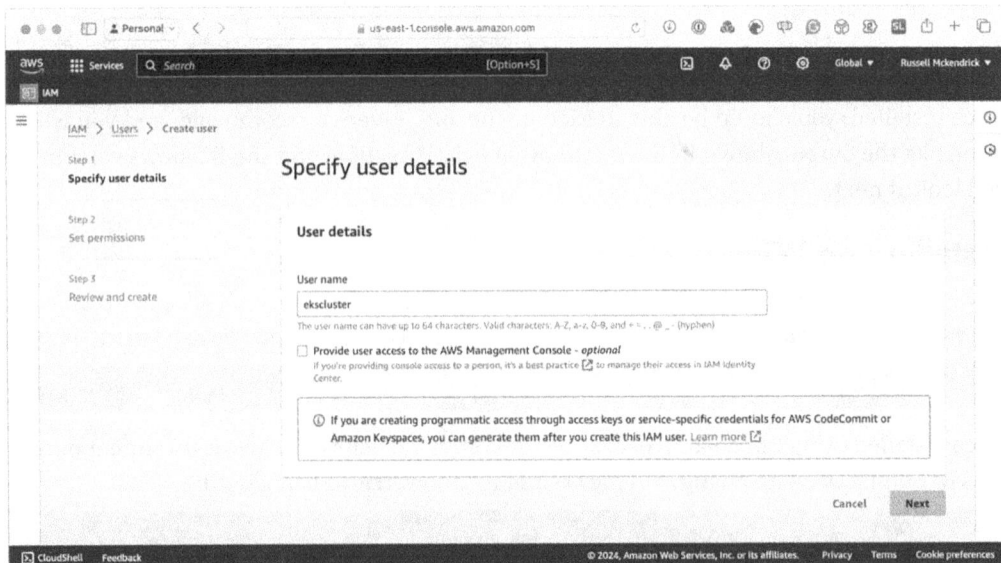

Figure 16.2: Adding a user

4. Rather than create a group, we will grant our user an existing policy. To do this, select **Attach existing policies directly**, select the **AdministratorAccess** policy, and then click **Next**.

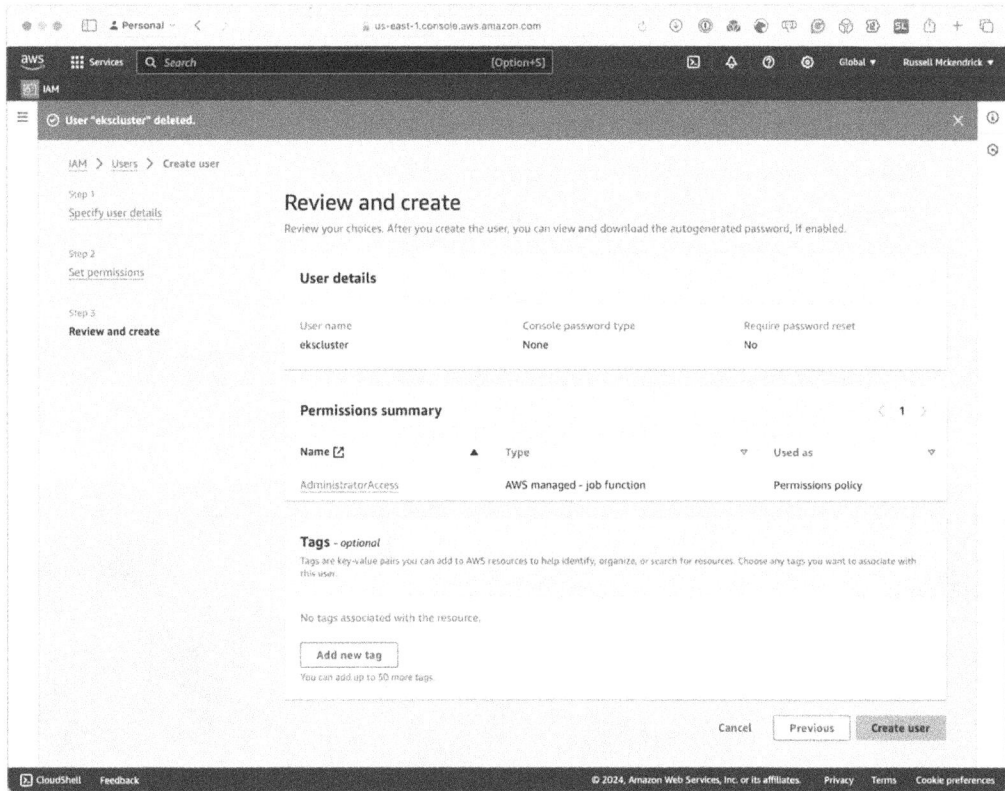

Figure 16.3: Assigning permissions

> As this is a test in a non-production AWS account and we are going to remove this user at the end of the chapter, I am using quite a permissive policy. If you are going to be deploying this into a more production-like environment, I recommend referring to the AWS documentation for a more detailed guide on setting the correct permissions and policies.

5. Once you have reviewed the information, click on the **Create user** button. Once your user has been created, select the **ekscluster** user from the list, select the **Security credentials** tab, and click the **Create access key** button in the **Access keys** section. Select **Command Line Interface (CLI)** and proceed to create the access key by clicking the **Next** and then **Create access key** buttons; once the key has been created, click on **Download .csv file** and, finally, the **Done** button.

> Keep the file you have downloaded safe, as it contains valid credentials for accessing your AWS account.

Return to your terminal and then run the following command to create a default profile:

```
$ aws configure
```

This will ask for a few bits of information, as follows:

- **AWS access key identifier** (**ID**): This is the access key ID from the **comma-separated values** (**CSV**) file we downloaded.
- **AWS secret access key:** This is the key from the CSV file.
- **Default region name:** I entered us-west-2.
- **Default output format:** I left this blank.

To test that everything is working, you can run the following command:

```
$ aws ec2 describe-regions
```

This will list the AWS Regions that are available, and the output should look something like this:

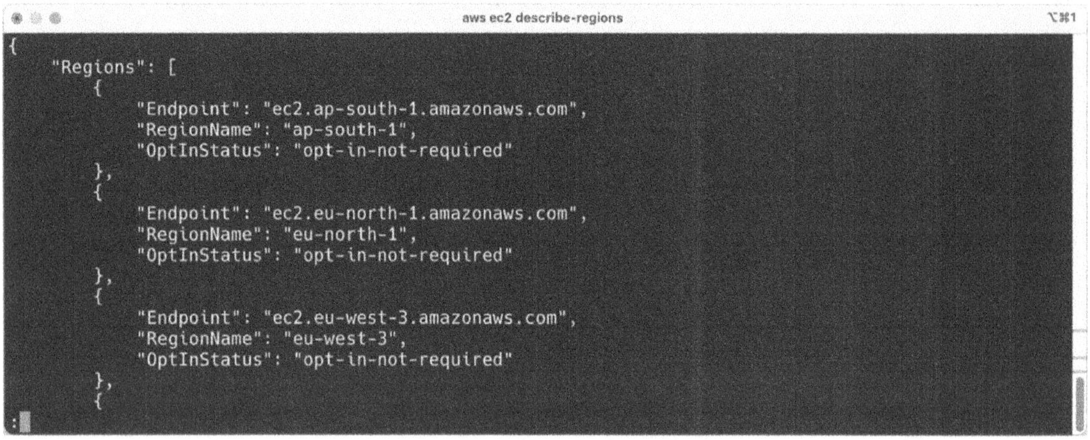

Figure 16.4: Testing the AWS CLI

Now that we have the AWS CLI installed and configured for our account, we need to install the second command-line tool, which we'll use to launch the Amazon EKS cluster.

Installing eksctl, the official CLI for Amazon EKS

While it is possible to launch an Amazon EKS cluster using the AWS CLI, it is complicated and has many steps. To get around this, Weaveworks created a simple command-line tool that generates an AWS CloudFormation template and then launches your cluster.

Unfortunately, Weaveworks stopped commercial operations in early 2024, but before ceasing operations, they passed control of the project to the AWS team.

> AWS CloudFormation is Amazon's **Infrastructure-as-Code** (**IaC**) definition language. It lets you define your AWS resources so that they can be deployed across multiple accounts or repeatedly in the same one. This is useful if you have to keep spinning up an environment, for example, as part of a **continuous integration** (**CI**) build.

As you may have already guessed, installing on macOS and Windows follows the same pattern we have been using; macOS users can run the following command:

```
$ brew install eksctl
```

Likewise, on Windows, you can run:

```
$ choco install eksctl
```

Installing `eksctl` on Linux, like the other tools, is slightly different, with the commands being:

```
$ PLATFORM=$(uname -s)_$(uname -m)
$ curl -sLO "https://github.com/eksctl-io/eksctl/releases/latest/download/
eksctl_$PLATFORM.tar.gz"
$ tar -xzf eksctl_$PLATFORM.tar.gz -C /tmp && rm eksctl_$PLATFORM.tar.gz
$ sudo mv /tmp/eksctl /usr/local/bin
```

Once installed, you should be able to run the command below to get the version number:

```
$ eksctl version
```

So, we are now ready to launch our Amazon EKS cluster.

Launching your Amazon Elastic Kubernetes Service cluster

With all of the prerequisites installed, we can finally start deploying our Amazon EKS cluster. Once deployed, we will be able to start interacting with it to launch a workload like we did in *Chapter 15, Kubernetes Clusters on Google Kubernetes Engine*.

To do this, we will use the defaults built into the `eksctl` command, as this is just a sandbox Amazon EKS against which we can run some commands. This will launch an Amazon EKS cluster with the following attributes:

*   In the **us-west-1** Region
*   With two worker nodes, using the **m5.large** instance type
*   Uses the official AWS EKS **Amazon Machine Image** (**AMI**)
*   Uses **Amazon's Virtual Private Cloud** (**VPC**) for its networking service
*   With an automatically generated random name

So, without further ado, let's launch our cluster by running the following command:

```
$ eksctl create cluster
```

You should go and make a drink or catch up on emails, as this process can take up to 30 minutes to complete. If you are not deploying an Amazon EKS cluster, here is my output when running the command.

First of all, some basic information is displayed about the version of `eksctl` and which Region will be used:

```
[i]   eksctl version 0.180.0-dev+763027060.2024-05-29T21:36:10Z
[i]   using region us-west-2
```

Next up, it will give some information on the networking and AZs it will be deploying resources into, as illustrated in the following code snippet:

```
[i]   setting availability zones to [us-west-2d us-west-2b us-west-2c]
[i]   subnets for us-west-2d - public:192.168.0.0/19 private:192.168.96.0/19
[i]   subnets for us-west-2b - public:192.168.32.0/19 private:192.168.128.0/19
[i]   subnets for us-west-2c - public:192.168.64.0/19 private:192.168.160.0/19
```

It will now give details of which version of the AMI it is going to use, along with the Kubernetes version that the image supports, as follows:

```
[i]   nodegroup "ng-11c87ff4" will use "[AmazonLinux2/1.29]"
[i]   using Kubernetes version 1.29
```

Now it knows all the elements, it is going to create a cluster. Here, you can see it making a start on the deployment:

```
[i]   creating EKS cluster "hilarious-wardrobe-1717847351" in "us-west-2" region
with managed nodes
[i]   will create 2 separate CloudFormation stacks for cluster itself and the
initial managed nodegroup
[i]   if you encounter any issues, check CloudFormation console or try
'eksctl utils describe-stacks --region=us-west-2 --cluster=hilarious-
wardrobe-1717847351'
[i]   Kubernetes API endpoint access will use default of {publicAccess=true,
privateAccess=false} for cluster "hilarious-wardrobe-1717847351" in "us-west-2"
```

As you can see, it has called my cluster `hilarious-wardrobe-1717847351`; this will be referenced throughout the build. By default, logging is not enabled, as we can see here:

```
[i]   CloudWatch logging will not be enabled for cluster "hilarious-
wardrobe-1717847351" in "us-west-2"
[i]   you can enable it with 'eksctl utils update-cluster-logging --enable-
types={SPECIFY-YOUR-LOG-TYPES-HERE (e.g. all)} --region=us-west-2
--cluster=hilarious-wardrobe-1717847351'
```

Now is the point where we wait as the control plane and cluster are deploying:

```
[i]2 sequential tasks: { create cluster control plane "hilarious-
wardrobe-1717847351",2 sequential sub-tasks: {wait for control plane to become
```

```
ready, create managed nodegroup "ng-11c87ff4",}}
[i]  building cluster stack "eksctl-hilarious-wardrobe-1717847351-cluster"
[i]  waiting for CloudFormation stack "eksctl-hilarious-wardrobe-1717847351-
cluster"
[i]  building managed nodegroup stack "eksctl-hilarious-wardrobe-1717847351-
nodegroup-ng-11c87ff4"
[i]  deploying stack "eksctl-hilarious-wardrobe-1717847351-nodegroup-ng-
11c87ff4"
[i]  waiting for CloudFormation stack "eksctl-hilarious-wardrobe-1717847351-
nodegroup-ng-11c87ff4"
[i]  waiting for the control plane to become ready
```

Once deployed, it will download the cluster credentials and configure kubectl, as follows:

```
[✔]  saved kubeconfig as "/Users/russ.mckendrick/.kube/config"
[i]  no tasks
[✔]  all EKS cluster resources for "hilarious-wardrobe-1717847351" have been
created
```

The final step is to wait for the nodes to become available, as is happening here:

```
2024-06-08 13:03:34 [✔]  created 0 nodegroup(s) in cluster "hilarious-
wardrobe-1717847351"
2024-06-08 13:03:35 [i]  node "ip-192-168-34-120.us-west-2.compute.internal" is
ready
2024-06-08 13:03:35 [i]  node "ip-192-168-67-233.us-west-2.compute.internal" is
ready
2024-06-08 13:03:35 [i]  waiting for at least 2 node(s) to become ready in "ng-
11c87ff4"
2024-06-08 13:03:35 [i]  nodegroup "ng-11c87ff4" has 2 node(s)
2024-06-08 13:03:35 [i]  node "ip-192-168-34-120.us-west-2.compute.internal" is
ready
2024-06-08 13:03:35 [i]  node "ip-192-168-67-233.us-west-2.compute.internal" is
ready
2024-06-08 13:03:35 [✔]  created 1 managed nodegroup(s) in cluster "hilarious-
wardrobe-1717847351"
```

Now that we have both nodes online and ready, it is time to display a message confirming that everything is ready, as follows:

```
2024-06-08 13:03:36 [i]  kubectl command should work with "/Users/russ.
mckendrick/.kube/config", try 'kubectl get nodes'
2024-06-08 13:03:36 [✔]  EKS cluster "hilarious-wardrobe-1717847351" in "us-
west-2" region is ready
```

Now that the cluster is ready, let's do as the output suggests and run `kubectl get nodes`. As expected, this gives us details on the two nodes that make up our cluster, as illustrated in the following screenshot:

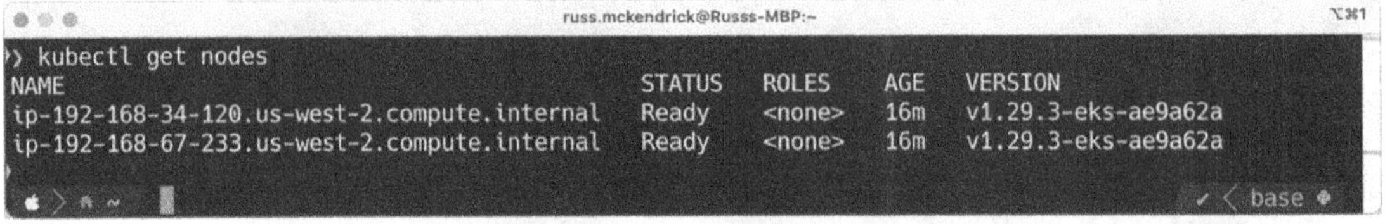

```
) kubectl get nodes
NAME                                           STATUS   ROLES    AGE   VERSION
ip-192-168-34-120.us-west-2.compute.internal   Ready    <none>   16m   v1.29.3-eks-ae9a62a
ip-192-168-67-233.us-west-2.compute.internal   Ready    <none>   16m   v1.29.3-eks-ae9a62a
```

Figure 16.5: Viewing the two Amazon EKS cluster nodes

Now that we have a cluster up and running, let's deploy the same workload we launched when we worked with our **Google Kubernetes Engine** (**GKE**) cluster.

Deploying a workload and interacting with your cluster

In *Chapter 15, Kubernetes Clusters on Google Kubernetes Engine*, we used the Guestbook example from the GCP GKE examples GitHub repository. In this section, first we will deploy the workload before exploring the web-based AWS console. So now let's start on our Guestbook deployment.

Deploying the workload

Even though our cluster runs in AWS using Amazon EKS, we will use the same set of YAML files we used to launch our workload in GKE using our local `kubectl`; to do this, follow the steps below:

1. As before, our first step is launching the Redis Leader deployment and service by running the two commands below:

```
$ kubectl apply -f https://raw.githubusercontent.com/GoogleCloudPlatform/
kubernetes-engine-samples/main/quickstarts/guestbook/redis-leader-
deployment.yaml
$ kubectl apply -f https://raw.githubusercontent.com/GoogleCloudPlatform/
kubernetes-engine-samples/main/quickstarts/guestbook/redis-leader-
service.yaml
```

2. Once the Redis Leader is deployed, we need to launch the Redis Follower deployment and service, as follows:

```
$ kubectl apply -f https://raw.githubusercontent.com/GoogleCloudPlatform/
kubernetes-engine-samples/main/quickstarts/guestbook/redis-follower-
deployment.yaml
$ kubectl apply -f https://raw.githubusercontent.com/GoogleCloudPlatform/
kubernetes-engine-samples/main/quickstarts/guestbook/redis-follower-
service.yaml
```

3. Once the Redis Leader and Follower are up and running, it's time to launch the frontend deployment and service using the following commands:

```
$ kubectl apply -f https://raw.githubusercontent.com/GoogleCloudPlatform/
kubernetes-engine-samples/main/quickstarts/guestbook/frontend-deployment.
yaml
$ kubectl apply -f https://raw.githubusercontent.com/GoogleCloudPlatform/
kubernetes-engine-samples/main/quickstarts/guestbook/frontend-service.
yaml
```

4. After a few minutes, we will be able to run the following command to get information on the service we have just launched, which should include details on where to access our workload:

```
$ kubectl get service frontend
```

You will notice that, this time, the output is slightly different from the output we got when running the workload on GKE, as we can see in the following screenshot:

Figure 16.6: Getting information on the frontend service

As you can see, rather than an **Internet Protocol** (**IP**) address, we get a **Uniform Resource Locator** (**URL**). Copy that into your browser.

Once you have opened the URL, given that we have used the same commands and workload configuration, you won't be surprised to see the **Guestbook** application, as shown in the following screenshot:

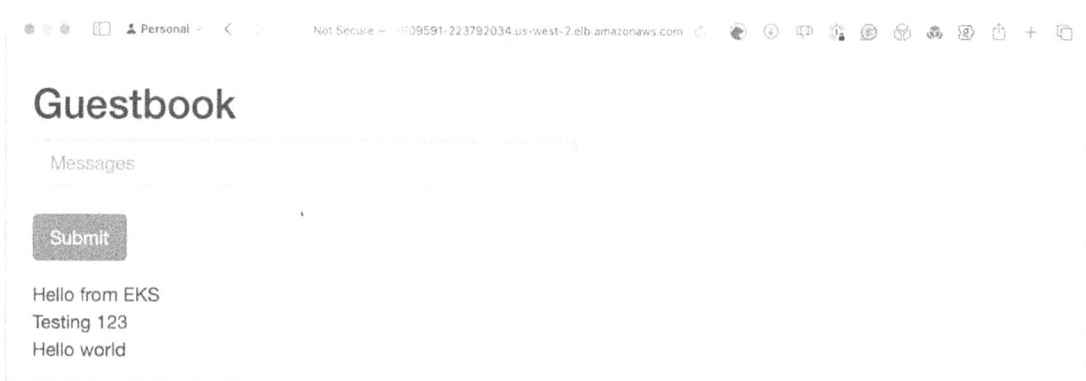

Figure 16.7: The Guestbook application with a few examples

Now that our workload is up and running, let's explore what we can see about our cluster within the AWS console.

Exploring the AWS console

This section will examine our newly deployed workload using the AWS console. First, log in to the AWS console at `https://console.aws.amazon.com/`. Once logged in, select the **US West (Oregon) us-west-2** Region in the Region selector at the top right of the screen, next to your username.

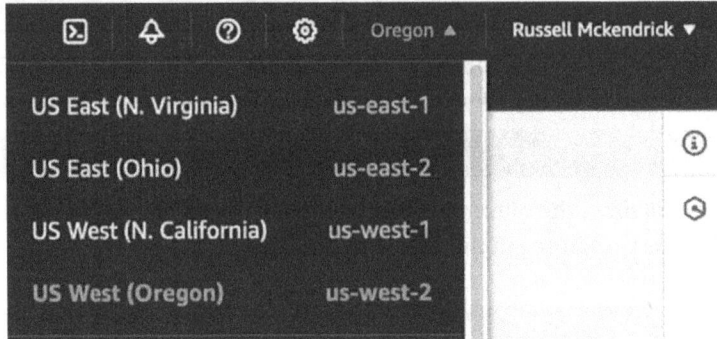

Figure 16.8: Selecting the correct Region

Once the correct Region is selected, search for **Elastic Kubernetes Service** in the search bar on the top right and select the service, which should be the first result, to be taken to the EKS page in the Oregon Region.

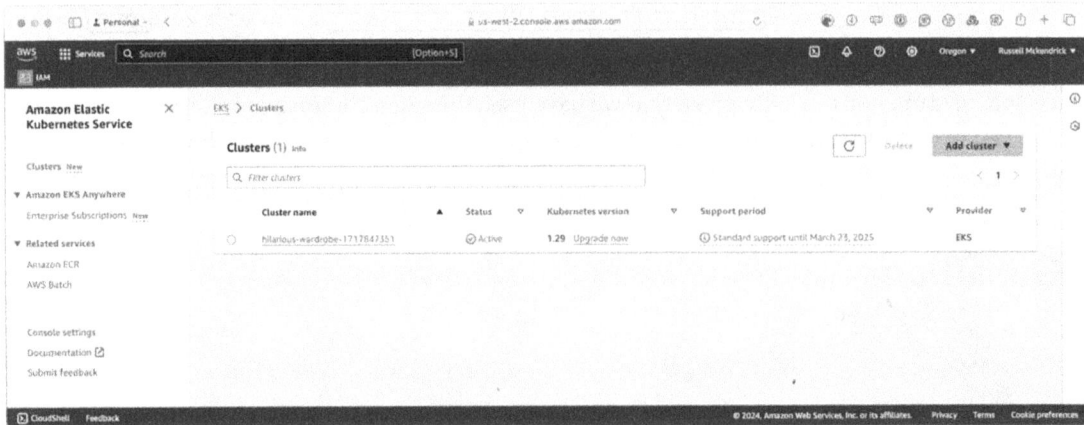

Figure 16.9: Our first look at our EKS cluster in the AWS console

So far, so good; well, we will talk about it in a moment – click on your cluster name, and you will be presented with something that looks like the following page:

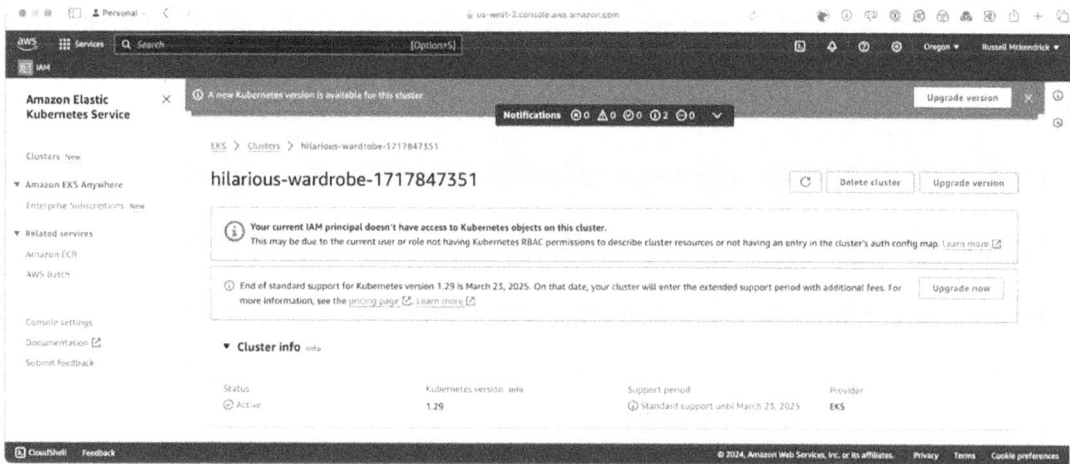

Figure 16.10: Access denied!

So, let's unpack what is happening here. You may think, *"This is my main user, and surely it has full access?".*

There is a good reason for this; when `eksctl` launched our cluster, it granted the **ekscluster** user we created earlier in the chapter permission to interact with the cluster using AWS services as we configured the AWS CLI to connect using this user and not the main user we are currently logged in as.

This means that to view workloads and the like within the AWS console, we need to log in as the user we created earlier. To do this, return to IAM in the AWS console, go to the **Users** page, and select the **ekscluster** user; go to the **Security Credentials** tab and then click on the **Enable console access** button:

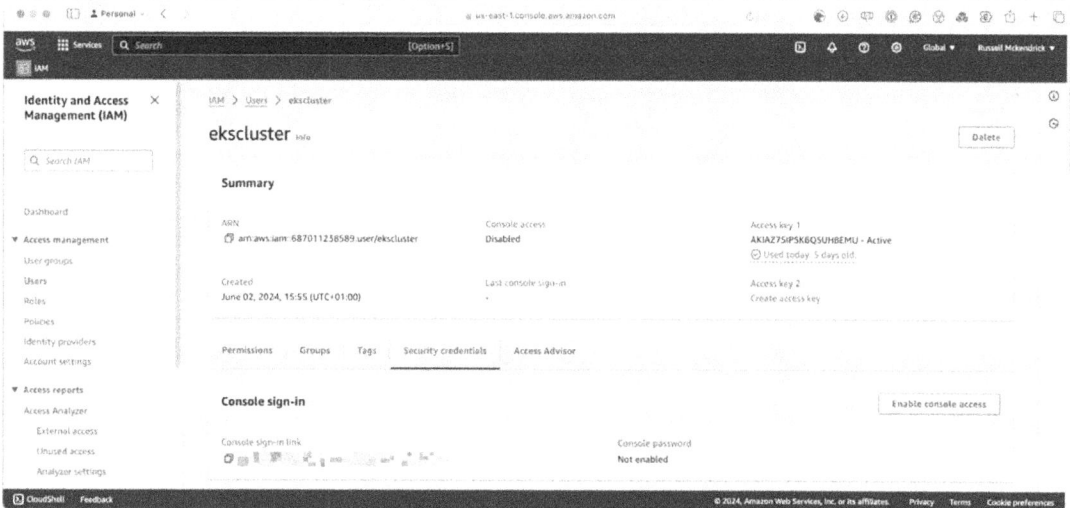

Figure 16.11: Enabling console access for the ekscluster user

Select the **Autogenerated password** option and enable access; finally, download the CSV file containing the credentials as before.

Once downloaded, sign out of the AWS console and open the CSV file you downloaded. Go to the console sign-in URL. This is a URL that allows IAM users, like the one we created, to sign in to your account; use the username and password in the CSV file.

Once you have signed in, return to the EKS page and select your cluster; this time, you will not see any complaints about permissions.

When you first open the cluster, you will see several tabs. These are:

- **Overview**: Displays various cluster details, such as the version of Kubernetes running, endpoint information, the cluster status, creation time and date, etc.
- **Resources**: Provides information on nodes, Pods, namespaces, and workloads
- **Compute**: Shows node information, node groups, and details on any Fargate profiles associated with the cluster
- **Networking**: Details the VPC configuration
- **Add-ons**: Lists installed and available add-ons for the cluster
- **Access**: Displays IAM roles, AWS auth ConfigMap, and Kubernetes RBAC role bindings
- **Observability**: Configures and shows logging, monitoring, and recent events
- **Upgrade insights**: Lists available Kubernetes version upgrades and compatibility checks
- **Update history**: Provides a history of cluster and node group updates
- **Tags**: Lists and manages tags associated with the EKS cluster

Below, you can see details on the nodes:

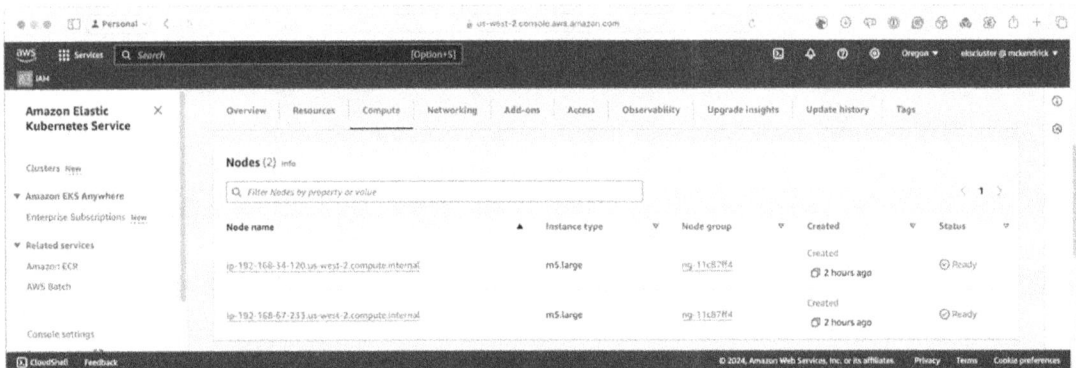

Figure 16.12: Viewing the two nodes in the cluster in the Compute tab

Clicking on **Resources**, selecting **Deployments**, and filtering down to the **default** workspace will show the workload we launched:

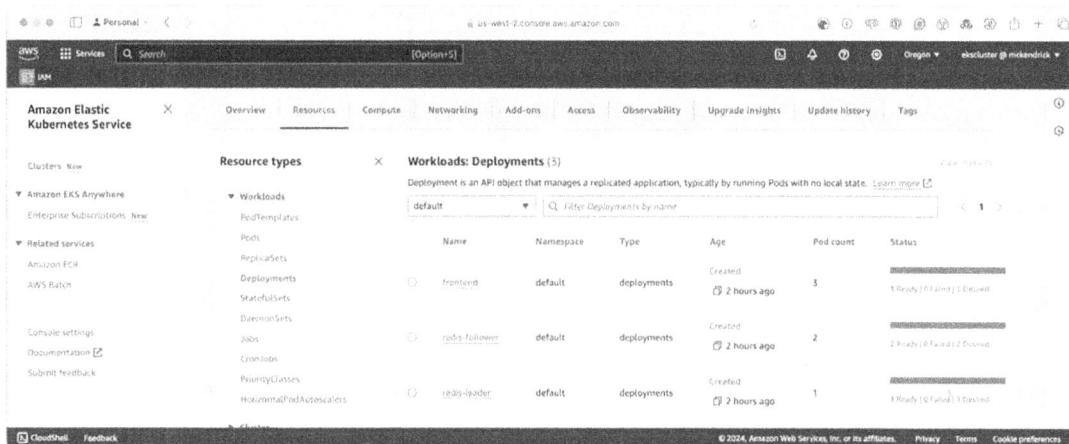

Figure 16.13: Viewing our workload

Clicking on one of the deployments will give you more information on the deployment – this includes details of the Pods, configuration, and so on. However, as you click around, you will find that all you can really do is view information on the services; there are no graphs, logging output, or anything that gives more than a basic overview of our workloads. This is because the AWS console is mostly just exposing information from Kubernetes itself.

Moving away from the EKS service page and going to the EC2 service section of the AWS console will display the two nodes, as illustrated in the following screenshot:

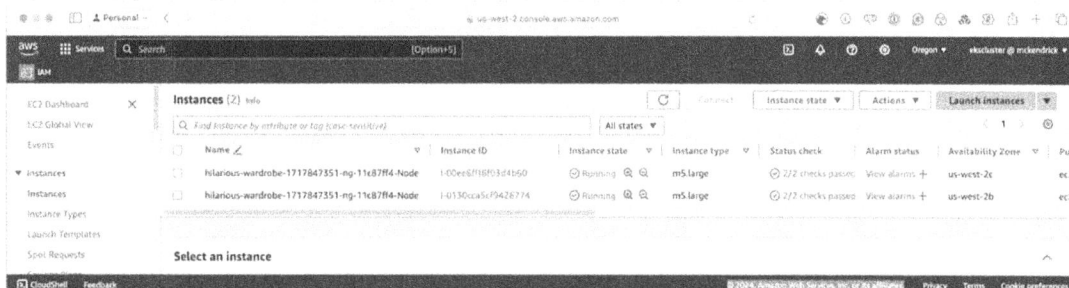

Figure 16.14: Looking at the raw EC2 compute resource

Here, you can drill down and find out more information on the instance, including CPU, RAM, and network utilization; however, this is only for the instance itself and not our Kubernetes workload.

Selecting **Load Balancers** from the **Load Balancing** section of the left-hand-side menu will show you the elastic load balancer that was launched and configured when we applied the frontend service, as illustrated in the following screenshot:

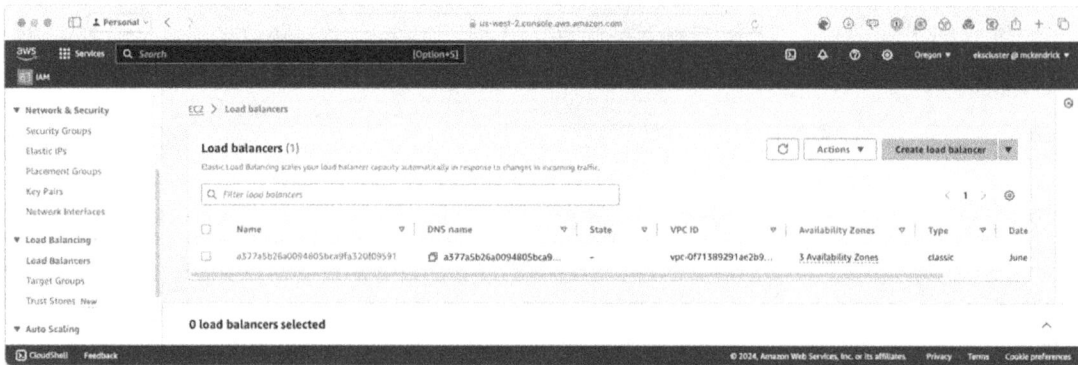

Figure 16.15: Looking at the raw load balancer resource

One final AWS service we are using is AWS CloudFormation, so entering **CloudFormation** in the Services menu and clicking on the link will take you to the CloudFormation service page.

Here, you will see two stacks: one for the EKS nodes, our two EC2 instances, and one for the EKS cluster, which is our Kubernetes management plane. These stacks are illustrated in the following screenshot:

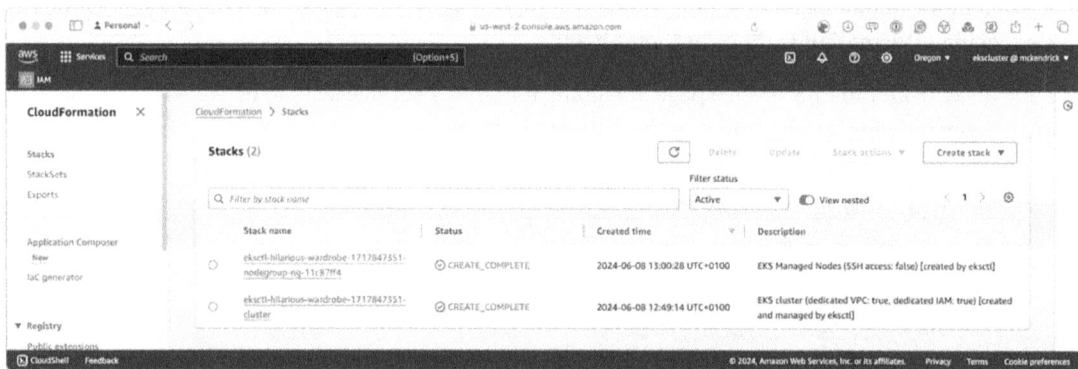

Figure 16.16: The two stacks that make up our cluster

Selecting one of the stacks will give you details on what happened when the stack was launched. It will list all the many resources created during the launch of the Amazon EKS cluster using eksctl.

You select a template and then view it in the designer; you can even see the CloudFormation template generated by eksctl, which is quite a complicated JSON file – if you click on the **View in Application Composer** button, you will be able to get a more digestible visual representation of the stack. A screenshot of this view can be seen below:

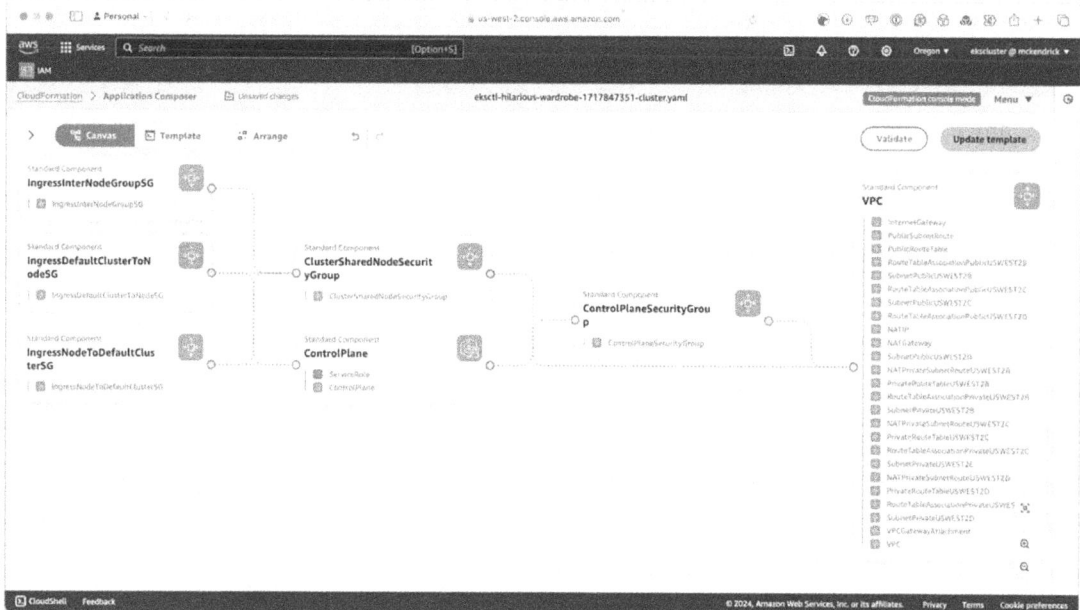

Figure 16.17: Reviewing the CloudFormation template in Application Composer

That is about all we can see in the AWS console. As we have seen, while Amazon EKS is relatively simple to launch using `eksctl`, its level of integration with the AWS console could be better compared to the GKE cluster we launched in the previous chapter.

While we were able to explore and view our workload, we could not interact with too much; also, the feedback on the cluster is tied to the basic monitoring offered by the Amazon EC2 service.

Once you have finished with your Amazon EKS cluster, you can delete it.

Deleting your Amazon Elastic Kubernetes Service cluster

You can delete your cluster by running the following command, making sure to replace the cluster name with that of your own:

```
$ eksctl delete cluster --name hilarious-wardrobe-1717847351
```

Deleting the cluster takes less time to run than when we launched it; in fact, it takes around 5 minutes.

As before, `eksctl` gives you details on what it is doing as it deletes the resources:

```
[i]  deleting EKS cluster "hilarious-wardrobe-1717847351"
[i]  will drain 0 unmanaged nodegroup(s) in cluster "hilarious-
wardrobe-1717847351"
[i]  starting parallel draining, max in-flight of 1
[i]  deleted 0 Fargate profile(s)
```

The first thing that is updated is the local kubectl configuration, as we can see here:

```
[✔]   kubeconfig has been updated
```

Then, any resources that have been launched as part of deploying workloads into our cluster are terminated:

```
[i]   cleaning up AWS load balancers created by Kubernetes objects of Kind
Service or Ingress
```

Then, the two AWS CloudFormation stacks are removed, which in turn removes all of the resources they created and configured, as illustrated in the following code snippet:

```
[i]   2 sequential tasks: { delete nodegroup "ng-11c87ff4", delete cluster
control plane "hilarious-wardrobe-1717847351" [async] }
[i]   will delete stack "eksctl-hilarious-wardrobe-1717847351-nodegroup-ng-
11c87ff4"
[i]   waiting for stack "eksctl-hilarious-wardrobe-1717847351-nodegroup-ng-
11c87ff4" to get deleted
[i]   waiting for CloudFormation stack "eksctl-hilarious-wardrobe-1717847351-
nodegroup-ng-11c87ff4"
[i]   will delete stack "eksctl-hilarious-wardrobe-1717847351-cluster"
[✔]   all cluster resources were deleted
```

At this point, our cluster has been completely deleted.

> Please double-check the EC2, EKS, and CloudFormation sections in the AWS console to ensure that all services have been correctly deleted, as you will be charged for any orphaned or idle resources left behind. While this is an unlikely scenario, it is best to double-check now rather than receive an unexpected bill at the end of the month.

So, how much would our Amazon EKS cluster have cost us to run for a month?

There are two sets of costs that we need to consider:

- The first is for the Amazon EKS cluster itself. It is US Dollars (USD) 0.10 per hour for each Amazon EKS cluster you create; however, each Amazon EKS cluster can run multiple node groups, so you shouldn't have to launch more than one per Region. This means that the Amazon EKS cluster costs around $73 per month.

- The next consideration is the AWS resources used by the cluster, for example, the EC2 cluster nodes, in our case, would have cost around $70 each, and the total cost to run our cluster for a month would be around $213. I say around because there are charges for bandwidth and the AWS **Elastic Load Balancing** (**ELB**) service, which will increase the cost of our workload further.

A link to the pricing overview page can be found in the *Further reading* section at the end of this chapter.

Summary

In this chapter, we discussed the origins of AWS and Amazon EKS before walking through how to sign up for an account and how to install and configure the two command-line tools required to easily launch an Amazon EKS cluster.

Once our cluster was up and running, we deployed the same workload as when we launched our GKE cluster. We did not have to make any allowances for the workload running on a different cloud provider – it just worked, even deploying a load balancer using the AWS native load balancing service without us having to instruct it to do so.

However, we did find that Amazon EKS is less integrated with the AWS console than the Google service we looked at. We also learned that we had to install a second command-line tool to easily launch our cluster due to the complications of trying to do so using the AWS CLI. This would have been around eight steps, assuming the Amazon VPC configuration and IAM roles had been created and deployed.

This lack of integration and complexity in launching and maintaining clusters compared to other providers would put me off running my Kubernetes workloads on Amazon EKS – it all feels a little disjointed and not as slick as the Google offering.

In the next chapter, we will examine launching an **Azure Kubernetes Service (AKS)** cluster on Microsoft Azure, the last of the three public providers we will cover.

Further reading

Here are links to more information on some of the topics and tools we have covered in this chapter:

- AWS: `https://aws.amazon.com/`
- Amazon EKS: `https://aws.amazon.com/eks/`
- The AWS CLI: `https://aws.amazon.com/cli/`
- eksctl: `https://eksctl.io/`
- eksctl support status update: `https://github.com/aws/containers-roadmap/issues/2280`
- Official documentation: `https://docs.aws.amazon.com/eks/latest/userguide/what-is-eks.html`
- Amazon EKS pricing: `https://aws.amazon.com/eks/pricing/`

Join our community on Discord

Join our community's Discord space for discussions with the authors and other readers:

`https://packt.link/cloudanddevops`

17

Kubernetes Clusters on Microsoft Azure with Azure Kubernetes Service

The final public cloud Kubernetes service we'll examine is **Azure Kubernetes Service (AKS)**, hosted in Microsoft Azure, one of the "big three" public cloud providers – the other two of the big three we have already covered in *Chapter 15, Kubernetes Clusters on Google Kubernetes Engine,* and *Chapter 16, Launching a Kubernetes Cluster on Amazon Web Services with Amazon Elastic Kubernetes Service.*

By the end of the chapter, you will have configured your local environment with the tools needed to interact with your Microsoft Azure account and launch your AKS cluster.

After deploying the cluster, we'll launch the same workload as in the previous chapters and explore the integration level between your AKS cluster and the Microsoft Azure portal.

Finally, at the end of the chapter, we will discuss the three services we covered in this and the previous two chapters, as well as which one I would recommend.

To do this, we will be covering the following topics:

- What are Microsoft Azure and Azure Kubernetes Service?
- Preparing your local environment
- Launching your Azure Kubernetes Service cluster
- Deploying a workload and interacting with your cluster
- Deleting your Azure Kubernetes Service cluster

Technical requirements

If you plan on following along with the examples covered in this chapter, you need a Microsoft Azure account with a valid payment method attached.

Following the examples in this chapter will incur a cost, and it is essential to terminate any resources you launch once you have finished with them to prevent unwanted expenses. All prices quoted in this chapter are correct at the time of print, and we recommend that you review the current costs before you launch any resources.

What are Microsoft Azure and Azure Kubernetes Service?

Before we start to look at installing the supporting tools, let's quickly discuss the origins of the services we'll be looking at, starting with Microsoft Azure.

Microsoft Azure

In 2008, Microsoft formally announced that it had a new service called Windows Azure.

This service was part of a project known internally as Project Red Dog, which had been in development since 2004. This project aimed to deliver data center services using core Windows components.

The five core components that Microsoft announced at their 2008 developer conference were as follows:

- **Microsoft SQL Data Services:** This was a cloud version of the Microsoft SQL Database service running as a **Platform as a Service (PaaS)**, which aimed to remove the complexity of hosting your own SQL services.
- **Microsoft .NET Services:** Another PaaS service that allowed developers to deploy their .NET-based applications into a Microsoft-managed .NET runtime.
- **Microsoft SharePoint:** A **Software as a Service (SaaS)** version of the popular intranet product.
- **Microsoft Dynamics:** A SaaS version of Microsoft's CRM product.
- **Windows Azure:** An **Infrastructure as a Service (IaaS)** offering, like other cloud providers, enabling users to spin up virtual machines, storage, and the networking services needed to support their compute workloads.

All these services were built on top of the Red Dog operating system, from which the project took its name; this was a specialized version of the Windows operating system with a built-in cloud layer.

In 2014, Windows Azure was renamed Microsoft Azure, reflecting the name of the underlying operating system powering the cloud services and that Azure was running many Linux-based workloads. As part of this announcement, newly appointed Microsoft CEO Satya Nadella showed the now famous (or infamous, depending on your point of view) "Microsoft loves Linux" slide, with "loves" represented by a heart emoji.

I say "famous" and "infamous" because one-time Microsoft CEO Steve Ballmer was once quoted as saying the following:

> *"Linux is a cancer that attaches itself in an intellectual property sense to everything it touches."*

So, this was seen as quite a U-turn, which took many by surprise.

In 2020, it was revealed that more than 50% of virtual machine cores are running Linux workloads, and 60% of Azure Marketplace images are now Linux-based. This is primarily attributed to Microsoft's embrace of Linux and open-source projects like Kubernetes, which has led us to their native Kubernetes service.

Azure Kubernetes Service

Originally, Microsoft launched a container-based service called **Azure Container Service (ACS)**. This allowed users to deploy their container workloads and choose to have them powered by one of three different orchestrators: Docker Swarm, DC/OS, or Kubernetes. All of these provided a container-based cluster solution.

It soon became apparent that Kubernetes was the most popular of the three orchestrators, so ACS was gradually replaced by AKS. AKS is a CNCF-compliant, purely Kubernetes-based service. The transition took about two years, with AKS becoming generally available in 2018 and ACS being retired in early 2020.

The AKS service is closely integrated with Azure Active Directory, Policies, and other key Microsoft Azure services. Alongside AKS, Microsoft also offers other container services; the newest is called Azure Container Apps.

Azure Container Apps is a serverless platform that runs containerized applications without the end user managing infrastructure, with features such as dynamic scaling based on traffic, event, CPU, or memory load, HTTPS or TCP ingress, Dapr integration, and autoscaling. It allows the use of various Azure native tools for management, secure secret handling, internal service discovery, and traffic splitting for deployment strategies. Applications can run containers from any registry and integrate with other Azure services. Confusingly, you can also choose to launch your container-based workloads in Azure App Services.

However, rather than discussing all the services in Microsoft Azure that you can use to run your container-based workloads, I have always found it easier to roll up your sleeves and get hands-on with a service, so without further delay, let's look at installing the tools we will need to launch and manage our AKS cluster.

Preparing your local environment

Before you launch your cluster, there are a few tasks you need to complete. First, you will need somewhere to launch the cluster, so if you don't already have one, you will need to sign up for an Azure account.

Creating a free Microsoft Azure account

If you don't already have an account, head to `https://azure.microsoft.com/free/`, where you can sign up for a free account:

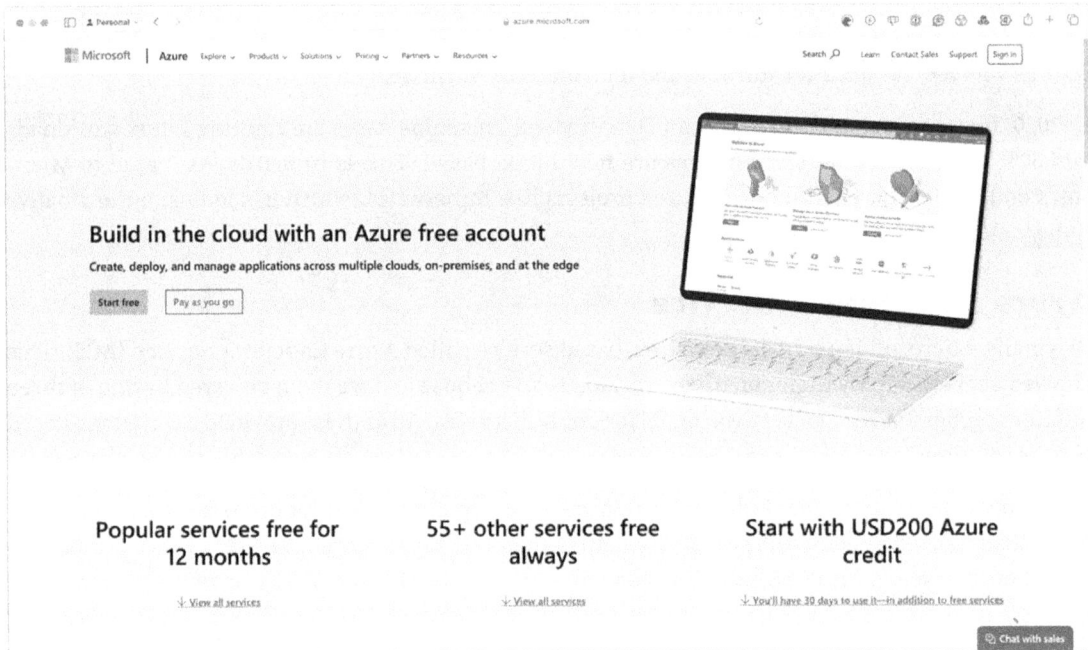

Figure 17.1: Reviewing what Microsoft Azure services you can get for free

At the time of writing, your free account includes 12 months of popular services, $200 of credit that can be used to explore and test the different Azure services, and access to over 55 services that will always be free.

Click on the **Start free** button and follow the onscreen instructions. The sign-up process will take about 15 minutes, and you will need to provide valid credit or debit card information to complete it and gain access to your free account.

Once you can access your account, the next step is installing the Azure CLI.

The Azure CLI

Microsoft provides a powerful cross-platform command-line tool for managing your Microsoft Azure resources. Installing it on macOS, Linux, and, of course, Windows couldn't be more straightforward.

Installing on macOS

If you have been following along with the previous two chapters, you may have already guessed that we will use Homebrew to install the Azure CLI on macOS.

To do this, run the following command:

```
$ brew install azure-cli
```

Once the Azure CLI has been installed, run the following command:

```
$ az --version
```

This should return something like the following screenshot:

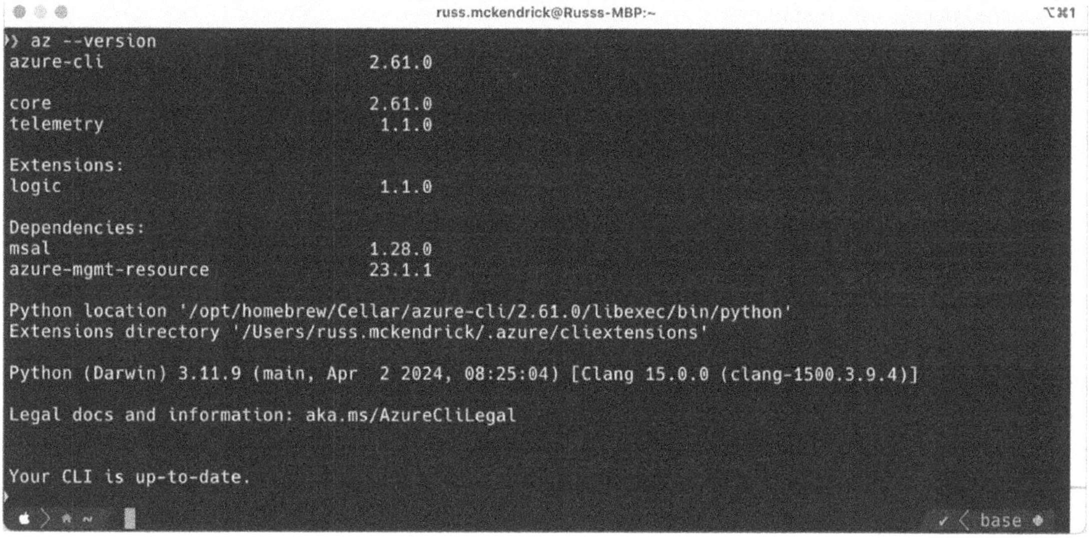

```
>> az --version
azure-cli                          2.61.0

core                               2.61.0
telemetry                           1.1.0

Extensions:
logic                               1.1.0

Dependencies:
msal                               1.28.0
azure-mgmt-resource                23.1.1

Python location '/opt/homebrew/Cellar/azure-cli/2.61.0/libexec/bin/python'
Extensions directory '/Users/russ.mckendrick/.azure/cliextensions'

Python (Darwin) 3.11.9 (main, Apr  2 2024, 08:25:04) [Clang 15.0.0 (clang-1500.3.9.4)]

Legal docs and information: aka.ms/AzureCliLegal


Your CLI is up-to-date.
```

Figure 17.2: Checking the Azure CLI version on macOS

Once the Azure CLI has been installed, you can move on to the *Configuring the Azure CLI* section.

Installing on Linux

Microsoft provides an installation script that covers the most common Linux distributions. To run the script, use the following command:

```
$ curl -L https://aka.ms/InstallAzureCli | bash
```

This will download, install, and configure everything required to run the Azure CLI on your Linux distribution of choice. Once it has been installed, you will need to restart your session.

You can do this by logging out and then back in, or, on some distributions, by running the following command:

```
$ source ~/.profile
```

Once you have restarted your session, run the following command:

```
$ az --version
```

This will return output almost exactly the same as the macOS output we covered when installing the CLI on that operating system, with the only difference being information on the operating system. Once the Azure CLI has been installed, move on to the *Configuring the Azure CLI* section.

Installing on Windows

You can install the Azure CLI on Windows machines in a few ways. Your first option is to download a copy of the installer from `https://aka.ms/installazurecliwindows` and then run the installer by double-clicking it.

Your next option is to use the following PowerShell command, which will download the installer from the preceding URL and install it:

```
$ProgressPreference = 'SilentlyContinue'; Invoke-WebRequest -Uri https://aka.
ms/installazurecliwindows -OutFile .\AzureCLI.msi; Start-Process msiexec.exe
-Wait -ArgumentList '/I AzureCLI.msi /quiet'; Remove-Item .\AzureCLI.msi
```

The third option is to use the **Chocolatey** package manager by running the following command:

```
$ choco install azure-cli
```

Whichever way you choose to install the package, run the following command once it has been installed to find out the version number:

```
$ az --version
```

As you may have guessed, this will also show you something similar to the output we saw when running the command on macOS. Now that we have the Azure CLI installed, we can configure it.

Configuring the Azure CLI

Configuring the Azure CLI is a straightforward process; you need to run the following command:

```
$ az login
```

This will open your default browser, asking you to log in. Once you are logged in, the Azure CLI will be configured to work the account associated with the user you logged in as.

If you are having problems or running a command-line-only installation of the Azure CLI (on a remote Linux server, for example), then running the following command will give you a URL and unique sign-in code to use:

```
$ az login --use-device-code
```

Once you are logged in, your command-line session should return some information on your Azure account. You can view this again by using the following command:

```
$ az account show
```

If you cannot install the Azure CLI locally for any reason, you will not lose anything, as there is a web-based terminal with the Azure CLI you can use in the Azure portal. We will look at this next.

Accessing Azure Cloud Shell

To access Azure Cloud Shell, open `https://portal.azure.com/` and log in with your credentials. Once you're logged in, click on the Cloud Shell icon in the menu bar, which is found at the top of the page; it is the first icon next to the central search box.

When launching a cloud shell, you have the option of attaching storage. Given that we will be using kubectl, we want our configuration to persist, so select the **Mount storage account** option and choose your subscription, as shown in the following screenshot:

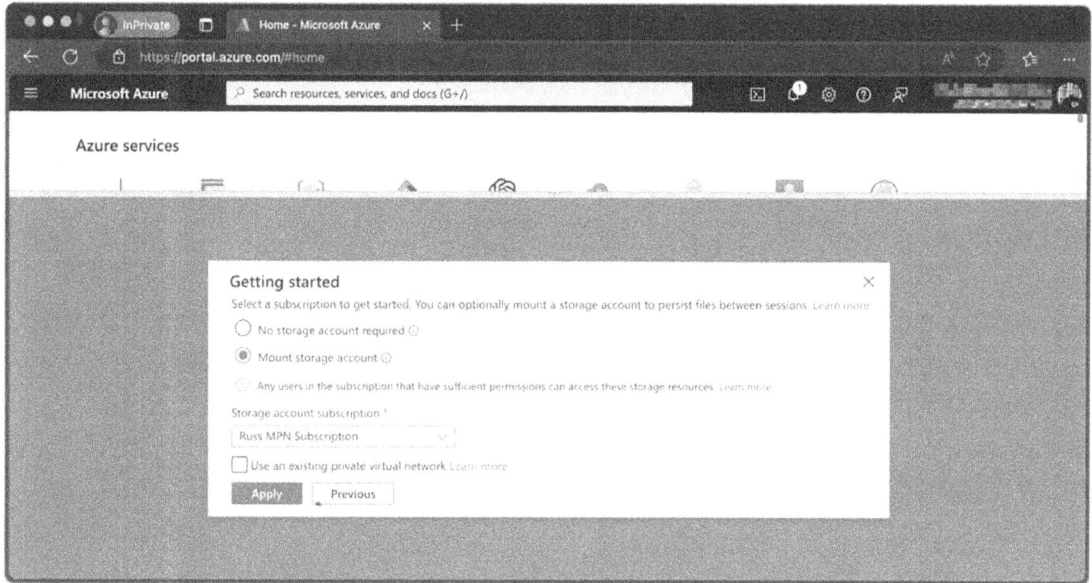

Figure 17.3: Choosing a storage option for your cloud shell session

Clicking on the **Apply** button will then present you with three options, as you can see in the following screenshot. Choose the **We will create a storage account for you** option:

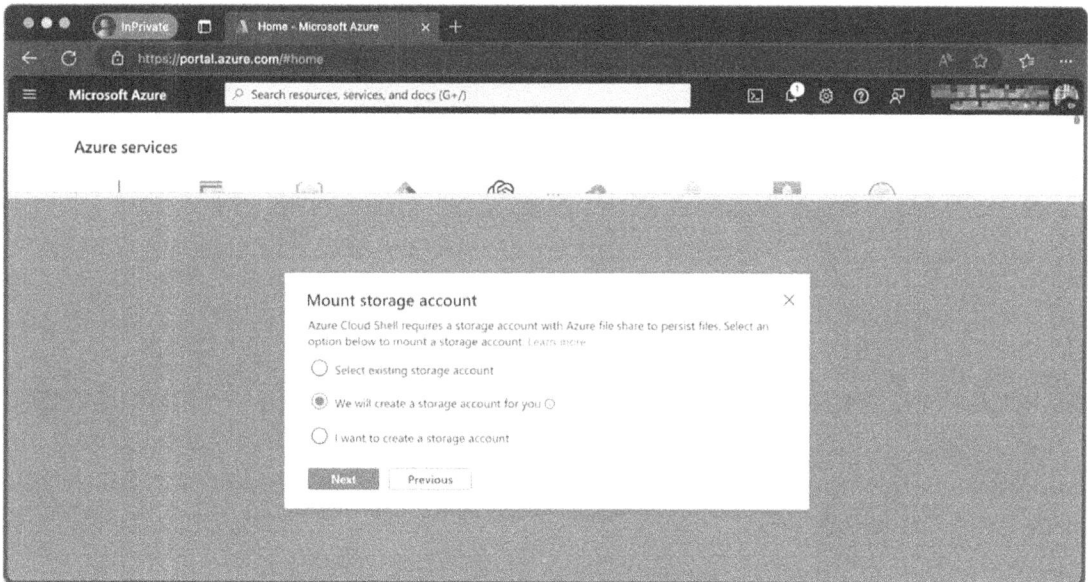

Figure 17.4: How do you want to create a storage account?

After about a minute, **Cloud Shell** should open, and you will be presented with a command prompt:

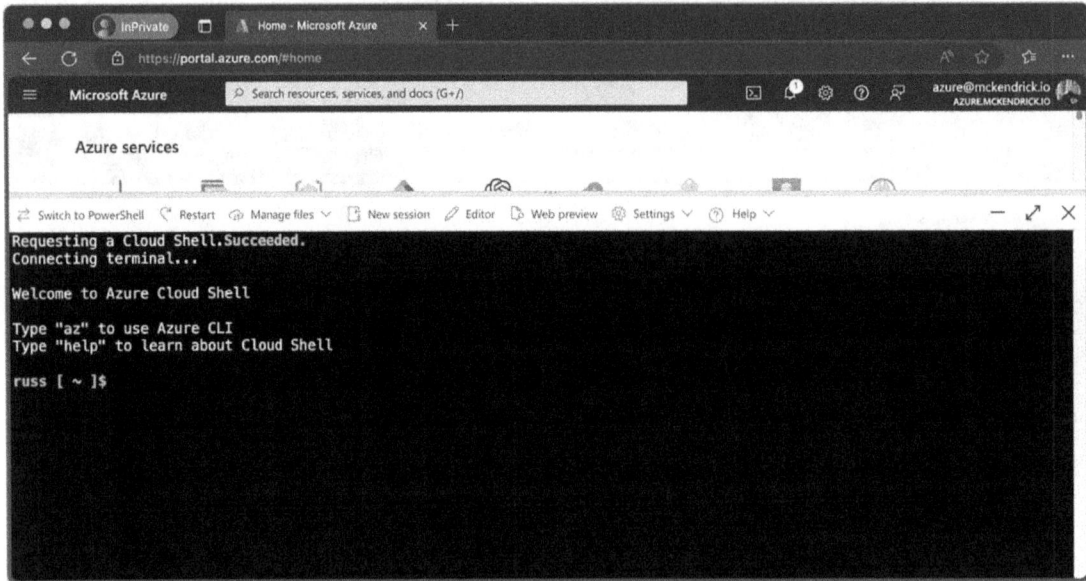

Figure 17.5: Logged in and ready to use

Now that you have a command prompt, running the following command, as we did on the local Azure CLI installation, will give you information on the version of the Azure CLI installed:

```
$ az --version
```

You will not need to run the az login command as the Azure portal took care of that for you in the background when your **Cloud Shell** instance was launched.

Now that you have access to a configured Azure CLI, in one form or another, we can look at launching our AKS cluster.

Launching your Azure Kubernetes Service cluster

With all of the prerequisites in place, we can now launch our AKS cluster. To do this, we will need to run just two commands.

The first of the commands creates an Azure resource group:

```
$ az group create --name rg-myfirstakscluster-eus --location eastus -o table
```

In the preceding command, we are creating a resource group called rg-myfirstakscluster-eus in the eastus region and setting the output to be formatted as a table rather than the JSON, which is the default output type for the Azure CLI.

> A **resource group** is a logical container used to group related Azure resources. Services launched within the resource group can inherit settings such as role-based access control, locks, and regions in which the resources are launched.

Once the resource group has been created, you should see confirmation formatted in a table like the output below:

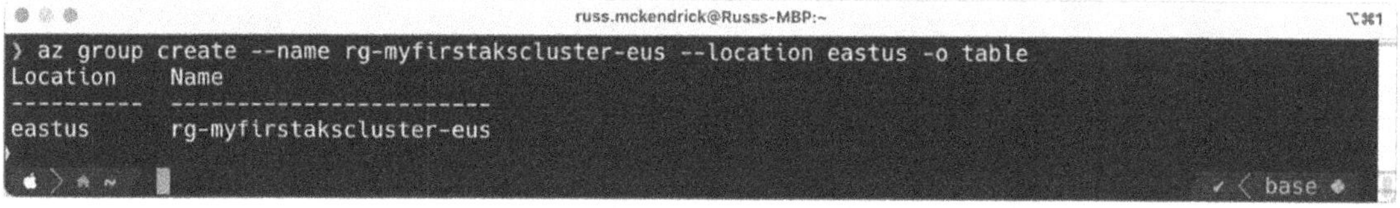

```
) az group create --name rg-myfirstakscluster-eus --location eastus -o table
Location    Name
----------  ------------------------
eastus      rg-myfirstakscluster-eus
)
```

Figure 17.6: Creating the resource group

Now that we have a container for our resources with the resource group, we can launch our AKS cluster by running the command below. As you can see, it references the resource group we just created:

```
$ az aks create --resource-group rg-myfirstakscluster-eus --name aks-
myfirstakscluster-eus --node-count 2 --enable-addons monitoring --generate-ssh-
keys -o yaml
```

Launching and configuring the cluster will take about five minutes, so while that deploys, I will work through the options we passed to the preceding `az aks create` command:

- `--resource-group`: As you may have guessed, this is the resource group where you would like to launch your AKS cluster. The cluster will inherit the resource group's location. In our example, we are using the `rg-myfirstakscluster-eus` resource group we created in the command before last, and the cluster will be created in `eastus`.

- `--name`: This passes the name of the cluster you are launching. We are calling ours `aks-myfirstakscluster-eus`.

- `--node-count`: Here, you set the number of nodes you want to launch. We are launching two. At the time of writing, the default instance type for nodes is Standard_DS2_v2, meaning that each node will have 2 x vCPUs and 7 GB of RAM.

- `--enable-addons`: This flag is used to supply a list of add-ons to enable while the cluster is being launched – we are just enabling the monitoring add-on.

- `--generate-ssh-keys`: This will generate SSH public and private key files for the cluster.

- `-o`: As we mentioned when discussing the previous command, this determines the command output format. This time, we output the results returned as YAML when we run the command because the output is more readable than the JSON and table options.

Once your cluster has been launched, you should see something like the following output:

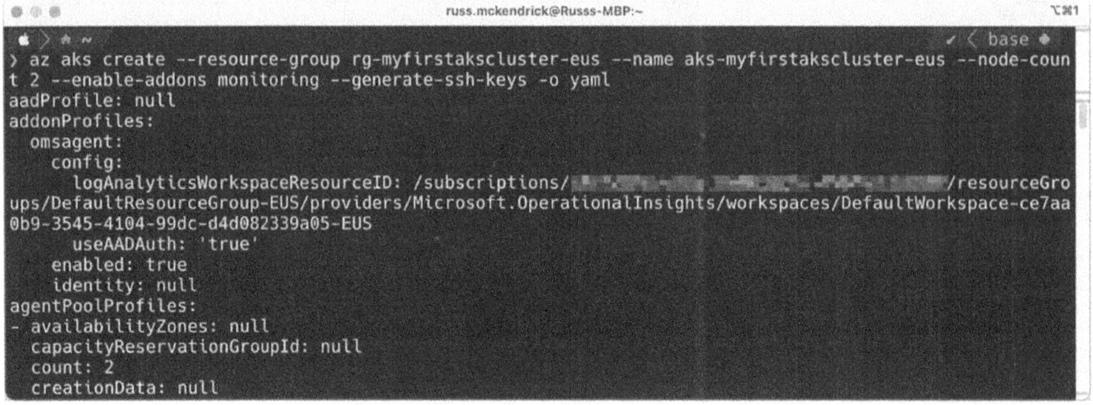

Figure 17.7: Viewing the output of the cluster launch

As you can see, there is a lot of information. We will not worry about this, though, as we will use the Azure CLI and portal to interact with the cluster rather than handcrafting API requests to the Azure Resource Manager API.

Now that our cluster is up and running, the last task we need to do before deploying our example workload is to configure our local kubectl client to interact with the cluster.

To do this, run the following command:

```
$ az aks get-credentials --resource-group rg-myfirstakscluster-eus --name aks-
myfirstakscluster-eus
```

Once this command has been run, you should see something like the following:

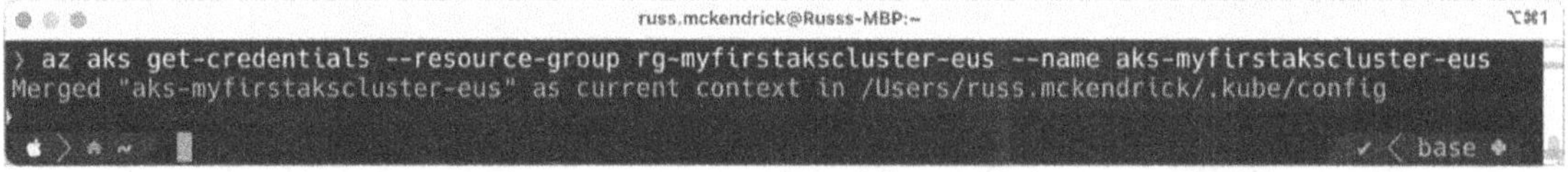

Figure 17.8: Downloading the cluster credentials and configuring kubectl

With our cluster launched and the local kubectl configured, we can now start issuing commands against the cluster, and if you have been through the previous two chapters, you will already know that the command is:

```
$ kubectl get nodes
```

This will return the nodes within the cluster, as shown in the following screenshot:

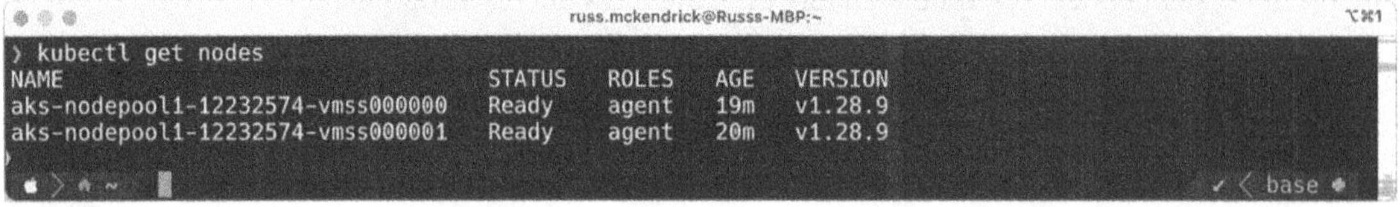

Figure 17.9: Checking the nodes are up and running

We are now able to launch the example guestbook workload that we have been using in the previous two chapters to test our cluster.

Deploying a workload and interacting with your cluster

We are going to be using the same workload we launched in *Chapter 15, Kubernetes Clusters on Google Kubernetes Engine*, and *Chapter 16, Launching a Kubernetes Cluster on Amazon Web Services with Amazon Elastic Kubernetes Service*, so I am not going to go into detail here other than to cover the commands.

Launching the workload

We start with the Redis leader deployment and service:

```
$ kubectl apply -f https://raw.githubusercontent.com/GoogleCloudPlatform/
kubernetes-engine-samples/main/quickstarts/guestbook/redis-leader-deployment.
yaml
$ kubectl apply -f https://raw.githubusercontent.com/GoogleCloudPlatform/
kubernetes-engine-samples/main/quickstarts/guestbook/redis-leader-service.yaml
```

Followed by the Redis follower:

```
$ kubectl apply -f https://raw.githubusercontent.com/GoogleCloudPlatform/
kubernetes-engine-samples/main/quickstarts/guestbook/redis-follower-deployment.
yaml
$ kubectl apply -f https://raw.githubusercontent.com/GoogleCloudPlatform/
kubernetes-engine-samples/main/quickstarts/guestbook/redis-follower-service.
yaml
```

Finally, we can launch the frontend deployment and service using the following commands:

```
$ kubectl apply -f https://raw.githubusercontent.com/GoogleCloudPlatform/
kubernetes-engine-samples/main/quickstarts/guestbook/frontend-deployment.yaml
$ kubectl apply -f https://raw.githubusercontent.com/GoogleCloudPlatform/
kubernetes-engine-samples/main/quickstarts/guestbook/frontend-service.yaml
```

Then, after a few minutes, we will be able to run the following command to get information on the frontend service:

```
$ kubectl get service frontend
```

Like the previous times, we have deployed the example workload. This will give us the public IP address we can use to access the guestbook application:

Figure 17.10: Getting information on the frontend service

Entering the IP address into a browser, making sure to use `http://<ipaddress>` as we have not configured an SSL certificate, will show the Guestbook application:

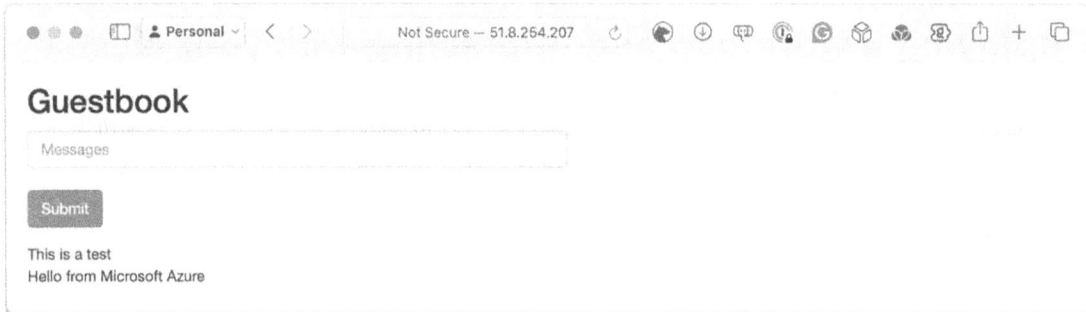

Figure 17.11: Viewing the Guestbook application

We can move to the Azure portal now that the workload is running.

Exploring the Azure portal

If you haven't already, log in to the Azure portal at `https://portal.azure.com/`. Once you are logged in, start typing **Kubernetes** into the Search resources, services, and docs search box at the top of the page.

In the list of services, you will see **Kubernetes services**. Click on this service, and you will be presented with a list of Kubernetes services running within the subscriptions your user has access to.

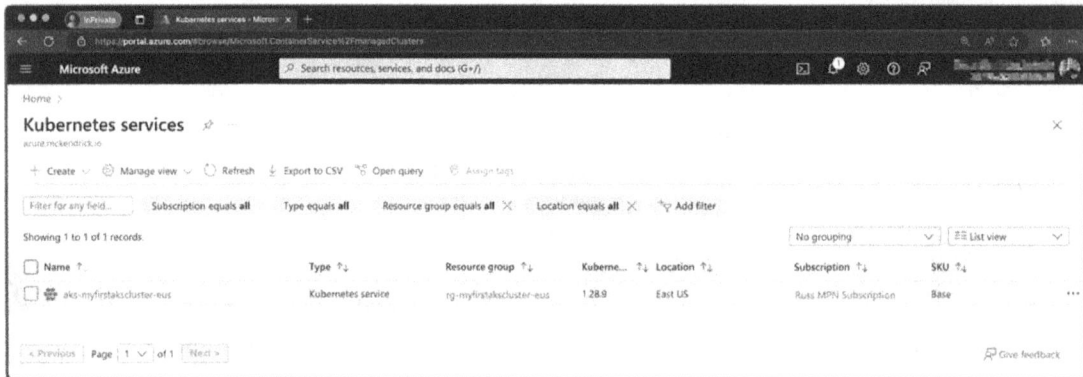

Figure 17.12: Listing the Kubernetes services

Clicking on `aks-myfirstakscluster-eus` will take you to an overview page. This will be our jumping-off point for viewing our workload and cluster information.

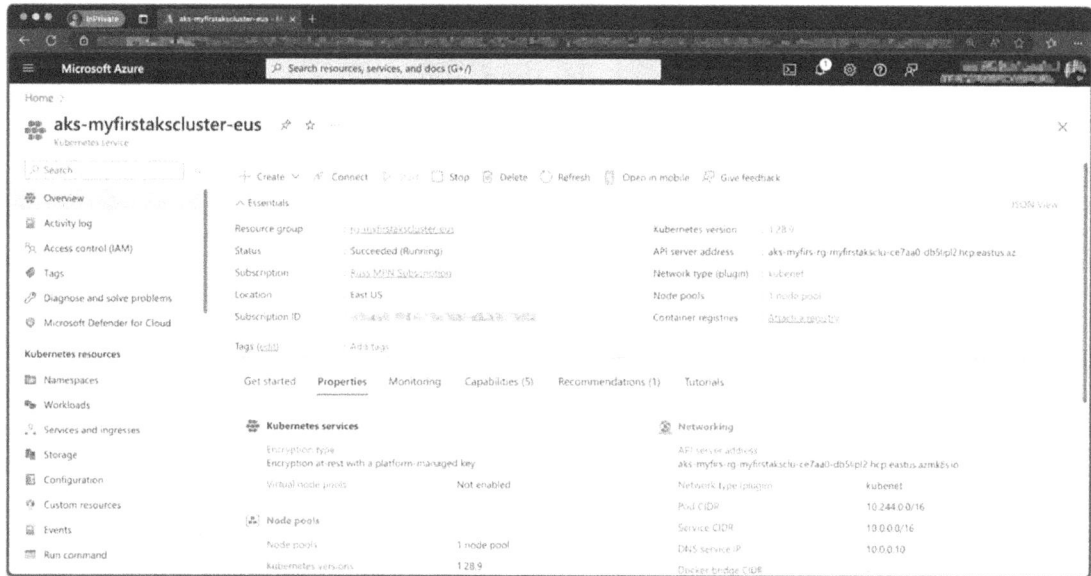

Figure 17.13: The cluster overview page

You will see several options under the Kubernetes resources menu on the left. Let's work through them one by one.

Namespaces (Kubernetes resources)

Here, you will find all the namespaces active within the cluster. As we didn't define a custom namespace when we launched our workload, our deployments and services will be listed under the `default` namespace.

In addition to the `default` namespace, there are also the ones deployed as part of the cluster: `kube-node-lease`, `kube-public`, and `kube-system`. I recommend leaving these alone.

You will be presented with the **Overview** page if you click on the default namespace. Here, you can edit the YAML that defines the namespaces, view an events log, and configure any service meshes that may have been deployed; in our test, there aren't any.

Workloads (Kubernetes resources)

As you may have already guessed, you can view information on your workload here. In the following screenshot, I have filtered the list only to show the workloads in the default namespace:

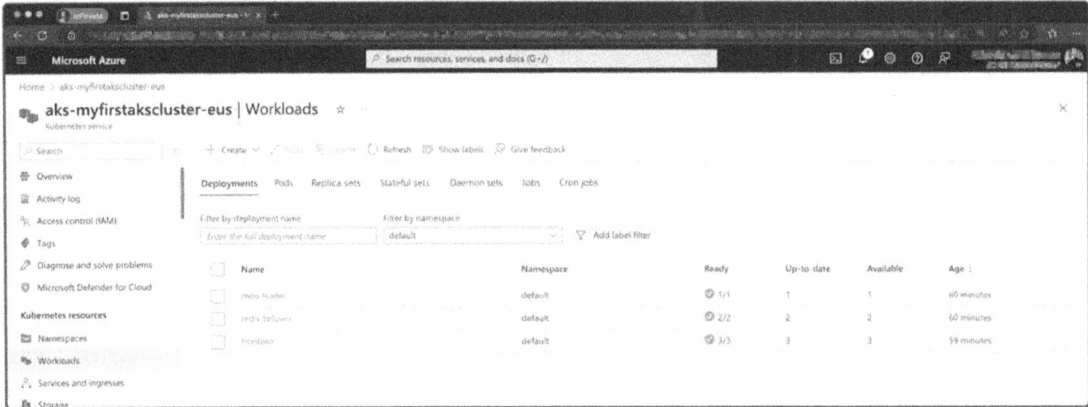

Figure 17.14: Viewing the workloads

Clicking on one of the deployments will give you a more detailed view of the deployment. For example, selecting the **frontend** deployment shows the following:

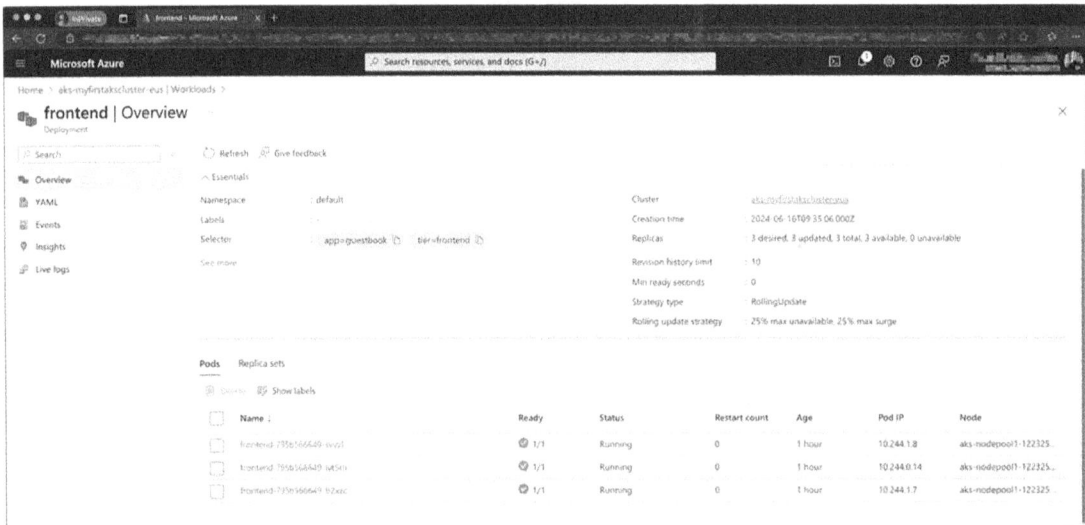

Figure 17.15: Drilling down into a deployment

As you can see from the menu on the left, there are a few additions to the options: in addition to **YAML** and **Events,** we now have the option to view **Insights.** We will cover insights in more detail at the end of this section.

The next option is **Live logs.** Here, you can select a pod and stream the logs in real time:

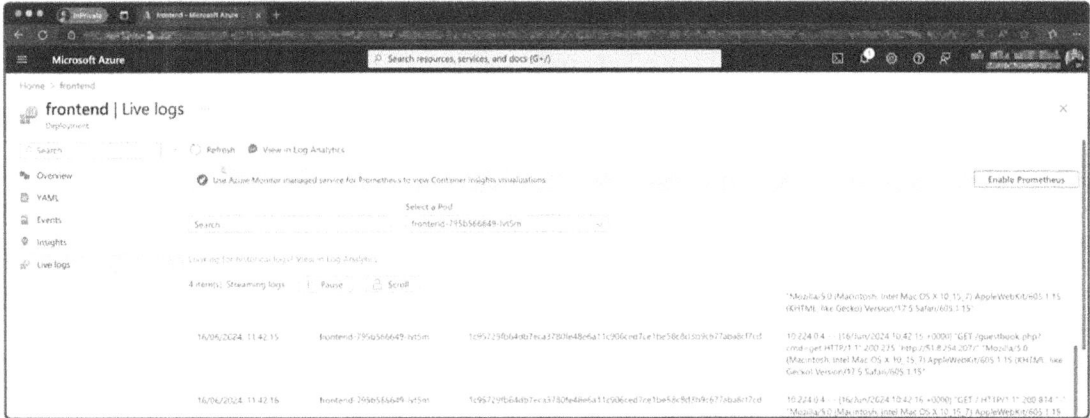

Figure 17.16: Viewing the pod logs in real time

Returning to the **Workloads** screen and selecting the **Pods** tab will give you a list of the pods that make up your workload. The IP address and the node the pod is active on are listed. This is good for getting a quick overview of your running pods.

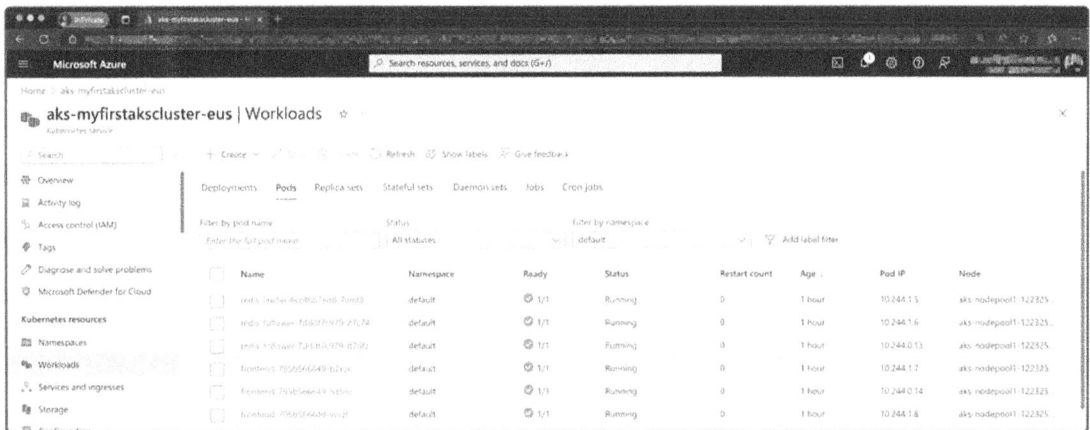

Figure 17.17: Listing all the running pods

Clicking on one of the pods will give you an overview and show you the YAML and any events for the pod.

The next tab on the workload screen is **Replica sets,** which provides useful way to see the replica sets deployed as part of your workload at a glance. Again, clicking on one of the listed replica sets gives you the now-familiar options of **Overview, YAML,** and **Events.**

Figure 17.18: Listing all the replica sets

The next tab in **Workloads** is **Stateful sets;** we don't have any stateful sets in our namespace, nor does Microsoft in the other namespaces, so there isn't much for us to see here. However, if there were, and if you were to select it, you would see the same information as we have seen on the other tabs in the **Workloads** section.

Next up, we have the **Daemon sets** tab. Again, there are no daemon sets in our workspace, but there are ones launched as part of the cluster by Microsoft that you can explore.

Finally, we have the last two tabs, **Jobs** and **Cron jobs;** here, you will find details of any jobs and cron jobs you have deployed within the cluster.

Services and ingresses (Kubernetes resources)

Here, you can find a list of all the services you have deployed in your cluster. As you can see from the following screenshot, you can get an overview of the cluster IPs used for the service along with any external IPs you have configured:

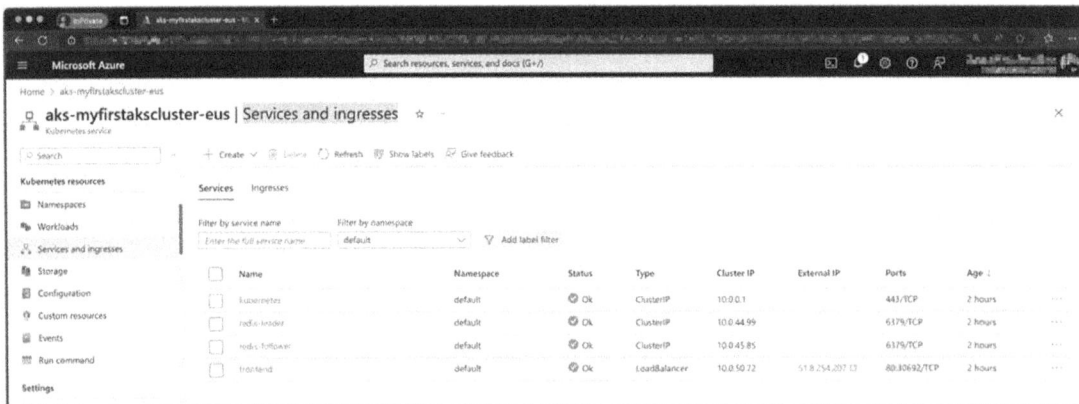

Figure 17.19: Viewing the services

Clicking on one of the services listed will provide the now-familiar view and allow you to drill deeper into the configuration of the services.

Storage (Kubernetes resources)

If we have any persistent storage configured within the cluster, you can view the details and manage them here; in our example workload, we don't, so there will be little to see.

Configuration (Kubernetes resources)

Here, you can view and edit any ConfigMaps or Secrets you have configured within the cluster. As we don't have any of these configured in our workload, the ones listed are for the cluster itself, so I wouldn't recommend making any changes to the items present.

Custom resources (Kubernetes resources)

Here, you can manage any custom resources attached to your cluster.

Events (Kubernetes resources)

Here are all the real-time events for the cluster; these events can help you monitor and troubleshoot any health issues within the cluster and application workloads.

Run command (Kubernetes resources)

This is a helpful addition; here, you can run any kubectl commands directly from the Azure portal without having to launch a cloud shell or have your local copy of kubectl configured:

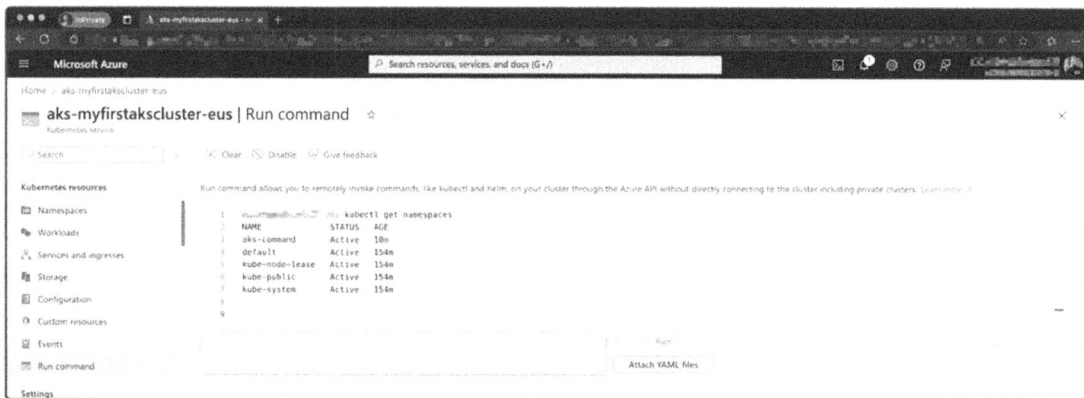

Figure 17.20: Running commands against the cluster from the portal

Moving onto the **Settings** section of the menu, we have the following.

Node pools (Settings)

Here, you will find details of your node pool and the option to upgrade the Kubernetes version running within the pool. This option is only available if you upgrade the version of Kubernetes running on the control plane.

You can also scale the pool and have the option to add a node pool. In the following screenshot, we can see what the scale option looks like:

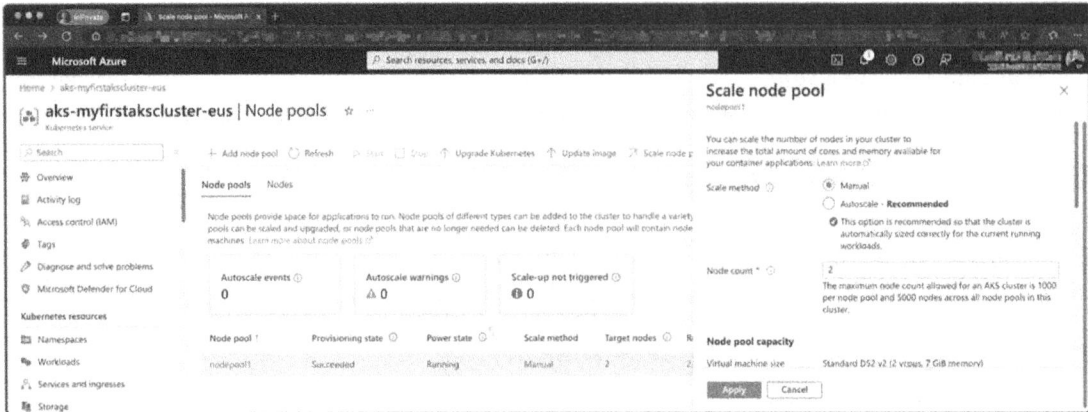

Figure 17.21: Reviewing the scaling options

You can also view information on each of the nodes, as seen on the screen below:

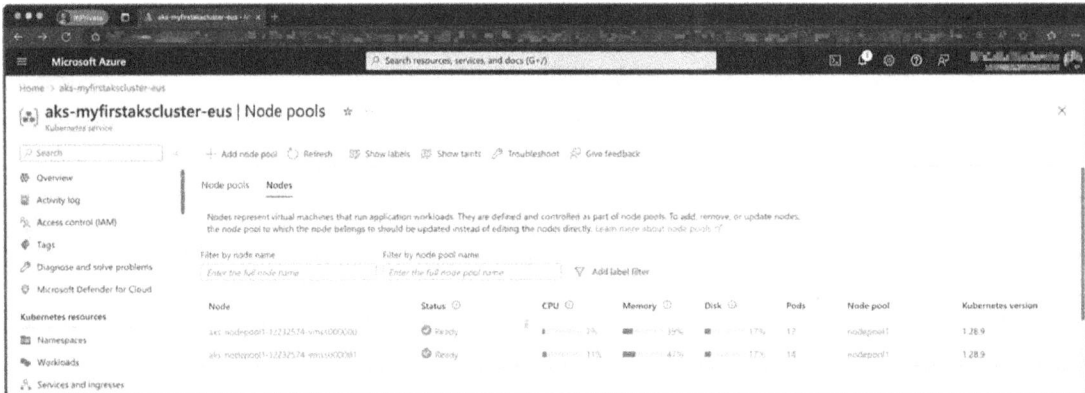

Figure 17.22: Reviewing the nodes

Cluster configuration (Settings)

In the last point, I mentioned that you can only upgrade the Kubernetes version running within your node pools if you upgrade the control plane, and this option is where you do that. Microsoft manages the control plane and separates it from your node pools.

The Kubernetes control plane provides backward compatibility for up to three releases, so you can usually only upgrade within three releases of the version you are currently running.

Application scaling (Settings)

Here, you can enable **Kubernetes Event-Driven Autoscaler** (**KEDA**), which dynamically adjusts workloads in response to events from external sources. At the time of writing, there are some limitations in the scaling events that can be supported when configuring the service through the Azure portal. It supports scaling with the following sources: Azure Service Bus, Cron, Memory, and CPU.

Networking (Settings)

In this section, you can view and manage your cluster's network settings.

Extensions + applications (Settings)

As part of Azure Marketplace, Microsoft provides first- and third-party applications you can deploy into your AKS cluster. In this section, you can manage these deployments.

Backup (Settings)

The Azure native backup services now support backing up your Kubernetes workloads and application data; this can all be managed from here.

Other options (Settings)

Several other options in the **Settings** menu allow you to configure various parts of your cluster and connect your cluster to other Azure native services. If you require any further information, links to the AKS documentation can be found in the **Further reading** section of this chapter at the end.

Insights (Monitoring)

The last part of the Azure portal we are going to look at is the **Insights** option found in the **Monitoring** menu in the cluster view. As you may recall, when we deployed our cluster, we enabled the monitoring add-on using the `--enable-addons` monitoring flag.

This enabled Microsoft's native monitoring to ship data from a resource to the Azure Log Analytics service. Once the data has been shipped to this service, Microsoft presents this information back to you, most commonly as insights. Most Azure services have an **Insights** option, and the data here can be used by Azure Monitor to create and generate alerts.

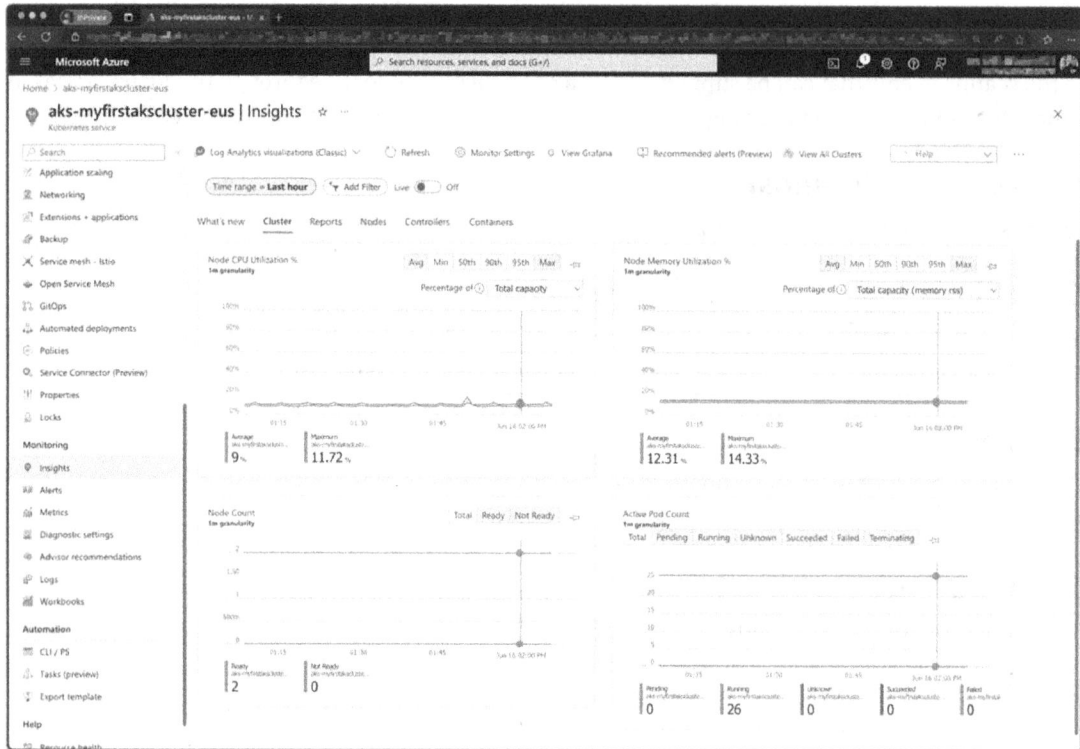

Figure 17.23: The Insights page

There are a few tabs on the **Insights** page; let's look:

- **Cluster:** This is shown in the preceding screenshot and gives you a quick view of the CPU and memory utilization across the whole cluster. It also shows both node and pod counts.

- **Reports:** Here, you can find pre-written reports on node monitoring (performance), resource monitoring (availability), billing, and networking. As the service matures, more reports will be added.

- **Nodes:** Here, you can get an overview of your nodes.

- **Controllers:** This is where you will find details on the controllers launched within your cluster – for example, replica sets and daemon sets.

- **Containers:** Here, you can find details of all the containers running on your deployed pods.

Now, you may think there is a lot of repetition in the preceding sections – and there is a little bit of that; however, if you need to see what is going on within your cluster quickly, you now have a way of getting that information without having to drill through a lot of pages to find it.

I recommend looking around and clicking on as many options as possible to explore the level of integration of your cluster and the Azure portal. Once you have finished, it is time for us to remove the cluster.

Deleting your Azure Kubernetes Service cluster

The final thing we will look at is how to delete the cluster. Moving back to the Azure CLI, all we need to run to delete the cluster is the following command:

```
$ az aks delete --resource-group rg-myfirstakscluster-eus --name aks-
myfirstakscluster-eus
```

You will be asked if you are sure, and answering yes will proceed to delete the cluster.

The process takes about five minutes. The preceding command only deletes the cluster and not the resource group. To delete the latter, run the following command:

```
$ az group delete --name rg-myfirstakscluster-eus
```

Again, you will be asked if you want to delete the group – just answer yes.

So, how much will our cluster cost to run?

Unlike the other two cloud services we examined in the previous two chapters, cluster management is free for non-production workloads, and you only pay for the compute resource.

In our case, two Standard_DS2_v2 instances in the US East region would cost around $213 per month, or if we choose the standard cluster management, it will cost $286 per month.

Other options, such as new generations of instances, could give us a similar-sized cluster for less money. For example, we could launch a different cluster using the following commands:

```
$ az group create --name rg-myfirstakscluster-eus --location eastus -o table
$ az aks create --resource-group rg-myfirstakscluster-eus --name aks-
myfirstakscluster-eus --node-count 2 --enable-addons monitoring --generate-ssh-
keys --node-vm-size standard_ds3_v2 -o yaml
```

This will give us a two-node cluster with four vCPUs and 16 GB of RAM for around $140 per month if we go for the non-production option.

Comparing the three public cloud offerings

Before we finish the chapter, let's quickly compare the three public cloud offerings:

Feature	Google Kubernetes Engine (GKE)	Amazon Elastic Kubernetes Service (EKS)	Microsoft AKS
Kubernetes version support	Latest versions, frequent updates	Slightly behind in version support	Latest versions, frequent updates

Automatic updates	Automatic for control plane and nodes	On-demand for control plane, manual for nodes	On-demand for control plane and nodes
Ease of use	High – intuitive interface	Medium – complex setup	High – intuitive interface
Integration with cloud services	Strong integration with GCP services	Strong integration with AWS services	Strong integration with Azure services
Scalability	Good, supports auto-scaling	Good, supports auto-scaling	Good, supports auto-scaling
Security features	Strong, integrates with GCP security tools	Strong, integrates with AWS security tools	Strong, integrates with Azure security tools
Pricing	Free control plane, pay for nodes	$0.10/hour for control plane and node costs	Free control plane, pay for nodes
Multi-zone cluster support	Yes	Yes	Yes
Private cluster support	Yes	Yes	Yes
Serverless compute option	Yes, using Cloud Run for Anthos	Yes, using Fargate for EKS	Yes, using AKS virtual nodes
Pricing	Free control plane, pay for nodes	Per hour cost for control plane and node costs	Free control plane, pay for nodes

Table 17.1: Comparison between the three public cloud offerings

There are some key points to note:

- GKE tends to lead in Kubernetes version support and automatic updates.
- AKS is considered by most to be the most user-friendly, especially for those already using Azure services.
- EKS charges for the control plane, while GKE and AKS only charge for the worker nodes.
- All three services offer strong integration with their respective cloud ecosystems.
- GKE is often praised for its advanced features and performance, leveraging Google's expertise as the original creator of Kubernetes.
- Each service has unique strengths: GKE in performance and features, EKS in AWS ecosystem integration, and AKS in ease of use and Azure integration.
- The best choice often depends on your existing cloud infrastructure, requirements, and familiarity with the cloud provider's ecosystem.

As you can see, your choice of which of the three services to use is very much connected with which cloud services you are already running workloads in; as we have learned in the last three chapters, once your clusters have been launched, the experience is pretty much the same.

Summary

In this chapter, we looked at how Microsoft Azure came to be, the history behind some of Microsoft's container services, and how they eventually settled on AKS.

We then signed up for an Azure account and installed and configured the Azure CLI before launching our own AKS cluster. Once it was launched, we deployed the same workload we deployed to our GKE and Amazon EKS clusters.

Once the workload was deployed, we moved onto the Azure portal and looked at the options for gaining insights into our workload and cluster and some of the cluster management options.

We finally deleted the resources we launched and discussed how much the cluster would cost to run.

Out of the three public cloud services we have examined over the last three chapters, I personally would put Microsoft Azure AKS first; it has the most rounded and feature-rich offerings, alongside its ease of use. I would put Google's offering, which we discussed in *Chapter 15, Kubernetes Clusters on Google Kubernetes Engine*, at a close second; it is good, but their pricing needs to be revised to rival Microsoft's offering.

This leaves Amazon's service, covered in *Chapter 16, Launching a Kubernetes Cluster on Amazon Web Services with Amazon Elastic Kubernetes Service*. AWS comes in as my least recommended service. It simply doesn't feel as polished as the offerings by Microsoft and Google, and it should feel like you are launching a service to complement other services offered by the cloud provider. Yet, it feels like you are running a Kubernetes cluster in AWS.

My personal opinion aside, the critical takeaway from having launched a Kubernetes cluster in three very different public cloud services is that once you have your cluster up and running and your kubectl client configured to interact with it, the experience is pretty much the same and your workload doesn't care where it is launched. You also don't have to consider the differences between the three providers – just a few years ago, this seemed like an unthinkable situation, and genuinely cloud-agnostic workloads were just a pipe dream.

In the next chapter, we are going to explore the security aspects of Kubernetes, including authentication and authorization, admission controllers, network policies, and other important topics.

Further reading

Here are links to more information on some of the topics and tools covered in this chapter:

- **Microsoft Azure:** https://azure.microsoft.com/
- **AKS:** https://azure.microsoft.com/services/kubernetes-service/
- **AKS official documentation:** https://docs.microsoft.com/en-us/azure/aks/
- **Microsoft Loves Linux:** https://www.microsoft.com/en-us/windows-server/blog/2015/05/06/microsoft-loves-linux/

- **Ballmer:** *"Linux is a cancer"*: `https://www.theregister.com/2001/06/02/ballmer_linux_is_a_cancer/`
- **Ballmer:** *I may have called Linux a cancer but now I love it*: `https://www.zdnet.com/article/ballmer-i-may-have-called-linux-a-cancer-but-now-i-love-it/`
- *Linux is Most Used OS in Microsoft Azure – over 50 percent of VM cores*: `https://build5nines.com/linux-is-most-used-os-in-microsoft-azure-over-50-percent-fo-vm-cores/`

Join our community on Discord

Join our community's Discord space for discussions with the authors and other readers:

`https://packt.link/cloudanddevops`

18

Security in Kubernetes

Authentication and authorization are the cornerstones of modern software systems in terms of providing the necessary identity management and access management, respectively. Many people confuse these two terms, despite the fact that they are quite different processes. Authentication has to do with the verification of the identity of a user, normally through some kind of mechanism like usernames and passwords, while authorization is all about what an authenticated user can access or do within a system. Authentication always comes first, after which authorization would take place in order for the system to be interacted with by verified users. Kubernetes extends this further with another model called **Role-Based Access Control (RBAC)**, which allows an administrator to define roles with certain privileges and then assign those roles to users, hence effectively implementing the principle of least privilege and allowing fine-grained access control.

Apart from Identity and Access Management, Kubernetes has a number of other security mechanisms to harden the rest of the components further. Being the most mature and widely adopted container orchestration platform, the design of Kubernetes places a lot of emphasis on the security of a wide range of components within clusters, nodes, containers, networks, and applications through the mitigation of risks at many layers.

Next, this chapter goes into some of the basic Kubernetes security concepts, from the different ways in which the system can flexibly authenticate-in X509 client certificates or tokens from OpenID Connect. In specialized cases, for example, the integration with LDAP, Kubernetes provides additional options. For example, the possibility of using an authenticating proxy or webhooks is also recommended. Then we will review the **RBAC** model from the platform that gives administrators control over access to resources in the cluster and allows them to manage users and groups along with ServiceAccounts.

We will also introduce one advanced feature in Kubernetes: Admission Controllers. An Admission Controller enforces security policies at the point of resource admission to validate and control resources before they enter the cluster. Admission controllers provide an additional layer of defense by governing resource requests through the enforcement of policies on the creation and modification of resources.

Pods and containers themselves need to be secured, as these are the runtimes of the workloads or applications that could interact with sensitive information. Kubernetes provides a set of `securityContext` options that enable administrators to declare particular security settings for containers; this includes forcing containers to run as non-root. Equally important will be network security, and we'll discuss how NetworkPolicies provide a mechanism to segregate and secure pod communication inside the cluster by controlling traffic flow at a granular level.

We'll then move to container runtime security. We will look at gVisor and Kata Containers as options for runtime, which introduce more security boundaries between either a user-space kernel to intercept system calls or a lightweight VM environment per container, respectively, which provides the speed of containers but the security of a VM.

Lastly, and most importantly, private registry credentials hold the key to guaranteeing security around container images inside the cluster. We will go through how Kubernetes handles these credentials safely – ensuring that only authorized components have access to them. By the end of this chapter, you will have a deeper understanding of these advanced security concepts and tools in Kubernetes. You will see precisely how to enhance your cluster's security posture, reduce risks, and have the best defense against possible vulnerabilities. With such measures, you will be able to secure your Kubernetes deployments at every layer, from identity management right through to runtime isolation, and reinforce the robustness of your containerized applications.

In this chapter, we will cover the following topics:

- Authentication and Authorization – User Access Control
- Admission Control – Security Policies and Checks
- Securing Pods and Containers
- Managing Secrets and Registry Credentials

Technical Requirements

For this chapter, you will need the following:

- A Kubernetes cluster to be deployed. We recommend using a multi-node or cloud-based Kubernetes cluster.
- The Kubernetes CLI (`kubectl`) installed on your local machine and configured to manage your Kubernetes cluster.

Basic Kubernetes cluster deployment (local and cloud-based) and `kubectl` installation were covered in *Chapter 3, Installing Your First Kubernetes Cluster*.

The previous chapters (*15*, *16*, and *17*) of this book have provided you with an overview of how to deploy a fully functional Kubernetes cluster on different cloud platforms and install the requisite CLIs to manage them.

You can download the latest code samples for this chapter from the official GitHub repository: `https://github.com/PacktPublishing/The-Kubernetes-Bible-Second-Edition/tree/main/Chapter18`.

Authentication and Authorization – User Access Control

It gives grounds for access control in ensuring that only authenticated and authorized users can use Kubernetes resources. Authentication verifies the identity of a user, while authorization decides what an authenticated user is allowed to do within the cluster. Kubernetes offers flexibility in authenticating via X509 certificates, OpenID Connect, token-based, and other approaches. Coupled with the verification process, RBAC does offer fine granular control over what users can do, thus helping administrators manage a wide range of permissions efficiently – a subject that will be dealt with in further detail in the following sections.

Let us start with authentication in the next section.

Authentication and User Management

The Kubernetes API server provides RESTful endpoints for managing the Kubernetes cluster and acts as the frontend to the shared state of the cluster. All interactions with the cluster, from users to internal components, are channeled through the Kubernetes API server, which acts as a frontend to the cluster's shared state.

Let us see how the authentication mechanism works in Kubernetes in the next sections.

The authentication workflow in Kubernetes

Just like a high-security facility, your Kubernetes cluster needs robust security measures to protect its resources. This involves a layered approach with several key components working together, as shown in the following figure:

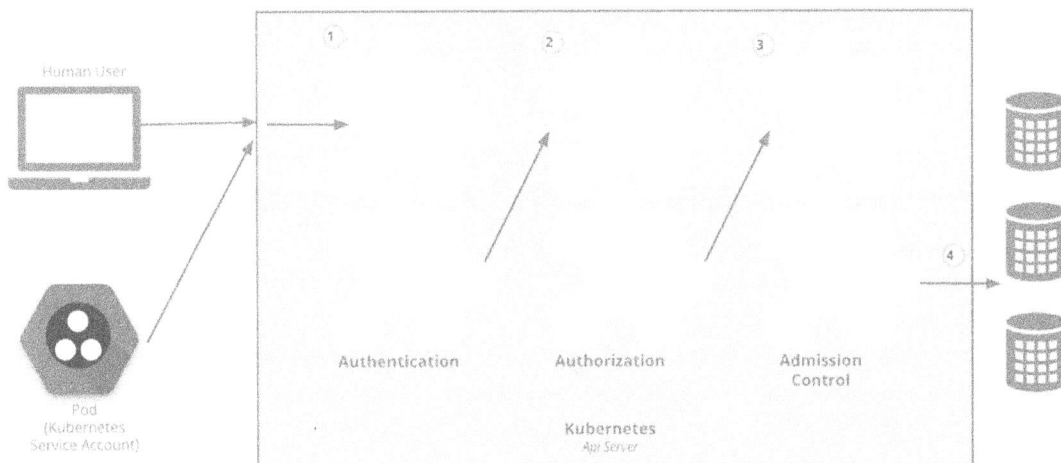

Figure 18.1: Request to Kubernetes API goes through several stages (source: https://kubernetes.io/ docs/concepts/security/controlling-access/)

- **Authentication:** This acts as the first line of defense, verifying the identity of anyone trying to access the Kubernetes API server. Imagine it like a security guard checking IDs at the entrance. Users might use passwords, tokens, or special certificates to prove they're authorized.

- **Authorization:** Once someone's identity is confirmed, authorization determines what they can actually do within the cluster. Think of it as granting specific access levels. Users might have permission to view resources, but not modify them, or they might be authorized to create new resources but only in specific areas.

- **Admission Control:** This stage adds an extra layer of scrutiny. Imagine it like a security scanner at the entrance. Admission control modules can inspect incoming requests, ensuring they comply with predefined security policies. They can even modify requests to enforce specific rules or reject them entirely if they pose a threat.

- **Auditing:** Just like keeping a log of who enters and exits a secure facility, auditing in Kubernetes keeps a record of all activity within the cluster. This includes actions taken by users, applications, and even the control plane itself. These logs are invaluable for monitoring suspicious activity and maintaining a secure environment.

By working together, these security measures create a layered defense system, ensuring that only authorized users can access your Kubernetes cluster and that their actions comply with established security policies.

We will learn some more details about the authentication mechanism in the next section.

Authentication to the Kubernetes API

Kubernetes API authentication ensures that only authorized users or services are allowed to talk to the resources running in a cluster. Each incoming request goes through an authentication setup, which is done in a chain of authenticator modules that are configured.

Requests to the API are always one of the following:

- Associated with an external, normal user or a **ServiceAccount** defined in the Kubernetes cluster.

- Treated as *anonymous* requests if the cluster has been configured to allow anonymous requests.

This is determined in the *authentication* process – the entire HTTP request is used as input to the process, but usually only request headers or the client certificate is analyzed. Authentication is carried out by authentication modules that depend on the cluster configuration. Your cluster may have multiple authenticator modules enabled, and then each of them is executed in sequence until one succeeds. If the request fails to authenticate, the API server will either respond with an HTTP status code of 401 (unauthorized) or, if anonymous requests are enabled, treat it as anonymous.

> Anonymous requests are essentially mapped to a special username called system:anonymous and a group called system:unauthenticated. This means that you can organize your authorization to resources for such requests, just as you can for other users or Service-Accounts.

Since all operations inside and outside the cluster must go through the Kubernetes API server, this means that all of them must go through the authentication process. This includes the operations of internal cluster components and Pods, which query the API server. For you, as an external user of the cluster, any requests that you make using kubectl commands or directly to the Kubernetes API server will also go through the authentication process:

- **Normal users:** Such users are managed *externally*, independent from the Kubernetes cluster. Currently, Kubernetes does not provide any objects to represent such users. The external management of users may be as simple (but *not* recommended) as static user-password files passed to the API server using the token-auth-file argument in the static Pod definition file /etc/kubernetes/manifests/kube-apiserver.yaml inside your control plane nodes (AKA master nodes) during startup. For production environments, leverage existing **identity providers (IdPs)** like **Google, GitHub, Azure Active Directory (AAD)**, or **AWS IAM** to manage users. Integrate your Kubernetes cluster with these IdPs using **OpenID Connect (OIDC** – https:// openid.net/connect/) tokens for a seamless authentication experience. Remember, regular user accounts in Kubernetes are global and don't have namespace restrictions.

- **Service accounts:** These are managed by the Kubernetes cluster and modeled as ServiceAccount objects. You can create and manage service accounts just like any other resource in Kubernetes, for example, using kubectl and YAML manifest files. This type of account is intended for processes in cluster components or running in Pods. The credentials for ServiceAccounts will be created as Secrets (manually or via TokenRequest API) that are mounted into Pods so that the container process can use them to talk to the Kubernetes API server. When a process authenticates using a ServiceAccount token, it is seen as a user called system:serviceaccount:<namespace>:<serviceAccountName>. Note that ServiceAccounts are namespaced.

As you can see, user management in Kubernetes is a mixture of different approaches that should fit all the needs of different organizations. The key takeaway here is that after the authentication process, the request will be either rejected (optionally treated as anonymous) or treated as coming from a particular user. The username attribute may be provided by the external user management system, as in the case of normal users, or it will be system:serviceaccount:<namespace>:<serviceAccountName> for ServiceAccounts. Additionally, the request will have more attributes associated with it, such as **User ID (UID)**, **groups**, and **extra fields**. This information is used for authorization processes based on RBAC, which we will explain in the next sections.

Now, let's look at the authentication methods that you can use with Kubernetes.

Authentication Methods in Kubernetes

The various authentications, in general, help in securely controlling access in Kubernetes to the API server. To validate users and services, a variety of authentication strategies can be enabled. Each is suited to different use cases and levels of security. These include tokens and certificates that verify the identities of both human users and applications interacting with the cluster. The good thing about the Kubernetes API server is that it provides support for multiple authentication mechanisms, so clusters can be configured using a combination of the previously-mentioned methods. In the following section, we will present some common authentication methods such as Static Token files, ServiceAccount tokens, X.509 client certificates, and OpenID Connect tokens.

Static token files

This method is the most basic one that Kubernetes offers for managing normal users. The approach somewhat resembles the /etc/shadow and /etc/passwd files in Unix/Linux systems. Note, however, that it is *not* recommended and is considered *unsecure* for production clusters.

In this method, you define a .csv file where each line has the following format:

```
token,user,uid,"group1,group2,group3"
```

Then, you pass the file when starting the Kubernetes API server process using the token-auth-file parameter in the static Pod definition file /etc/kubernetes/manifests/kube-apiserver.yaml inside your control plane nodes (AKA master nodes):

```
# /etc/kubernetes/manifests/kube-apiserver.yaml
...<removed for brevity>...
spec:
  containers:
  - command:
    - kube-apiserver
    - --advertise-address=192.168.59.154
    - --allow-privileged=true
    - --authorization-mode=Node,RBAC
    - --token-auth-file=/etc/kubernetes/user-tokens.csv
    - --client-ca-file=/var/lib/minikube/certs/ca.crt
...<removed for brevity>...
```

To authenticate against the API server, you need to use a standard HTTP **bearer authentication scheme** for your requests. This means that your requests will need to use an additional header that's in the following form:

```
Authorization: Bearer <token>
```

Based on this request information, the Kubernetes API server will match the token against the static token file and assign user attributes based on the matched record.

When using kubectl, you must modify your kubeconfig. You can do this using the kubectl command:

```
$ kubectl config set-credentials <contextUser> --token=<token>
```

After that, you need to create and use context with this user for your requests using the kubectl config use-context command.

> In Kubernetes versions prior to 1.19, there was a similar authentication method that allowed us to use an HTTP **basic authentication scheme** and a file passed by the basic-auth-file parameter to the API server. This method is no longer supported due to security reasons.

The following diagram visualizes the principles behind this method of authentication:

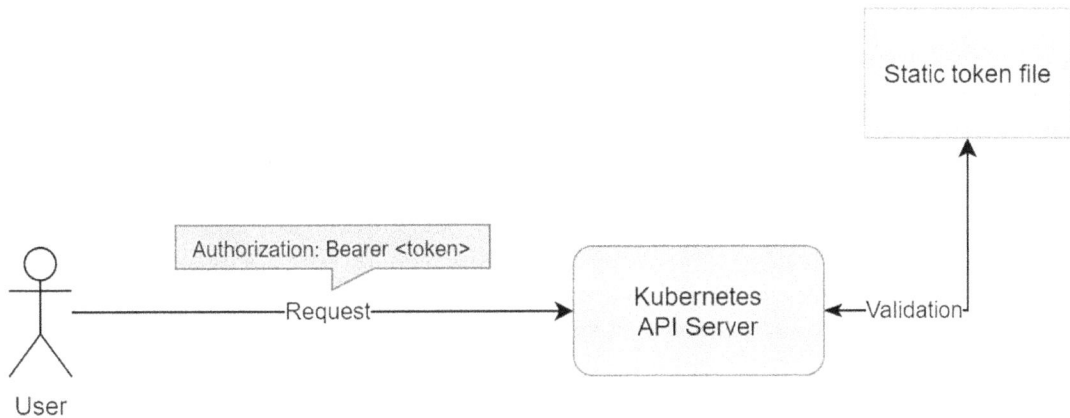

Figure 18.2: Static token file authentication in Kubernetes

We can now summarize the advantages and disadvantages of using the static token file method for authentication.

The advantages of the static token file method are as follows:

- It is easy to configure.
- It is easy to understand.

The disadvantages of the static token file method are as follows:

- It is unsecure; exposing a token file compromises all cluster users.
- It requires that we manually manage users.
- Adding new users or removing existing ones requires that we restart the Kubernetes API server.
- Rotating any tokens requires that we restart the Kubernetes API server.
- It takes extra effort to replicate the Token file content to every control plane node when you have a high availability control plane with multiple control plane nodes.

In short, this method is good for development environments and learning the principles behind authentication in Kubernetes, but it is not recommended for production use cases. Next, we will take a look at authenticating users using ServiceAccount tokens.

ServiceAccount tokens

As we mentioned in the introduction to this section, ServiceAccounts are meant for in-cluster identities for processes running in Pod containers or for cluster components. However, they can be used for authenticating external requests as well.

ServiceAccounts are Kubernetes objects and can be managed like any other resource in the cluster; that is, by using kubectl or raw HTTP requests to the API server. The tokens for ServiceAccounts are **JSON Web Tokens (JWTs)** and will be generated on-demand or using the kubectl create token command.

Every Kubernetes namespace has a pre-created ServiceAccount named `default`. Pods without a specified ServiceAccount automatically inherit this default account for authorization within the cluster. You can verify a Pod's ServiceAccount using `kubectl get pods/<podname> -o yaml` and checking the `spec.serviceAccountName` field.

Usually, when defining a Pod, you can specify what ServiceAccount should be used for processes running in the containers. You can do this using `.spec.serviceAccountName` in the Pod specification. The JWT token will be injected into the container; then, the process inside can use it in the HTTP bearer authentication scheme to authenticate to the Kubernetes API server. This is only necessary if it interacts with the API server in any way, for example, if it needs to discover other Pods in the cluster. We have summarized this authentication method in the following diagram:

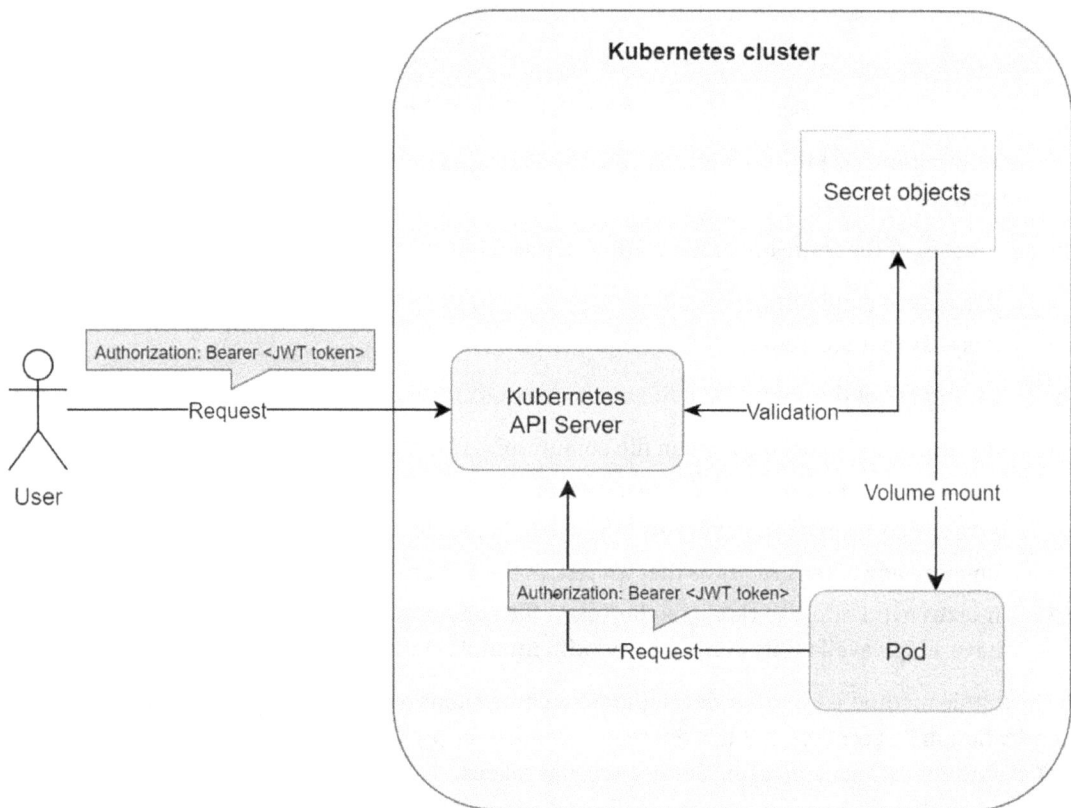

Figure 18.3: ServiceAccount authentication in Kubernetes

This also shows why ServiceAccount tokens can be used for external requests – the API server does not care about the origin of the request; all it is interested in is the bearer token that comes with the request header. Again, you can use this token in `kubectl` or in raw HTTP requests to the API server. Please note that this is generally not a recommended way to use ServiceAccounts, but it can be used in some scenarios, especially when you are unable to use an external authentication provider for normal users.

Prior to version 1.22, Kubernetes automatically generated API credentials for ServiceAccounts using Secrets. These Secrets contained tokens that Pods could mount for access. This approach had limitations:

- **Static Tokens:** Secrets stored tokens in plain text, posing a security risk if compromised.
- **Limited Control:** Token lifespans and permissions were not easily managed.

Starting from version 1.22, Kubernetes switched to a more secure approach. Pods now obtain tokens directly using the **TokenRequest** API. These tokens are as follows:

- **Short-lived:** Tokens have limited lifespans, reducing the impact of potential compromise.
- **Mounted into Pods:** Tokens are automatically mounted as volumes, eliminating the need for pre-stored Secrets.

While automatic mounting is preferred, you can still manually create Secrets for service account tokens. This might be useful for tokens requiring longer lifespans, but it's important to prioritize automatic token mounting for enhanced security in most scenarios.

> As we learned, Kubernetes automatically mounts Service Account API credentials within Pods for streamlined access. To disable this behavior and manage tokens differently, set `automountServiceAccountToken: false` either in the ServiceAccount manifest or within the Pod specification. This setting applies to all Pods referencing the ServiceAccount unless overridden by the specific Pod configuration. If both are defined, the Pod's setting takes precedence. Refer to the documentation for more details (`https://kubernetes.io/docs/tasks/configure-pod-container/configure-service-account/#opt-out-of-api-credential-automounting`).

We will now demonstrate how you can create and manage ServiceAccounts and how you can use JWT tokens to authenticate when using `kubectl`. This will also give a sneak peek into RBAC, which we are going to look at in more detail in the next section. Please follow these steps:

1. Create a YAML manifest for a new Namespace and a ServiceAccount as follows. We will configure RBAC for this account so that it can only read Pods in that namespace:

```yaml
# 01_serviceaccount/example-sa-ns.yaml
---
apiVersion: v1
kind: Namespace
metadata:
  name: example-ns


---
apiVersion: v1
kind: ServiceAccount
metadata:
  name: example-sa
  namespace: example-ns
```

Note that you can also use the *imperative* command `kubectl create serviceaccount example-sa` to create the resources.

2. Create a YAML manifest for a `Role` object named `pod-reader` in the `example-ns` namespace. This role will allow you to get, watch, and list Pods in this namespace. The `01_serviceaccount/pod-reader-role.yaml` YAML manifest file has the following contents:

```yaml
# 01_serviceaccount/pod-reader-role.yaml
apiVersion: rbac.authorization.k8s.io/v1
kind: Role
metadata:
  namespace: example-ns
  name: pod-reader
rules:
  - apiGroups: [""]
    resources: ["pods"]
    verbs: ["get", "watch", "list"]
```

3. Create a YAML manifest for `RoleBinding` named `reads-pods`. This is what *associates* the role that we created with our `example-sa` ServiceAccount – the account will now have the privilege of read-only access to Pods, and nothing more. The `01_serviceaccount/read-pods-rolebinding.yaml` YAML manifest file has the following contents:

```yaml
# 01_serviceaccount/read-pods-rolebinding.yaml
apiVersion: rbac.authorization.k8s.io/v1
kind: RoleBinding
metadata:
  name: read-pods
  namespace: example-ns
subjects:
  - kind: ServiceAccount
    name: example-sa
    namespace: example-ns
roleRef:
  kind: Role
  name: pod-reader
  apiGroup: rbac.authorization.k8s.io
```

4. Now, we can apply all the manifest files to the cluster at once using the `kubectl apply` command:

```
$ kubectl apply -f 01_serviceaccount/
namespace/example-ns created
serviceaccount/example-sa created
role.rbac.authorization.k8s.io/pod-reader created
```

```
rolebinding.rbac.authorization.k8s.io/read-pods created
```

5. Now, we will create a Token for the ServiceAccount as follows:

```
$ kubectl create token example-sa -n example-ns
```

Collect the JWT token from the command output, which you can use to authenticate as that ServiceAccount. If you are interested, you can inspect the contents of the JWT using `https://jwt.io/` as shown in the following figure:

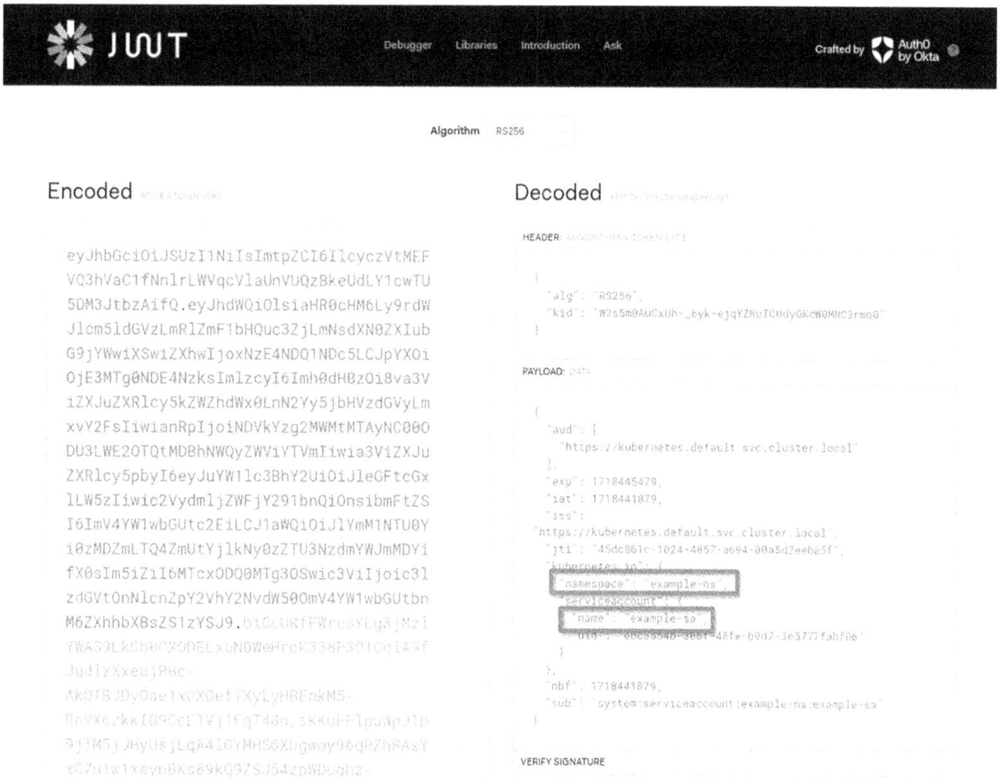

Figure 18.4: Inspecting a JWT for ServiceAccount

As you can see, the JWT maps to the `example-sa` ServiceAccount in the `example-ns` namespace. Additionally, you can identify that the actual username (marked as a `subject` in the payload) that will be mapped to in Kubernetes is `system:serviceaccount:example-ns:example-sa`, as we explained previously.

6. With this token, we can set up kubeconfig to test it. First, you need to create a user in your kubeconfig using the following command:

```
$ kubectl config set-credentials example-sa --token=<your-token>
User "example-sa" set.
```

Where the `example-sa` is the new ServiceAccount you have created and also replace `<your-token>` with the token string you collected earlier.

7. Create a new context that uses this user in the kubeconfig. You also need to know the cluster name that you are connecting to right now – you can check it using the kubectl config view command. Use the kubectl config set-context command to create a new context:

```
$ kubectl config set-context <new-context-name> --user=<new-user-created
--cluster=<clusterName>
```

For example, use the following command to create a new context named example-sa-context with minikube as the target cluster and example-sa as the user:

```
$ kubectl config set-context example-sa-context --user=example-sa
--cluster=minikube
Context "example-sa-context" created.
```

8. Before we switch to the newly created context, let us create a simple nginx Pod in the example-ns namespace. Copy the sample YAML Chapter18/references/sa-demo-nginx-pod.yaml to Chapter18/01_serviceaccount/nginx-pod.yaml and apply the configuration:

```
$ cp references/sa-demo-nginx-pod.yaml 01_serviceaccount/nginx-pod.yaml
$ kubectl apply -f 01_serviceaccount/nginx-pod.yaml
pod/nginx-pod created
$  kubectl get po -n example-ns
NAME           READY    STATUS     RESTARTS    AGE
nginx-pod      1/1      Running    0           12m
```

9. Also, before you switch to the new context, you may want to check the name of the context that you are currently using by utilizing the kubectl config current-context command. This will make it easier to go back to your old cluster admin context:

```
$ kubectl config current-context
minikube
```

10. Now, switch to the new context using the following command:

```
$ kubectl config use-context example-sa-context
Switched to context "example-sa-context".
```

11. You can also verify the identity of the credential you are currently using as follows:

```
$ kubectl auth whoami
ATTRIBUTE                                                VALUE
Username
system:serviceaccount:example-ns:example-sa
UID                                                      ebc5554b-306f-48fe-
b9d7-3e5777fabf06
Groups
[system:serviceaccounts system:serviceaccounts:example-ns
system:authenticated]
```

```
Extra: authentication.kubernetes.io/credential-id    [JTI=45dc861c-1024-
4857-a694-00a5d2eeba5f]
```

12. We are now ready to verify that our authentication works and that the RBAC roles allow read-only access to Pods in the `example-ns` namespace. First, try getting Pods:

```
$ kubectl get po -n example-ns
NAME         READY   STATUS    RESTARTS   AGE
nginx-pod    1/1     Running   0          18m
```

13. This worked as expected! Now, try getting Pods from the `kube-system` namespace:

```
$ kubectl get pods -n kube-system
Error from server (Forbidden): pods is forbidden: User
"system:serviceaccount:example-ns:example-sa" cannot list resource "pods"
in API group "" in the namespace "kube-system"
```

14. We have authenticated correctly, but the action was forbidden by RBAC authorization, which is what we expected. Lastly, let's try getting Service objects:

```
$ kubectl get svc -n example-ns
Error from server (Forbidden): services is forbidden: User
"system:serviceaccount:example-ns:example-sa" cannot list resource
"services" in API group "" in the namespace "example-ns"
```

This is also expected as the RBAC is not configured for the ServiceAccount to view or list the Service resources in the `example-ns` namespace.

As you can see, we have successfully used our ServiceAccount token for authentication and we have verified that our privileges work correctly. You can now switch back to your old `kubectl` context using the `kubectl config use-context <context-name>` command.

> The preceding procedure of configuring the `kubectl` context with a bearer token can be used for the static token file authentication method as well.

Let's summarize what the advantages and disadvantages of using ServiceAccount tokens for authentication are.

The advantages of using ServiceAccount tokens are as follows:

- Easy to configure and use, similar to static token files.
- Entirely managed by the Kubernetes cluster, so there's no need for external authentication providers.
- ServiceAccounts are namespaced.

The disadvantages of using ServiceAccount tokens are as follows:

- ServiceAccounts are intended for processes running in Pod containers to give them identity and let them use Kubernetes RBAC. *It is not a best practice for a user to use the ServiceAccount token.*

In general, using ServiceAccount tokens for external authentication is only good for development and test scenarios when you cannot integrate with external authentication providers. However, for production clusters, it is not the best option, mainly due to security concerns. Now, let's take a look at using X.509 client certificates for Kubernetes API authentication.

X.509 client certificates

Using X.509 client certificates is one of the industry standards for authentication processes. There is one important catch, however – you need to have good means of managing certificate signing, revoking, and rotation. Otherwise, you may hit very similar security issues as with using ServiceAccount tokens. You can learn more about X.509 certificates and the processes around them at `https://www.ssl.com/faqs/what-is-an-x-509-certificate/`.

This method works in Kubernetes as follows:

1. The Kubernetes API server starts with the `client-ca-file` argument. This provides **certificate authority** (CA) information to be used to validate client certificates presented to the API server. You can configure a custom CA certificate here or use the default CA created as part of the cluster deployment. For example, if you are using minikube, you can see a default CA file already configured in `kube-apiserver` as follows:

```
# /etc/kubernetes/manifests/kube-apiserver.yaml
...<removed for brevity>...
spec:
  containers:
  - command:
    - kube-apiserver
    - --advertise-address=192.168.59.154
    - --allow-privileged=true
    - --authorization-mode=Node,RBAC
    - --client-ca-file=/var/lib/minikube/certs/ca.crt
...<removed for brevity>...
```

2. Users who want to authenticate against the API server need to request an X.509 client certificate from the CA. This should be a secure and audited process. The subject common name (the `CN` attribute in the subject) of the certificate is used as the `username` attribute when authentication is successful. Note that as of Kubernetes 1.19, you can use the Certificates API to manage signing requests. More information is available in the official documentation: `https://kubernetes.io/docs/reference/access-authn-authz/certificate-signing-requests/`.

3. The user must present the client certificate during authentication to the API server, which validates the certificate against the CA. Based on that, the request goes through the authentication process successfully or is rejected. Again, if you are using a minikube cluster, you are already utilizing the certificate-based authentication, as shown in the following example:

```
$ kubectl config view -o json | jq '.users[]'
{
  "name": "example-sa",
  "user": {
    "token": "REDACTED"
  }
}
{
  "name": "minikube",
  "user": {
    "client-certificate": "/home/iamgini/.minikube/profiles/minikube/
client.crt",
    "client-key": "/home/iamgini/.minikube/profiles/minikube/client.key"
  }
}
```

While using the kubectl commands, users can configure this method of authentication in kubeconfig using the kubectl config set-credentials command, as we learned earlier. We have summarized this process in the following diagram:

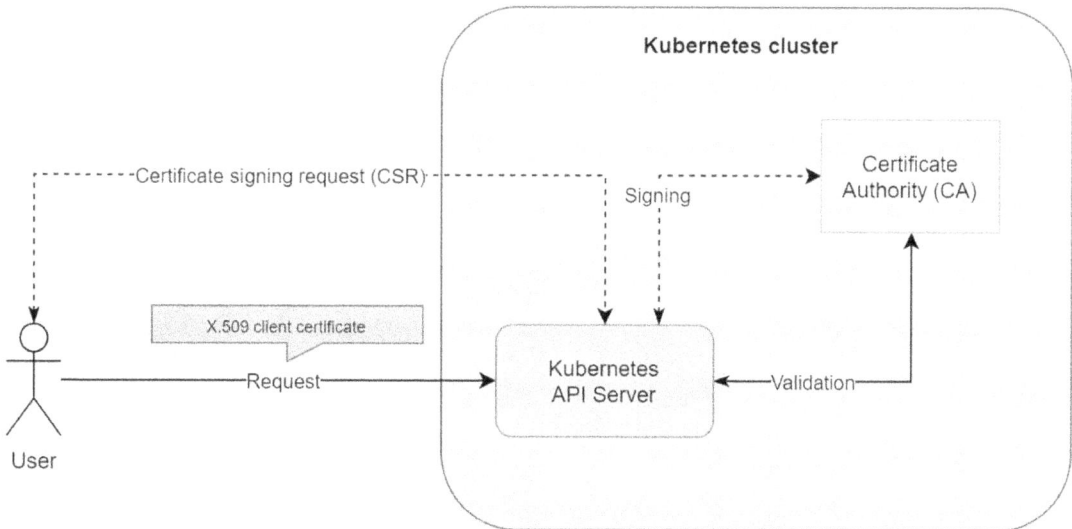

Figure 18.5: X.509 client certificate authentication in Kubernetes

Please note that this visualizes the case when initial CSR by the user is handled by the Certificate API in a Kubernetes cluster. This does not need to be the case as CA may be external to the cluster, and the Kubernetes API server can rely on a copy of the CA .pem file.

In the following hands-on exercise, we will generate and configure certificate-based authentication in Kubernetes:

1. Start with creating a private key using the `openssl` command:

    ```
    $ openssl genrsa -out iamgini.key 2048
    ```

2. Generate a **CertificateSigningRequest (CSR)**:

    ```
    $ openssl req -new -key iamgini.key -out iamgini.csr -subj "/CN=iamgini/
    O=web1/O=frontend"
    ```

3. Gather the CSR data and encode it using base64:

    ```
    $ cat iamgini.csr | base64 -w 0
    ```

4. Now, we need to create a `CertificateSigningRequest` resource with **Certificates API**; let us use the `csr.yaml` file as follows:

    ```yaml
    # csr.yaml
    apiVersion: certificates.k8s.io/v1
    kind: CertificateSigningRequest
    metadata:
      name: iamgini
    spec:
      request: <your encoded CSR content here from Step.3>
      signerName: kubernetes.io/kube-apiserver-client
      usages:
        - client auth
    ```

5. Create the `CertificateSigningRequest`:

    ```
    $ kubectl apply -f csr.yaml
    certificatesigningrequest.certificates.k8s.io/iamgini created
    ```

6. Now the administrators (or the users with the `certificatesigningrequests` privilege) can see the CSR resources:

    ```
    $ kubectl get csr
    NAME        AGE    SIGNERNAME                                REQUESTOR
    REQUESTEDDURATION    CONDITION
    iamgini     25s    kubernetes.io/kube-apiserver-client       minikube-user
    <none>              Pending
    ```

7. Check and approve the CSR as follows:

    ```
    $ kubectl certificate approve iamgini
    certificatesigningrequest.certificates.k8s.io/iamgini approved
    ```

8. Once the CSR is approved, gather the certificate data from the approved CSR resource as follows; the following command will extract the data to `iamgini.crt` file:

```
$ kubectl get csr iamgini -o json | jq -r '.status.certificate' | base64
--decode > iamgini.crt
```

9. Now, we have the private key and certificate as follows (you can delete the `.csr` file as it is not required anymore):

```
$ ls iamgini.*
iamgini.crt   iamgini.csr   iamgini.key
```

10. Now, we will configure the kubeconfig with our new user and context; create a new user entry in the kubeconfig as follows (remember to use the full path of the key and certificate file):

```
$ kubectl config set-credentials iamgini --client-key=/full-path/iamgini.
key --client-certificate=/full-path/iamgini.crt
User "iamgini" set.
```

11. Create a new context with the new user:

```
$ kubectl config set-context iamgini --cluster=minikube --user=iamgini
Context "iamgini" created.
```

12. Now, the kubeconfig is updated with the new user and context. Let us test the access. Change the kubeconfig context as follows:

```
$ kubectl config use-context iamgini
Switched to context "iamgini".
```

13. Verify the context and connection:

```
$ kubectl auth whoami
ATTRIBUTE     VALUE
Username      iamgini
Groups        [web1 frontend system:authenticated]
```

Congratulations; you have configured a new user with X509 certificate-based authentication. But remember, the user will not be able to do any kind of operation until you configure the appropriate RBAC resources.

Based on what we have learned, we can summarize the advantages of this method as follows:

- It's a much more secure process than using ServiceAccount tokens or static token files.
- Being unable to store certificates in the cluster means that it is not possible to compromise all certificates. X.509 client certificates can be used for high-privileged user accounts.
- X.509 client certificates can be revoked on demand. This is very important in case of security incidents.

The disadvantages of X.509 client certificate authentication are as follows:

- Certificates have an expiry date, which means they cannot be valid indefinitely. For simple use cases in development, this is a disadvantage. From a security perspective, in production clusters, this is a huge *advantage. But remember to ensure the certificate is stored safely as the file-based authentication mechanism is a security risk; the file could be stolen and used for unauthorized access.*

- Monitoring certificate expiration, revocation, and rotation must be handled. This should be an automated process so that we can quickly react in the case of security incidents.

- Using client certificates in the browser for authentication is troublesome, for example, when you would like to authenticate to Kubernetes Dashboard.

The key takeaway is that using X.509 client certificates is secure but requires sophisticated certificate management so that we have all the benefits. Now, we will take a look at OpenID Connect tokens, which is the recommended method for cloud environments.

OpenID Connect tokens

Using **OpenID Connect** (OIDC), you can achieve a **single sign-on** (SSO) experience for your Kubernetes cluster (and possibly other resources in your organization). OIDC is an authentication layer that's created on top of OAuth 2.0, which allows third-party applications to verify the identity of the end-user and obtain basic user profile information. OIDC uses JWTs, which you can obtain using flows that conform to the OAuth 2.0 specifications. The most significant issue with using OIDC for authenticating in Kubernetes is the limited availability of OpenID providers. But if you are deploying in a cloud environment, all tier 1 cloud service providers such as Microsoft Azure, Amazon Web Services, and Google Cloud Platform have their versions of OpenID providers. The beauty of *managed* Kubernetes cluster deployments in the cloud, such as AKS Amazon EKS, and Google Kubernetes Engine, is that they provide *integration* with their native OpenID provider out of the box or by a simple flip of a configuration switch. In other words, you do not need to worry about reconfiguring the Kubernetes API server and making it work with your chosen OpenID provider – you get it alongside the managed solution. If you are interested in learning more about the OIDC protocol, you can refer to the official web page at `https://openid.net`.

For more details and more specific flows, such as in the context of AAD please take a look at `https://docs.microsoft.com/en-us/azure/active-directory/develop/v2-protocols-oidc`.

In the following diagram, you can see the basics of the OIDC authentication flow on Kubernetes:

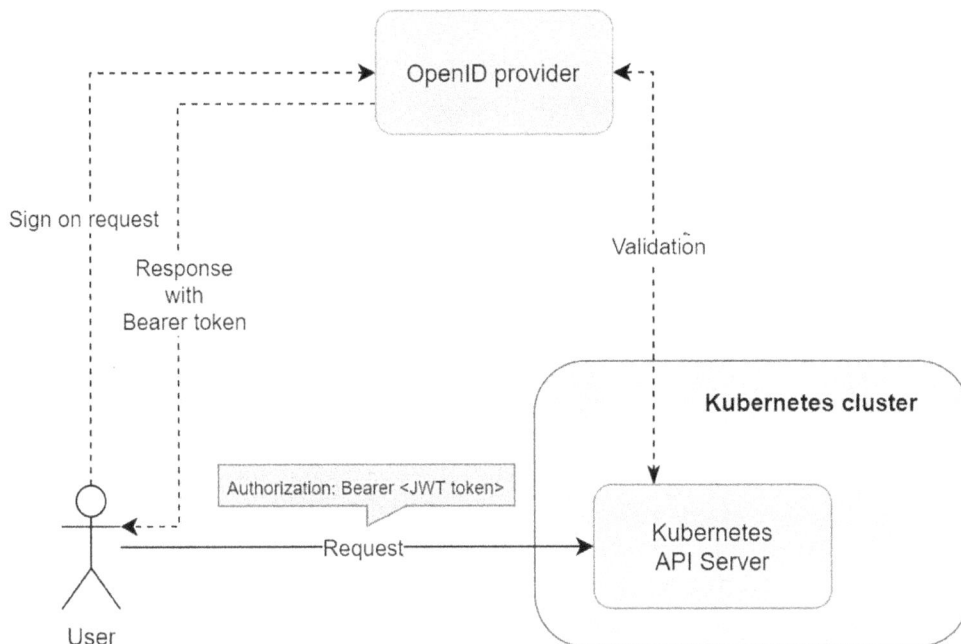

Figure 18.6: OIDC authentication in Kubernetes

The most important thing is that the OpenID provider is responsible for the SSO experience and managing the bearer tokens. Additionally, the Kubernetes API server must validate the bearer token that's received against the OpenID provider.

Using OIDC has the following advantages:

- You get SSO experience, which you can use with other services in your organization.
- Most of the cloud service providers have their own OpenID providers that easily integrate with their managed Kubernetes offerings.
- It can be also used with other OpenID providers and non-cloud deployments – this requires a bit more configuration though.
- It's a secure and scalable solution.

The disadvantages of the OIDC approach can be summarized as follows:

- Kubernetes has no web interface where you can trigger the authentication process. This means that you need to get the credentials by manually requesting them from the IdP. In managed cloud Kubernetes offerings, this is often solved by additional simple tooling to generate kubeconfig with credentials.
- OIDC tokens can be revoked by the IdP if it supports the token endpoint revocation feature. This allows you to invalidate tokens before their expiration time, for example, if a user's account is compromised. However, not all IdPs support this feature, and Kubernetes doesn't handle token revocation itself.

Using OIDC in Kubernetes

Kubernetes does not provide an integrated OpenID Connect Identity Provider. Thus, it relies on the external ones provided either by cloud providers or stand-alone tools. As we mentioned earlier in this section, the most popular cloud environments – like AWS, GCP, and Azure – natively provide OIDC integration in their managed Kubernetes offerings, which makes it pretty straightforward to enable SSO. Alternatively, the identity providers can also be set up independently for every organization using tools such as Dex, Keycloak, UAA, or OpenUnison for the non-cloud or self-managed Kubernetes clusters.

Identity Provider Requirements in Kubernetes

For an OIDC identity provider to work with Kubernetes, it has to satisfy a number of important pre-requisites:

- **Support OIDC Discovery**: OIDC discovery simplifies configuration efforts as through it all information about IdP endpoints and public keys are made available. Kubernetes reads the IdP's public keys from the discovery endpoint to validate OIDC tokens.

- **Transport Layer Security (TLS) Compliance**: The identity provider shall handle TLS to handle non-obsolete ciphers as sensitive authentication data handling is at stake.

- **CA-Signed Certificate**: Whether by using a commercial CA or a self-signed certificate, the certificate of the identity provider must have the CA flag set to TRUE. This is because Kubernetes uses Go's TLS client which strictly enforces this requirement so that Kubernetes can safely trust the identity provider's certificates during user token verification.

> For the self-deployer of an identity provider without a commercial CA, such tools as the Dex gencert script may be used to create a compliant CA certificate along with the signing key.

The following list contains some of the popular OIDC Identity Providers for Kubernetes:

- **Dex**: One of the more lightweight, open-source popular IdPs intended for use in a Kubernetes environment. it supports OIDC and works well with the authentication workflow that Kubernetes expects. Dex works by hooking into other external IdPs such as LDAP, GitHub, and Google, which would make it a good choice for organizations with more complicated identity scenarios.

- **Keycloak**: This is an open-source IdP that offers a more powerful feature set with extensive support for OIDC and SAML. Besides core functionality, Keycloak supports enterprise-grade features such as user federation and RBAC. Keycloak would be a good fit if you want to have more control or customization in your authentication setup.

- **OpenUnison**: Another IdP that is optimized for Kubernetes is OpenUnison, with features like natively integrating the Kubernetes RBAC and identity federation. It should be popular with enterprises that are ready to engage a prebuilt solution optimized to their own needs for securing Kubernetes.

- **Cloud Foundry User Account and Authentication (UAA):** This is an open-source multi-purpose IdP originating from Cloud Foundry. It supports OIDC and does an extremely strong job with cloud platform and enterprise authentication system integrations, making it suitable for more complex Kubernetes deployments in hybrid cloud environments.

Configuring OIDC with Kubernetes API Server

Enabling OIDC in Kubernetes will involve some configuration of the Kubernetes API server with certain OIDC-related flags. The major configurations include the following:

- `oidc-issuer-url`: The URL of the OIDC provider. It is used by Kubernetes for verification of token authenticity.
- `oidc-client-id string`: The client ID to use when authenticating with the IdP when Kubernetes is the client.
- `oidc-username-claim`: Specifies which claim in the token should map to the Kubernetes username.
- `oidc-groups-claim`: Maps the groups in the IdP to Kubernetes groups, in order to manage RBAC roles.

For further details on configuring specific OIDC identity providers, you can refer to the official resources such as Dex for Kubernetes Guide (`https://dexidp.io/docs/guides/kubernetes/`) or OpenID Connect Authentication in Kubernetes (`https://kubernetes.io/docs/reference/access-authn-authz/authentication/`).

The key takeaway about OIDC is that this is your best bet when configuring authentication for Kubernetes, especially if you are deploying production clusters in the cloud.

Other methods

Kubernetes offers a few other authentication methods that you can use. They are mainly intended for advanced use cases, such as integrating with LDAP or Kerberos. The first one is an **authenticating proxy**.

When you use an authenticating proxy in front of the Kubernetes API server, you can configure the API server to use certain HTTP headers to extract authentication user information from them. In other words, your authenticating proxy is doing the job of authenticating the user and passing down this information alongside the request in the form of additional headers.

You can find more information in the official documentation (`https://kubernetes.io/docs/reference/access-authn-authz/authentication/#authenticating-proxy`).

Another approach is known as **webhook token authentication**, whereby the Kubernetes API server uses an external service to verify the bearer tokens. The external service receives the information in the form of a TokenReview object from the API server via an HTTP POST request, performs verification, and sends back a TokenReview object with additional information about the result.

Find more information from the official documentation (`https://kuberntes.io/docs/reference/access-authn-authz/authentication/#webhook-token-authentication`).

Kubernetes also uses another common authentication method called **bootstrap tokens**. But bootstrap tokens are not used for general authentication but for the cluster node. Bootstrap tokens are a special type of secret in Kubernetes that simplify adding new nodes to a cluster. Stored in the `kube-system` namespace, these short-lived tokens allow the API server to authenticate kubelets (programs running on nodes) during the initial connection. This streamlines the bootstrapping process, making it easier to join new nodes or create new clusters from scratch. They can be used with or without the kubeadm tool and work seamlessly with Kubelet TLS Bootstrapping for secure communication. Refer to the documentation (`https://kubernetes.io/docs/reference/access-authn-authz/bootstrap-tokens`) to learn about authentication with bootstrap tokens and TLS bootstrapping.

In general, you need the authenticating proxy and webhook token authentication methods in special cases where you want to integrate with existing identity providers in your organization that are not supported by Kubernetes out of the box.

In the next section, we will look at authorization and RBAC in Kubernetes.

Authorization and introduction to RBAC

Security in Kubernetes relies on two crucial processes: **authentication** and **authorization**. Authentication verifies the identity of a user attempting to access the system, ensuring they are who they claim to be. This initial step typically involves checking credentials like usernames and passwords or tokens.

Following successful authentication, authorization comes into play. This process determines what actions a user can perform within the system. In Kubernetes, the API server evaluates a user's identity (derived from authentication) along with other request attributes, such as the specific API endpoint or action being requested. Based on pre-defined policies or external services, authorization modules decide whether to allow or deny the request.

Authentication is the first step in determining the identity of the user, whereas authorization is the next step when verifying if the user can perform the action they want to.

> Access controls based on specific object fields are handled by admission controllers, which occur after authorization and only if authorization allows the request. We will learn about admission controllers in the later sections of this chapter.

In the Kubernetes API server, authenticating a request results in a set of additional request attributes such as **user, group, API request verb, HTTP request verb**, and so on. These are then passed further to authorization modules that, based on these attributes, answer whether the user is allowed to do the action or not. If the request is denied by any of the modules, the user will be presented with an HTTP status code of `403` (`Forbidden`).

This is an important difference between HTTP status codes. If you receive `401`
(`Unauthorized`), this means that you have been not recognized by the system; for example, you have provided incorrect credentials or the user does not exist. If you receive
`403` (`Forbidden`), this means that authentication has been successful and you have been
recognized, but you are not *allowed* to do the action you requested. This is useful when
debugging issues regarding access to a Kubernetes cluster.

Kubernetes has a few authorization modes available that can be enabled by using the `authorization-`
`mode` argument when starting the Kubernetes API server, as follows:

```
# /etc/kubernetes/manifests/kube-apiserver.yaml
...<removed for brevity>...
spec:
  containers:
  - command:
    - kube-apiserver
    - --advertise-address=192.168.59.154
    - --allow-privileged=true
    - --authorization-mode=Node,RBAC
...<removed for brevity>...
```

The following are the authorization modes available in Kubernetes:

- **RBAC:** This allows you to organize access control and management with roles and privileges.
  RBAC is one of the industry standards for access management, also outside of Kubernetes. Roles
  can be assigned to users in the system, which gives them certain privileges and access. In this
  way, you can achieve very fine-grained access management that can be used to enforce the
  principle of least privilege. For example, you can define a role in the system that allows you
  to access certain files on a network share. Then, you can assign such roles to individual users
  in groups in the system to allow them to access these files. This can be done by associating the
  user with a role – in Kubernetes, you model this using the **RoleBinding** and **ClusterRoleBinding**
  objects. In this way, multiple users can be assigned a role, and a single user can have multiple
  roles assigned. Please note that in Kubernetes, RBAC is *permissive*, which means that there
  are no *deny* rules. Everything is denied by default, and you have to define *allow* rules instead.

- **Attribute-Based Access Control (ABAC):** This is part of the access control paradigm and is
  not only used in Kubernetes, which uses policies based on the attributes of the user, resource,
  and environment. This is a very fine-grained access control approach – you can, for example,
  define that the user can access a given file, but only if the user has clearance to access confidential data (user attribute), the owner of the file is Mike (resource attribute), and the user
  tries to access the file from an internal network (environment attribute). So, policies are sets
  of attributes that must be present together for the action to be performed. In Kubernetes, this
  is modeled using Policy objects. For example, you can define that the authenticated user, `mike`,
  can read any Pods in the `default` namespace. If you want to give the same access to user `bob`,
  then you need to create a new Policy for user bob.

- **Node:** This is a special-purpose authorization mode used for authorizing API requests made by kubelet in the cluster.

- **Webhook:** This mode is similar to webhooks for authentication. You can define an external service that needs to handle HTTP POST requests with a **SubjectAccessReview** object that's sent by the Kubernetes API server. This service must process the request and determine if the request should be allowed or denied. The response from the service should contain SubjectAccessReview, along with information, on whether the subject is allowed access. Based on that, the Kubernetes API server will either proceed with the request or reject it with an HTTP status code of 403. This approach is useful when you are integrating with existing access control solutions in the organization.

- **AlwaysAllow:** This grants unrestricted access to all requests, and is only suitable for testing environments due to security concerns.

- **AlwaysDeny:** This blocks all requests, and is useful solely for testing purposes to establish a baseline for authorization.

Currently, RBAC is considered an industry standard in Kubernetes due to its flexibility and ease of management. For this reason, RBAC is the only authentication mode we are going to describe in more detail.

RBAC mode in Kubernetes

Using RBAC in Kubernetes involves the following types of API resources that belong to the rbac. authorization.k8s.io API group:

- **Role** and **ClusterRole:** They define a set of permissions. Each rule in Role says which verb(s) are allowed for which API resource(s). The only difference between Role and ClusterRole is that Role is namespace-scoped, whereas ClusterRole is global.

- **RoleBinding** and **ClusterRoleBinding:** They associate users or a set of users (alternatively, groups or ServiceAccounts) with a given Role. Similarly, RoleBinding is namespace-scoped, while ClusterRoleBinding is cluster-wide. Please note that ClusterRoleBinding works with ClusterRole, but RoleBinding works with both ClusterRole and Role.

All these Kubernetes objects can be managed using kubectl and YAML manifests, just as you do with Pods, Services, and so on.

We will now demonstrate this in practice. In the previous section, we showed a basic RBAC configuration for a service account that was being used for authentication using kubectl. The example that we are going to use here will be a bit different and will involve creating a Pod that runs under a *dedicated* service account and periodically queries the Kubernetes API server for a list of Pods. In general, having dedicated service accounts for running your Pods is a good practice and makes it possible to ensure the principle of least privilege. For example, if your Pod needs to get the list of Pods in the cluster but does not need to create a new Pod, the ServiceAccount for this Pod should have a role assigned that allows you to list Pods, nothing more. Follow these steps to configure this example:

1. Begin by creating a dedicated namespace for the objects with the following YAML file:

```
# 02_rbac/rbac-demo-ns.yaml

---
apiVersion: v1
kind: Namespace
metadata:
  name: rbac-demo-ns
```

Create namespace by applying the YAML

```
$ kubectl apply -f 02_rbac/rbac-demo-ns.yaml
namespace/rbac-demo-ns created
```

2. To demonstrate, let us create a sample nginx Pod in the same namespace using the 02_rbac/nginx-pod.yaml definition:

```
$ kubectl apply -f 02_rbac/nginx-pod.yaml
pod/nginx-pod created
```

> Please note, that the nginx Pod is not doing anything here; we need the Pod pod-logger-app to fetch the nginx Pod details in the rbac-demo-ns namespace later.

3. Now, create a ServiceAccount named pod-logger. Create a YAML manifest named pod-logger-serviceaccount.yaml:

```
# 02_rbac/pod-logger-serviceaccount.yaml
apiVersion: v1
kind: ServiceAccount
metadata:
  name: pod-logger
  namespace: rbac-demo-ns
```

Apply the manifest to the cluster using the following command:

```
$ kubectl apply -f 02_rbac/pod-logger-serviceaccount.yaml
serviceaccount/pod-logger created
```

4. Create a role named `pod-reader`. This role will only allow the `get`, `watch`, and `list` verbs on pods resources in the Kubernetes RESTful API. In other words, this translates into an `/api/v1/namespaces/rbac-demo-ns/pods` endpoint in the API. Note that `apiGroups` specified as `""` mean the core API group. The structure of the `pod-reader-role.yaml` manifest file is as follows:

```yaml
# 02_rbac/pod-reader-role.yaml
apiVersion: rbac.authorization.k8s.io/v1
kind: Role
metadata:
  namespace: rbac-demo-ns
  name: pod-reader
rules:
  - apiGroups: [""]
    resources: ["pods"]
    verbs: ["get", "watch", "list"]
```

5. Apply the manifest to the cluster using the following command:

```
$ kubectl apply -f 02_rbac/pod-reader-role.yaml
role.rbac.authorization.k8s.io/pod-reader created
```

6. Now, we would normally create a RoleBinding object to associate the service account with the role. But to make this demonstration more interesting, we will create a Pod that's running under the `pod-logger` service account. This will essentially make the Pod unable to query the API for Pods because it will be *unauthorized* (remember that everything is denied by default in RBAC). Create a YAML manifest named `pod-logger-app.yaml` for a Pod called `pod-logger-app`, running without any additional controllers:

```yaml
# 02_rbac/pod-logger-app.yaml
apiVersion: v1
kind: Pod
metadata:
  name: pod-logger-app
  namespace: rbac-demo-ns
spec:
  serviceAccountName: pod-logger
  containers:
    - name: logger
      image: quay.io/iamgini/k8sutils:debian12
      command:
```

```
        - /bin/sh
        - -c
        - |
          SERVICEACCOUNT=/var/run/secrets/kubernetes.io/serviceaccount
          TOKEN=$(cat ${SERVICEACCOUNT}/token)
          while true
          do
            echo "Querying Kubernetes API Server for Pods in default
namespace..."
            curl --cacert $SERVICEACCOUNT/ca.crt --header "Authorization:
Bearer $TOKEN" -X GET https://kubernetes.default.svc.cluster.local/api/
v1/namespaces/rbac-demo-ns/pods
            sleep 10
          done
```

Here, the most important fields are .spec.serviceAccountName, which specifies the service account that the Pod should run under, and the command in the container definition, which we have overridden to periodically query the Kubernetes API.

7. Let us apply the 02_rbac/pod-logger-app.yaml to create the Pod as follows:

```
$ kubectl apply -f 02_rbac/pod-logger-app.yaml
pod/pod-logger-app created

$ kubectl get po -n rbac-demo-ns
NAME             READY   STATUS    RESTARTS   AGE
nginx-pod        1/1     Running   0          15m
pod-logger-app   1/1     Running   0          9s
```

8. Assigning the pod-logger service account, as explained in the previous section, will result in a Secret with a bearer JWT for this account to be mounted in the container filesystem under /var/run/secrets/kubernetes.io/serviceaccount/token. Let us verify this using kubectl exec, as follows:

```
$ kubectl exec -it -n rbac-demo-ns pod-logger-app -- bash
root@pod-logger-app:/# ls -l /var/run/secrets/kubernetes.io/
serviceaccount/
total 0
lrwxrwxrwx 1 root root 13 Jul 14 03:33 ca.crt -> ..data/ca.crt
lrwxrwxrwx 1 root root 16 Jul 14 03:33 namespace -> ..data/namespace
lrwxrwxrwx 1 root root 12 Jul 14 03:33 token -> ..data/token
```

9. The overridden commands run an infinite loop in a Linux shell (e.g., bash) in 10-second intervals. In each iteration, we query the Kubernetes API endpoint (`https://kubernetes/api/v1/namespaces/rbac-demo-ns/pods`) for Pods in the `rbac-demo-ns` namespace with the HTTP `GET` method using the `curl` command. To properly authenticate, we pass the contents of `/var/run/secrets/kubernetes.io/serviceaccount/token` as a **bearer** token in the `Authorization` header for the request. Additionally, we pass a CA certificate path to verify the remote server using the `cacert` argument. The certificate is injected into `/var/run/secrets/kubernetes.io/serviceaccount/ca.crt` by the Kubernetes runtime. When you inspect its logs, you should expect to see a bunch of messages with an HTTP status code of `403 (Forbidden)`. This is because the ServiceAccount does not have a RoleBinding type that associates it with the `pod-reader` Role yet.

10. Start following the logs of the `pod-logger-app` Pod using the following command:

```
$ kubectl logs -n rbac-demo-ns pod-logger-app -f
Querying Kubernetes API Server for Pods in rbac-demo-ns namespace...
  % Total    % Received % Xferd  Average Speed   Time    Time     Time
Current
                                 Dload  Upload   Total   Spent    Left
Speed
100   336  100   336    0     0  25611      0 --:--:-- --:--:-- --:--:--
25846
{
  "kind": "Status",
  "apiVersion": "v1",
  "metadata": {},
  "status": "Failure",
  "message": "pods is forbidden: User \"system:serviceaccount:rbac-demo-
ns:pod-logger\" cannot list resource \"pods\" in API group \"\" in the
namespace \"rbac-demo-ns\"",
  "reason": "Forbidden",
  "details": {
    "kind": "pods"
  },
  "code": 403
}
```

11. In a new console window (or by ending the logs with the *Ctrl + F* command), we will create and apply a RoleBinding that *associates* the `pod-logger` ServiceAccount with the `pod-reader` Role. Create a YAML manifest named `read-pods-rolebinding.yaml` that contains the following contents:

```
# 02_rbac/read-pods-rolebinding.yaml
apiVersion: rbac.authorization.k8s.io/v1
kind: RoleBinding
```

```
metadata:
  name: read-pods
  namespace: rbac-demo-ns
subjects:
- kind: ServiceAccount
  name: pod-logger
  namespace: rbac-demo-ns
roleRef:
  kind: Role
  name: pod-reader
  apiGroup: rbac.authorization.k8s.io
```

There are three key components in the RoleBinding manifest: name, which is used to identify the user; subjects, which reference the users, groups, or service accounts; and roleRef, which references the role.

12. Apply the RoleBinding manifest file using the following command:

```
$ kubectl apply -f 02_rbac/read-pods-rolebinding.yaml
rolebinding.rbac.authorization.k8s.io/read-pods created
```

13. Now check the pod-logger-app logs again; you will see that the Pod was able to successfully retrieve the list of Pods in the rbac-demo-ns namespace. In other words, the request was successfully authorized:

```
$ kubectl logs -n rbac-demo-ns pod-logger-app -f
...<removed for brevity>...
Querying Kubernetes API Server for Pods in default namespace...
  % Total    % Received % Xferd  Average Speed   Time    Time     Time
Current
                                 Dload  Upload   Total   Spent    Left
Speed
{
  "kind": "PodList",
  "apiVersion": "v1",
  "metadata": {
    "resourceVersion": "4889"
  },
  "items": [
    {
      "metadata": {
        "name": "nginx-pod",
        "namespace": "rbac-demo-ns",
        "uid": "b62b2bdb-2677-4809-a134-9d6cfa07ecad",
...<removed for brevity>...
```

14. Lastly, you can delete the RoleBinding type using the following command:

```
$ kubectl delete rolebinding read-pods -n rbac-demo-ns
```

15. Now, if you inspect the logs of the pod-logger-app Pod again, you will see that the requests are denied with an HTTP status code of 403 again.

Congratulations! You have successfully used RBAC in Kubernetes to be able to read the Pods in the cluster for a Pod running under ServiceAccount. To clean up the Kubernetes environment, you can delete the rbac-demo-ns namespace so that the resources you created will be removed as part of the namespace removal.

As we have explored authentication and authorization, in the next section, let us learn about another security feature in Kubernetes called admission controllers.

Admission Control – Security Policies and Checks

Imagine a security checkpoint at a critical facility. Admission controllers in Kubernetes function similarly for your cluster. They act as gatekeepers, intercepting requests to the Kubernetes API server before resources are created, deleted, or modified. These controllers can validate or modify the requests based on predefined rules, ensuring that only authorized and properly configured resources enter the system. Also note that admission controllers do not (and cannot) block requests to read (get, watch, or list) objects.

> Several key features of Kubernetes rely on specific admission controllers to function correctly. Therefore, a Kubernetes API server without the appropriate admission controllers is incomplete and will not support all expected features.

There are two types of admission controllers:

- **Validation controllers:** These controllers meticulously examine incoming requests. If they find anything suspicious or non-compliant with set policies, they reject the request entirely.

- **Mutation controllers:** These controllers have the power to modify requests before they are stored permanently. They can, for instance, add missing security annotations or adjust resource limits.

Now, let's get introduced to the two-phase admission process.

The Two-Phase Admission Process

Admission control in Kubernetes operates in a two-step process, ensuring only compliant resources enter your cluster.

The high-level flow taken by the mutation and validation phases of Kubernetes admission control is represented by the following figure. This flow takes incoming requests into Kubernetes to process in a manner that first does the appropriate mutations for modifying or enriching requests before doing any actual validation to see if the request meets all required security and policy validations that are required.

This sequence is exposed to the flow, showing the way, Kubernetes enforces consistency, security, and policy compliance before allowing any changes to the cluster state:

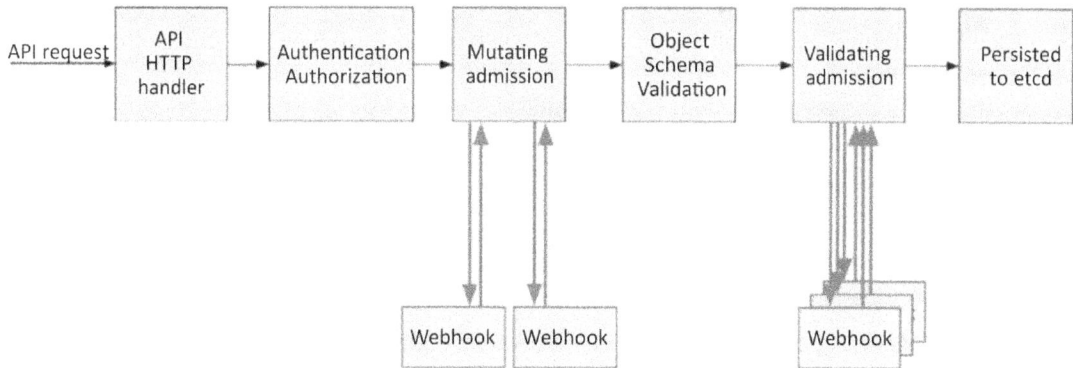

Figure 18.7: Admission controllers in the API request processing flow (image source: https://kubernetes. io/blog/2019/03/21/a-guide-to-kubernetes-admission-controllers/)

Here's a breakdown of each phase.

Mutation Phase

The Mutation phase is a step in admission control in Kubernetes, where controllers running in the role of mutation controllers will change the incoming API requests to make them compliant with cluster policies before further processing. Such controllers basically act like "molders," which not only see to it that the requests are consistent with established settings but can also automatically add or adjust settings, for example, defaults or security labels. This makes the system maintain the policy compliance and alignment of configurations without manual input.

There are a few examples presented here for this phase:

- Adding missing security annotations to pods.
- Adjusting resource requests and limits for pods based on pre-defined rules.
- Injecting sidecar containers for specific functionalities.

Validation Phase

It is during the Validation phase, in this sequence, that Kubernetes admission control completes a controller's doing in the preceding Mutation phase. Controllers then closely scrutinize incoming requests that might have been modified by some controller. Often referred to as the "guardians," these controllers check requests for adherence to cluster policies and security standards. It is an important phase in the prevention of misconfiguration and unauthorized changes that maintain cluster integrity and security by rejecting requests not meeting set criteria.

Some of the example actions are listed here:

- Approve the request if it adheres to set criteria (e.g., resource quotas, security standards).
- Reject the request if it violates any policies, providing informative error messages.

In the next section, we will learn how to turn off and turn on admission controllers in Kubernetes.

Enabling and disabling Admission controllers

To check which Admission controllers are enabled, you typically need to inspect the configuration of the **Kubernetes API server**. This is often done by accessing the configuration file where the API server is defined, usually located in the system's configuration directories or managed through a configuration management tool. Look for the `--enable-admission-plugins` flag, which specifies the list of admission controllers currently active.

For example, in a minikube environment, you can SSH into the minikube VM using the `minikube ssh` command. Once inside, you can locate and inspect the `kube-apiserver.yaml` file, typically found in `/etc/kubernetes/manifests/`. Use `sudo cat /etc/kubernetes/manifests/kube-apiserver.yaml` to view its contents and look for the `--enable-admission-plugins` flag:

```
$ minikube ssh 'sudo grep -- '--enable-admission-plugins' /etc/kubernetes/
manifests/kube-apiserver.yaml'
    - --enable-admission-plugins=NamespaceLifecycle,LimitRanger,ServiceAcco
unt,DefaultStorageClass,DefaultTolerationSeconds,NodeRestriction,MutatingA
dmissionWebhook,ValidatingAdmissionWebhook,ResourceQuota
```

To modify the list of enabled plugins, edit this file with a text editor like nano or vi, adjust the plugins as needed, and then save your changes. The kubelet watches the manifest files and will automatically restart the API server (recreate the Pod) if it detects any changes to the manifest file.

It is also possible to turn off the default admissions controllers as follows:

```
kube-apiserver --disable-admission-plugins=PodNodeSelector,AlwaysDeny ...
```

In the following section, we will learn the list of admission controllers available in Kubernetes.

Common Admission Controllers

In Kubernetes, the admission controllers are built into the `kube-apiserver` and should only be configured by the cluster administrator. Among these controllers, two are particularly notable: **MutatingAdmissionWebhook** and **ValidatingAdmissionWebhook**. These controllers execute the respective mutating and validating admission control webhooks that are configured through the API:

- **Basic Controls:** `AlwaysAdmit` (deprecated), `AlwaysDeny` (deprecated), `AlwaysPullImages`
- **Defaults:** `DefaultStorageClass`, `DefaultTolerationSeconds`
- **Security:** `DenyEscalatingExec`, `DenyServiceExternalIPs`, `PodSecurityPolicy`, `SecurityContextDeny`
- **Resource Management:** `LimitRanger`, `ResourceQuota`, `RuntimeClass`
- **Object Lifecycle:** `NamespaceAutoProvision`, `NamespaceExists`, `NamespaceLifecycle`, `PersistentVolumeClaimResize`, `StorageObjectInUseProtection`
- **Node Management:** `NodeRestriction`, `TaintNodesByCondition`
- **Webhooks:** `MutatingAdmissionWebhook`, `ValidatingAdmissionWebhook`

- **Others:** `EventRateLimit, LimitPodHardAntiAffinityTopology, OwnerReferencesPermissio nEnforcement, PodNodeSelector` (deprecated), `Priority, ServiceAccount`

There are numerous advantages to using admission controllers in a Kubernetes cluster. Let us learn about a few in the next section.

Benefits of Admission Controllers

There are multiple advantages of using admission controllers in your Kubernetes clusters, including the following:

- **Enhanced Security:** By enforcing security policies like pod security standards, admission controllers help keep your cluster safe from unauthorized or vulnerable deployments.
- **Policy Enforcement:** You can define rules for resource usage, image pulling, and more, which admission controllers will automatically enforce.
- **Consistency and Standardization:** Admission controllers ensure that resources across your cluster adhere to established best practices and configurations.

To summarize, the admission controllers section has emphasized that admission controllers play a very important role in ensuring the security of Kubernetes, and this is through the Mutation and Validation phases. We learned how the mutation controllers run modifications on the requests to make sure they comply with cluster policies, while validation controllers ensure none other than those meeting the security standards are processed. In all, the foregoing processes improve the overall state of the security of the Kubernetes clusters by assuring compliance and prohibiting unauthorized changes.

In the next sections, let us learn about how to secure workloads in Kubernetes using Security Context and NetworkPolicies.

Securing Pods and Containers

Securing Pods and containers is essential to keeping your Kubernetes environment in a healthy state, since these directly interact with workloads and sensitive data. In the next sections, we are going to talk about how the securityContext settings and NetworkPolicies can enforce strict access controls and isolation in place to strengthen the security of Pods and containers in your cluster.

Securing Pods and Containers in Kubernetes Using Security Context

In Kubernetes, a **securityContext** defines a set of security settings that determine how a Pod or container operates within the cluster. This allows you to enforce security best practices and minimize the attack surface by restricting privileges and controlling access.

The primary purpose of securityContext is to enhance the security of your Kubernetes clusters by defining how a pod or container should run within the cluster. By specifying security settings, you can ensure that your applications adhere to the principle of least privilege, reducing the potential for malicious activities and accidental misconfigurations.

A typical use case for securityContext is to run containers as non-root users. This prevents containers from having unnecessary permissions, thereby limiting the potential damage if a container is compromised. Additionally, you can configure other security settings such as read-only filesystems and fine-grained capabilities to further strengthen your cluster's security posture.

Key Components of SecurityContext

Here's a breakdown of the key components of a securityContext along with illustrative examples.

User and Group

This security context specifies the user and group ID, under which processes inside the container will run. By enforcing the principle of least privilege, it grants containers only the minimum permissions necessary to function. The following code snippet shows a typical example of a Pod definition with the securityContext configured:

```
apiVersion: v1
kind: Pod
metadata:
  name: my-app
spec:
  containers:
  - name: app-container
    securityContext:
      runAsUser: 1000   # Run container processes as user ID 1000
      runAsGroup: 1000  # Run container processes as group ID 1000
```

Linux Capabilities

Capabilities are special privileges that can be granted to containers beyond the limitations of a user. securityContext allows you to define which capabilities a container should have, enabling specific functionalities without providing full root access, as shown in the following example:

```
apiVersion: v1
kind: Pod
metadata:
  name: privileged-container
spec:
  containers:
  - name: app-container
    securityContext:
      capabilities:
        add:
          - CAP_NET_ADMIN  # Grant network management capabilities
```

Refer to the Linux capabilities documentation to learn more (https://linux-audit.com/kernel/capabilities/linux-capabilities-hardening-linux-binaries-by-removing-setuid/).

Privileged Mode

By default, containers run in an unprivileged mode. securityContext offers the option to run a container in privileged mode, granting it access to capabilities usually reserved for the root user. However, due to the increased security risk, this should be used with extreme caution. Refer to the following code snippet for en example Pod YAML with privileged mode securityContext:

```yaml
apiVersion: v1
kind: Pod
metadata:
  name: privileged-container
spec:
  containers:
  - name: app-container
    securityContext:
      privileged: true   # Run container in privileged mode (use cautiously)
```

Read-Only Root Filesystem

This securityContext allows you to configure the container to have a read-only root filesystem. This enhances security by preventing accidental or malicious modifications to the base system:

```yaml
apiVersion: v1
kind: Pod
metadata:
  name: read-only-container
spec:
  containers:
  - name: app-container
    securityContext:
      readOnlyRootFilesystem: true   # Mount root filesystem as read-only
```

There are a few more SecurityContext settings such as **Security Enhanced Linux (SELinux)**, **AppArmor** (https://kubernetes.io/docs/tutorials/security/apparmor/), **Seccomp**, and so on. Refer to https://kubernetes.io/docs/tasks/configure-pod-container/security-context/ to learn more.

You also need to know where is the best place in the configuration you can apply SecurityContext; let us learn that in the next section.

Applying SecurityContext at Pod and Container Levels

In Kubernetes, the securityContext can be applied at both the pod level and the container level, offering flexibility in defining security settings for your applications.

Pod-Level SecurityContext

When applied at the pod level, the securityContext settings are inherited by all containers within the pod. This is useful for setting default security configurations that should apply uniformly across all containers in the pod:

```yaml
apiVersion: v1
kind: Pod
metadata:
  name: my-pod
spec:
  securityContext:
    runAsUser: 1000  # All containers in the pod run as user ID 1000
    runAsGroup: 1000 # All containers in the pod run as group ID 1000
  containers:
  - name: app-container
    image: my-app-image
  - name: sidecar-container
    image: my-sidecar-image
```

Container-Level SecurityContext

When applied at the container level, the securityContext settings only affect the specific container. This allows for more granular control, where different containers within the same pod can have different security configurations.

```yaml
apiVersion: v1
kind: Pod
metadata:
  name: my-pod
spec:
  containers:
  - name: app-container
    image: my-app-image
    securityContext:
      runAsUser: 1000  # This container runs as user ID 1000
      capabilities:
        add: ["CAP_NET_ADMIN"]  # Grant specific capabilities
  - name: sidecar-container
    image: my-sidecar-image
    securityContext:
      runAsUser: 2000  # This container runs as user ID 2000
      readOnlyRootFilesystem: true  # This container has a read-only root
filesystem
```

In the following section, let us demonstrate the Security Context with an example Pod.

Applying Security Context to a Pod

The following example creates a Pod with a container that runs with a `read-only` root filesystem and specifies non-root user and group IDs:

```yaml
# pod-with-security-context.yaml
apiVersion: v1
kind: Pod
metadata:
  name: security-context-demo
spec:
  containers:
    - name: app-container
      image: nginx:latest
      securityContext:
        runAsUser: 1000                 # Run container processes as user ID
1000
        runAsGroup: 1000                # Run container processes as group ID
1000
        readOnlyRootFilesystem: true  # Mount the root filesystem as read-only
      volumeMounts:
        - name: html-volume
          mountPath: /usr/share/nginx/html
  volumes:
    - name: html-volume
      emptyDir: {}                      # Volume to provide writable space
```

In the preceding YAML, note the following:

- `runAsUser` and `runAsGroup`: These settings ensure that the container runs with a specific non-root user ID and group ID, following the principle of least privilege.
- `readOnlyRootFilesystem`: This setting mounts the container's root filesystem as read-only, preventing any accidental or malicious modifications to the base system.

Create the Pod using the YAML as follows:

```
$ kubectl apply -f pod-with-security-context.yaml
pod/security-context-demo created
```

Once the Pod is created, let us test a few commands inside the container to verify the securityContext we applied:

```
$ kubectl exec -it security-context-demo -- /bin/sh
~ $ id
```

```
uid=1000 gid=1000 groups=1000
~ $ touch /testfile
touch: /testfile: Read-only file system
```

You can see the Read-only filesystem error; this is expected.

Refer to `https://kubernetes.io/docs/tasks/configure-pod-container/security-context/` to learn more about the security context in Kubernetes.

The next section introduces the control of network flow in Kubernetes using the **NetworkPolicy** object. You will see that you can build a kind of network firewall directly in Kubernetes so that you can prevent Pods from being able to reach one another.

Securing Pods using the NetworkPolicy object

The **NetworkPolicy** object is the last resource kind we need to discover as part of this chapter to have an overview of services in this chapter. NetworkPolicy will allow you to define network firewalls directly implemented in your cluster.

Why do you need NetworkPolicy?

When you have to manage a real Kubernetes workload in production, you'll have to deploy more and more applications onto it, and it is possible that these applications will have to communicate with each other.

Achieving communication between applications is really one of the fundamental objectives of a microservice architecture. Most of this communication will be done through the network, and the network is forcibly something that you want to secure by using firewalls.

Kubernetes has its own implementation of network firewalls called NetworkPolicy. Say that you want one nginx resource to be accessible on port 80 from a particular IP address and to block any other traffic that doesn't match these requirements. To do that, you'll need to use NetworkPolicy and attach it to that Pod.

NetworkPolicy brings three benefits, as follows:

- You can build egress/ingress rules based on **Classless Inter-Domain Routing (CIDR)** blocks.
- You can build egress/ingress rules based on Pods labels and selectors (just as we've seen before with services' and Pods' association).
- You can build egress/ingress rules based on namespaces (a notion we will discover in the next chapter).

Lastly, keep in mind that for NetworkPolicy to work, you'll need to have a Kubernetes cluster with a CNI plugin installed. CNI plugins are generally not installed by default on Kubernetes. If you're using minikube for learning purposes, the good news is that it has an integration with Calico, which is a CNI plugin with NetworkPolicy support implemented out of the box. You just need to recreate the minikube cluster this way:

```
$ minikube start --network-plugin=cni --cni=calico --container-
runtime=containerd
```

If you're using Kubernetes on top of a cloud platform, we suggest you read the documentation of your cloud provider in order to verify which CNI options your cloud platform offers and whether it implements NetworkPolicy support.

Understanding Pods are not isolated by default

By default, in Kubernetes, Pods are not isolated and any Pod can be reached by any other Pod without any constraint.

If you don't use NetworkPolicy, Pods will remain just like that: accessible by everything without any constraint. Once you attach the NetworkPolicy to a Pod, the rules described on the NetworkPolicy will be applied to the Pod.

To establish communication between two Pods associated with network policies, both sides must be open. It means Pod *A* must have an egress rule to Pod *B*, and Pod *B* must have an ingress rule from Pod *A*; otherwise, the traffic will be denied. The following figure illustrates this:

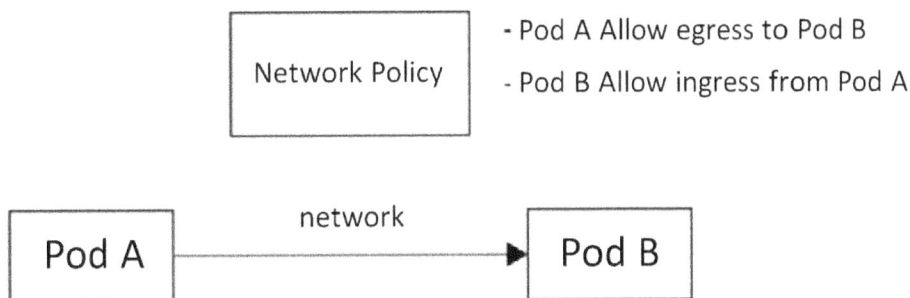

Figure 18.8: One of the Pods is broken but the service will still forward traffic to it

Keep in mind that you'll have to troubleshoot NetworkPolicy because it can be the root cause of a lot of issues. Let's now configure a NetworkPolicy between two Pods by using labels and selectors.

Configuring NetworkPolicy with labels and selectors

First, let's create two nginx Pods to demonstrate our example. To demonstrate the isolation, we will use two separate namespaces in this example. You will learn more about Kubernetes namespace a in *Chapter 6, Namespaces, Quotas, and Limits for Multi-Tenancy in Kubernetes.*

> Implementing complete communication isolation within a namespace can be complex and have unintended consequences. Carefully evaluate your needs and potential impacts before applying any restrictions.

Let's create the namespaces and two Pods with two distinct labels so that they become easier to target with the NetworkPolicy.

Our `web1` namespace with `nginx1` pod will be created as follows:

```
# web1-app.yaml
---
apiVersion: v1
kind: Namespace
metadata:
  labels:
    project: web1
  name: web1


---
apiVersion: v1
kind: Pod
metadata:
  name: nginx1
  namespace: web1
  labels:
    app: nginx1
spec:
  containers:
    - name: nginx1
      image: nginx
```

Also, the `web2` namespace with a `nginx2` pod will be created as follows:

```
# web2-app.yaml
---
apiVersion: v1
kind: Namespace
metadata:
  labels:
    project: web2
  name: web2
---
apiVersion: v1
kind: Pod
metadata:
  name: nginx2
  namespace: web2
  labels:
    app: nginx2
spec:
```

```
    containers:
      - name: nginx2
        image: nginx
```

In the previous code snippets, we used namespaces (web1 and web2) instead of deploying to the default namespace.

Let's create the resources and verify the Pods as follows:

```
$ kubectl apply -f web-app1.yaml
namespace/web1 created
pod/nginx1 created

$ kubectl apply -f web-app2.yaml
namespace/web2 created
pod/nginx2 created

$ kubectl get po -o wide -n web1
NAME        READY    STATUS      RESTARTS    AGE    IP                NODE
NOMINATED NODE    READINESS GATES
nginx1    1/1      Running     0           3m     10.244.120.71     minikube    <none>
<none>
$ kubectl get po -o wide -n web2
NAME        READY    STATUS      RESTARTS    AGE      IP                NODE
NOMINATED NODE    READINESS GATES
nginx2    1/1      Running     0           2m53s    10.244.120.72     minikube    <none>
<none>
```

Now that the two Pods are created with distinct labels inside different namespaces, we use the -o wide flag to get the IP address of both Pods. Run a curl command from the nginx1 Pod to reach the nginx2 Pod, to confirm that by default, network traffic is allowed because no NetworkPolicy is created at this point. The code is illustrated here; 10.244.120.72 is the IP address of the nginx2 Pod in the web2 namespace:

```
$ kubectl -n web1 exec nginx1 -- curl 10.244.120.72
  % Total     % Received % Xferd  Average Speed   Time     Time     Time  Current
                                  Dload  Upload   Total    Spent    Left  Speed
100    615  100    615     0       0    698k        0 --:--:-- --:--:-- --:--:--   600k
<!DOCTYPE html>
<html>
<head>
<title>Welcome to nginx!</title>
...<removed for brevity>...
```

As you can see, we correctly received the nginx home page from the nginx2 Pod.

Now, let's block all the ingress traffic to the web2 namespace explicitly. To do that, we can create a default policy as follows:

```yaml
# default-deny-ingress.yaml
---
apiVersion: networking.k8s.io/v1
kind: NetworkPolicy
metadata:
  name: default-deny-ingress
spec:
  podSelector: {}
  policyTypes:
    - Ingress
  ingress: []
```

In the preceding YAML snippet, note the following:

- `podSelector: {}`: Selects pods to which the NetworkPolicy applies. In this case, {} selects all pods in the namespace. This means that the rules defined in the NetworkPolicy *will apply to all pods in the namespace*, regardless of their labels.

- `policyTypes: - Ingress`: Specifies the type of policy being applied, which is "Ingress" in this case. This means that the NetworkPolicy will control incoming (ingress) traffic to the selected pods.

- `ingress: []`: Defines the list of ingress rules for the NetworkPolicy. In this case, the list is empty ([ ]), indicating that there are no specific ingress rules defined. Therefore, all incoming traffic to the selected pods will be denied by default.

Let's apply this deny policy to our web2 namespace to block all incoming (ingress) traffic as follows:

```
$ kubectl apply -f default-deny-ingress.yaml -n web2
networkpolicy.networking.k8s.io/default-deny-ingress created
```

We will try to access the nginx2 pod from nginx1 pod now and see the output:

```
$ kubectl -n web1 exec nginx1 -- curl 10.244.120.72
  % Total    % Received % Xferd  Average Speed   Time    Time     Time  Current
                                 Dload  Upload   Total   Spent    Left  Speed
  0     0    0     0    0     0      0      0 --:--:--  0:02:15 --:--:--     0
curl: (28) Failed to connect to 10.244.120.72 port 80 after 135435 ms: Couldn't
connect to server
command terminated with exit code 28
```

It is clear from the previous output that the traffic to web2 namespace and Pods are denied with the default-deny-ingress NetworkPolicy resource.

Now, we will add NetworkPolicy to `nginx2` in the `web2` namespace to explicitly allow traffic coming from the Pod `nginx1` in the `web1` namespace. Here is how to proceed with the YAML code:

```yaml
# allow-from-web1-netpol.yaml
apiVersion: networking.k8s.io/v1
kind: NetworkPolicy
metadata:
  name: allow-from-web1-netpol
  namespace: web2
spec:
  podSelector:
    matchLabels:
      app: nginx2
  policyTypes:
    - Ingress
  ingress:
    - from:
        - namespaceSelector:
            matchLabels:
              project: web1
        - podSelector:
            matchLabels:
              app: nginx1
      ports:
        - protocol: TCP
          port: 80
```

Please note the `namespaceSelector.matchLabels` here with the `project: web1` label, which we used for `web1` namespace explicitly for this purpose. Let's apply this NetworkPolicy, as follows:

```
$ kubectl apply -f nginx2-networkpolicy.yaml
networkpolicy.networking.k8s.io/nginx2-networkpolicy created
```

Now, let's run the same `curl` command we did before, as follows:

```
$ kubectl -n web1 exec nginx1 -- curl 10.244.120.72
  % Total    % Received % Xferd  Average Speed   Time    Time     Time  Current
                                 Dload  Upload   Total   Spent    Left  Speed
100   615  100   615    0     0  1280k      0 --:--:-- --:--:-- --:--:--  600k
<!DOCTYPE html>
<html>
<head>
<title>Welcome to nginx!</title>
...<removed for brevity>...
```

As you can see, it works just like it did before. Why? For the following two reasons:

- `nginx2` now explicitly allows ingress traffic on port `80` from `nginx1` in the `web1` namepsace; everything else is denied.
- `nginx1` has no NetworkPolicy, and thus, egress traffic to everything is allowed for it.

Keep in mind that if no NetworkPolicy is set on the Pod, the default behaviour applies—everything is allowed for the Pod.

We strongly encourage you to make a habit of using NetworkPolicy along with your Pod. Lastly, please be aware that NetworkPolicy can also be used to build firewalls based on CIDR blocks. It might be useful, especially if your Pods are called from outside the cluster. Otherwise, when you need to configure firewalls between Pods, it is recommended to proceed with labels and selectors as you already did with the services' configuration.

Next, we will focus on yet another important aspect of securing Kubernetes, namely Securing Communication via TLS Certificates between Kubernetes components. In this section, we will talk about how the TLS certificate helps in securing the data in transit and ensures secure interactions among various components that make up the ecosystem of Kubernetes.

Securing Communication — TLS Certificates Between Kubernetes Components

In Kubernetes, secure communication between various components is critical. **Transport Layer Security (TLS)**, and **Secure Sockets Layer (SSL)**, play a crucial role in encrypting data transmissions and establishing trust between services.

By implementing **TLS with mutual authentication (mTLS)**, both the client and server involved in communication can verify each other's identities using digital certificates issued by a trusted CA. This adds a layer of security by preventing unauthorized access and ensuring data integrity.

Here are some examples of how TLS certificates are used in Kubernetes:

- **API Server to etcd:** The API server, the central control plane component, communicates with etcd, the distributed key-value store, to manage cluster state. Utilizing mTLS between these components safeguards sensitive cluster data from interception or tampering.
- **Ingress Controller to Services:** The ingress controller, acting as a single entry point for external traffic, routes requests to backend services. Implementing mTLS between the ingress controller and services ensures that only authorized services receive traffic, mitigating potential security breaches.
- **Internal Service Communication:** Services within the cluster can also leverage mTLS for secure communication. This is particularly important for services that handle sensitive data or require strong authentication.
- **Service Meshes – for instance, Istio:** These types of service mesh have a variety of advanced traffic management and security capabilities, such as automatic mTLS between microservices. This makes the process of securing service-to-service communication easier without having to embed these communications with TLS configuration in the code that developers manage.

- **Load Balancers:** Applications deployed behind a load balancer can also be used to secure communication between the load balancer and backend services with the use of TLS. In a configuration like this, the data will remain encrypted along the entire path.

- Another security mechanism would be enabling IPSec within a Kubernetes cluster to encrypt network traffic between nodes. This may be useful in the protection of the traffic in cloud environments or between various data centers.

Through the deployment of TLS certificates with mTLS, Kubernetes administrators significantly bolster the security of their clusters. This approach encrypts communication paths, verifies the identities of communicating components, and mitigates risks associated with unauthorized data access or tampering.

In the next section, we will learn how to enable container security using special containers such as gVisor and Kata Containers.

Container Security – gVisor and Kata Containers

Traditional containers share the host operating system kernel with other applications running on the machine, which can pose security risks if a container vulnerability allows access to the underlying system. **gVisor** and **Kata Containers** emerge as alternative container runtime technologies that prioritize security. Let us learn about them in the next sections.

gVisor (Guest Virtual Machine Supervisor)

gVisor is a lightweight virtual machine implemented in user space. It acts as a sandbox for each container, isolating it from the host kernel and other containers.

The following figure shows the high-level architecture of gVisor.

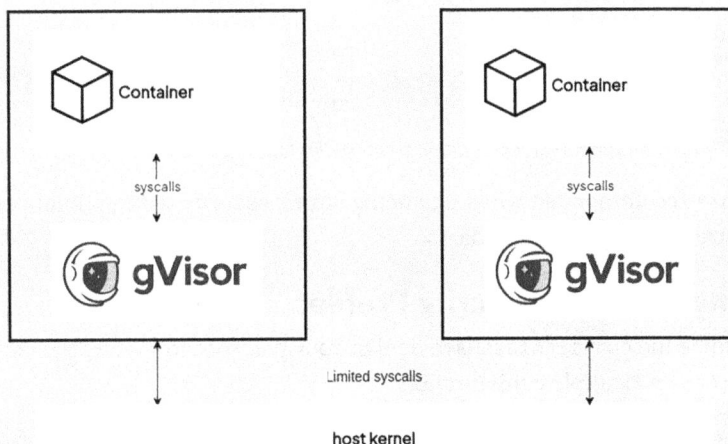

Figure 18.9: gVisor architecture

By virtualizing kernel functionalities for each container, gVisor ensures that container vulnerabilities cannot directly compromise the host system. It establishes a robust isolation boundary, even in compromised container scenarios. gVisor is best suited for environments requiring the highest level of security isolation, despite potentially higher resource overhead.

Kata Containers

Kata Containers utilize lightweight VMs that are similar to traditional VMs but optimized for container workloads. Kata Containers offers a secure execution environment by isolating containers within lightweight VMs. This enhanced isolation strengthens security compared to standard containers while maintaining performance efficiency.

The following figure demonstrates how Kata Containers are different from traditional container technologies.

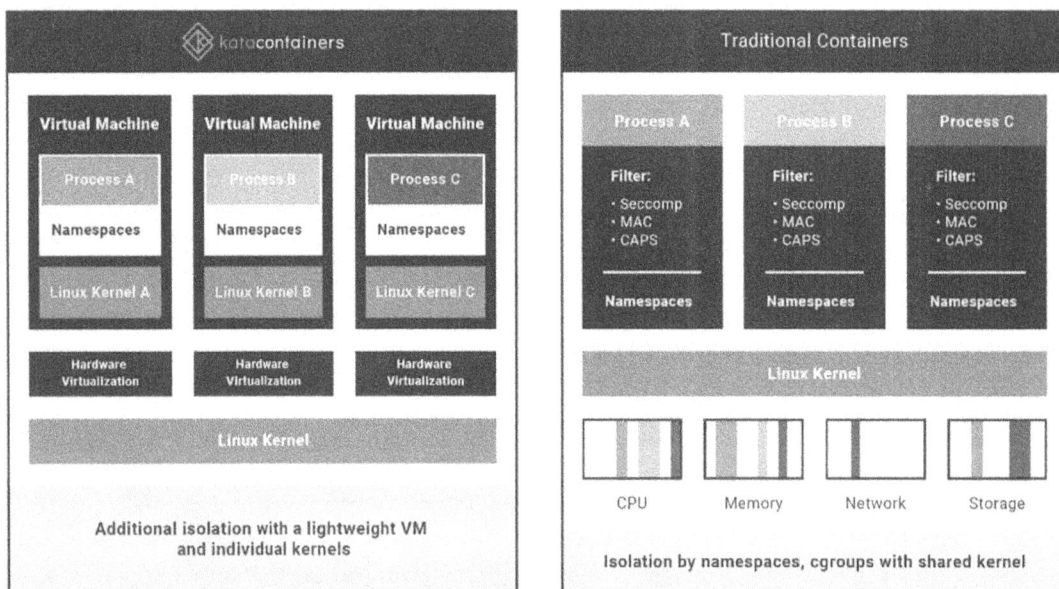

Figure 18.10: Kata Containers versus traditional containers (source: https://katacontainers.io/learn/)

Kata Containers are recommended when balancing strong security with optimal performance, particularly for resource-intensive workloads.

Using RuntimeClass for Security Profiles

In Kubernetes, utilize the `runtimeClassName` field in your pod spec to specify the container runtime environment. Here is an example configuration:

```
apiVersion: v1
kind: Pod
metadata:
  name: secure-pod
```

```
spec:
  runtimeClassName: kata-containers  # Specifies Kata Containers runtime
  containers:
  - name: my-app
    image: my-secure-image
```

This setup directs Kubernetes to use the Kata Containers runtime for enhanced security isolation.

We learned several important things about Kubernetes security in this chapter. Before we conclude the chapter, let us learn about another security topic in the next section, which is about accessing private registries and container images.

Managing Secrets and Registry Credentials

In Kubernetes, registry credentials are necessary for securely pulling container images from private registries that require authentication. Without these credentials, Kubernetes pods cannot access images stored in private repositories. Managing these credentials securely is crucial to ensure that only authorized pods can retrieve and use specific container images.

Using kubectl create secret docker-registry simplifies the management of container registry credentials in Kubernetes. It ensures security by encrypting secrets at rest, making them accessible only to authorized nodes. This approach reduces complexity compared to manual methods, minimizing errors and improving operational efficiency. Moreover, it seamlessly integrates with Kubernetes pod specifications, allowing straightforward configuration of imagePullSecrets to authenticate pod access to private container registries.

Using kubectl to create a Docker registry secret

To illustrate, here's how you can create a Docker registry secret and integrate it into a Kubernetes pod configuration:

```
$ kubectl create secret docker-registry my-registry-secret \
    --docker-server=your-registry.com \
    --docker-username=your_username \
    --docker-password=your_password \
    --docker-email=your-email@example.com
```

Replace your-registry.com, your_username, your_password, and your-email@example.com with your actual registry details.

Update your Pod YAML to use the newly created secret for pulling images from the private registry:

```
apiVersion: v1
kind: Pod
metadata:
  name: my-pod
spec:
```

```
  containers:
  - name: my-container
    image: your-registry.com/your-image:tag
  imagePullSecrets:
  - name: my-registry-secret
```

Ensure `my-registry-secret` matches the name used when creating the Docker registry secret.

When Kubernetes creates the Pod, the image will be pulled from the private registry using the image-PullSecrets as the authentication credentials.

Congratulations, you have reached the end of this long chapter about Kubernetes security.

Summary

This chapter covered *authentication* and *authorization* in Kubernetes. First, we provided an overview of the available authentication methods in Kubernetes and explained how you can use ServiceAccount tokens for external user authentication. Next, we focused on RBAC in Kubernetes. You learned how to use Roles, ClusterRoles, RoleBindings, and ClusterRoleBindings to manage authorization in your cluster. We demonstrated a practical use case of RBAC for ServiceAccounts by creating a Pod that can list Pods in the cluster using the Kubernetes API (respecting the principle of least privilege).

After that, we learned about Admission Controllers in Kubernetes and what controllers are available to secure your Kubernetes cluster. We also learned about SecurityContext and different samples for securityContext configurations. We also discovered how to control traffic flow between Pods by using an object called NetworkPolicy that behaves like a networking firewall within the cluster. As part of the container security, we explored the alternative container runtimes such as Kata Containers and gVisor options. Finally, we learned how to configure the credentials for the private container registries. In the next chapter, we are going to dive deep into advanced techniques for scheduling Pods.

Further reading

- **Controlling Access to the Kubernetes API:** https://kubernetes.io/docs/concepts/security/controlling-access
- **Managing Service Accounts:** https://kubernetes.io/docs/reference/access-authn-authz/service-accounts-admin/
- **Configure Service Accounts for Pods:** https://kubernetes.io/docs/tasks/configure-pod-container/configure-service-account/
- **Certificates and Certificate Signing Requests:** https://kubernetes.io/docs/reference/access-authn-authz/certificate-signing-requests/
- **Authorization:** https://kubernetes.io/docs/reference/access-authn-authz/authorization/
- **What is OpenID Connect:** https://openid.net/developers/how-connect-works/
- **Admission controller:** https://kubernetes.io/docs/reference/access-authn-authz/admission-controllers

- **Security Context:** `https://kubernetes.io/docs/tasks/configure-pod-container/security-context/`

Join our community on Discord

Join our community's Discord space for discussions with the authors and other readers:

`https://packt.link/cloudanddevops`

19

Advanced Techniques for Scheduling Pods

At the beginning of the book, in *Chapter 2, Kubernetes Architecture – from Container Images to Running Pods*, we explained the principles behind the Kubernetes scheduler (kube-scheduler) control plane component and its crucial role in the cluster. In short, its responsibility is to schedule container workloads (Kubernetes Pods) and assign them to healthy nodes that fulfill the criteria required for running a particular workload.

This chapter will cover how you can control the criteria for scheduling Pods in the cluster. We will pay particular attention to Node **affinity**, **taints**, and **tolerations** for Pods. We will also take a closer look at **scheduling policies**, which give kube-scheduler flexibility in how it prioritizes Pod workloads. You will find all of these concepts important in running production clusters at the cloud scale.

In this chapter, we will cover the following topics:

- Refresher – What is kube-scheduler?
- Managing Node affinity
- Using Node taints and tolerations
- Understanding Static Pods in Kubernetes
- Extended Scheduler Configurations in Kubernetes

Technical requirements

For this chapter, you will need the following:

- A *multi-node* Kubernetes cluster is required. Having a multi-node cluster will make understanding Node affinity, taints, and tolerations much easier.
- The Kubernetes CLI (kubectl) installed on your local machine and configured to manage your Kubernetes cluster.

Basic Kubernetes cluster deployment (local and cloud-based) and `kubectl` installation have been covered in *Chapter 3, Installing Your First Kubernetes Cluster*. The previous chapters *15*, *16*, and *17* have provided you an overview of how to deploy a fully functional Kubernetes cluster on different cloud platforms.

You can download the latest code samples for this chapter from the official GitHub repository: `https://github.com/PacktPublishing/The-Kubernetes-Bible-Second-Edition/tree/main/Chapter19`.

Refresher — What is kube-scheduler?

In Kubernetes clusters, kube-scheduler is a critical component of the control plane. The main responsibility of this component is scheduling container workloads (Pods) and **assigning** them to healthy compute nodes (also known as worker nodes) that fulfill the criteria required for running a particular workload. To recap, a Pod is a group of one or more containers with a shared network and storage and is the smallest **deployment unit** in the Kubernetes system. You usually use different Kubernetes controllers, such as Deployment objects and StatefulSet objects, to manage your Pods, but it is kube-scheduler that eventually assigns the created Pods to particular Nodes in the cluster.

> For managed Kubernetes clusters in the cloud, such as **Azure Kubernetes Service** (AKS) or Amazon **Elastic Kubernetes Service** (EKS), you typically do not have access to the control plane or controller nodes as they are managed by the cloud service provider. This means you won't have direct access to components like `kube-scheduler` or control over its configuration, such as scheduling policies. However, you can still control all the parameters of Pods that influence their scheduling.

kube-scheduler queries the **Kubernetes API Server** (`kube-apiserver`) at regular intervals in order to list the Pods that have not been *scheduled*. At creation, Pods are marked as *not* scheduled – this means no Node was elected to run them. A Pod that is not scheduled will be registered in the `etcd` cluster state but without any Node assigned to it and, thus, no running kubelet will be aware of this Pod. Ultimately, no container described in the Pod specification will run at this point.

Internally, the Pod object, as it is stored in `etcd`, has a property called `nodeName`. As the name suggests, this property should contain the name of the Node that will host the Pod. When this property is set, we say the Pod is in a `scheduled` state; otherwise, the Pod is in a `pending` state.

We need to find a way to fill this `nodeName` value, and this is the role of kube-scheduler. For this, kube-scheduler polls kube-apiserver at regular intervals. It looks for Pod resources with an empty `nodeName` property. Once it finds such Pods, it will execute an algorithm to elect a Node and will update the `nodeName` property in the Pod object by issuing a request to kube-apiserver. When selecting a Node for the Pod, `kube-scheduler` will take into account its internal scheduling policies and criteria that you defined for the Pods. Finally, the kubelet that is responsible for running Pods on the selected Node will notice that there is a new Pod in the `scheduled` state for the Node and will attempt to start the Pod. These principles have been visualized in the following diagram:

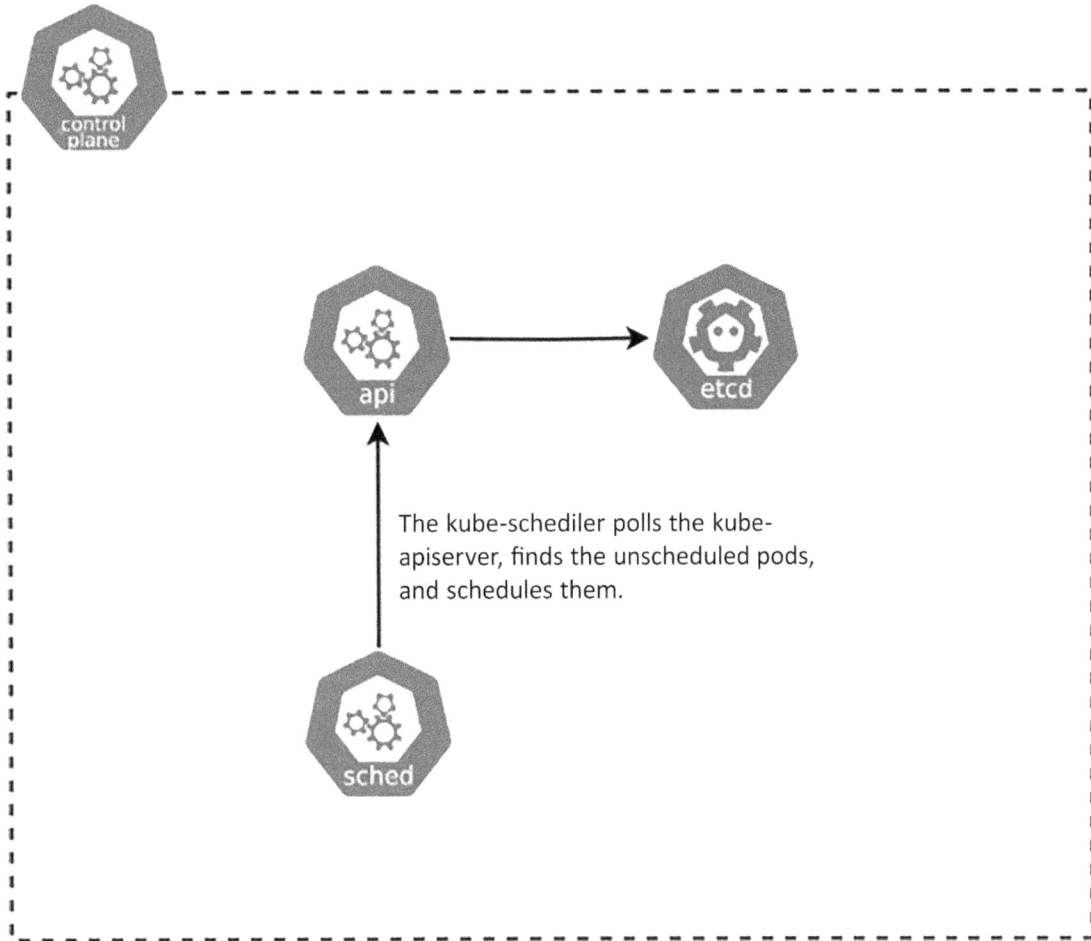

Figure 19.1: Interactions of kube-scheduler and kube-apiserver

The scheduling process for a Pod is performed in two phases:

- **Filtering**: kube-scheduler determines the set of Nodes that are capable of running a given Pod. This includes checking the actual state of the Nodes and verifying any resource requirements and criteria specified by the Pod definition. At this point, if there are no Nodes that can run a given Pod, the Pod cannot be scheduled and remains pending.

- **Scoring**: kube-scheduler assigns scores to each Node based on a set of **scheduling policies**. Then, the Pod is assigned by the scheduler to the Node with the highest score. We will cover scheduling policies in a later section of this chapter.

`kube-scheduler` will consider criteria and configuration values you can optionally pass in the Pod specification. By using these configurations, you can control precisely how kube-scheduler will elect a Node for the Pod. To control where a Pod runs, you can set constraints to restrict it to specific nodes or indicate preferred nodes. We have learned that, typically, Kubernetes will handle Pod placement effectively without any manual constraints, ensuring Pods are spread across Nodes to prevent resource shortages. However, there are times when you might need to influence Pod placement, such as ensuring a Pod runs on a Node with an SSD or co-locating Pods that communicate frequently within the same availability zone.

> The decisions of kube-scheduler are valid precisely at the point in time when the Pod is scheduled. Once the Pod is scheduled and running, kube-scheduler will not perform any rescheduling operations while it is running (which can be for days or even months). So, even if the Pod no longer matches the Node according to your rules, it will remain running. Rescheduling will only happen if the Pod is terminated, and a new Pod needs to be scheduled.

You can use the following methods to influence Pod scheduling in Kubernetes:

- Use the `nodeSelector` field to match against node labels.
- Set affinity and anti-affinity rules.
- Specify the `nodeName` field.
- Define Pod topology spread constraints.
- Taints and tolerations.

In the next sections, we will discuss these configurations to control the scheduling of Pods. Before we jump into the hands-on practices, make sure you have a multi-node Kubernetes cluster to experience the node scheduling scenarios.

In our case, we are using a multi-node Kubernetes cluster using `minikube` as follows (you may change the driver to kvm2, Docker, or Podman):

```
$ minikube start \
   --driver=virtualbox \
   --nodes 3 \
   --cni calico \
   --cpus=2 \
   --memory=2g \
   --kubernetes-version=v1.30.0 \
   --container-runtime=containerd
```

The `--nodes=3` argument will trigger minikube to deploy a Kubernetes cluster with the first node as a controller node (or Master node) and the second and third nodes as compute nodes (or worker nodes), as shown below:

```
$ kubectl get nodes
NAME            STATUS    ROLES          AGE     VERSION
minikube        Ready     control-plane  3m34s   v1.30.0
minikube-m02    Ready     <none>         2m34s   v1.30.0
minikube-m03    Ready     <none>         87s     v1.30.0
```

If you are using any other Kubernetes cluster for learning, then you may skip this minikube cluster setup.

Now, let's take a look at Node affinity, together with Node name and Node selector.

Managing Node affinity

To understand how **Node affinity** works in Kubernetes, we first need to take a look at the most basic scheduling options, which use **Node name** and **Node selector** for Pods.

Using nodeName for Pods

As we mentioned before, each Pod object has a nodeName field, which is usually controlled by kube-scheduler. Nevertheless, it is possible to set this property directly in the YAML manifest when you create a Pod or create a controller that uses a Pod template. This is the simplest form of statically scheduling Pods on a given Node and is generally *not recommended* – it is not flexible and does not scale at all. The names of Nodes can change over time, and you risk running out of resources on the Node.

> You may find setting nodeName explicitly useful in debugging scenarios when you want to run a Pod on a specific Node.

We are going to demonstrate all scheduling principles on an example Deployment object that we introduced in *Chapter 11, Using Kubernetes Deployments for Stateless Workloads*. This is a simple Deployment that manages *five* Pod replicas of an nginx webserver.

Before we use nodeName in the Deployment manifest, we need to know what Nodes we have in the cluster so that we can understand how they are scheduled and how we can influence the scheduling of Pods. You can get the list of Nodes using the kubectl get nodes command, as follows:

```
$ kubectl get nodes
NAME            STATUS    ROLES          AGE     VERSION
minikube        Ready     control-plane  3m34s   v1.30.0
minikube-m02    Ready     <none>         2m34s   v1.30.0
minikube-m03    Ready     <none>         87s     v1.30.0
```

In our example, we are running a three-Node cluster (remember to refer to your correct cluster Node names in the manifest later). For simplicity, let's refer to minikube as Node1, minikube-m02 as Node2, and minikube-m02 as Node3.

For the demonstration, we want to schedule all five nginx Pods to `minikube-m02`. Create the following YAML manifest named `n01_nodename/nginx-deployment.yaml`:

```yaml
# 01_nodename/nginx-deployment.yaml
apiVersion: apps/v1
kind: Deployment
metadata:
  name: nginx-app
spec:
  replicas: 5
  selector:
    matchLabels:
      app: nginx
      environment: test
  template:
    metadata:
      labels:
        app: nginx
        environment: test
    spec:
      containers:
        - name: nginx
          image: nginx:1.17
          ports:
            - containerPort: 80
```

Apply the Deployment YAML as usual:

```
$ kubectl apply -f 01_nodename/nginx-deployment.yaml
deployment.apps/nginx-app created
```

The Deployment object will create five Pod replicas. Use `kubectl get pods -o wide` to see the Pods and Node names. Let's use a customized output as follows:

```
$ kubectl get pods --namespace default --output=custom-columns="NAME:.metadata.
name,STATUS:.status.phase,NODE:.spec.nodeName"
NAME                          STATUS    NODE
nginx-app-7b547cfd87-4g9qx    Running   minikube
nginx-app-7b547cfd87-m76l2    Running   minikube-m02
nginx-app-7b547cfd87-mjf78    Running   minikube-m03
nginx-app-7b547cfd87-vvrgk    Running   minikube-m02
nginx-app-7b547cfd87-w7jcw    Running   minikube-m03
```

As you can see, by default, the Pods have been distributed uniformly – Node1 has received one Pod, Node2 two Pods, and Node3 two Pods. This is a result of the default scheduling policies enabled in `kube-scheduler` for filtering and scoring.

> If you are running a **non-managed** Kubernetes cluster, you can inspect the logs for the kube-scheduler Pod using the `kubectl logs` command, or even directly in the control plane nodes in `/var/log/kube-scheduler.log`. This may also require increased verbosity of logs for the kube-scheduler process. You can read more at `https://kubernetes.io/docs/reference/command-line-tools-reference/kube-scheduler/`.

At this point, the Pod template in `.spec.template.spec` does not contain any configurations that affect the scheduling of the Pod replicas.

We will now **forcefully** assign all Pods in the Deployment to **Node2** (`minikube-m02` in our case) in the cluster using the `nodeName` field in the Pod template. Change the `nginx-deployment.yaml` YAML manifest so that it has this property set with the correct Node name for *your* cluster:

```
# 01_nodename/nginx-deployment.yaml
apiVersion: apps/v1
kind: Deployment
metadata:
  name: nginx-app
spec:
  ...
  template:
    ...
    spec:
      nodeName: minikube-m02
...<removed for brevity>...
```

Notice the line `nodeName: minikube-m02`; we are explicitly stating that `minikube-m02` should be used as the Node for deploying our nginx Pods.

Apply the manifest to the cluster using the `kubectl apply -f ./nginx-deployment.yaml` command:

```
$ kubectl apply -f 01_nodename/nginx-deployment.yaml
deployment.apps/nginx-app created
```

Now, inspect the Pod status and Node assignment again:

```
$ kubectl get pods --namespace default --output=custom-columns="NAME:.metadata.name,STATUS:.status.phase,NODE:.spec.nodeName"
NAME                          STATUS    NODE
nginx-app-85b577894f-8tqj7    Running   minikube-m02
nginx-app-85b577894f-9c6hd    Running   minikube-m02
```

```
nginx-app-85b577894f-fldxx    Running    minikube-m02
nginx-app-85b577894f-jrnjc    Running    minikube-m02
nginx-app-85b577894f-vs7c5    Running    minikube-m02
```

As expected, *all five* Pods are now running on Node2 (`minikube-m02`). These are all new Pods – when you change the Pod template in the Deployment specification, it causes an internal rollout using a new ReplicaSet object, while the old ReplicaSet object is scaled down, as explained in *Chapter 11*, *Using Kubernetes Deployments for Stateless Workloads*.

> In this way, we have actually *bypassed* `kube-scheduler`. If you inspect events for one of the Pods using the `kubectl describe` pod command, you will see that it lacks any events with `Scheduled` as a reason.

Next, we are going to take a look at another basic method of scheduling Pods, which is `nodeSelector`.

Using nodeSelector for Pods

Pod specification has a special field, `.spec.nodeSelector`, that gives you the ability to schedule your Pod only on Nodes that have certain label values. This concept is similar to **label selectors** for Deployments and StatefulSets, but the difference is that it allows only simple *equality-based* comparisons for labels. You cannot do advanced *set-based* logic.

This is especially useful in:

- **Hybrid clusters**: Ensure Windows containers run on Windows nodes and Linux containers run on Linux nodes by specifying the operating system as a scheduling criterion.
- **Resource allocation**: Target Pods to nodes with specific resources (CPU, memory, storage) to optimize resource utilization.
- **Hardware requirements**: Schedule Pods requiring special hardware (e.g., GPUs) only on nodes with those capabilities.
- **Security zones**: Define security zones with labels and use `nodeSelector` to restrict Pods to specific zones for enhanced security.

Every Kubernetes Node comes by default with a set of labels, which include the following:

- `kubernetes.io/arch`: Describes the Node's processor architecture, for example, `amd64` or `arm`. This is also defined as `beta.kubernetes.io/arch`.
- `kubernetes.io/os`: Has a value of `linux` or `Windows`. This is also defined as `beta.kubernetes.io/os`.
- `node-role.kubernetes.io/control-plane`: The role of the node in the Kubernetes cluster.

If you inspect the labels for one of the Nodes, you will see that there are plenty of them. In our case, some of them are specific to `minikube` clusters:

```
$ kubectl describe nodes minikube
Name:              minikube
```

```
    Roles:                  control-plane
    Labels:                 beta.kubernetes.io/arch=amd64
                            beta.kubernetes.io/os=linux
                            kubernetes.io/arch=amd64
                            kubernetes.io/hostname=minikube
                            kubernetes.io/os=linux
                            minikube.k8s.io/
commit=5883c09216182566a63dff4c326a6fc9ed2982ff
                            minikube.k8s.io/name=minikube
                            minikube.k8s.io/primary=true
                            minikube.k8s.io/updated_at=2024_07_21T16_40_25_0700
                            minikube.k8s.io/version=v1.33.1
                            node-role.kubernetes.io/control-plane=
                            node.kubernetes.io/exclude-from-external-load-balancers=
...<removed for brevity>...
```

Of course, you can define your *own* labels for the Nodes and use them to control scheduling. Please note that in general you should use semantic labeling for your resources in Kubernetes, rather than give them special labels just for the purpose of scheduling. Let's demonstrate how to do that by following these steps:

1. Use the kubectl label nodes command to add a node-type label with a superfast value to Node 1 and Node 2 in the cluster:

```
$ kubectl label nodes minikube-m02 node-type=superfast
node/minikube-m02 labeled
$ kubectl label nodes minikube-m03 node-type=superfast
node/minikube-m03 labeled
```

2. Verify the node labels as follows:

```
$ kubectl get nodes --show-labels |grep superfast
minikube-m02    Ready      <none>          2m59s    v1.31.0
beta.kubernetes.io/arch=amd64,beta.kubernetes.io/
os=linux,kubernetes.io/arch=amd64,kubernetes.io/
hostname=minikube-m02,kubernetes.io/os=linux,minikube.k8s.io/
commit=5883c09216182566a63dff4c326a6fc9ed2982ff,minikube.k8s.io/
name=minikube,minikube.k8s.io/primary=false,minikube.k8s.io/updated_
at=2024_10_13T15_26_06_0700,minikube.k8s.io/version=v1.33.1,node-
type=superfast
minikube-m03    Ready      <none>          2m30s    v1.31.0
beta.kubernetes.io/arch=amd64,beta.kubernetes.io/
os=linux,kubernetes.io/arch=amd64,kubernetes.io/
hostname=minikube-m03,kubernetes.io/os=linux,minikube.k8s.io/
commit=5883c09216182566a63dff4c326a6fc9ed2982ff,minikube.k8s.io/
```

```
name=minikube,minikube.k8s.io/primary=false,minikube.k8s.io/updated_
at=2024_10_13T15_26_34_0700,minikube.k8s.io/version=v1.33.1,node-
type=superfast
```

3. Edit the ./nginx-deployment.yaml Deployment manifest (or create another one called 02_nodeselector/nginx-deployment.yaml) so that nodeSelector in the Pod template is set to node-type: superfast, as follows:

```yaml
# 02_nodeselector/nginx-deployment.yaml
apiVersion: apps/v1
kind: Deployment
metadata:
  name: nginx-app
spec:
  ...
  template:
    ...
    spec:
      nodeSelector:
        node-type: superfast
...<removed for brevity>
```

4. Apply the manifest to the cluster using the kubectl apply -f 02_nodeselector/nginx-deployment.yaml command and inspect the Pod status and Node assignment again. You may need to wait a while for the Deployment rollout to finish:

```
$ kubectl get pods --namespace default --output=custom-columns="NAME:.
metadata.name,STATUS:.status.phase,NODE:.spec.nodeName"
NAME                          STATUS    NODE
nginx-app-6c5b8b758-2dcsc     Running   minikube-m02
nginx-app-6c5b8b758-48c5t     Running   minikube-m03
nginx-app-6c5b8b758-pfmvg     Running   minikube-m03
nginx-app-6c5b8b758-v6rhj     Running   minikube-m02
nginx-app-6c5b8b758-zqvqm     Running   minikube-m02
```

As you can see in the preceding output, Pods are now assigned to minikube-m02 and minikube-m03 (minikube-m02 has been assigned with three Pods and minikube-m02 with two Pods). The Pods have been distributed among Nodes that have the node-type=superfast label.

5. In contrast, if you change the ./nginx-deployment.yaml manifest so that nodeSelector in the Pod template is set to node-type: slow, which no Node in the cluster has assigned, we will see that Pods could not be scheduled and the Deployment will be stuck. Edit the manifest (or copy to a new file called 02_nodeselector/nginx-deployment-slow.yaml):

```yaml
# 02_nodeselector/nginx-deployment.yaml
apiVersion: apps/v1
```

```
kind: Deployment
metadata:
  name: nginx-app
spec:
  ...
  template:
    ...
    spec:
      nodeSelector:
        node-type: slow
...<removed for brevity>
```

6. Apply the manifest to the cluster as follows:

```
$ kubectl apply -f  02_nodeselector/nginx-deployment-slow.yaml
```

Inspect the Pod status and Node assignment again:

```
$ kubectl get pods --namespace default --output=custom-columns="NAME:.
metadata.name,STATUS:.status.phase,NODE:.spec.nodeName"
NAME                         STATUS     NODE
nginx-app-6c5b8b758-48c5t    Running    minikube-m03
nginx-app-6c5b8b758-pfmvg    Running    minikube-m03
nginx-app-6c5b8b758-v6rhj    Running    minikube-m02
nginx-app-6c5b8b758-zqvqm    Running    minikube-m02
nginx-app-9cc8544f4-7dwcd    Pending    <none>
nginx-app-9cc8544f4-cz947    Pending    <none>
nginx-app-9cc8544f4-lfqqj    Pending    <none>
```

The reason why three new Pods are pending and four old Pods are still running is the default configuration of rolling updates in the Deployment object. By default, maxSurge is set to 25% of Pod replicas (the absolute number is *rounded up*), so in our case, two Pods are allowed to be created above the desired number of five Pods. In total, we now have seven Pods. At the same time, maxUnavailable is also 25% of Pod replicas (but the absolute number is *rounded down*), so in our case, one Pod out of five cannot be available. In other words, four Pods must be Running. And because the new Pending Pods cannot get a Node in the process of scheduling, the Deployment is stuck waiting and not progressing. Normally, in this case, you need to either perform a rollback to the previous version for the Deployment or change nodeSelector to one that matches existing Nodes properly. Of course, there is also an alternative of adding a new Node with matching labels or adding missing labels to the existing ones without performing a rollback.

We will now continue the topic of scheduling Pods by looking at the first of some more advanced techniques: **Node affinity**.

Using the nodeAffinity configuration for Pods

The concept of Node affinity expands the `nodeSelector` approach and provides a richer language for defining which Nodes are preferred or avoided for your Pod. In everyday life, the word "affinity" is defined as "*a natural liking for and understanding of someone or something,*" and this best describes the purpose of Node affinity for Pods. That is, you can control which Nodes your Pod will be *attracted* to or *repelled* by.

With Node affinity, represented in `.spec.affinity.nodeAffinity` for the Pod, you get the following enhancements over simple `nodeSelector`:

- You get a richer language for expressing the rules for matching Pods to Nodes. For example, you can use the `In`, `NotIn`, `Exists`, `DoesNotExist`, `Gt`, and `Lt` operators for labels.
- Similar to `nodeAffinity`, it is possible to do scheduling using `inter-Pod` affinity (`podAffinity`) and additionally `anti-affinity` (`podAntiAffinity`). Anti-affinity has the opposite effect of affinity – you can define rules that repel Pods. In this way, you can make your Pods be attracted to Nodes that *already run* certain Pods. This is especially useful if you want to collocate Pods to decrease latency.
- It is possible to define **soft** affinity and anti-affinity rules that represent a *preference* instead of a **hard** rule. In other words, the scheduler can still schedule the Pod, even if it cannot match the soft rule. Soft rules are represented by the `preferredDuringSchedulingIgnoredDuringE xecution` field in the specification, whereas hard rules are represented by the `requiredDuri ngSchedulingIgnoredDuringExecution` field.
- Soft rules can be **weighted**, and it is possible to add multiple rules with different weight values. The scheduler will consider this weight value together with the other parameters to make a decision on the affinity.

> Even though there is no Node anti-affinity field provided by a separate field in the spec, as in the case of inter-Pod anti-affinity you can still achieve similar results by using the `NotIn` and `DoesNotExist` operators. In this way, you can make Pods be repelled from Nodes with specific labels, also in a soft way. Refer to the documentation to learn more: `https://kubernetes.io/docs/concepts/scheduling-eviction/assign-pod-node/#affinity-and-anti-affinity`.

The use cases and scenarios for defining the Node affinity and inter-Pod affinity/anti-affinity rules are *unlimited*. It is possible to express all kinds of requirements in this way, provided that you have enough labeling on the Nodes. For example, you can model requirements like scheduling the Pod only on a Windows Node with an Intel CPU and premium storage in the West Europe region but currently not running Pods for MySQL, or try not to schedule the Pod in availability zone 1, but if it is not possible, then availability zone 1 is still OK.

To demonstrate Node affinity, we will try to model the following requirements for our Deployment: "*Try* to schedule the Pod only on Nodes with a `node-type` label with a `fast` or `superfast` value, but if this is not possible, use any Node but *strictly* not with a `node-type` label with an `extremelyslow` value." For this, we need to use:

- A **soft Node affinity** rule of type `preferredDuringSchedulingIgnoredDuringExecution` to match `fast` and `superfast` Nodes.
- A **hard Node affinity** rule of type `requiredDuringSchedulingIgnoredDuringExecution` to repel the Pod strictly from Nodes with `node-type` as `extremelyslow`. We need to use the `NotIn` operator to get the anti-affinity effect.

In our cluster, we are going to first have the following labels for Nodes:

- Node1: `slow`
- Node2: `fast`
- Node3: `superfast`

As you can see, according to our requirements, the Deployment Pods should be scheduled on Node2 and Node3, unless something is preventing them from being allocated there, like a lack of CPU or memory resources. In that case, Node1 would also be allowed as we use the soft affinity rule.

Next, we will relabel the Nodes in the following way:

- Node1: `slow`
- Node2: `extremelyslow`
- Node3: `extremelyslow`

Subsequently, we will need to redeploy our Deployment (for example, scale it down to zero and up to the original replica count, or use the `kubectl rollout restart` command) to reschedule the Pods again. After that, looking at our requirements, kube-scheduler should assign all Pods to Node1 (because it is still allowed by the soft rule) but avoid *at all costs* Node2 and Node3. If, by any chance, Node1 has no resources to run the Pod, then the Pods will be stuck in the `Pending` state.

> To solve the issue of rescheduling already running Pods (in other words, to make kube-scheduler consider them again), there is an incubating Kubernetes project named **Descheduler**. You can find out more here: `https://github.com/kubernetes-sigs/descheduler`.

To do the demonstration, please follow these steps:

1. Use the `kubectl label nodes` command to add a `node-type` label with a `slow` value for Node1, a `fast` value for Node2, and a `superfast` value for Node3:

```
$ kubectl label nodes --overwrite minikube node-type=slow
node/minikube labeled

$ kubectl label nodes --overwrite minikube-m02 node-type=fast
node/minikube-m02 labeled

$ kubectl label nodes --overwrite minikube-m03 node-type=superfast
node/minikube-m03 not labeled
# Note that this label was already present with this value
```

2. Edit the `03_affinity/nginx-deployment.yaml` Deployment manifest and define the soft Node affinity rule as follows:

```
# 03_affinity/nginx-deployment.yaml
...
spec:
  ...
  template:
    ...
    spec:
      affinity:
        nodeAffinity:
          requiredDuringSchedulingIgnoredDuringExecution:
            nodeSelectorTerms:
            - matchExpressions:
              - key: node-type
                operator: NotIn
                values:
                - extremelyslow
          preferredDuringSchedulingIgnoredDuringExecution:
          - weight: 1
            preference:
              matchExpressions:
              - key: node-type
                operator: In
                values:
                - fast
                - superfast
...<removed for brevity>...
```

As you can see, we have used `nodeAffinity` (not `podAffinity` or `podAntiAffinity`) with `preferredDuringSchedulingIgnoredDuringExecution` set so that it has only one soft rule: `node-type` should have a `fast` value or a `superfast` value. This means that if there are no resources on such Nodes, they can still be scheduled on other Nodes. Additionally, we specify one hard anti-affinity rule in `requiredDuringSchedulingIgnoredDuringExecution`, which says that `node-type` *must not* be `extremelyslow`. You can find the full specification of Pod's `.spec.affinity` in the official documentation: `https://kubernetes.io/docs/concepts/scheduling-eviction/assign-pod-node/#affinity-and-anti-affinity`.

3. Apply the manifest to the cluster using the `kubectl apply -f 03_affinity/nginx-deployment.yaml` command and inspect the Pod status and Node assignment again. You may need to wait a while for the Deployment rollout to finish:

```
$ kubectl get pods --namespace default --output=custom-columns="NAME:.
metadata.name,STATUS:.status.phase,NODE:.spec.nodeName"
NAME                             STATUS    NODE
nginx-app-7766c596cc-4d4sl       Running   minikube-m02
nginx-app-7766c596cc-4h6k6       Running   minikube-m03
nginx-app-7766c596cc-ksld5       Running   minikube-m03
nginx-app-7766c596cc-nw9hx       Running   minikube-m02
nginx-app-7766c596cc-tmwhm       Running   minikube-m03
```

Our Node affinity rules were defined to prefer Nodes that have node-type set to either fast or superfast, and indeed the Pods were scheduled for Node2 and Node3 only.

Now, we will perform an experiment to demonstrate how the soft part of Node affinity works together with the hard part of Node anti-affinity. We will relabel the Nodes as described in the introduction, redeploy the Deployment, and observe what happens. Please follow these steps:

1. Use the kubectl label nodes command to add a node-type label with a slow value for Node 0, an extremelyslow value for Node1, and an extremelyslow value for Node2:

```
$ kubectl label nodes --overwrite minikube node-type=slow
node/minikube not labeled
# Note that this label was already present with this value

$ kubectl label nodes --overwrite minikube-m02 node-type=extremelyslow
node/minikube-m02 labeled

$ kubectl label nodes --overwrite minikube-m03 node-type=extremelyslow
node/minikube-m03 labeled
```

2. At this point, if you were to check Pod assignments using kubectl get pods, there would be no difference. This is because, as we explained before, a Pod's assignment to Nodes is valid only at the time of scheduling, and after that, it is not changed unless they are restarted. To force the restart of Pods, we could scale the Deployment down to zero replicas and then back to five. But there is an easier way, which is to use an imperative kubectl rollout restart command. This approach has the benefit of not making the Deployment unavailable, and it performs a rolling restart of Pods without a decrease in the number of available Pods. Execute the following command:

```
$ kubectl rollout restart deployment nginx-app
deployment.apps/nginx-app restarted
```

3. Inspect the Pod status and Node assignment again. You may need to wait a while for the Deployment rollout to finish:

```
$ kubectl get pods --namespace default --output=custom-columns="NAME:.
metadata.name,STATUS:.status.phase,NODE:.spec.nodeName"
```

```
NAME                          STATUS     NODE
nginx-app-7d8c65464c-5d9cc    Running    minikube
nginx-app-7d8c65464c-b97g8    Running    minikube
nginx-app-7d8c65464c-cqwh5    Running    minikube
nginx-app-7d8c65464c-kh8bm    Running    minikube
nginx-app-7d8c65464c-xhpss    Running    minikube
```

The output shows that, as expected, all Pods have been scheduled to Node1, which is labeled with node-type=slow. We allow such Nodes if there is nothing better, and in this case, Node2 and Node3 have the node-type=extremelyslow label, which is prohibited by the hard Node anti-affinity rule.

> To achieve even higher granularity and control of Pod scheduling, you can use *Pod topology spread constraints*. More details are available in the official documentation: https://kubernetes.io/docs/concepts/workloads/pods/pod-topology-spread-constraints/.

Congratulations, you have successfully configured Node affinity for our Deployment Pods! We will now explore another way of scheduling Pods – *Taints and tolerations*.

Using Node taints and tolerations

Using the Node and inter-Pod affinity mechanism for scheduling Pods is very powerful, but sometimes you need a simpler way of specifying which Nodes should *repel* Pods. Kubernetes has two slightly older and simpler features for this purpose – **taints** and **tolerations**. You apply a taint to a given Node (which describes some kind of limitation) and the Pod must have a specific toleration defined to be schedulable on the tainted Node. If the Pod has a toleration, it does not mean that the taint is *required* on the Node. The definition of *taint* is "a trace of a bad or undesirable substance or quality," and this reflects the idea pretty well – all Pods will *avoid* a Node if there is a taint set for them, but we can instruct Pods to *tolerate* a specific taint.

> If you look closely at how taints and tolerations are described, you can see that you can achieve similar results with Node labels and Node hard and soft affinity rules with the NotIn operator. There is one catch – you can define taints with a NoExecute effect, which will result in the termination of the Pod if it cannot tolerate it. You cannot get similar results with affinity rules unless you restart the Pod manually.

Taints for Nodes have the following structure: <key>=<value>:<effect>. The **key** and **value** pair *identifies* the taint and can be used for more granular tolerations, for example, tolerating all taints with a given key and any value. This is similar to labels, but please remember that taints are separate properties, and defining a taint does not affect Node labels. In our example demonstration, we will use our taint with a machine-check-exception key and a memory value. This is, of course, a theoretical example where we want to indicate that there is a hardware issue with memory on the host, but you could also have a taint with the same key and instead a cpu or disk value.

In general, your taints should *semantically* label the type of issue that the Node is experiencing. There is nothing preventing you from using any keys and values for creating taints, but if they make semantic sense, it is much easier to manage them and define tolerations.

The taint can have different effects:

- `NoSchedule` – kube-scheduler *will not schedule* Pods to this Node. Similar behavior can be achieved using a hard Node affinity rule.
- `PreferNoSchedule` – kube-scheduler *will try to not schedule* Pods to this Node. Similar behavior can be achieved using a soft Node affinity rule.
- `NoExecute` – kube-scheduler *will not schedule* Pods to this Node and *evict* (terminate and reschedule) running Pods from this Node. You cannot achieve similar behavior using Node affinity rules. Note that when you define a toleration for a Pod for this type of taint, it is possible to control how long the Pod will tolerate the taint before it gets evicted, using `tolerationSeconds`.

Kubernetes manages quite a few `NoExecute` taints automatically by monitoring the Node hosts. The following taints are built in and managed by **NodeController** or the `kubelet`:

- `node.kubernetes.io/not-ready`: Added when NodeCondition `Ready` has a `false` status.
- `node.kubernetes.io/unreachable`: Added when NodeCondition `Ready` has an `Unknown` status. This happens when `NodeController` cannot reach the Node.
- `node.kubernetes.io/memory-pressure`: Node is experiencing memory pressure.
- `node.kubernetes.io/disk-pressure`: Node is experiencing disk pressure.
- `node.kubernetes.io/network-unavailable`: Network is currently down on the Node.
- `node.kubernetes.io/unschedulable`: Node is currently in an `unschedulable` state.
- `node.cloudprovider.kubernetes.io/uninitialized`: Intended for Nodes that are prepared by an external cloud provider. When the Node gets initialized by `cloud-controller-manager`, this taint is removed.

To add a taint on a Node, you use the `kubectl taint node` command in the following way:

```
$ kubectl taint node <nodeName> <key>=<value>:<effect>
```

So, for example, if we want to use key `machine-check-exception` and a `memory` value with a `NoExecute` effect for Node1, we will use the following command:

```
$ kubectl taint node minikube machine-check-exception=memory:NoExecute
node/minikube tainted
```

To remove the same taint, you need to use the following command (bear in mind the - character at the end of the taint definition):

```
$ kubectl taint node minikube machine-check-exception=memory:NoExecute-
node/minikube untainted
```

You can also remove all taints with a specified key:

```
$ kubectl taint node minikube machine-check-exception:NoExecute-
```

To counteract the effect of the taint on a Node for specific Pods, you can define tolerations in their specification. In other words, you can use tolerations to ignore taints and still schedule the Pods to such Nodes. If a Node has multiple taints applied, the Pod must tolerate all of its taints. Tolerations are defined under .spec.tolerations in the Pod specification and have the following structure:

```
tolerations:
- key: <key>
  operator: <operatorType>
  value: <value>
  effect: <effect>
```

The operator can be either Equal or Exists. Equal means that the key and value of the taint must match exactly, whereas Exists means that just key must match and value is not considered. In our example, if we want to ignore the taint, the toleration will need to look like this:

```
tolerations:
- key: machine-check-exception
  operator: Equal
  value: memory
  effect: NoExecute
```

> Please note, you can define multiple tolerations for a Pod to ensure the correct Pod placement.

In the case of NoExecute tolerations, it is possible to define an additional field called tolerationSeconds, which specifies how long the Pod will tolerate the taint until it gets evicted. So, this is a way of having partial toleration of taint with a timeout. Please note that if you use NoExecute taints, you usually also need to add a NoSchedule taint. In this way, you can prevent any **eviction loops** from happening when the Pod has a NoExecute toleration with tolerationSeconds set. This is because the taint has no effect for a specified number of seconds, which also includes *not* preventing the Pod from being scheduled for the tainted Node.

> When Pods are created in the cluster, Kubernetes automatically adds two Exists tolerations for node.kubernetes.io/not-ready and node.kubernetes.io/unreachable with tolerationSeconds set to 300.

Now that we have learned about taints and tolerations, we will put this knowledge into practice with a few demonstrations. Please follow the next steps to go through the taints and tolerations exercise:

1. If you have the `nginx-app` Deployment with Node affinity defined still running from the previous section, it will currently have all Pods running on Node1 (minikube). The Node affinity rules are constructed in such a way that the Pods cannot be scheduled on Node2 and Node3. Let's see what happens if you taint Node1 with `machine-check-exception=memory:NoExecute`:

```
$ kubectl taint node minikube machine-check-exception=memory:NoExecute
node/minikube tainted
```

2. Check the Pod status and Node assignment:

```
$ kubectl get pods --namespace default --output=custom-columns="NAME:.
metadata.name,STATUS:.status.phase,NODE:.spec.nodeName"
NAME                           STATUS    NODE
nginx-app-7d8c65464c-5j69n     Pending   <none>
nginx-app-7d8c65464c-c8j58     Pending   <none>
nginx-app-7d8c65464c-cnczc     Pending   <none>
nginx-app-7d8c65464c-drpdh     Pending   <none>
nginx-app-7d8c65464c-xss9b     Pending   <none>
```

All Deployment Pods are now in the `Pending` state because kube-scheduler is unable to find a Node that can run them.

3. Edit the `./nginx-deployment.yaml` Deployment manifest (or check `04_taints/nginx-deployment.yaml`) and remove `affinity`. Instead, define taint toleration for `machine-check-exception=memory:NoExecute` with a timeout of 60 seconds as follows:

```
# 04_taints/nginx-deployment.yaml
apiVersion: apps/v1
kind: Deployment
metadata:
  name: nginx-app
spec:
  ...
  template:
    ...
    spec:
      tolerations:
        - key: machine-check-exception
          operator: Equal
          value: memory
          effect: NoExecute
          tolerationSeconds: 60
...<removed for brevity>...
```

When this manifest is applied to the cluster, the old Node affinity rules which prevented scheduling to Node2 and Node3 will be gone. The Pods will be able to schedule on Node2 and Node3, but Node1 has taint `machine-check-exception=memory:NoExecute`. So, the Pods should *not* be scheduled to Node0, as `NoExecute` implies `NoSchedule`, *right*? Let's check that.

4. Apply the manifest to the cluster using the `kubectl apply -f 04_taints/nginx-deployment.yaml` command and inspect the Pod status and Node assignment again. You may need to wait a while for the Deployment rollout to finish:

```
$ kubectl get pods --namespace default --output=custom-columns="NAME:.
metadata.name,STATUS:.status.phase,NODE:.spec.nodeName"
NAME                          STATUS     NODE
nginx-app-84d755f746-4zkjd    Running    minikube
nginx-app-84d755f746-58qmh    Running    minikube-m02
nginx-app-84d755f746-5h5vk    Running    minikube-m03
nginx-app-84d755f746-psmgf    Running    minikube-m02
nginx-app-84d755f746-zkbc6    Running    minikube-m03
```

This result may be a bit surprising. As you can see, we got Pods scheduled on Node2 and Node3, but at the same time Node1 has received Pods, and they are in an eviction loop every 60 seconds! The explanation for this is that `tolerationSeconds` for the `NoExecute` taint implies that the whole taint is ignored for 60 seconds. So, kube-scheduler can schedule the Pod on Node1, even though it will get evicted later.

5. Let's fix this behavior by applying a recommendation to use a `NoSchedule` taint whenever you use a `NoExecute` taint. In this way, the evicted Pods will have no chance to be scheduled on the tainted Node again, unless, of course, they start tolerating this type of taint too. Execute the following command to taint Node0:

```
$ kubectl taint node minikube machine-check-exception=memory:NoSchedule
node/minikube tainted
```

6. Inspect the Pod status and Node assignment again:

```
$ kubectl get pods --namespace default --output=custom-columns="NAME:.
metadata.name,STATUS:.status.phase,NODE:.spec.nodeName"
NAME                          STATUS     NODE
nginx-app-84d755f746-58qmh    Running    minikube-m02
nginx-app-84d755f746-5h5vk    Running    minikube-m03
nginx-app-84d755f746-psmgf    Running    minikube-m02
nginx-app-84d755f746-sm2cm    Running    minikube-m03
nginx-app-84d755f746-zkbc6    Running    minikube-m03
```

In the output, you can see that the Pods are now distributed between Node2 and Node3 – exactly as we wanted.

7. Now, remove *both* taints from Node1:

```
$ kubectl taint node minikube machine-check-exception-
node/minikube untainted
```

8. Restart the Deployment to reschedule the Pods using the following command:

```
$ kubectl rollout restart deployment nginx-app
deployment.apps/nginx-app restarted
```

9. Inspect the Pod status and Node assignment again:

```
$ kubectl get pods --namespace default --output=custom-columns="NAME:.
metadata.name,STATUS:.status.phase,NODE:.spec.nodeName"
NAME                          STATUS    NODE
nginx-app-5bdd957558-fj7bk    Running   minikube-m02
nginx-app-5bdd957558-mrddn    Running   minikube-m03
nginx-app-5bdd957558-mz2pz    Running   minikube-m02
nginx-app-5bdd957558-pftz5    Running   minikube-m03
nginx-app-5bdd957558-vm6k9    Running   minikube
```

The Pods are again distributed evenly between all three Nodes.

10. And finally, let's see how the combination of the NoExecute and NoSchedule taints work, with tolerationSeconds for NoExecute set to 60. Apply two taints to Node1 again:

```
$ kubectl taint node minikube machine-check-exception=memory:NoSchedule
node/minikube tainted
$ kubectl taint node minikube machine-check-exception=memory:NoExecute
node/minikube tainted
```

11. Immediately after that, start watching Pods with their Node assignments. Initially, you will see that the Pods are still running on Node1 for some time. But after 60 seconds, you will see:

```
$ kubectl get pods --namespace default --output=custom-columns="NAME:.
metadata.name,STATUS:.status.phase,NODE:.spec.nodeName"
NAME                          STATUS    NODE
nginx-app-5bdd957558-7n42p    Running   minikube-m03
nginx-app-5bdd957558-fj7bk    Running   minikube-m02
nginx-app-5bdd957558-mrddn    Running   minikube-m03
nginx-app-5bdd957558-mz2pz    Running   minikube-m02
nginx-app-5bdd957558-pftz5    Running   minikube-m03
```

As we expected, the Pods were evicted after 60 seconds and there were no eviction-schedule loops.

This has demonstrated a more advanced use case for taints that you cannot easily substitute with Node affinity rules.

Let's learn about static Pods in the next section.

Understanding Static Pods in Kubernetes

Static Pods offer a different way to manage Pods within a Kubernetes cluster. Unlike regular Pods, which are controlled by the cluster's Pod schedulers and API server, static Pods are managed directly by the kubelet daemon on a specific node. The API server isn't aware of these Pods, except through mirror Pods that the kubelet creates for them.

Key characteristics:

- **Node-specific:** Static Pods are tied to a single node and can't be moved elsewhere in the cluster.
- **Kubelet management:** The kubelet on the designated node handles starting, stopping, and restarting static Pods.
- **Mirror Pods:** The kubelet creates mirror Pods on the API server to reflect the state of static Pods, but these mirror Pods can't be controlled through the API.

Static Pods can be created using two main methods. The first method is the filesystem-hosted configuration, where you place Pod definitions in YAML or JSON format in a specific directory on the node. The kubelet scans this directory regularly and manages Pods based on the files present. The second method is the web-hosted configuration, where the Pod definition is hosted in a YAML file on a web server. The kubelet is configured with the URL of this file and periodically downloads it to manage the static Pods.

Static Pods are often used for bootstrapping essential cluster components, like the API server or controller manager, on each node. However, for running Pods on every node in a cluster, DaemonSets are usually recommended. Static Pods have limitations, such as not being able to reference other Kubernetes objects like Secrets or ConfigMaps and not supporting ephemeral containers. Understanding static Pods can be useful in scenarios where tight control over Pod placement on individual nodes is needed.

So far, we have covered different mechanisms to control the Pod scheduling and placement. In the next section, we will give a short overview of other scheduler configurations and features.

Extended scheduler configurations in Kubernetes

In addition to the scheduler customizations, Kubernetes also supports some advanced scheduling configurations, which we will discuss in this section.

Scheduler configuration

You can customize this scheduling behavior using a configuration file. This file defines how the scheduler prioritizes Nodes for Pods based on various criteria.

Key concepts:

- **Scheduling profiles:** The configuration file can specify multiple scheduling profiles. Each profile has a distinct name and can be configured with its own set of plugins.

- **Scheduling plugins:** Plugins are like building blocks that perform specific tasks during the scheduling process. They can filter Nodes based on resource availability, hardware compatibility, or other factors.

- **Extension points:** These are stages within the scheduling process where plugins can be hooked in. Different plugins are suited for different stages, such as filtering unsuitable Nodes or scoring suitable Nodes.

> **IMPORTANT: Scheduling Policies (Pre-v1.23 Kubernetes)**
>
> Kubernetes versions before v1.23 allowed specifying scheduling policies via kube-scheduler flags or ConfigMaps. These policies defined how the scheduler selected nodes for Pods using predicates (filtering criteria) and priorities (scoring functions). As of v1.23, this functionality is replaced by scheduler configuration. This new approach allows more flexibility and control over scheduling behavior.

The Kubernetes scheduler configuration file provides several benefits for managing Pod placement within your cluster. It provides the flexibility to tailor scheduling behavior to your needs. For instance, you can prioritize Pods requiring GPUs to land on Nodes with those resources. Additionally, you can develop custom plugins to handle unique scheduling requirements not addressed by default plugins. Finally, the ability to define multiple profiles allows you to create granular controls by assigning different scheduling profiles to different types of Pods.

Before we conclude the chapter, let's look at the Node restrictions feature in Kubernetes scheduling.

Node isolation and restrictions

Kubernetes allows you to isolate Pods on specific Nodes using node labels. These labels can define properties like security requirements or regulatory compliance. This ensures Pods are only scheduled on Nodes that meet these criteria. To prevent a compromised node from manipulating labels for its own benefit, the `NodeRestriction` admission plugin restricts the `kubelet` from modifying labels with a specific prefix (e.g., `node-restriction.kubernetes.io/`). To leverage this functionality, you'll need to enable the `NodeRestriction` plugin and `Node authorizer`. Then, you can add labels with the restricted prefix to your Nodes and reference them in your Pod's `nodeSelector` configuration. This ensures your Pods only run on pre-defined, isolated environments.

Tuning Kubernetes scheduler performance

In large Kubernetes clusters, efficient scheduler performance is crucial. Let's explore a key tuning parameter: `percentageOfNodesToScore`.

The `percentageOfNodesToScore` setting determines how many Nodes the scheduler considers when searching for a suitable Pod placement. A higher value means the scheduler examines more Nodes, potentially finding a better fit but taking longer. Conversely, a lower value leads to faster scheduling but might result in suboptimal placement. You can configure this value in the kube-scheduler configuration file. The valid range is 1% to 100%, with a default calculated based on cluster size (50% for 100 nodes, 10% for 5,000 nodes).

To set `percentageOfNodesToScore` to 50% for a cluster with hundreds of nodes, you'd include the following configuration in the scheduler file:

```
apiVersion: kubescheduler.config.k8s.io/v1alpha1
kind: KubeSchedulerConfiguration
algorithmSource:
  provider: DefaultProvider
...
percentageOfNodesToScore: 50
```

The optimal value depends on your priorities. If fast scheduling is critical, a lower value might be acceptable. However, if ensuring the best possible placement outweighs speed concerns, a higher value is recommended. Avoid setting it too low to prevent the scheduler from overlooking potentially better Nodes.

With this, we have finished this chapter, and we have learned about different mechanisms and strategies available in Kubernetes to control Pod placement on Nodes, including `nodeName`, `nodeSelector`, `nodeAffinity`, taints and tolerations, and also other useful advanced scheduler configurations.

Summary

This chapter has given an overview of advanced techniques for Pod scheduling in Kubernetes. First, we recapped the theory behind kube-scheduler implementation and explained the process of scheduling Pods. Next, we introduced the concept of Node affinity in Pod scheduling. We discussed the basic scheduling methods, which use Node names and Node selectors, and based on that, we explained how more advanced Node affinity works. We also explained how to use the affinity concept to achieve anti-affinity, and what inter-Pod affinity/anti-affinity is. After that, we discussed taints for Nodes and tolerations specified by Pods. You learned about some different effects of taints, and put that knowledge into practice in an advanced use case involving `NoExecute` and `NoSchedule` taints on a Node. Lastly, we discussed some advanced scheduling features in Kubernetes such as scheduler configurations, Node isolation, and static pods.

In the next chapter, we are going to discuss the **autoscaling** of Pods and Nodes in Kubernetes. This is a topic that shows how flexibly Kubernetes can run workloads in cloud environments.

Further reading

- Kubernetes Scheduler: https://kubernetes.io/docs/concepts/scheduling-eviction/kube-scheduler/
- Assigning Pods to Nodes: https://kubernetes.io/docs/concepts/scheduling-eviction/assign-pod-node/
- Taints and Tolerations: https://kubernetes.io/docs/concepts/scheduling-eviction/taint-and-toleration/
- Scheduler Configuration: https://kubernetes.io/docs/reference/scheduling/config/

For more information regarding Pod scheduling in Kubernetes, please refer to the following PacktPub books:

- *The Complete Kubernetes Guide*, by *Jonathan Baier, Gigi Sayfan, Jesse White* (`https://www.packtpub.com/en-in/product/the-complete-kubernetes-guide-9781838647346`)
- *Getting Started with Kubernetes – Third Edition*, by *Jonathan Baier, Jesse White* (`https://www.packtpub.com/en-in/product/getting-started-with-kubernetes-9781788997263`)
- *Kubernetes for Developers*, by *Joseph Heck* (`https://www.packtpub.com/en-in/product/kubernetes-for-developers-9781788830607`)

You can also refer to official documents:

- Kubernetes documentation (`https://kubernetes.io/docs/home/`), which is always the most up-to-date source of knowledge about Kubernetes in general.
- Node affinity is covered at `https://kubernetes.io/docs/concepts/scheduling-eviction/assign-pod-node/`.
- Taint and tolerations are covered at `https://kubernetes.io/docs/concepts/scheduling-eviction/taint-and-toleration/`.
- Pod priorities and preemption (which we have not covered in this chapter) are described at `https://kubernetes.io/docs/concepts/configuration/pod-priority-preemption/`.
- Advanced kube-scheduler configuration using scheduling profiles (which we have not covered in this chapter) is described at `https://kubernetes.io/docs/reference/scheduling/config`.

Join our community on Discord

Join our community's Discord space for discussions with the authors and other readers:

`https://packt.link/cloudanddevops`

20

Autoscaling Kubernetes Pods and Nodes

Needless to say, having **autoscaling** capabilities for your cloud-native application is considered the holy grail of running applications in the cloud. In short, by autoscaling, we mean a method of automatically and dynamically adjusting the amount of computational resources, such as CPU and RAM, available to your application. The goal of autoscaling is to add or remove resources based on the **activity and demand** of end users. So, for example, an application might require more CPU and RAM during daytime hours, when users are most active, but much less during the night. Similarly, for example, if you are supporting an e-commerce business infrastructure, you can expect a huge spike in demand during so-called *Black Friday*. In this way, you can not only provide a better, highly available service to users but also reduce your **cost of goods sold (COGS)** for the business. The fewer resources you consume in the cloud, the less you pay, and the business can invest the money elsewhere – this is a *win-win* situation. There is, of course, no single rule that fits all use cases, hence good autoscaling needs to be based on critical usage metrics and should have **predictive features** to anticipate the workloads based on history.

Kubernetes, as the most mature container orchestration system available, comes with a variety of built-in autoscaling features. Some of these features are natively supported in every Kubernetes cluster and some require installation or specific type of cluster deployment. There are also multiple *dimensions* of scaling that you can have:

- **Vertical for Pods:** This involves adjusting the amount of CPU and memory resources available to a Pod. Pods can run under limits specified for CPU and memory, to prevent excessive consumption, but these limits may require automatic adjustment rather than a human operator guessing. This is implemented by a **VerticalPodAutoscaler (VPA)**.
- **Horizontal for Pods:** This involves dynamically changing the number of Pod replicas for your Deployment or StatefulSet. These objects come with nice scaling features out of the box, but adjusting the number of replicas can be automated using a **HorizontalPodAutoscaler (HPA)**.

- **Horizontal for Nodes:** Another dimension of horizontal scaling (scaling *out*), but this time at the level of a Kubernetes Node. You can scale your whole cluster by adding or removing the Nodes. This requires, of course, a Kubernetes Deployment that runs in an environment that supports the dynamic provisioning of machines, such as a cloud environment. This is implemented by a **Cluster Autoscaler** (CA), available for some cloud vendors.

In this chapter, we will cover the following topics:

- Pod resource requests and limits
- Autoscaling Pods vertically using a VerticalPodAutoscaler
- Autoscaling Pods horizontally using a HorizontalPodAutoscaler
- Autoscaling Kubernetes Nodes using a Cluster Autoscaler
- Alternative autoscalers for Kubernetes

Technical requirements

For this chapter, you will need the following:

- A Kubernetes cluster deployed. We recommend using a multi-node Kubernetes cluster.
- A multi-node **Google Kubernetes Engine** (GKE) cluster. This is a prerequisite for VPA and cluster autoscaling.
- A Kubernetes CLI (kubectl) installed on your local machine and configured to manage your Kubernetes cluster.

Basic Kubernetes cluster deployment (local and cloud-based) and kubectl installation have been covered in *Chapter 3, Installing Your First Kubernetes Cluster*.

Chapters *15*, *16*, and *17* of this book have provided you with an overview of how to deploy a fully functional Kubernetes cluster on different cloud platforms and install the requisite CLIs to manage them.

The latest code samples for this chapter can be downloaded from the official GitHub repository at https://github.com/PacktPublishing/The-Kubernetes-Bible-Second-Edition/tree/main/ Chapter20.

Pod resource requests and limits

Before we dive into the topics of autoscaling in Kubernetes, we need to get a bit more of an understanding of how to control the CPU and memory resource (known as **compute resources**) usage by using Pod containers in Kubernetes. Controlling the use of compute resources is important since, in this way, it is possible to enforce **resource governance** – this allows better planning of the cluster capacity and, most importantly, prevents situations when a single container can consume all compute resources and prevent other Pods from serving the requests.

When you create a Pod, it is possible to specify how much of the compute resources its containers **require** and what the **limits** are in terms of permitted consumption. The Kubernetes resource model provides an additional distinction between two classes of resources: **compressible** and **incompressible**. In short, a compressible resource can be easily throttled, without severe consequences.

A perfect example of such a resource is the CPU – if you need to throttle CPU usage for a given container, the container will operate normally, just slower. On the other hand, we have incompressible resources that cannot be throttled without severe consequences – memory allocation is an example of such a resource. If you do not allow a process running in a container to allocate more memory, the process will crash and result in a container restart.

To control the resources for a Pod container, you can specify two values in its specification:

- `requests`: This specifies the guaranteed amount of a given resource provided by the system. You can also think about this the other way around – this is the amount of a given resource that the Pod container requires from the system in order to function properly. This is important as Pod scheduling is dependent on the `requests` value (not `limits`), namely, the `PodFitsResources` predicate and the `BalancedResourceAllocation` priority.
- `limits`: This specifies the **maximum** amount of a given resource that is provided by the system. If specified together with `requests`, this value must be greater than or equal to `requests`. Depending on whether the resource is compressible or incompressible, exceeding the limit has different consequences – compressible resources (CPU) will be throttled, whereas incompressible resources (RAM) *might* result in container kill and restart.

You can allow the overcommitment of resources by setting different values for requests and limits. The system will then be able to handle brief periods of high resource usage more gracefully while it optimizes overall resource utilization. This works because it's fairly unlikely that all containers on a Node will reach their resource limits simultaneously. Hence, Kubernetes can use the available resources more effectively most of the time. It's kind of like overprovisioning in the case of virtual machines or airlines overbooking because not everybody uses all of their allocation at the same time. This means you can actually get more Pods on each Node, which improves overall resource utilization.

If you do not specify `limits` at all, the container can consume as much of the resource on a Node as it wants. This can be controlled by namespace **resource quotas** and **limit ranges**, which we explored in *Chapter 6, Namespaces, Quotas, and Limits for Multi-Tenancy in Kubernetes.*

> In more advanced scenarios, it is also possible to control huge pages and ephemeral storage `requests` and `limits`.

Before we dive into the configuration details, we need to look at the units for measuring CPU and memory in Kubernetes:

- For CPU, the base unit is **Kubernetes CPU (KCU)**, where 1 is equivalent to, for example, 1 vCPU on Azure, 1 core on GCP, or 1 hyperthreaded core on a bare-metal machine.
- Fractional values are allowed: `0.1` can be also specified as `100m` (*milliKCUs*).
- For memory, the base unit is a **byte**; you can, of course, specify standard unit prefixes, such as `M`, `Mi`, `G`, or `Gi`.

To enable com pute resource `requests` and `limits` for Pod containers in the `nginx` Deployment that we used in the previous chapters, make the following changes to the YAML manifest, `resource-limit/nginx-deployment.yaml`:

```yaml
# resource-limit/nginx-deployment.yaml
apiVersion: apps/v1
kind: Deployment
metadata:
  name: nginx-deployment-example
spec:
  replicas: 5

...<removed for brevity>...
    spec:
      containers:
        - name: nginx
          image: nginx:1.17
          ports:
            - containerPort: 80
          resources:
            limits:
              cpu: 200m
              memory: 60Mi
            requests:
              cpu: 100m
              memory: 50Mi
```

For each container that you have in the Pod, specify the `.spec.template.spec.containers[*].resources` field. In this case, we have set `limits` at `200m` KCU and `60Mi` for RAM, and `requests` at `100m` KCU and `50Mi` for RAM.

When you apply the manifest to the cluster using `kubectl apply -f resource-limit/nginx-deployment.yaml`, describe one of the Nodes in the cluster that run Pods for this Deployment, and you will see the detailed information about compute resources quotas and allocation:

```
$ kubectl describe node minikube-m03
...<removed for brevity>...
Non-terminated Pods:              (5 in total)
  Namespace                 Name                                            CPU
Requests  CPU Limits  Memory Requests  Memory Limits  Age
  ---------                 ----                                            ----
-------   ---------   ---------------  -------------  ---
  default                   nginx-deployment-example-6d444cfd96-f5tnq       100m
(5%)      200m (10%)  50Mi (2%)        60Mi (3%)      23s
```

```
   default                      nginx-deployment-example-6d444cfd96-k6j9d      100m
(5%)     200m (10%)   50Mi (2%)         60Mi (3%)      23s
   default                      nginx-deployment-example-6d444cfd96-mqxxp      100m
(5%)     200m (10%)   50Mi (2%)         60Mi (3%)      23s
   kube-system                  calico-node-92bdc                             250m
(12%)    0 (0%)       0 (0%)            0 (0%)         6d23h
   kube-system                  kube-proxy-5cd4x                              0
(0%)         0 (0%)   0 (0%)            0 (0%)         6d23h
Allocated resources:
  (Total limits may be over 100 percent, i.e., overcommitted.)
  Resource             Requests     Limits
  --------             --------     ------
  cpu                  550m (27%)   600m (30%)
  memory               150Mi (7%)   180Mi (9%)
  ephemeral-storage    0 (0%)       0 (0%)
  hugepages-2Mi        0 (0%)       0 (0%)
Events:                <none>
```

Now, based on this information, you could experiment and set CPU `requests` for the container to a value higher than the capacity of a single Node in the cluster; in our case, we modify the value as follows in `resource-limit/nginx-deployment.yaml`:

```
...
        resources:
          limits:
            cpu: 2000m
            memory: 60Mi
          requests:
            cpu: 2000m
            memory: 50Mi
...
```

Apply the configuration as follows:

```
$ kubectl apply -f resource-limit/nginx-deployment.yaml
deployment.apps/nginx-deployment-example configured
```

Check the Pod status as follows, and you will notice that new Pods hang in the `Pending` state because they cannot be scheduled on a matching Node:

```
$ kubectl get pod
NAME                                        READY   STATUS    RESTARTS   AGE
nginx-deployment-example-59b669d85f-cdptx   1/1     Running   0          52s
nginx-deployment-example-59b669d85f-hdzdf   1/1     Running   0          54s
nginx-deployment-example-59b669d85f-ktn59   1/1     Running   0          54s
```

```
nginx-deployment-example-59b669d85f-vdn87    1/1    Running    0    52s
nginx-deployment-example-69bd6d55b4-n2mzq    0/1    Pending    0    3s
nginx-deployment-example-69bd6d55b4-qb62p    0/1    Pending    0    3s
nginx-deployment-example-69bd6d55b4-w7xng    0/1    Pending    0    3s
```

Investigate the Pending state by describing the Pod as follows:

```
$ kubectl describe pod nginx-deployment-example-69bd6d55b4-n2mzq
...
Events:
  Type     Reason            Age              From               Message
  ----     ------            ----             ----               -------
  Warning  FailedScheduling  23m (x21 over 121m)  default-scheduler  0/3 nodes
are available: 1 node(s) had untolerated taint {machine-check-exception:
memory}, 2 Insufficient cpu. preemption: 0/3 nodes are available: 1 Preemption
is not helpful for scheduling, 2 No preemption victims found for incoming pod.
```

> Some of the autoscaling mechanisms discussed in this chapter are currently in alpha or beta versions, which might not be fully stable and thus might not be suitable for production environments. For more mature autoscaling solutions, please refer to the Alternative autoscalers for Kubernetes section of this chapter.

In the preceding output, there were no Nodes that could accommodate a Pod that has a container requiring 2000m KCU, and therefore the Pod cannot be scheduled at this time.

Knowing now how to manage compute resources, we will move on to autoscaling topics: first, we will explain the vertical autoscaling of Pods.

Autoscaling Pods vertically using a VerticalPodAutoscaler

In the previous section, we managed requests and limits for compute resources manually. Setting these values correctly requires some accurate human *guessing*, observing metrics, and performing benchmarks to adjust. Using overly high requests values will result in a waste of compute resources, whereas setting requests too low might result in Pods being packed too densely and having performance issues. Also, in some cases, the only way to scale the Pod workload is to do it **vertically** by increasing the amount of compute resources it can consume. For bare-metal machines, this would mean upgrading the CPU hardware and adding more physical RAM. For containers, it is as simple as allowing them more of the compute resource quotas. This works, of course, only up to the capacity of a single Node. You cannot scale vertically beyond that unless you add more powerful Nodes to the cluster.

To help resolve these issues, you can use a VPA, which can increase and decrease CPU and memory resource requests for Pod containers dynamically.

The following diagram shows the vertical scaling of Pods.

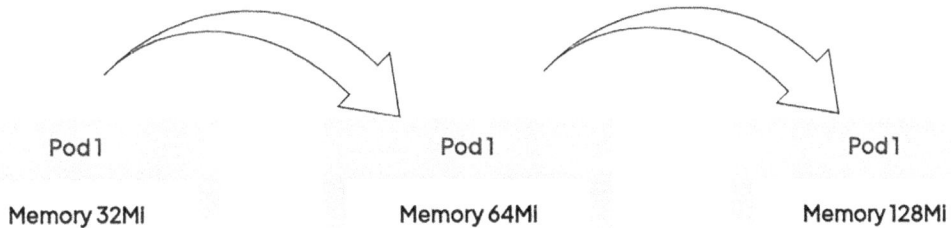

Figure 20.1: Vertical scaling of Pods

The goal is to better match the *actual* usage of the container rather than relying on hardcoded, pre-defined values resources request and limit values. Controlling `limits` within specified ratios is also supported.

The VPA is created by a **Custom Resource Definition (CRD)** object named `VerticalPodAutoscaler`. This means the object is not part of standard Kubernetes API groups and must be installed in the cluster. The VPA is developed as part of an **autoscaler** project (`https://github.com/kubernetes/autoscaler`) in the Kubernetes ecosystem.

There are three main components of a VPA:

- **Recommender:** Monitors the current and past resource consumption and provides recommended CPU and memory request values for a Pod container.
- **Updater:** Checks for Pods with incorrect resources and **deletes** them, so that the Pods can be recreated with the updated `requests` and `limits` values.
- **Admission plugin:** Sets the correct resource `requests` and `limits` on new Pods created or recreated by their controller, for example, a Deployment object, due to changes made by the updater.

The reason that the updater needs to terminate Pods, and the VPA has to rely on the admission plugin, is that Kubernetes does not support dynamic changes to the resource `requests` and `limits`. The only way is to terminate the Pod and create a new one with new values.

> A VPA can run in a recommendation-only mode where you see the suggested values in the VPA object, but the changes are not applied to the Pods. A VPA is currently considered *experimental* and using it in a mode that recreates the Pods may lead to downtime for our application. This should change when in-place updates of Pod `requests` and `limits` are implemented.

Some Kubernetes offerings come with one-click or operator support for installing a VPA. Two good examples are OpenShift and GKE. Refer to the *Automatically adjust pod resource levels with the vertical pod autoscaler* article (`https://docs.openshift.com/container-platform/4.16/nodes/pods/nodes-pods-vertical-autoscaler.html`) to learn about VPA implementation in OpenShift.

Enabling InPlacePodVerticalScaling

In-place pod resizing, an alpha feature introduced in Kubernetes 1.27, allows for the dynamic adjustment of pod resources without requiring restarts, potentially improving application performance and resource efficiency.

Alpha Feature Warning

In-place pod resizing is an alpha feature as of Kubernetes 1.27 and may be changed in future versions without notice. It should not be deployed on production clusters because of potential instability; more generally, an alpha feature may not be subject to a stable version and can change at any time.

To activate this capability, the `InPlacePodVerticalScaling` feature gate must be enabled across all cluster nodes.

For Kubernetes clusters, enable the feature gate using the following method:

1. Update `/etc/kubernetes/manifests/kube-apiserver.yaml` (or appropriate configuration for your Kubernetes cluster).

2. Add `feature-gates` as follows:

```
# /etc/kubernetes/manifests/kube-apiserver.yaml
...
  - command:
    - kube-apiserver
  ...<removed for brevity>...
    - --feature-gates=InPlacePodVerticalScaling=true
```

For minikube environments, incorporate the feature gate during cluster startup as follows:

```
$ minikube start --feature-gates=InPlacePodVerticalScaling=true
```

We will now quickly explain how to enable VPA if you are running a GKE cluster.

Enabling a VPA in GKE

Enabling a VPA in **Google Kubernetes Engine** (**GKE**) is as simple as running the following command:

```
$ gcloud container clusters update <cluster-name> --enable-vertical-pod-
autoscaling
```

Note that this operation causes a restart to the Kubernetes control plane.

If you want to enable a VPA for a new cluster, use the additional argument `--enable-vertical-pod-autoscaling`:

```
$ gcloud container clusters create k8sforbeginners --num-nodes=2 --zone=us-
central1-a --enable-vertical-pod-autoscaling
```

The GKE cluster will have a VPA CRD available, and you can use it to control the vertical autoscaling of Pods.

Let's learn how to enable VPA for standard Kubernetes clusters in the next section. If you are using a different type of Kubernetes cluster, follow the specific instructions for your setup.

Enabling a VPA for other Kubernetes clusters

In the case of different platforms such as AKS or EKS (or even local deployments for testing), you need to install a VPA manually by adding a VPA CRD to the cluster. The exact, most recent steps are documented in the corresponding GitHub repository: https://github.com/kubernetes/autoscaler/tree/master/vertical-pod-autoscaler#installation.

To install a VPA in your cluster, please perform the following steps:

1. Clone the Kubernetes autoscaler repository (https://github.com/kubernetes/autoscaler):

    ```
    $ git clone https://github.com/kubernetes/autoscaler
    ```

2. Navigate to the VPA component directory:

    ```
    $ cd autoscaler/vertical-pod-autoscaler
    ```

3. Begin installation using the following command. This assumes that your current kubectl context is pointing to the desired cluster:

    ```
    $ ./hack/vpa-up.sh
    ```

4. This will create a bunch of Kubernetes objects. Verify that the main component Pods are started correctly using the following command:

    ```
    $ kubectl get pods -n kube-system | grep vpa
    vpa-admission-controller-5b64b4f4c4-vsn9j    1/1      Running    0
    5m34s
    vpa-recommender-54c76554b5-m7wnk             1/1      Running    0
    5m34s
    vpa-updater-7d5f6fbf9b-rkwlb                 1/1      Running    0
    5m34s
    ```

The VPA components are running, and we can now proceed to test a VPA on real Pods.

Using a VPA

For demonstration purposes, we need a Deployment with Pods that cause actual consumption of CPU. The Kubernetes autoscaler repository has a good, simple example that has **predictable** CPU usage: https://github.com/kubernetes/autoscaler/blob/master/vertical-pod-autoscaler/examples/hamster.yaml. We are going to modify this example a bit and do a step-by-step demonstration.

> **WARNING**
>
> VPA utilization heavily depends on the distribution and maturity of the underlying Kubernetes. Sometimes, the Pods are not rescheduled as expected, which may lead to application downtime. Therefore, if full automation of the VPA is enabled, that may result in cascading issues related to resource overcommitment and cluster instability if monitoring is not performed.

Let's prepare the Deployment first:

1. First of all, enable the metric server for your Kubernetes cluster. You can use the default metric server (https://github.com/kubernetes-sigs/metrics-server) and deploy it within your Kubernetes cluster. If you are using a minikube cluster, enable the metric server as follows:

    ```
    $ minikube addons enable metrics-server
    ```

2. Create a new Namespace for this:

    ```yaml
    # vpa/vpa-demo-ns.yaml

    ---
    apiVersion: v1
    kind: Namespace
    metadata:
      labels:
        project: vpa-demo
      name: vpa-demo
    ```

    Create the Namespace using the `kubectl apply` command as follows:

    ```
    $ kubectl apply -f vpa/vpa-demo-ns.yaml
    namespace/vpa-demo created
    ```

3. Create the `hamster-deployment.yaml` YAML manifest file (check vpa/hamster-deployment.yaml for sample):

    ```yaml
    # vpa/hamster-deployment.yaml
    apiVersion: apps/v1
    kind: Deployment
    metadata:
      name: hamster
      namespace: vpa-demo
    spec:
      selector:
        matchLabels:
    ```

```
      app: hamster
  replicas: 5
  template:
    metadata:
      labels:
        app: hamster
    spec:
      containers:
        - name: hamster
          image: ubuntu:20.04
          resources:
            requests:
              cpu: 100m
              memory: 50Mi
          command:
            - /bin/sh
            - -c
            - while true; do timeout 0.5s yes >/dev/null; sleep 0.5s;
done
```

It's a real hamster! The `command` that is used in the Pod's ubuntu container repeatedly consumes the maximum available CPU for 0.5 seconds and does nothing for 0.5 seconds. This means that the actual CPU usage will stay, on average, at around 500m KCU. However, the `requests` value for resources specifies that it requires 100m KCU. This means that the Pod will consume more than it declares, but since there are no `limits` set, Kubernetes will not throttle the container CPU. This could potentially lead to incorrect scheduling decisions by the Kubernetes Scheduler.

4. Apply the manifest to the cluster using the following command:

```
$ kubectl apply -f vpa/hamster-deployment.yaml
deployment.apps/hamster created
```

5. Check the Pods in the vpa-demo Namespace:

```
$  kubectl get po -n vpa-demo
NAME                       READY   STATUS    RESTARTS   AGE
hamster-7fb7dbff7-hmzt5    1/1     Running   0          8s
hamster-7fb7dbff7-lbk9f    1/1     Running   0          8s
hamster-7fb7dbff7-ql6gd    1/1     Running   0          8s
hamster-7fb7dbff7-qmxd8    1/1     Running   0          8s
hamster-7fb7dbff7-qtrpp    1/1     Running   0          8s
```

6. Let's verify what the CPU usage of the Pod is. The simplest way is to use the `kubectl top` command:

```
$ kubectl top pod -n vpa-demo
NAME                       CPU(cores)    MEMORY(bytes)
hamster-7fb7dbff7-hmzt5    457m          0Mi
hamster-7fb7dbff7-1bk9f    489m          0Mi
hamster-7fb7dbff7-ql6gd    459m          0Mi
hamster-7fb7dbff7-qmxd8    453m          0Mi
hamster-7fb7dbff7-qtrpp    451m          0Mi
```

As we expected, the CPU consumption for each Pod in the deployment oscillates at around 500m KCU.

With that, we can move on to creating a VPA for our Pods. VPAs can operate in four **modes** that you specify by means of the `.spec.updatePolicy.updateMode` field:

- `Recreate`: Pod container `limits` and `requests` are assigned on Pod creation and dynamically updated based on calculated recommendations. To update the values, the Pod must be restarted. Please note that this may be disruptive to your application.
- `Auto`: Currently equivalent to `Recreate`, but when in-place updates for Pod container `requests` and `limits` are implemented, this can automatically switch to the new update mechanism.
- `Initial`: Pod container `limits` and `requests` values are assigned on Pod creation only.
- `Off`: A VPA runs in recommendation-only mode. The recommended values can be inspected in the VPA object, for example, by using `kubectl`.

First, we are going to create a VPA for `hamster` Deployment, which runs in `Off` mode, and later we will enable `Auto` mode. To do this, please perform the following steps:

1. Create a VPA YAML manifest named vpa/hamster-vpa.yaml:

```
# vpa/hamster-vpa.yaml
apiVersion: autoscaling.k8s.io/v1
kind: VerticalPodAutoscaler
metadata:
  name: hamster-vpa
  namespace: vpa-demo
spec:
  targetRef:
    apiVersion: apps/v1
    kind: Deployment
    name: hamster
  updatePolicy:
    updateMode: 'Off'
  resourcePolicy:
    containerPolicies:
```

```
    - containerName: '*'
      minAllowed:
        cpu: 100m
        memory: 50Mi
      maxAllowed:
        cpu: 1
        memory: 500Mi
      controlledResources:
        - cpu
        - memory
```

This VPA is created for a Deployment object with the name hamster, as specified in .spec. targetRef. The mode is set to "Off" in .spec.updatePolicy.updateMode ("Off" needs to be specified in quotes to avoid being interpreted as a Boolean) and the container resource policy is configured in .spec.resourcePolicy.containerPolicies. The policy that we used allows Pod container requests for CPU to be adjusted automatically between 100m KCU and 1000m KCU, and for memory between 50Mi and 500Mi.

2. Apply the manifest file to the cluster:

```
$ kubectl apply -f vpa/hamster-vpa.yaml
verticalpodautoscaler.autoscaling.k8s.io/hamster-vpa created
```

3. You need to wait a while for the recommendation to be calculated for the first time. Then, check what the recommendation is by describing the VPA:

```
$ kubectl describe vpa hamster-vpa -n vpa-demo
...<removed for brevity>...
Status:
  Conditions:
    Last Transition Time:  2024-08-11T09:20:44Z
    Status:                True
    Type:                  RecommendationProvided
  Recommendation:
    Container Recommendations:
      Container Name:  hamster
      Lower Bound:
        Cpu:     461m
        Memory:  262144k
      Target:
        Cpu:     587m
        Memory:  262144k
      Uncapped Target:
        Cpu:     587m
```

```
        Memory:    262144k
      Upper Bound:
        Cpu:       1
        Memory:    500Mi
Events:            <none>
```

The VPA has recommended allocating a bit more than the expected 500m KCU and 262144k memory. This makes sense, as the Pod should have a safe buffer for CPU consumption.

4. Now we can check the VPA in practice and change its mode to Auto. Modify vpa/hamster-vpa.yaml:

```
# vpa/hamster-vpa.yaml
apiVersion: autoscaling.k8s.io/v1
kind: VerticalPodAutoscaler
metadata:
  name: hamster-vpa
  namespace: vpa-demo
spec:
...
  updatePolicy:
    updateMode: Auto
...
```

5. Apply the manifest to the cluster:

```
$ kubectl apply -f vpa/hamster-vpa.yaml
verticalpodautoscaler.autoscaling.k8s.io/hamster-vpa configured
```

6. After a while, you will notice that the Pods for the Deployment are being restarted by the VPA:

```
$ kubectl get po -n vpa-demo -w
NAME                      READY   STATUS              RESTARTS   AGE
hamster-7fb7dbff7-24p89   0/1     ContainerCreating   0          2s
hamster-7fb7dbff7-6nz8f   0/1     ContainerCreating   0          2s
hamster-7fb7dbff7-hmzt5   1/1     Running             0          20m
hamster-7fb7dbff7-lbk9f   1/1     Running             0          20m
hamster-7fb7dbff7-ql6gd   1/1     Terminating         0          20m
hamster-7fb7dbff7-qmxd8   1/1     Terminating         0          20m
hamster-7fb7dbff7-qtrpp   1/1     Running             0          20m
hamster-7fb7dbff7-24p89   1/1     Running             0          2s
hamster-7fb7dbff7-6nz8f   1/1     Running             0          2s
```

7. We can inspect one of the restarted Pods to see the current `requests` for resources:

```
$ kubectl describe pod hamster-7fb7dbff7-24p89 -n vpa-demo
...
Annotations:       ...<removed for brevity>...
                   vpaObservedContainers: hamster
                   vpaUpdates: Pod resources updated by hamster-vpa:
container 0: memory request, cpu request
...
Containers:
  hamster:
    ...
    Requests:
      cpu:        587m
      memory:     262144k
...<removed for brevity>...
```

As you can see, the newly started Pod has CPU and memory `requests` set to the values recommended by the VPA!

> A VPA should not be used with an HPA running on CPU/memory metrics at this moment. However, you can use a VPA in conjunction with an HPA running on custom metrics.

Next, we are going to discuss how to autoscale Pods horizontally using an HPA.

Autoscaling Pods horizontally using a HorizontalPodAutoscaler

While a VPA acts like an optimizer of resource usage, the true scaling of your Deployments and Stateful-Sets that run multiple Pod replicas can be done using an HPA. At a high level, the goal of the HPA is to automatically scale the number of replicas in Deployment or StatefulSets depending on the current CPU utilization or other custom metrics (including multiple metrics at once). The details of the algorithm that determines the target number of replicas based on metric values can be found here: https://kubernetes.io/docs/tasks/run-application/horizontal-Pod-autoscale/#algorithm-details.

> Not all applications will work equally efficiently with HPAs and VPAs. Some of them might work better using one method, but others might either not support autoscaling or even suffer from the method. Always analyze your application behavior prior to using any autoscaling approach.

A high-level diagram to demonstrate the difference between vertical and horizontal scaling is given below:

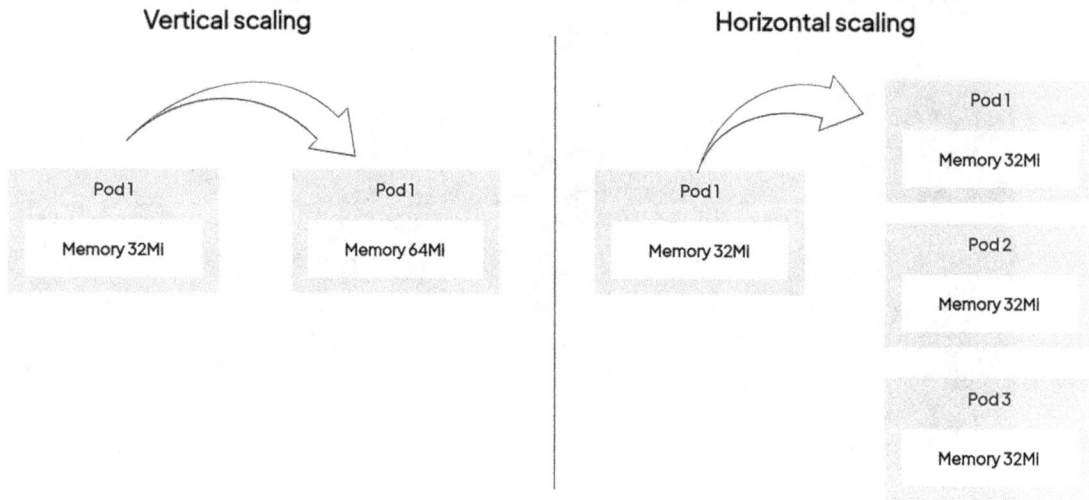

Figure 20.2: Vertical scaling vs. horizontal scaling for the Pods

HPAs are highly configurable, and, in this chapter, we will cover a standard scenario in which we would like to autoscale based on target CPU usage.

> The HPA is an API resource in the Kubernetes autoscaling API group. The current stable version is `autoscaling/v2`, which includes support for scaling based on memory and custom metrics. When using `autoscaling/v1`, the new fields introduced in `autoscaling/v2` are preserved as annotations.

The role of the HPA is to monitor a configured metric for Pods, for example, CPU usage, and determine whether a change to the number of replicas is needed. Usually, the HPA will calculate the average of the current metric values from all Pods and determine whether adding or removing replicas will bring the metric value closer to the specified target value. For example, say you set the target CPU usage to be 50%. At some point, increased demand for the application causes the Deployment Pods to have 80% CPU usage. The HPA would decide to add more Pod replicas so that the average usage across all replicas will fall and be closer to 50%. And the cycle repeats. In other words, the HPA tries to maintain the average CPU usage to be as close to 50% as possible. This is like a continuous, closed-loop controller – a thermostat reacting to temperature changes in a building is a good example.

The following figure shows a high-level diagram of the Kubernetes HPA components:

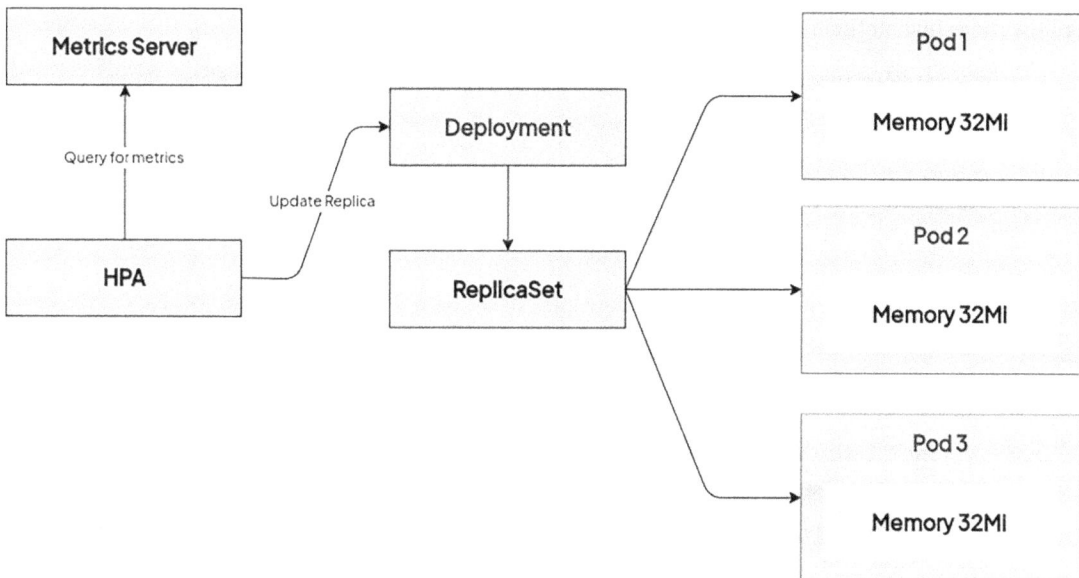

Figure 20.3: HPA overview in Kubernetes

HPA additionally uses mechanisms such as a **stabilization window** to prevent the replicas from scaling down too quickly and causing unwanted replica **flapping**.

> GKE has beta functionality for multidimensional Pod autoscaling, which combines horizontal scaling using CPU metrics and vertical scaling based on memory usage at the same time. Read more about this feature in the official documentation: `https://cloud.google.com/kubernetes-engine/docs/how-to/multidimensional-pod-autoscaling`. Please note that this feature is subject to the Pre-GA Offerings Terms in the General Service Terms and is provided "as is" with limited support; refer to the launch stage descriptions for more details.

As an HPA is a built-in feature of Kubernetes, there is no need to perform any installation. We just need to prepare a Deployment for testing and create a `HorizontalPodAutoscaler` Kubernetes resource.

Deploying the app for HPA demonstration

To test an HPA, we are going to rely on the standard CPU usage metric. This means that we need to configure `requests` for CPU on the Deployment Pods; otherwise, autoscaling is not possible as there is no absolute number that is needed to calculate the percentage metric. On top of that, we again need a Deployment that can consume a predictable amount of CPU resources. Of course, in real use cases, the varying CPU usage would be coming from actual demand for your application from end users.

First of all, enable the metric server for your Kubernetes cluster. You can use the default metric server (`https://github.com/kubernetes-sigs/metrics-server`) and deploy it within your Kubernetes cluster. If you are using a minikube cluster, enable the metric server as follows:

```
$ minikube addons enable metrics-server
```

Follow these instructions to learn how to implement HPA:

1. To isolate our resources for this demonstration, create a new Namespace as follows:

    ```yaml
    # hpa/hpa-demo-ns.yaml

    ---
    apiVersion: v1
    kind: Namespace
    metadata:
      labels:
        project: hpa-demo
      name: hpa-demo
    ```

2. Apply the YAML and create the Namespace:

    ```
    $ kubectl apply -f hpa/hpa-demo-ns.yaml
    namespace/hpa-demo created
    ```

3. For the demonstration, we will use a simple web server container based on a custom image, `quay.io/iamgini/one-page-web:1.0`. The following YAML contains a simple Deployment definition that will create one replica of the Pod:

    ```yaml
    ---
    # hpa/todo-deployment.yaml
    apiVersion: apps/v1
    kind: Deployment
    metadata:
      name: todo-app
      namespace: hpa-demo
    spec:
      replicas: 1  # Adjust as needed
      selector:
        matchLabels:
          app: todo
      template:
        metadata:
          labels:
            app: todo
        spec:
          containers:
            - name: todoapp
              image: quay.io/ginigangadharan/todo-app:2.0
              ports:
                - containerPort: 3000
    ```

```
      resources:
        requests:
          memory: "50Mi"    # Request 50 MiB of memory
          cpu: "50m"        # Request 0.05 CPU core
        limits:
          memory: "100Mi"   # Request 100 MiB of memory
          cpu: "100m"       # Request 0.1 CPU core
```

4. Apply the configuration and ensure the Pods are running as expected:

```
$ kubectl apply -f hpa/todo-deployment.yaml
deployment.apps/todo-app created

$ kubectl get po -n hpa-demo
NAME                        READY   STATUS    RESTARTS   AGE
todo-app-5cfb496d77-16r69   1/1     Running   0          8s
```

5. To expose the application, let us create a Service using the following YAML:

```
# hpa/todo-service.yaml
apiVersion: v1
kind: Service
metadata:
  name: todo-app
  namespace: hpa-demo
spec:
  type: ClusterIP
  selector:
    app: todo
  ports:
    - port: 8081          # Port exposed within the cluster
      targetPort: 3000    # containerPort on the pods
```

6. Apply the configuration and verify the Service resource:

```
$ kubectl apply -f hpa/todo-service.yaml
service/todo-app created

$ kubectl get svc -n hpa-demo
NAME       TYPE        CLUSTER-IP      EXTERNAL-IP   PORT(S)    AGE
todo-app   ClusterIP   10.96.171.71    <none>        8081/TCP   15s
```

7. Now the application is running and exposed via a `ClusterIP` Service, let us use a `kubectl` `port-forward` command to access the application outside of the cluster:

```
$ kubectl port-forward svc/todo-app -n hpa-demo 8081:8081
Forwarding from 127.0.0.1:8081 -> 3000
Forwarding from [::1]:8081 -> 3000
```

8. Open a web browser and launch `http://localhost:8081`. You will see the Todo application is running as follows:

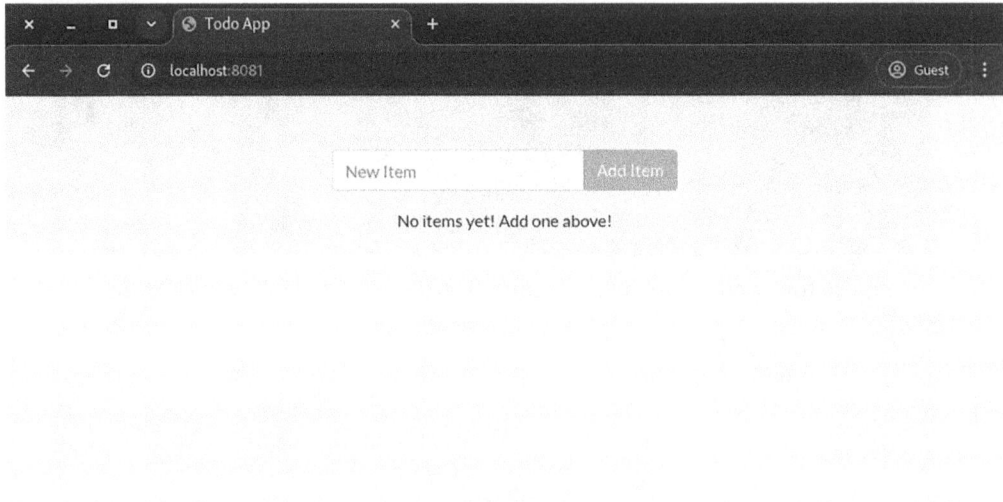

Figure 20.4: The Todo app is running on Kubernetes

9. On the console, press *Ctrl+C* to end the `kubectl` `port-forward` task.

Now we have Todo application Deployment running in the cluster and it is time to learn how HPA works. In the next section, we will learn how to create the HPA and apply load to the deployment to see the autoscaling.

Implementing an HPA

You have already learned that you can scale the number of Pods by using the `kubectl` `scale` command (e.g., `kubectl scale deployment one-page-web -n hpa-demo --replicas 3`), but in this case, we want to learn how an HPA helps us with automated scaling based on the workload.

As we learned earlier in this section, an HPA triggers scaling based on metrics, and so we should give a workload to the Pods. There are several tools available to generate a simulated workload for a web application, in the interests of stress testing and load testing. In this demonstration, we will use a tiny program called hey for the load testing. hey is a lightweight HTTP load-testing tool written in Go. It was designed to make it easy to generate traffic to web applications in order to measure performance under load. It does this by simplifying benchmarking, since users can quickly send a slew of requests and view things like response times and request throughput.

It is also possible to increase the load using other methods. For instance, you can run another container to access application pods with commands like:

```
$ kubectl run -i --tty load-generator --rm --image=busybox:1.28
--restart=Never -- /bin/sh -c "while sleep 0.01; do wget -q -O-
http://todo-app; done"
```

However, this method may not be efficient for controlling the workload precisely.

The hey application is available for Linux, macOS, and Windows (https://github.com/rakyll/hey) and installation is pretty simple:

1. Download the hey package for your operating system.

2. Set executable permission and copy the file to an executable path (eg., `ln -s ~/Downloads/hey_linux_amd64 ~/.local/bin/`).

Now, create HPA resources to scale the one-page web Deployment based on the workload:

1. Prepare the HPA YAML as follows:

```yaml
# hpa/todo-hpa.yaml
apiVersion: autoscaling/v2
kind: HorizontalPodAutoscaler
metadata:
  name: todo-hpa
  namespace: hpa-demo
spec:
  scaleTargetRef:
    apiVersion: apps/v1
    kind: Deployment
    name: todo-app
  minReplicas: 1
  maxReplicas: 5
  metrics:
    - resource:
        name: cpu
        target:
          averageUtilization: 80
          type: Utilization
      type: Resource
```

2. Apply the configuration and create the HPA:

```
$ kubectl apply -f hpa/todo-hpa.yaml
horizontalpodautoscaler.autoscaling/todo-hpa created
```

```
$ kubectl get hpa -n hpa-demo
NAME          REFERENCE               TARGETS            MINPODS   MAXPODS
REPLICAS      AGE
todo-hpa      Deployment/todo-app     cpu: <unknown>/80%  1         5
0             6s
```

3. Let us use a kubectl port-forward command again to access the Todo application outside of the cluster:

```
$ kubectl port-forward svc/todo-app -n hpa-demo 8081:8081
Forwarding from 127.0.0.1:8081 -> 3000
Forwarding from [::1]:8081 -> 3000
```

4. Nobody is using the **todo** application, hence the Pod replica remains 1. Let us simulate the workload by using the hey utility now. On another console, execute the hey workload command as follows:

```
$ hey -z 4m -c 25 http://localhost:8081
```

Please see the parameters and details of the preceding command below:

- -z 4m: Runs for 4 minutes to sustain the load for a longer period
- -c 25: Uses 15 concurrent connections to generate higher load, aiming to push CPU usage closer to 80%
- http://localhost:8081: The URL to access the todo appliaction (enabled by the kubectl port-forward command)

You will find a lot of connection entries in your kubectl port-forward console as hey is simulating the load on the one-page web application now.

5. Open the third console (without waiting for hey to finish the execution) and check the Pod resource utilization:

```
$ watch 'kubectl get po -n hpa-demo;kubectl top pods -n hpa-demo'
Every 2.0s: kubectl get po -n hpa-demo;kubectl top pods -n hpa-demo

NAME                           READY    STATUS     RESTARTS   AGE
todo-app-5cfb496d77-5kc27      1/1      Running    0          76s
todo-app-5cfb496d77-l6r69      1/1      Running    0          10m
todo-app-5cfb496d77-pb7tx      1/1      Running    0          76s
NAME                           CPU(cores)   MEMORY(bytes)
todo-app-5cfb496d77-5kc27      10m          14Mi
todo-app-5cfb496d77-l6r69      100m         48Mi
todo-app-5cfb496d77-pb7tx      7m           14Mi
```

You can see that there are three Pods (or more) created now because hey is applying more workload to the todo application, which triggers the HPA to create mode replicas.

6.   Also, check the deployment details and confirm the replica count and the events to see the scaling events:

```
$ kubectl describe deployments.apps todo-app -n hpa-demo
Name:                   todo-app
...<removed for brevity>...
Replicas:               3 desired | 3 updated | 3 total | 3 available | 0
unavailable
StrategyType:           RollingUpdate
...<removed for brevity>...
Events:
  Type    Reason           Age     From                   Message
  ----    ------           ----    ----                   -------
  Normal  ScalingReplicaSet  16m     deployment-controller  Scaled up
replica set todo-app-749854577d to 1
  Normal  ScalingReplicaSet  13m     deployment-controller  Scaled up
replica set todo-app-5cfb496d77 to 1
  Normal  ScalingReplicaSet  13m     deployment-controller  Scaled down
replica set todo-app-749854577d to 0 from 1
  Normal  ScalingReplicaSet  4m9s    deployment-controller  Scaled up
replica set todo-app-5cfb496d77 to 3 from 1
```

Congratulations! You have successfully configured horizontal autoscaling for your Deployment using an HPA. As part of housekeeping, delete the resources by deleting the hpa-demo namespace (e.g., kubectl delete namespaces hpa-demo). In the next section, we will take a look at autoscaling Kubernetes Nodes using a CA, which gives even more flexibility when combined with an HPA.

Autoscaling Kubernetes Nodes using a Cluster Autoscaler

So far, we have discussed scaling at the level of individual Pods, but this is not the only way in which you can scale your workloads on Kubernetes. It is possible to scale the cluster itself to accommodate changes in demand for compute resources – at some point, we will need more Nodes to run more Pods. You can configure a fixed number of nodes to manage Node-level capacity manually. This approach is still applicable even if the process of setting up, managing, and decommissioning these nodes is automated.

This is solved by the CA, which is part of the Kubernetes autoscaler repository (https://github.com/kubernetes/autoscaler/tree/master/cluster-autoscaler). The CA must be able to provision and deprovision Nodes for the Kubernetes cluster, so this means that vendor-specific plugins must be implemented. You can find the list of supported cloud service providers here: https://github.com/kubernetes/autoscaler/tree/master/cluster-autoscaler#deployment.

The CA periodically checks the status of Pods and Nodes and decides whether it needs to take action:

- If there are Pods that cannot be scheduled and are in the Pending state because of insufficient resources in the cluster, CA will add more Nodes, up to the predefined maximum size.
- If Nodes are under-utilized and all Pods can be scheduled even with a smaller number of Nodes in the cluster, the CA will remove the Nodes from the cluster, unless it has reached the pre-defined minimum size. Nodes are gracefully drained before they are removed from the cluster.
- For some cloud service providers, the CA can also choose between different SKUs for VMs to better optimize the cost of operating the cluster.

> Pod containers must specify requests for the compute resources to make the CA work properly. Additionally, these values should reflect real usage; otherwise, the CA will not be able to make the correct decisions for your type of workload.

The CA can complement HPA capabilities. If the HPA decides that there should be more Pods for a Deployment or StatefulSet, but no more Pods can be scheduled, then the CA can intervene and increase the cluster size.

Before we explore more about the CA, let us note some of the limitations involved in CA-based Kubernetes autoscaling.

CA limitations

The CA has several constraints that can impact its effectiveness:

- There's a delay between the CA requesting a new node from the cloud provider and the node becoming available. This delay, often several minutes, can impact application perform ance during periods of high demand.
- CA's scaling decisions are based solely on pod resource requests and limits, not actual CPU or memory utilization. This can lead to underutilized nodes and resource inefficiency if pods over-request resources.
- CA is primarily designed for cloud environments. While it can be adapted for on-premises or other infrastructures, it requires additional effort. This involves custom scripts or tools to manage node provisioning and deprovisioning, as well as configuring the autoscaler to interact with these mechanisms. Without cloud-based autoscaling features, managing the cluster's size becomes more complex and requires closer monitoring.

Enabling the CA entails different steps depending on your cloud service provider. Additionally, some configuration values are specific for each of them. We will first take a look at GKE in the next section.

> **WARNING – Resource Consumption Notice**
>
> Be very cautious with CA configurations, as many such configurations can easily lead to very high resource consumption and impact system instability or unexpected scaling behaviors. Always monitor and fine-tune your configuration to avoid resource exhaustion or performance degradation.

Enabling the CA in GKE

For GKE, it is easiest to create a cluster with CA enabled from scratch. To do that, you need to run the following command to create a cluster named k8sbible:

```
$ gcloud container clusters create k8sbible \
  --enable-autoscaling \
  --num-nodes 3 \
  --min-nodes 2 \
  --max-nodes 10 \
  --region=us-central1-a
...<removed for brevity>...
Creating cluster k8sbible in us-central1-a... Cluster is being health-checked
(master is healthy)...done.
Created [https://container.googleapis.com/v1/projects/k8sbible-project/zones/
us-central1-a/clusters/k8sbible].

To inspect the contents of your cluster, go to: https://console.cloud.google.
com/kubernetes/workload_/gcloud/us-central1-a/k8sbible?project=k8sbible-project

kubeconfig entry generated for k8sbible.
NAME       LOCATION       MASTER_VERSION      MASTER_IP      MACHINE_TYPE   NODE_
VERSION          NUM_NODES  STATUS
k8sbible  us-central1-a  1.29.7-gke.1008000  <removed>      e2-medium
1.29.7-gke.1008000  3          RUNNING
```

In the preceding command:

- cloud container clusters create k8sbible: Creates a new Kubernetes cluster named k8sbible.
- --enable-autoscaling: Enables autoscaling for the cluster's node pools.
- --num-nodes 3: Sets the initial number of nodes to 3.
- --min-nodes 2: Sets the minimum number of nodes to 2.
- --max-nodes 10: Sets the maximum number of nodes to 10.
- --region=us-central1-a: Specifies the region as us-central1-a.

You should have configured your GCP account with appropriate configurations and permission including **Virtual Private Cloud (VPC)**, Networks, Security, etc.

In the case of an existing cluster, you need to enable the CA on an existing Node pool. For example, if you have a cluster named k8sforbeginners with one Node pool named nodepool1, then you need to run the following command:

```
$ gcloud container clusters update k8sforbeginners --enable-autoscaling --min-
nodes=2 --max-nodes=10 --zone=us-central1-a --node-pool=nodepool1
```

The update will take a few minutes.

Verify the autoscaling feature using the gcloud CLI as follows:

```
$ gcloud container node-pools describe default-pool --cluster=k8sdemo |grep
autoscaling -A 1
autoscaling:
  enabled: true
```

Learn more about autoscaling in GKE in the official documentation: https://cloud.google.com/kubernetes-engine/docs/concepts/cluster-autoscaler.

Once configured, you can move on to *Using the CA*.

Enabling a CA in Amazon Elastic Kubernetes Service

Setting up a CA in Amazon EKS cannot currently be done in a one-click or one-command action. You need to create an appropriate IAM policy and role, deploy the CA resources to the Kubernetes cluster, and undertake manual configuration steps. For this reason, we will not cover this in the book and we request that you refer to the official instructions: https://docs.aws.amazon.com/eks/latest/userguide/cluster-autoscaler.html.

Once configured, move on to *Using the CA*.

Enabling a CA in Azure Kubernetes Service

AKS provides a similar CA setup experience to GKE – you can use a one-command procedure to either deploy a new cluster with CA enabled or update the existing one to use the CA. To create a new cluster named k8sforbeginners-aks from scratch in the k8sforbeginners-rg resource group, execute the following command:

```
$ az aks create --resource-group k8sforbeginners-rg \
  --name k8sforbeginners-aks \
  --node-count 1 \
  --enable-cluster-autoscaler \
  --min-count 1 \
```

```
    --max-count 10 \
    --vm-set-type VirtualMachineScaleSets \
    --load-balancer-sku standard \
    --generate-ssh-keys
```

You can control the minimum number of Nodes in autoscaling by using the `--min-count` parameter, and the maximum number of Nodes by using the `--max-count` parameter.

To enable the CA on an existing AKS cluster named `k8sforbeginners-aks`, execute the following command:

```
$ az aks update --resource-group k8sforbeginners-rg --name k8sforbeginners-aks
  --enable-cluster-autoscaler --min-count 2 --max-count 10
```

The update will take a few minutes.

Learn more in the official documentation: `https://docs.microsoft.com/en-us/azure/aks/cluster-autoscaler`. Additionally, the CA in AKS has more parameters that you can configure using the **autoscaler profile**. Further details are provided in the official documentation at `https://docs.microsoft.com/en-us/azure/aks/cluster-autoscaler#using-the-autoscaler-profile`.

Now, let's take a look at how to use a CA in a Kubernetes cluster.

Using the CA

We have just configured the CA for the cluster and it might take a bit of time for the CA to perform its first actions. This depends on the CA configuration, which may be vendor-specific. For example, in the case of AKS, the cluster will be evaluated every 10 seconds (`scan-interval`), to check whether it needs to be scaled up or down. If scaling down needs to happen after scaling up, there is a 10-minute delay (`scale-down-delay-after-add`). Scaling down will be triggered if the sum of requested resources divided by capacity is below 0.5 (`scale-down-utilization-threshold`).

As a result, the cluster may automatically scale up, scale down, or remain unchanged after the CA is enabled.

For the demonstration, we are using a GKE cluster with two nodes:

```
$ kubectl get nodes -o custom-columns=NAME:.metadata.name,CPU_ALLOCATABLE:.
  status.allocatable.cpu,MEMORY_ALLOCATABLE:.status.allocatable.memory
NAME                                      CPU_ALLOCATABLE   MEMORY_ALLOCATABLE
gke-k8sdemo-default-pool-1bf4f185-6422    940m              2873304Ki
gke-k8sdemo-default-pool-1bf4f185-csv0    940m              2873312Ki
```

- Based on this, we have a computing capacity of 1.88 cores CPU and 5611.34 Mi memory in total in the GKE cluster.
- Remember, there is a bit of KCU consumed by the kube-system namespace Pods.
- Check the exact number of CPU and memory usage using the `kubectl top nodes` command in your cluster.

Unfortunately, there is no simple way to have predictable and varying CPU usage in a container out of the box. So, we need to set up a Deployment with a Pod template to achieve this for our demonstration. We'll use another hamster Deployment to create an `elastic-hamster` Deployment (refer to the `Chapter20/ca` directory in the GitHub repo). The `hamster.sh` shell script running continuously in the container will operate in a way that increases the workload based on the `TOTAL_HAMSTER_USAGE` value. We'll set the total desired work for all hamsters across all Pods. Each Pod will query the Kubernetes API to determine the number of currently running replicas for the Deployment. Then, we'll divide the total desired work by the number of replicas to determine the workload for each hamster.

For instance, if we set the total work for all hamsters to 1.0, which represents the total KCU consumption in the cluster, and deploy five replicas, each hamster will do 1.0/5 = 0.2 work. This means they will work for 0.2 seconds and rest for 0.8 seconds. If we scale the Deployment to 10 replicas, each hamster will then do 0.1 seconds of work and rest for 0.9 seconds. Thus, the hamsters collectively always work for 1.0 seconds, regardless of the number of replicas. This mimics a real-world scenario where end users generate traffic that needs to be managed, and this load is distributed among the Pod replicas. The more Pod replicas there are, the less traffic each has to handle, resulting in lower average CPU usage.

> You may use alternative methods to increase the workload using tools you are familiar with. However, to avoid introducing additional tools in this context, we are employing a workaround to demonstrate the workload increase and scaling.

Follow these steps to implement and test the cluster autoscaling in the cluster:

1. To isolate the testing, we will use a `ca-demo` Namespace:

```
# ca/ca-demo-ns.yaml

---
apiVersion: v1
kind: Namespace
metadata:
  labels:
    project: ca-demo
  name: ca-demo
```

2. To query Deployments via the Kubernetes API, you'll need to set up additional RBAC permissions. More details can be found in *Chapter 18*, *Security in Kubernetes*. Prepare a `Role` definition as follows:

```
# ca/deployment-reader-role.yaml
apiVersion: rbac.authorization.k8s.io/v1
kind: Role
metadata:
  namespace: ca-demo
  name: deployment-reader
```

```
rules:
- apiGroups: ["apps"]
  resources: ["deployments"]
  verbs: ["get", "watch", "list"]
```

3. Prepare a `ServiceAccount` for the hamster Pods to use:

```
# ca/elastic-hamster-serviceaccount.yaml
apiVersion: v1
kind: ServiceAccount
metadata:
  name: elastic-hamster
  namespace: ca-demo
```

4. Also, prepare a `RoleBinding` YAML:

```
# ca/read-deployments-rolebinding.yaml
apiVersion: rbac.authorization.k8s.io/v1
kind: RoleBinding
metadata:
  name: read-deployments
  namespace: ca-demo
subjects:
- kind: ServiceAccount
  name: elastic-hamster
  namespace: default
roleRef:
  kind: Role
  name: deployment-reader
  apiGroup: rbac.authorization.k8s.io
```

5. The hamster deployment is very simple, as follows, but with a special container image (refer to `ca/elastic-hamster-deployment.yaml`):

```
...
    spec:
      serviceAccountName: elastic-hamster
      containers:
        - name: hamster
          image: quay.io/iamgini/elastic-hamster:1.0
          resources:
            requests:
              cpu: 500m
```

```
        memory: 50Mi
      env:
        - name: TOTAL_HAMSTER_USAGE
          value: "1.0"
```

We have created a custom container image `elastic-hammer` with `hamster.sh` script inside (refer to the `ca/Dockerfile` and `ca/hamster.sh` in the `Chaper20` folder).

6. Finally, create an HPA to autoscale the Pods:

```yaml
# elastic-hamster-hpa.yaml
apiVersion: autoscaling/v1
kind: HorizontalPodAutoscaler
metadata:
  name: elastic-hamster-hpa
  namespace: ca-demo
spec:
  minReplicas: 1
  maxReplicas: 25
  metrics:
    - resource:
        name: cpu
        target:
          averageUtilization: 75
          type: Utilization
      type: Resource
  scaleTargetRef:
    apiVersion: apps/v1
    kind: Deployment
    name: elastic-hamster
```

7. Instead of applying YAML files one by one, let us apply them together; apply all the YAML files under `ca` directory as follows:

```
$ kubectl apply -f ca/
namespace/ca-demo created
role.rbac.authorization.k8s.io/deployment-reader created
deployment.apps/elastic-hamster created
horizontalpodautoscaler.autoscaling/elastic-hamster-hpa created
serviceaccount/elastic-hamster created
rolebinding.rbac.authorization.k8s.io/read-deployments created
```

Now, based on the calculation, we have `maxReplicas: 25` configured in the HPA. As per the shell script calculation, HPA will try to schedule 25 Pods with a `cpu: 500m` request. Indeed, the cluster doesn't have enough capacity to schedule those Pods and the CA will start scaling the Kubernetes nodes.

8. Check the Pods, as we will find that several Pods have a Pending status due to capacity issues:

```
$ kubectl get po -n ca-demo
NAME                                READY   STATUS    RESTARTS   AGE
elastic-hamster-87d4db7fd-4tmxn     0/1     Pending   0          7m20s
elastic-hamster-87d4db7fd-591lcd    1/1     Running   0          8m4s
elastic-hamster-87d4db7fd-5d2gf     0/1     Pending   0          7m20s
elastic-hamster-87d4db7fd-5m27q     0/1     Pending   0          8m4s
elastic-hamster-87d4db7fd-7nc48     0/1     Pending   0          7m19s
...<removed for brevity>...
elastic-hamster-87d4db7fd-st7r5     0/1     Pending   0          7m34s
elastic-hamster-87d4db7fd-twb86     1/1     Running   0          8m48s
elastic-hamster-87d4db7fd-xrppp     0/1     Pending   0          7m34s
```

9. Check the nodes now; you will find a total of 10 nodes in the cluster now (which is the maximum number we configured using the `--max-nodes 10` parameter):

```
$ kubectl top pod -n ca-demo
NAME                                    CPU(cores)   CPU%
MEMORY(bytes)    MEMORY%
gke-k8sdemo-default-pool-1bf4f185-6422  196m         20%     1220Mi
43%
gke-k8sdemo-default-pool-1bf4f185-csv0  199m         21%     1139Mi
40%
gke-k8sdemo-default-pool-1bf4f185-fcsd  751m         79%     935Mi
33%
gke-k8sdemo-default-pool-1bf4f185-frq6  731m         77%     879Mi
31%
gke-k8sdemo-default-pool-1bf4f185-h8hw  742m         78%     846Mi
30%
gke-k8sdemo-default-pool-1bf4f185-j99r  733m         77%     923Mi
32%
gke-k8sdemo-default-pool-1bf4f185-k6xq  741m         78%     986Mi
35%
...<removed for brevity>...
```

This shows how the CA has worked together with the HPA to seamlessly scale the Deployment and cluster at the same time to accommodate the workload (not a full workload in our case due to the maximum node limit). We will now show what automatic scaling down looks like. Perform the following steps:

1. To decrease the load in the cluster, let us reduce the number of maximum replicas in the HPA. It is possible to edit the YAML and apply it in the system, but let us use a `kubectl patch` command here:

```
$ kubectl patch hpa elastic-hamster-hpa -n ca-demo -p '{"spec":
{"maxReplicas": 2}}'
horizontalpodautoscaler.autoscaling/elastic-hamster-hpa patch
```

2. The Pod count will be adjusted now based on the updated HPA:

```
$ kubectl get pod -n ca-demo
NAME                               READY   STATUS    RESTARTS   AGE
elastic-hamster-87d4db7fd-2qghf    1/1     Running   0          20m
elastic-hamster-87d4db7fd-mdvpx    1/1     Running   0          19m
```

3. Since the capacity demand is less, the CA will start scaling down the nodes as well. But when scaling down, the CA allows a 10-minute grace period to reschedule the Pods from a node onto other nodes before it forcibly terminates the node. So, check the nodes after 10 minutes and you will see the unwanted nodes have been removed from the cluster.

```
$ kubectl get nodes
NAME                                      STATUS   ROLES    AGE    VERSION
gke-k8sdemo-default-pool-1bf4f185-6422    Ready    <none>   145m
v1.29.7-gke.1008000
gke-k8sdemo-default-pool-1bf4f185-csv0    Ready    <none>   145m
v1.29.7-gke.1008000
```

This shows how efficiently the CA can react to a decrease in the load in the cluster when the HPA has scaled down the Deployment. Earlier, without any intervention, the cluster scaled to 10 Nodes for a short period of time and then scaled down to just two Nodes. Imagine the cost difference between having an eight-node cluster running all the time and using the CA to cleverly autoscale on demand!

> To ensure that you are not charged for any unwanted cloud resources, you need to clean up the cluster or disable cluster autoscaling to be sure that you are not running too many Nodes.

This demonstration concludes our chapter about autoscaling in Kubernetes. But before we go to the summary, let us touch on some other Kubernetes autoscaling tools in the next section.

Alternative autoscalers for Kubernetes

Compared to the basic Kubernetes autoscaler, other autoscalers such as **Kubernetes Eventdriven Autoscaling (KEDA)** and **Karpenter** offer more flexibility and efficiency by managing resource scaling based on application-specific metrics and workloads. KEDA permits autoscaling based on events originating outside a cluster and at custom metrics. This is well suited for event-driven applications.

On the other hand, Karpenter simplifies node provisioning and scaling by automatically adapting the node count based on workload demands, using your cluster resources efficiently and cost-effectively. Together, these tools enable fine-grained scaling control so that applications can adequately perform under variable load conditions.

Let us learn about these two common Kubernetes autoscaler tools in the coming sections.

KEDA

KEDA (`https://keda.sh`) is designed to enable event-driven scaling in Kubernetes by allowing you to scale the number of pod replicas based on custom metrics or external events. Unlike traditional autoscalers, which rely on CPU or memory usage, KEDA can trigger scaling based on metrics from various event sources, such as message queues, HTTP request rates, and custom application metrics. This makes it particularly useful for workloads that are driven by specific events or metrics rather than general resource usage.

KEDA integrates seamlessly with the existing Kubernetes HPA and can scale applications up or down based on dynamic workloads. By supporting a wide range of event sources, it offers flexibility and precision in scaling decisions. KEDA helps ensure that resources are allocated efficiently in response to real-time demand, which can optimize costs and improve application performance.

The following diagram shows the architecture and components of KEDA:

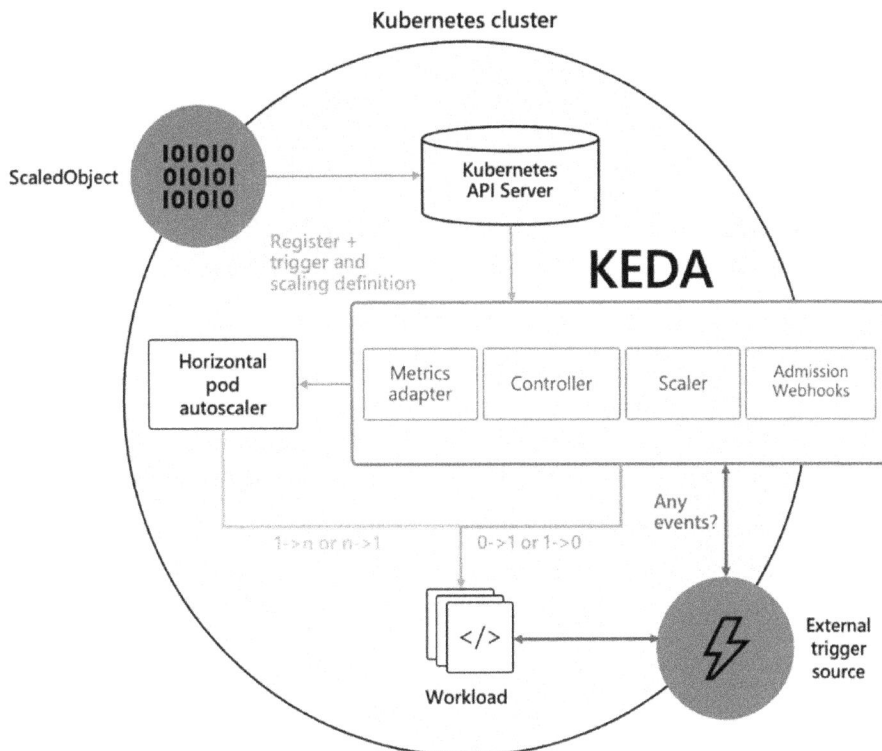

Figure 20.5: KEA architecture (image source: https://keda.sh/docs/2.15/concepts/)

KEDA is an open source project hosted by the CNCF and provides best-effort support via GitHub for filing bugs and feature requests. There are several different vendors that include KEDA as part of their offering and support, including Azure Container Apps, Red Hat OpenShift Autoscaler with custom metrics, and KEDA Add-On for Azure Kubernetes Service.

Karpenter

Karpenter (`https://karpenter.sh`) is an advanced Kubernetes CA that focuses on optimizing the provisioning and scaling of nodes within a cluster. It automates the process of scaling compute resources by dynamically adjusting the number of nodes based on the needs of your workloads. Karpenter is designed to rapidly adapt to changes in demand and optimize the cluster's capacity, thereby improving both performance and cost efficiency.

The following diagram shows how Karpenter works in a Kubernetes cluster.

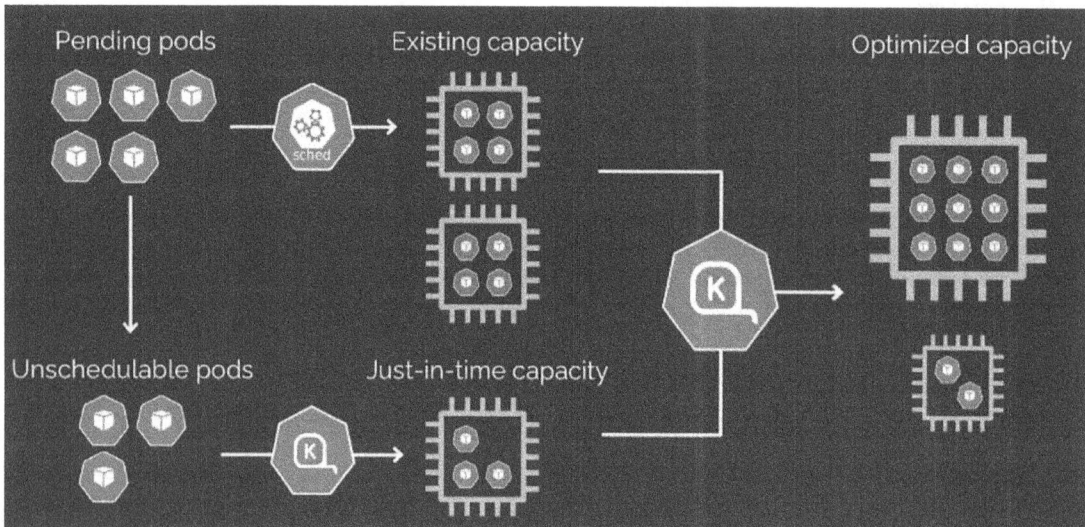

Figure 20.6: Workings of Karpenter (image source: https://karpenter.sh)

Karpenter offers fast and efficient node scaling with capabilities like capacity optimization and intelligent provisioning. It ensures that the right types and amounts of nodes are available to meet workload requirements, minimizing waste and cost. By providing sophisticated scaling and provisioning features, Karpenter helps maintain cluster performance while keeping operational costs in check.

Implementing autoscaling using KEDA or Karpenter is beyond the scope of this book; please refer to the documentation (`https://keda.sh/docs/latest`) to learn more.

Now, let's summarize what we have learned in this chapter.

Summary

In this chapter, you have learned about autoscaling techniques in Kubernetes clusters. We first explained the basics behind Pod resource requests and limits and why they are crucial for the autoscaling and scheduling of Pods.

Next, we introduced the VPA, which can automatically change requests and limits for Pods based on current and past metrics. After that, you learned about the HPA, which can be used to automatically change the number of Deployment or StatefulSet replicas. The changes are done based on CPU, memory, or custom metrics. Lastly, we explained the role of the CA in cloud environments. We also demonstrated how to efficiently combine the HPA with the CA to achieve the scaling of your workload together with the scaling of the cluster.

There is much more that can be configured in the VPA, HPA, and CA, so we have just scratched the surface of powerful autoscaling in Kubernetes! We also mentioned alternative Kubernetes autoscalers such as KEDA and Karpenter.

In the next chapter, we will explain advanced Kubernetes topics such as traffic management using ingress, multi-cluster strategies, and emerging technologies.

Further reading

- Horizontal Pod Autoscaling: `https://kubernetes.io/docs/tasks/run-application/horizontal-pod-autoscale/`
- Autoscaling Workloads: `https://kubernetes.io/docs/concepts/workloads/autoscaling/`
- HorizontalPodAutoscaler Walkthrough: https://kubernetes.io/docs/tasks/run-application/horizontal-pod-autoscale-walkthrough/
- Cluster Autoscaling: `https://kubernetes.io/docs/concepts/cluster-administration/cluster-autoscaling/`.

For more information regarding autoscaling in Kubernetes, please refer to the following Packt books:

- *The Complete Kubernetes Guide*, by *Jonathan Baier, Gigi Sayfan, Jesse White* (`https://www.packtpub.com/en-in/product/the-complete-kubernetes-guide-9781838647346`)
- *Getting Started with Kubernetes – Third Edition*, by *Jonathan Baier, Jesse White* (`https://www.packtpub.com/en-in/product/getting-started-with-kubernetes-9781788997263`)
- *Kubernetes for Developers*, by *Joseph Heck* (`https://www.packtpub.com/en-in/product/kubernetes-for-developers-9781788830607`)
- *Hands-On Kubernetes on Windows*, by *Piotr Tylenda* (`https://www.packtpub.com/product/hands-on-kubernetes-on-windows/9781838821562`)

You can also refer to the official Kubernetes documentation:

- Kubernetes documentation (`https://kubernetes.io/docs/home/`). This is always the most up-to-date source of knowledge regarding Kubernetes in general.
- General installation instructions for the VPA are available here: `https://github.com/kubernetes/autoscaler/tree/master/vertical-pod-autoscaler#installation`.
- EKS' documentation offers its own version of the instructions: `https://docs.aws.amazon.com/eks/latest/userguide/vertical-pod-autoscaler.html`.

Join our community on Discord

Join our community's Discord space for discussions with the authors and other readers:

`https://packt.link/cloudanddevops`

21

Advanced Kubernetes: Traffic Management, Multi-Cluster Strategies, and More

Advanced topics in Kubernetes, beyond those covered in the earlier parts of this book, will be discussed in this final chapter. We will start by looking into the advanced use of Ingress for some really sophisticated routing to your Pods, followed by effective methodologies for troubleshooting Kubernetes and hardening Kubernetes security, as well as best practices for optimizing a Kubernetes setup.

This final chapter will introduce you to advanced Kubernetes traffic routing in Kubernetes using Ingress resources. In a nutshell, Ingress allows exposing your Pods running behind a Service object to the external world using HTTP and HTTPS routes. So far, we have introduced ways to expose your application using Service objects directly, especially the LoadBalancer Service. But this approach only works well in cloud environments where you have the cloud-controller-manager running. It works by configuring external load balancers to be used with this type of Service. Moreover, each LoadBalancer Service requires a separate instance of the cloud load balancer, which brings additional costs and maintenance overhead. Next, we are going to introduce Ingress and Ingress Controller, which can be used in any type of environment to provide routing and load-balancing capabilities for your application. You are also going to learn how to use the nginx web server as an Ingress Controller and how you can configure the dedicated Azure **Application Gateway Ingress Controller** (**AGIC**) for your AKS cluster.

Further, we are going to review some of the recent Kubernetes projects that include KubeVirt for virtualization and serverless solutions, such as Knative and OpenFaaS. You will also learn about ephemeral containers and how they are used in real-time troubleshooting, the role of different Kubernetes plugins, and multi-cluster management. Although we will be giving an overview of most of them, kindly note that some of these topics are only at a high level because they go beyond the detailed scope of this book.

In this chapter, we will cover the following topics:

- Advanced Traffic Routing with Ingress

- Gateway API
- Modern Advancements with Kubernetes
- Maintaining Kubernetes Clusters – Day 2 tasks
- Securing a Kubernetes Cluster – Best Practices
- Troubleshooting Kubernetes

Technical Requirements

For this chapter, you will need the following:

- A Kubernetes cluster deployed. We recommend using a multi-node, cloud-based Kubernetes cluster. It is also possible to use Ingress in `minikube` after enabling the required add-ons.
- An AKS cluster is required to follow the section about the Azure AGIC.
- The Kubernetes CLI (`kubectl`) needs to be installed on your local machine and configured to manage your Kubernetes cluster.

Basic Kubernetes cluster deployment (local and cloud-based) and `kubectl` installation have been covered in *Chapter 3, Installing Your First Kubernetes Cluster*.

The previous chapters of this book, *15*, *16*, and *17*, have provided you with an overview of how to deploy a fully functional Kubernetes cluster on different cloud platforms and install the requisite CLIs to manage them.

You can download the latest code samples for this chapter from the official GitHub repository at `https://github.com/PacktPublishing/The-Kubernetes-Bible-Second-Edition/tree/main/Chapter21`.

Advanced Traffic Routing with Ingress

This section will explain how Ingress can be used to supply advanced networking and mechanisms of traffic routing in Kubernetes. Fundamentally, an Ingress is a reverse proxy Kubernetes resource. It will route incoming requests from outside of the cluster to services inside of the cluster based on rules specified in the ingress configuration. A single entry may be used to allow external users to access applications deployed within the cluster.

Before we look at Ingress and its resources, let's do a quick recap of the various Kubernetes service types that we have used to access applications.

Refresher — Kubernetes Services

In *Chapter 8, Exposing Your Pods with Services*, you learned about the Service objects that can be used to expose Pods to load-balanced traffic, both internal as well as external. Internally, they are implemented as virtual IP addresses managed by kube-proxy at each of the Nodes. We are going to do a quick recap of different types of services:

- `ClusterIP`: Exposes Pods using internally visible, virtual IP addresses managed by `kube-proxy` on each Node. This means that the Service will only be reachable from within the cluster.

- `NodePort`: Like the `ClusterIP` service, it can be accessed via any node's IP address and a specified port. Kube-proxy exposes a port in the 30000-32767 range – by default, this is configurable – on each node and sets up forwarding rules so that connections to this port are directed to the corresponding `ClusterIP` service.

- `LoadBalancer`: Usually used in cloud environments where you have **software-defined networking (SDN)**, and you can configure load balancers on demand that redirect traffic to your cluster. In cloud-controller-manager, the automatic provisioning of load balancers in the cloud is driven by vendor-specific plugins. This type of service combines the approach of the `NodePort` Service with an additional external load balancer in front of it, which routes traffic to NodePorts.

You can still, of course, use the service internally via its `ClusterIP`.

It might sound tempting to always use Kubernetes services for enabling external traffic to the cluster, but there are a couple of disadvantages to using them all the time. We will now introduce the Ingress object and discuss why it is needed, and when it should be used instead of Services to handle external traffic.

Overview of the Ingress object

In the previous section, we briefly reviewed the Service objects in Kubernetes and the role they play in routing traffic. From the viewpoint of incoming traffic, the most important ones are the NodePort service and the LoadBalancer service. Generally, the NodePort service is only used along with another component of routing and load balancing, because exposing several external endpoints on all Kubernetes nodes is not secure. Now, this leaves us with the LoadBalancer service, which relies, under the hood, on NodePort. However, there are some limitations to using this type of service in certain use cases:

- The Layer-4 `LoadBalancer` service is based on OSI layer 4, routing the traffic on the basis of the TCP/UDP protocol. Most HTTP/HTTPS-based applications demand L7 load-balancing, which is associated with OSI layer 7 applications.

- HTTPS traffic cannot be terminated and offloaded in an L4 load balancer.

- It is not possible to use the same L4 load balancer for name-based virtual hosting across several domain names.

- Path-based routing could be implemented if you had an L7 load balancer. In fact, you cannot at all configure an L4 load balancer to proxy requests like `https://<loadBalancerIp>/service1` to the Kubernetes service named `service1` and `https://<loadBalancerIp>/service2` to the ones proxied to the Kubernetes service named `service2`, since an L4 load balancer is completely unaware of the HTTP(S) protocol.

- Some features, like sticky sessions or cookie affinity, require an L7 load balancer.

In Kubernetes, these problems can be solved by using an Ingress object, which can be used to implement and model L7 load balancing. Ingress object is only used for defining the routing and balancing rules; for example, what path shall be routed to what Kubernetes Service.

Let's take a look at an example YAML manifest file, `ingress/portal-ingress.yaml`, for Ingress:

```yaml
# ingress/portal-ingress.yaml
apiVersion: networking.k8s.io/v1
kind: Ingress
metadata:
  name: portal-ingress
  namespace: ingress-demo
  annotations:
    nginx.ingress.kubernetes.io/rewrite-target: /
spec:
  rules:
    - host: k8sbible.local
      http:
        paths:
          - path: /video
            pathType: Prefix
            backend:
              service:
                name: video-service
                port:
                  number: 8080
          - path: /shopping
            pathType: Prefix
            backend:
              service:
                name: shopping-service
                port:
                  number: 8080
```

We can visualize what is happening behind the Ingress Controller in the following diagram:

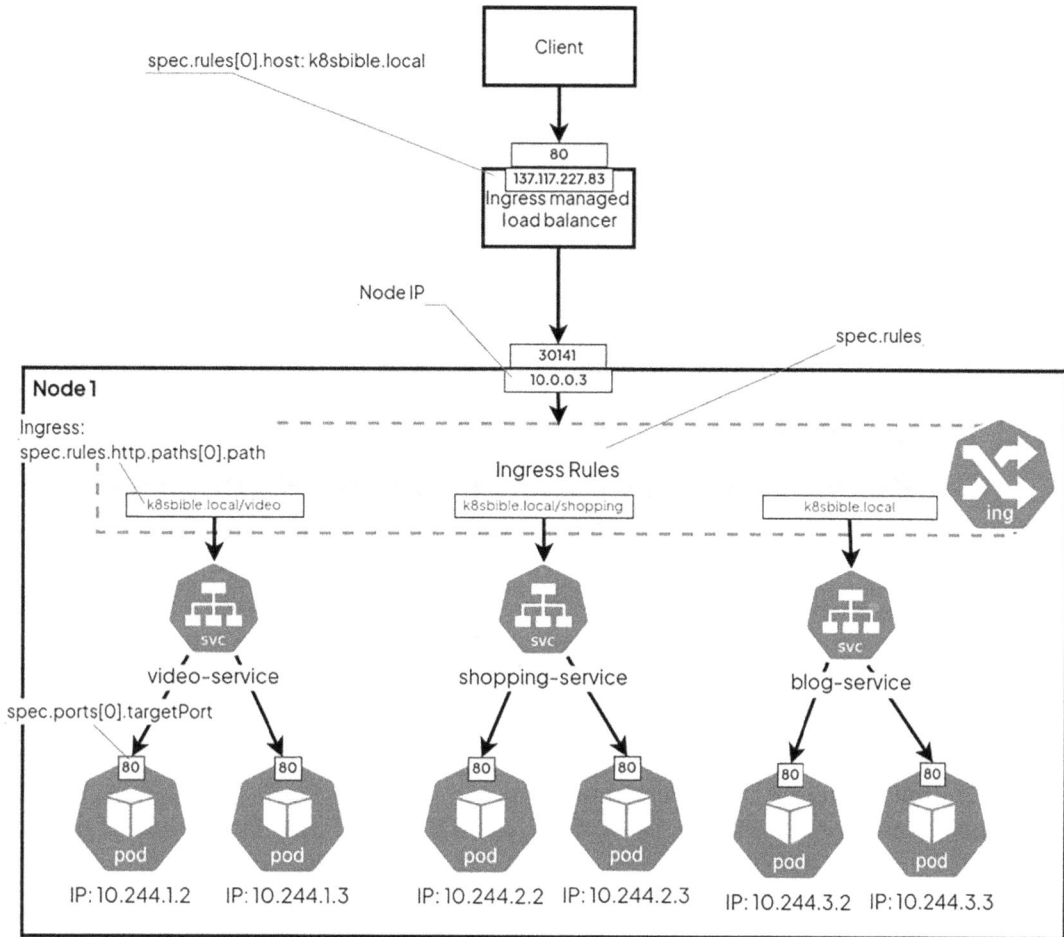

Figure 21.1: Using nginx as an Ingress Controller in a cloud environment

Simply said, Ingress is an abstract definition of routing rules for your Services. Alone, it is not doing anything; it needs Ingress Controller to actually process and implement these rules—you can apply the manifest file, but at this point, it will have no effect. But first, we're going to explain how Ingress HTTP routing rules are built. Each of these rules in the specification contains the following:

- **Optional host:** We are not using this field in our example; hence, the rule we have defined here applies to all incoming traffic. If the field value is provided, then the rule applies only to requests that have this host as a destination—you can have multiple hostnames resolve to the same IP address. The host field supports wildcards.

- **Listing of the path routings:** Each of the paths has an associated Ingress backend, which you define by providing serviceName and servicePort. In the preceding example, all requests arriving at the path with the prefix /video will be routed to the Pods of the video-service Service and all requests arriving at the path with the prefix /shopping will be routed to the Pods of the shopping-service Service. The path fields support prefixes and exact matching, and you can also use implementation-specific matching, which is carried out by the underlying Ingress Controller.

This way, you will be able to configure complex routing rules that involve multiple Services in the cluster, but externally, they will be visible as a single endpoint with multiple paths available.

In order to materialize these Ingress objects, we need to have an Ingress Controller installed in the cluster, which we will learn about in the next section.

Using nginx as an Ingress Controller

The Ingress Controller is a Kubernetes controller that one deploys manually to the cluster, most often as a DaemonSet or a Deployment object running dedicated Pods handling incoming traffic load balancing and smart routing. It is responsible for the processing of the Ingress objects; that is, it's responsible for those specifying that they want to use the Ingress Controller and dynamic configuration of real routing rules.

> Unlike other types of controllers that run as part of the `kube-controller-manager` binary, Ingress controllers are not started automatically with a cluster. The Kubernetes project maintains a number of Ingress controllers, including AWS, GCE, and ngnix Ingress controllers. For third-party Ingress controller projects, see the documentation for a detailed list:
>
> `https://kubernetes.io/docs/concepts/services-networking/ingress-controllers/#additional-controllers`

One of the commonly used Ingress controllers for Kubernetes is nginx. The correct term is **Nginx Ingress Controller**. Ingress Controller (`https://www.f5.com/products/nginx/nginx-ingress-controller`) is installed in the cluster as a Deployment with a set of rules for handling Ingress API objects. The Ingress Controller is exposed as a Service with a type that depends on the installation – in cloud environments, this will be `LoadBalancer`.

> You will frequently encounter dedicated Ingress Controllers in cloud environments, which utilize special features provided by the cloud provider to allow the external load balancer to communicate directly with the Pods. There is no extra Pod overhead in this case, and even `NodePort` Services might not be needed. Such routing is done at the level of SDN and CNI, whereas the load balancer may use the private IPs of the Pods. We will review an example of such an approach in the next section when we discuss the **Application Gateway ingress controller for AKS**.

The installation of `ingress-nginx` is described for different environments in the official documentation: `https://kubernetes.github.io/ingress-nginx/deploy/`.

Note that while using Helm is the preferred deployment method, some environments might require specific instructions. For cloud environments, the installation of ingress-nginx is usually very simple and involves applying a single YAML manifest file (or enabling the ingress-nginx while creating the cloud based managed Kubernetes clusters), which creates multiple Kubernetes objects. For example, it is possible to deploy the required ingress controller components in AKS or GKE using a single command as follows:

```
$ kubectl apply -f https://raw.githubusercontent.com/kubernetes/ingress-nginx/
controller-v1.11.2/deploy/static/provider/cloud/deploy.yaml
```

This was true when writing and when this deployment was tested. For the most current stable version, please refer to the documentation of the Ingress Controller. Also, note that different Kubernetes distributions might have different prerequisites to implement such features; for example, you should have cluster-admin permission on the cluster to enable ingress-nginx in a GKE cluster. Refer to the documentation (`https://kubernetes.github.io/ingress-nginx/`) to learn more.

In AWS, a **Network Load Balancer** (**NLB**) is used to expose the Nginx Ingress Controller by configuring it with a Service of Type LoadBalancer:

```
$ kubectl apply -f https://raw.githubusercontent.com/kubernetes/ingress-nginx/
controller-v1.11.2/deploy/static/provider/aws/deploy.yaml
```

The YAML contains several resources to set up ingress in the cluster, including `Roles`, `RoleBinding`, `Namespace`, `ConfigMap`, and so on.

If you do not have a cloud environment or cloud-based Kubernetes deployment, then refer to the following section to deploy the Ingress Controller in the minikube cluster.

Deploying the NGINX Ingress Controller in minikube

A multi-node `minikube` Kubernetes cluster can be deployed using the following command:

```
$ minikube start --cni calico --nodes 3 --kubernetes-version=v1.31.0
```

Once the Kubernetes cluster is up and running, enable Ingress in the `minikube` cluster using the following command:

```
$ minikube addons enable ingress
  ingress is an addon maintained by Kubernetes. For any concerns contact
minikube on GitHub.

...<removed for brevity>...
  Verifying ingress addon...
  The 'ingress' addon is enabled
```

Verify the Pods in the `ingress-nginx` Namespace as follows:

```
$ kubectl get pods -n ingress-nginx
NAME                                        READY   STATUS      RESTARTS   AGE
ingress-nginx-admission-create-rsznt        0/1     Completed   0          78s
ingress-nginx-admission-patch-4c7xh         0/1     Completed   0          78s
ingress-nginx-controller-6fc95558f4-zdhp7   1/1     Running     0          78s
```

Now, the Ingress controller is ready to monitor Ingress resources. In the following section, we will learn how to deploy Ingress resources in our Kubernetes cluster.

Deploying Ingress Resources in Kubernetes

Now, we are ready to deploy our application; refer to the `Chapter21/ingress` directory in the repository where we have prepared the following YAML files:

- `00-ingress-demo-ns.yaml`: Create the `ingress-demo` Namespace.
- `video-portal.yaml`: Create a video portal with ConfigMap, Deployment and Service.
- `blog-portal.yaml`: Create a blog portal with ConfigMap, Deployment and Service.
- `shopping-portal.yaml`: Create a shopping portal with ConfigMap, Deployment, and Service.
- `portal-ingress.yaml`: Create ingress resource to create path-based ingress for our website (`k8sbible.local`).

Inside the `portal-ingress.yaml` file, the following rule tells ingress to serve `video-service` when users access `k8sbible.local/video`:

```
# ingress/portal-ingress.yaml
...
spec:
  rules:
    - host: k8sbible.local
      http:
        paths:
          - path: /video
            pathType: Prefix
            backend:
              service:
                name: video-service
                port:
                  number: 8080
  ...
```

The following rule tells ingress to serve `shopping-service` when users access `k8sbible.local/shopping`:

```
  ...
          - path: /shopping
            pathType: Prefix
            backend:
              service:
                name: shopping-service
                port:
                  number: 8080
```

```
...
Finally, the following rule tells the ingress to serve the blog-service when
users access k8sbible.local/ or k8sbible.local:

...
            - path: /
              pathType: Prefix
              backend:
                service:
                  name: blog-service
                  port:
                    number: 8080
```

> Since we already learned about Deployment, ConfigMaps, and Services, we are going to skip explaining those items here; you may refer to the YAML files in the repository for more information.

Apply the YAML files inside the ingress directory as follows:

```
$ kubectl apply -f ingress/
namespace/ingress-demo created
configmap/blog-configmap created
deployment.apps/blog created
service/blog-service created
ingress.networking.k8s.io/portal-ingress created
configmap/shopping-configmap created
deployment.apps/shopping created
service/shopping-service created
configmap/video-configmap created
deployment.apps/video created
service/video-service created
```

Check the Pods, Services, and Ingress resources:

```
$ kubectl get po,svc,ingress -n ingress-demo
NAME                            READY   STATUS    RESTARTS   AGE
pod/blog-675df44d5-5s8sg        1/1     Running   0          88s
pod/shopping-6f88c5f485-lw6ts   1/1     Running   0          88s
pod/video-7d945d8c9f-wkxc5      1/1     Running   0          88s


NAME                  TYPE        CLUSTER-IP      EXTERNAL-IP   PORT(S)
AGE
service/blog-service  ClusterIP   10.111.70.32    <none>        8080/TCP
```

```
88s
service/shopping-service    ClusterIP    10.99.103.137    <none>           8080/TCP
88s
service/video-service       ClusterIP    10.109.3.177     <none>           8080/TCP
88s


NAME                                      CLASS    HOSTS          ADDRESS
PORTS     AGE
ingress.networking.k8s.io/portal-ingress  nginx    k8sbible.local
192.168.39.18    80         88s
```

If you are using a cloud-based Kubernetes cluster, then k8sbible.local or whatever host you have used in the ingress configuration should point to the cloud LoadBalancer IP address. If you do not have any actual domain name registered, then you can simulate the same using local /etc/hosts entries (C:\windows\system32\drivers\etc\hosts in Windows machines).

For example, assume we have deployed a minikube cluster and used the following VM IP addresses which we fetched using minikube ip command inside the /etc/hosts file:

```
# k8sbible minikube
192.168.39.18 k8sbible.local
```

Now, you can access your portal using http://k8sbible.local. Open a browser (or use the curl command) and test different services, as shown in the following figure.

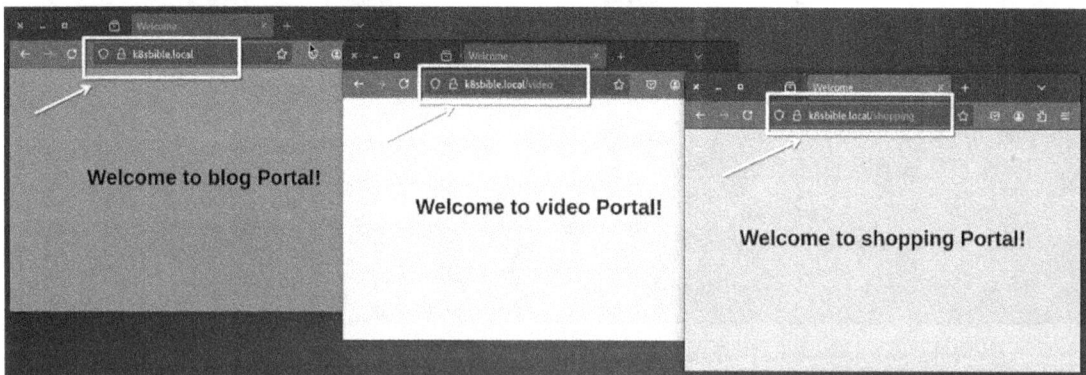

Figure 21.2: Ingress serving different services.

When you perform an HTTP request to http://k8sbible.local/video, the traffic will be routed by nginx to video-service. Similarly, when you use the /shopping path, the traffic will be routed to shopping-service. Note that you are only using one cloud load balancer (or a public IP/hostname) for this operation and that the actual routing to Kubernetes Services is performed by the Ingress Controller Pods using path-based routing.

In practice, you should set up SSL certificates for your HTTP endpoints if you want proper security. It is possible to set up SSL for ingress, but you need a domain name or local environment alternatives—local domain names. We are not setting up a local domain name for our examples for simplicity and clarity. Refer to the documentation of the cert-manager to learn more about this:

```
https://cert-manager.io/docs/tutorials/acme/nginx-ingress/
```

Congratulations! You have successfully configured the Ingress and Ingress Controller in your cluster.

As we mentioned at the beginning of this section on Ingress, there are multiple Ingress controllers and methods available to use. Before we learn about another Ingress method, let us learn about ingressClass in the next section.

ingressClass and Multiple Ingress Controllers

In some situations, we may need different configurations for the Ingress controller. With a single Ingress controller, you may not be able to implement it as the customized configuration may impact other Ingress objects in the Kubernetes cluster. In such cases, you can deploy multiple Ingress controllers within a single Kubernetes cluster by using the ingressClass mechanism. Some of the scenarios are listed here:

- **Different classes of Ingress for different requirements**: Kubernetes ingress controllers can be annotated with specific Ingress classes, such as nginx-public and nginx-private. This can help to direct various types of traffic; for instance, public traffic can be served by a performance-optimized controller while your internal services remain behind tighter access controls.
- **Multi-protocol support**: Different applications will require support for multiple protocols, including HTTP/HTTPS and TCP/UDP. This can be handled by having different ingress controllers for each protocol. In this way, applications with different protocol requirements will be supported on the same Kubernetes cluster without relying on an ingress controller for all types. The performance will be enhanced along with reducing the complexity of configuration.

It's important to note the .metadata.name of your ingressClass resource because this name is needed when specifying the ingressClassName field on your Ingress object. This ingressClassName field replaces the older method of using annotations to link an Ingress to a specific controller, as outlined in the **IngressSpec** v1 documentation.

If you don't specify an IngressClass when creating an Ingress, and your cluster has exactly one IngressClass marked as default, Kubernetes will automatically apply that default IngressClass to the Ingress. To mark an IngressClass as the default, you should set the ingressclass.kubernetes.io/is-default-class annotation on that IngressClass, with the true value. While this is the intended specification, it's important to note that different Ingress controllers may have slight variations in their implementation of these features.

Now, let us check the nginx ingress controller we used in the previous hands-on lab to identify the ingressClass:

```
$ kubectl get IngressClass -o yaml
apiVersion: v1
items:
- apiVersion: networking.k8s.io/v1
  kind: IngressClass
  metadata:
    name: nginx
    annotations:
      ingressclass.kubernetes.io/is-default-class: "true"
...<removed for brevity>...
  spec:
    controller: k8s.io/ingress-nginx
```

In the preceding snippet, the following is true:

- The ingressClass name is nginx (.metadata.name)
- You can see the ingressclass.kubernetes.io/is-default-class: "true"

In the following section, we are going to explore a special type of Ingress Controller for AKS named Azure Application Gateway Ingress Controller.

Azure Application Gateway Ingress Controller for AKS

As discussed in detail in the preceding section, the use of the nginx Ingress Controller is a rather flexible approach to traffic routing within a Kubernetes cluster. Though this approach generally serves well, it can be a bit complex when one moves to opt for cloud providers such as **Azure Kubernetes Service (AKS)** due to multiple layers of load balancing. Those layers can introduce unnecessary complexity and raise the number of failure points.

To solve these problems, AKS offers a native L7 load balancer service called **Azure Application Gateway Ingress Controller (AGIC)**. AGIC works in tandem with the networking services of Azure to support much more efficient and reliable traffic routing, enabling direct communications with Pods via their private IP addresses. Such functionality is made possible through some Azure SDN features, such as VNet Peering.

Why Choose AGIC for AKS?

The reasons for choosing AGIC for AKS are as follows:

- **Streamlined Load Balancing:** AGIC eliminates the need to use a separate Azure Load Balancer that would then proxy requests to the nginx Ingress Controller Pods using NodePorts. Instead, it forwards the traffic directly to the Pods. This reduces the layers involved in load balancing and minimizes the possibility of failure points.
- **Direct Pod Communication:** AGIC leverages the Azure SDN capability to enable direct communications with the Pods, without having kube-proxy manage the routing of services.

This high-level design of AGIC is shown in the following diagram:

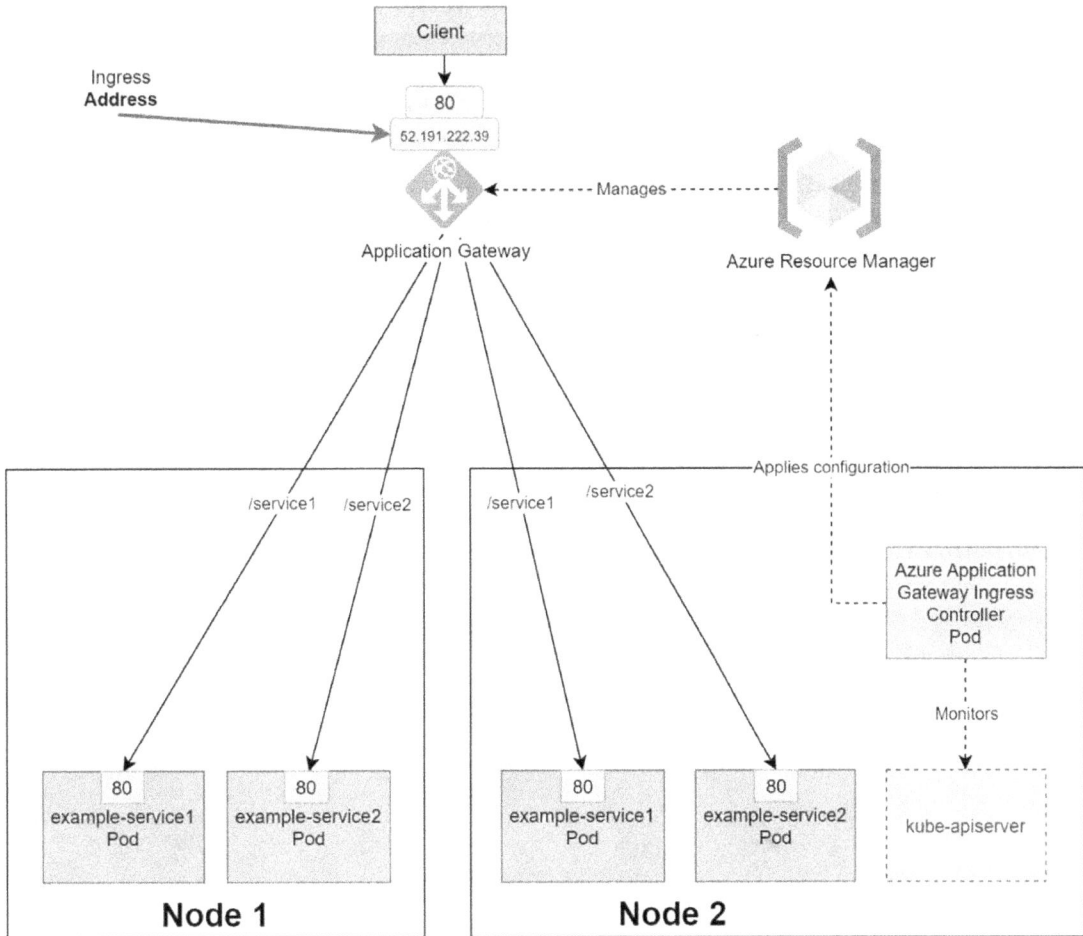

Figure 21.3: Application Gateway Ingress Controller in AKS

It is possible to configure AGIC on an existing AKS cluster, and that is described in the official documentation: `https://docs.microsoft.com/en-us/azure/application-gateway/tutorial-ingress-controller-add-on-existing`.

For ease and simplicity, we will be creating a new AKS cluster with AGIC enabled, using a single command. To deploy the two-node cluster named `k8sforbeginners-aks-agic` in the `k8sforbeginners-rg` resource group, execute the following command:

```
$ az aks create --resource-group myResourceGroup --name myAKSCluster --node-
count 2 --network-plugin azure --enable-managed-identity -a ingress-appgw
--appgw-name MyAppGateway --appgw-subnet-cidr "10.2.0.0/16" --generate-ssh-keys
```

This will create an Azure Application Gateway named `AksApplicationGateway` with the subnet CIDR `10.2.0.0/16`.

When the cluster finishes deploying, we need to generate `kubeconfig` to use it with `kubectl`. Run the following command (it will switch to a new context so you will still have the old context available later):

```
$ az aks get-credentials --resource-group k8sforbeginners-rg --name
k8sforbeginners-aks-agic
Merged "k8sforbeginners-aks-agic" as current context in .kube/config
```

Now we can use the same YAML manifests for Deployments and Services in the `ingress` directory—in the Book repo, the same as in the preceding section. But we need to make some changes in the YAML for AGIC; for better clarity, we copy the content of the `ingress` directory to `aks_agic` directory and modify it there. Modify the Ingress resource definition as follows:

```
# aks-agic/portal-ingress.yaml

...

spec:
  ingressClassName: azure-application-gateway

...
```

We also renamed the namespace to `agic-demo` to isolate the testing. Apply the YAML definitions from `aks_agic` directory as follows:

```
$ kubectl apply -f aks_agic/
```

Wait a few moments for the Application Gateway to update its configuration. To retrieve the external IP address of the Ingress, run the following:

```
$ kubectl get ingress
NAME              CLASS    HOSTS   ADDRESS          PORTS   AGE
example-ingress   <none>   *       52.191.222.39    80      36m
```

In our case, the IP address is 52.191.222.39.

Test the configuration by navigating to `/video` and `/shopping` paths using the retrieved IP address:

- **Service 1**: `http://<external-IP>/video` will be served by the `video-service` Pods.
- **Service 2**: `http://<external-IP>/shopping` will be served by the `shopping-service` Pods.
- **Default service**: `http://<external-IP>/` will be served by the `blog-service` Pods.

With this setup, you've successfully configured and tested the AGIC in AKS.

In the following section, we will learn about the Gateway API in Kubernetes, which is a relatively new and powerful approach to managing traffic routing within a cluster.

Gateway API

The Kubernetes Gateway API is an evolving set of resources that offers a more expressive and extensible way of defining network traffic routing in the cluster.

It's designed to eventually replace the Ingress API with a more powerful and flexible mechanism for configuring load balancing, HTTP routing, and other network-related features.

The three main API resources comprising the Gateway API are as follows:

- `GatewayClass` represents a class of Gateways that share a common set of configurations and are operated by the same controller implementing this resource.
- `Gateway` is an instance of an environment where traffic is being controlled through a controller, for example, a cloud load balancer.
- `HTTPRoute` defines HTTP-specific rules for routing traffic from a Gateway listener to backend network endpoints, typically represented as Services.

The following figure shows the high-level flow with Gateway API resources:

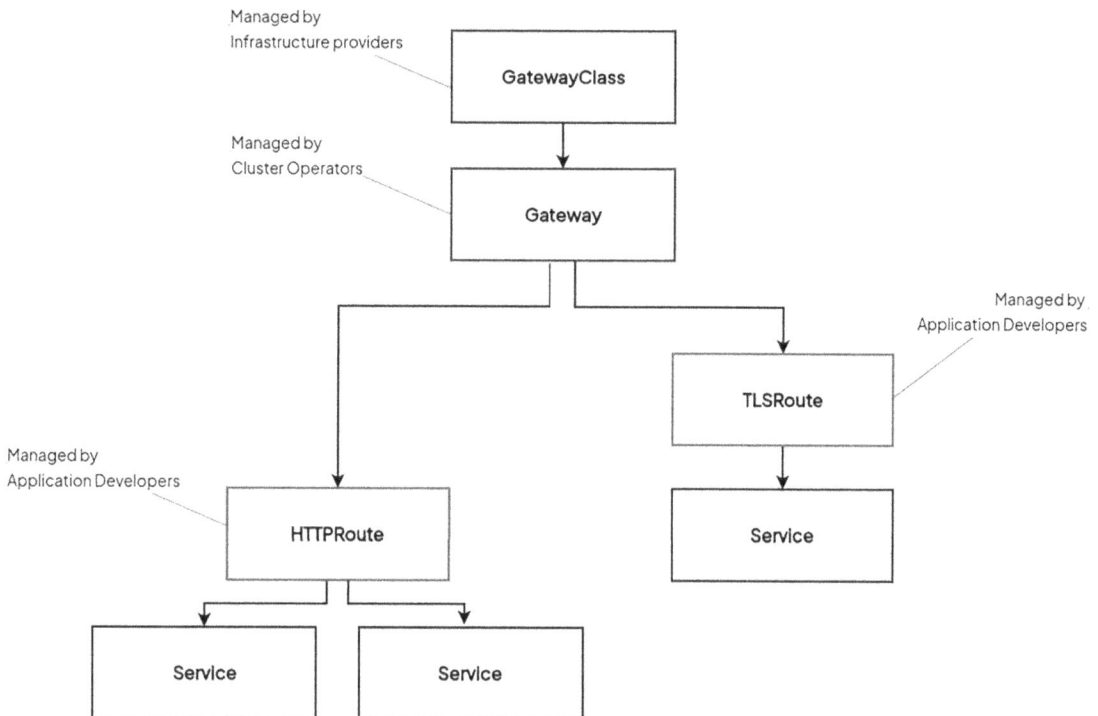

Figure 21.4: Resource model of Gateway API

Among these, the major benefits of using Gateway API over Ingress API include flexibility for complex routing scenarios like multi-level, cross-namespace routing. In addition, the design emphasizes extensibility, where third-party developers can write their own Gateway controllers that will interact seamlessly with Kubernetes. Furthermore, the Gateway API allows more fine-grained control over routing rules, traffic policies, and load-balancing management tasks.

A typical `GatewayClass` is provided here for reference:

```yaml
# gateway_api/gatewayclass.yaml
apiVersion: gateway.networking.k8s.io/v1
kind: GatewayClass
metadata:
  name: dev-cluster-gateway
spec:
  controllerName: "example.net/gateway-controller"
```

`gateway-api/gateway_api/gateway.yaml` contains a typical `Gateway` resource pointing to dev-cluster-gateway as gatewayClassName:

```yaml
apiVersion: gateway.networking.k8s.io/v1
kind: Gateway
metadata:
  name: dev-gateway
  namespace: gateway-api-demo
spec:
  gatewayClassName: dev-cluster-gateway
  listeners:
    - name: http
      protocol: HTTP
      port: 80
```

Finally, we have `HTTPRoute` (similar to Ingress) with rules pointing to different services:

```yaml
# gateway-api/httproute.yaml
apiVersion: gateway.networking.k8s.io/v1
kind: HTTPRoute
metadata:
  name: dev-httproute
  namespace: gateway-api-demo
spec:
  parentRefs:
    - name: dev-cluster-gateway
      kind: Gateway
      namespace: gateway-api-demo
  hostnames:
    - "k8sbible.local"
  rules:
    - matches:
        - path:
```

```
        type: PathPrefix
        value: /video
    backendRefs:
      - name: video-service
        port: 80
```

The following diagram explains the components involved in the Gateway API workflow:

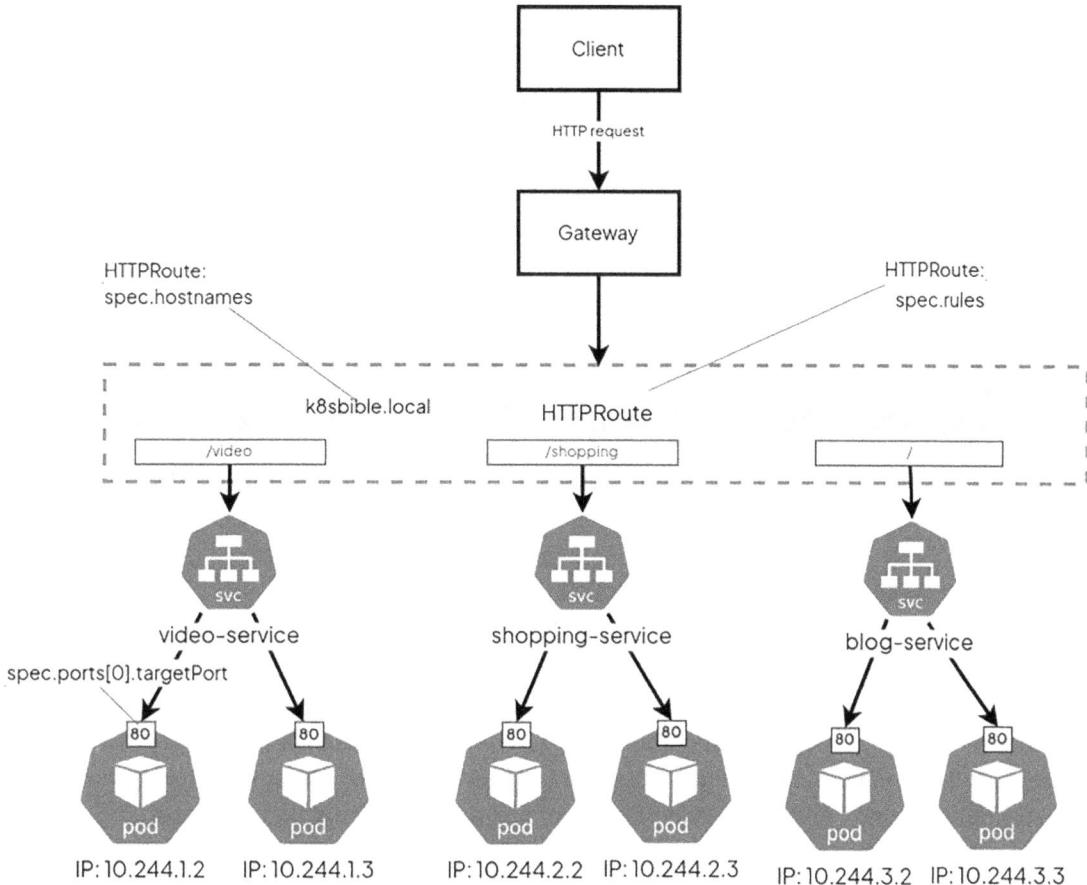

Figure 21.5: Gateway API components.

If you want to explore further, refer to the documentation and implement Gateway API in your cluster. The Gateway API is designed to replace the Ingress API, but it does not directly support the Ingress resource type. Therefore, you'll need to convert your existing Ingress resources to Gateway API resources as a one-time migration. For guidance on how to perform this migration, consult the Ingress migration guide (https://gateway-api.sigs.k8s.io/guides/migrating-from-ingress).

Before we conclude the advanced routing, Ingress, and Gateway API topics, let us get a quick introduction to EndPointSlices in the next section.

Understanding Endpoints and EndpointSlices

Traditionally, Kubernetes has managed the deployment of applications by means of Pods, where the Service objects serve as reliable networking intermediaries. The Services would act as some sort of doorway into the Pods and maintain a record of a corresponding Endpoints object that listed the active, healthy Pods matching the selector criteria of a Service. This does not scale when size grows.

Suppose a Service represents several Pods. The corresponding Endpoints object carries the IP and port for each of the Pods, which gets disseminated across the cluster and used in networking configurations. In the case of any update to this object, it would always affect the nodes across the whole cluster, even in cases of minor changes, resulting in heavy network traffic and intensive node processing.

Addressing this challenge, **EndpointSlices** slice up the monolithic Endpoints object into smaller pieces. Each EndpointSlice, by default, accommodates 100 endpoints that represent network details for pods.

Updates would be done much more surgically with EndpointSlices. Instead of re-downloading the whole Endpoints object, only a slice containing this exact Pod would be updated. This reduces network traffic and Node workload and, most importantly, enables higher scalability and performance, which has proven to be an exciting prospect in the evolution of Kubernetes.

Refer to the EndPointSlices documentation to learn more (`https://kubernetes.io/docs/concepts/services-networking/endpoint-slices/`).

In the next section, we'll explore the world of advanced technologies such as Serverless Computing, Machine Learning, Virtualization, and how they integrate with Kubernetes.

Modern Advancements with Kubernetes

Kubernetes plays at the forefront of integrating and supporting a set of advanced technologies that reshape the IT landscape. Consequently, Kubernetes provides a flexible and scalable platform that can easily integrate modern and cutting-edge solutions for serverless computing such as **Knative**; function-as-a-service like **OpenFaas**; virtual machine management like **KubeVirt**; or machine learning workflows like **Kubeflow**. Such solutions extend the functionality of Kubernetes and, in turn, help organizations innovate and move toward the adoption of new paradigms with a greater degree of efficiency and speed.

In this chapter, we will delve into the details of two of the most powerful frameworks, Knative and OpenFaaS, along with their primary use cases.

Serverless with Knative and OpenFaaS

Serverless computing is changing the paradigm for building and deploying applications, and this frees up developers to just write code while the infrastructure management goes to automated platforms. With **Knative** and **OpenFaaS** in a Kubernetes environment, serious serverless capabilities will be able to deploy, scale, and manage functions-as-a-service.

Knative is a Kubernetes-based platform that abstracts much of the underlying complexity associated with managing containerized applications. It provides automation for tasks such as autoscaling, traffic management, and event-driven function execution. Furthermore, this also makes Knative very effective in applications requiring the processing of variable workloads or event-driven tasks in an efficient manner. You can use Knative to scale microservices up during peak times, handle background jobs like user upload processing, or build super-responsive event-driven systems.

Another flexible framework is **OpenFaaS**, which offers a whole lot of ease while deploying functions on Kubernetes. OpenFaaS allows deploying lightweight, serverless functions in containers to ensure easy scaling and easy management. That will be very useful in a microservice architecture, where you scale each function separately based on demand. OpenFaaS is ideal for use cases involving real-time data processing, functions triggered by events, or building APIs to resize images or transform data without the overhead of the entire application stack. With Knative on top of OpenFaaS, an organization can better utilize Kubernetes in a mission to reduce complexity and scale with ever more efficient applications.

Kubeflow — Machine Learning on Kubernetes

Kubeflow is an open-source platform that enables easy and smooth deployment, scaling, and management of machine learning workflows on Kubernetes. It ties together all types of tools and frameworks into one system and enables data scientists and developers to focus on the creation and experimentation with ML models without being worried about managing infrastructure.

Kubeflow (`https://www.kubeflow.org`) can automate the whole machine learning cycle, starting from data preparation through model training to deployment and monitoring. It works with most popular ML frameworks such as **TensorFlow**, **PyTorch**, and **XGBoost**, so these tools should seamlessly integrate into your current workflow. Running on top of Kubernetes, Kubeflow gets its scalability and resiliency from the Kubernetes layer, meaning your ML workloads will be able to scale up if needed and automatically recover from failures.

In particular, Kubeflow is an effective solution for managing large ML projects where model training needs to be done on distributed datasets, when deploying a model into production, or when repeatedly retraining models with new data. In turn, this would mean the realization of a truly powerful and flexible platform that accelerates the development and deployment of machine learning applications atop Kubernetes.

In the next section, we will learn what KubeVirt is.

KubeVirt — Virtual Machines on Kubernetes

KubeVirt (`https://kubevirt.io`) is an open-source project that extends Kubernetes with the management of VMs besides containerized workloads. The integration lets an organization run VMs inside a Kubernetes cluster, letting traditional applications using VMs be deployed side by side on one managed platform with modern, containerized applications.

KubeVirt allows the smooth coexistence of VMs with containers. It enables one to take advantage of the powerful orchestration and scaling of Kubernetes for all workloads. This will be very helpful in organizations that are moving to cloud-native environments but still need support for legacy applications running on VMs. In such cases, KubeVirt can manage, scale, and orchestrate them just like containerized applications within the same Kubernetes environment.

For those using Red Hat OpenShift, this is the productized version of KubeVirt called OpenShift Virtualization. This will bring all these capabilities, giving powers to run and manage VMs directly from within OpenShift next to their containerized workloads. It will reduce operations and complexity, unlock flexible and efficient use of resources, and make it easier to modernize the IT infrastructure while continuing to support existing applications based on VMs.

We discussed new cluster builds, and most of the time, we talked about Kubernetes for development environments, such as minikube clusters. In reality, there are running Kubernetes clusters that house production critical applications and it is of the utmost importance to ensure all kinds of cluster maintenance tasks are taken care of as part of day-2 operations.

In the following sections, we will explore some of the Kubernetes maintenance tasks.

Maintaining Kubernetes Clusters — Day 2 Tasks

In the following sections, we will highlight the standard Kubernetes maintenance tasks such as backup, upgrade, multi-cluster management, and so on.

Kubernetes Cluster Backup and Restore

Kubernetes backup and restore is a significant concern for ensuring data integrity and business continuity in any production environment. Of all the crucial elements that must be part of the backup scope in a Kubernetes cluster, the most essential one is the `etcd`, or the key-value store where all the critical configurations and states of the cluster are stored. `etcd` backup for on-premise or self-managed clusters involves taking snapshots and securely storing them.

Taking Backup of etcd

Backing up an `etcd` cluster is essential to the integrity of all Kubernetes objects because the `etcd` stores your entire Kubernetes cluster state. Regular backups let you restore your cluster if you lose all control plane nodes. The backup process creates a snapshot file with all the Kubernetes state and other critical data. Since this data contains potentially sensitive information, it is a good idea to encrypt the snapshot files.

> Mechanism through the use of `etcdctl` backup is purely at the `etcd` project level. In other Kubernetes distributions, there will be adequate tools or a mechanism available to perform `etcd` backup. For example, this `cluster-backup.sh` script is part of the `etcd` `Cluster Operator` in `OpenShift` and wraps the execution of `etcdctl snapshot save`, simplifying operating and executing snapshots against `etcd` clusters.

The etcdctl tool allows you to create a snapshot of your etcd cluster directly from a live etcd member. This process doesn't impact the performance of your etcd instance.

The etcdctl and etcdutl tools can be installed from the etcd release page (https://github.com/etcd-io/etcd/releases/):

```
$ ETCDCTL_API=3 etcdctl \
  --endpoints=[https://127.0.0.1:2379] \
  --cacert=/etc/kubernetes/pki/etcd/ca.crt \
  --cert=/etc/kubernetes/pki/etcd/server.crt \
  --key=/etc/kubernetes/pki/etcd/server.key \
  snapshot save /tmp/snapshot-pre-patch.db
```

These files (trusted-ca-file, cert-file, and key-file) can typically be found in the etcd Pod's description (e.g., /etc/kubernetes/manifests/etcd.yaml).

After creating a snapshot, verify its integrity using the etcdutl tool:

```
$ etcdutl --write-out=table snapshot status snapshot.db
```

This command displays details such as the hash, revision, total keys, and snapshot size.

> If your etcd data is stored on a volume that supports snapshots (e.g., Amazon Elastic Block Store), you can back up the etcd data by taking a snapshot of the storage volume. This method is often used in cloud environments where storage snapshots can be automated.

In cloud-based clusters, managed services like **Google Kubernetes Engine** (GKE), Amazon EKS, or Azure AKS simplify the backup process. These platforms often provide integrated tools for automated backups and easy restoration. For example, you can use AWS Backup for EKS or Azure Backup for AKS to regularly back your cluster's state and configuration up without the need for manual intervention.

etcd Snapshot Restore with etcdutl

Restoring an etcd cluster from a snapshot is a critical and complex task, particularly in a multi-node setup where consistency across all nodes must be ensured. The process requires careful handling to avoid issues, especially if there are running API servers. Before initiating the restore, it's important to stop all API server instances to prevent inconsistencies. Once the restore is complete, you should restart the API servers, along with key Kubernetes components like kube-scheduler, kube-controller-manager, and kubelet, to ensure they don't rely on outdated data.

To perform the restore, use the etcdutl tool and specify the directory for the restored data:

```
$ etcdutl --data-dir <data-dir-location> snapshot restore snapshot.db
```

The specified <data-dir-location> will be created during the restoration process.

Reconfiguring the Kubernetes API Server

If the etcd cluster's access URLs change after restoration, you need to reconfigure and restart the Kubernetes API servers with the updated etcd server URLs (replace $NEW_ETCD_CLUSTER with the IP address):

```
. . .
--etcd-servers=$NEW_ETCD_CLUSTER
. . .
```

If a load balancer is used in front of the etcd cluster, update its configuration accordingly.

Leveraging Infrastructure as Code (IaC) and Configuration as Code (CaC) for Resilient Cluster Management

Backing up and restoring etcd is complex, considering the number of ways in which it could be performed to maintain data consistency and system stability. The most important thing is that you implement the IaC and CaC practices for your Kubernetes clusters and applications in order to avoid such challenges. That way, it would be quite easy to rebuild everything from scratch, having everything version-controlled, repeatable, and consistent in nature.

With the adoption of IaC and CaC practices, it is important to note that the four-eyes principle should be in place within Git workflows. That generally means all changes must undergo at least a review by two members of your team before merging. This practice will enhance code quality, ensure compliance, and minimize the chances of errors during backups and restorations.

To set this up robustly, treat your cluster as stateless and immutable. Keep YAML files for all configurations, such as namespaces, operators, **role-based access control** (**RBAC**) settings, NetworkPolicies, and so on. This should be versioned, committed to a repository, and automatically applied to your new cluster. This makes sure that the new cluster is the same as the old one, thus reducing downtime as far as possible and limiting human errors.

Extend this to your applications also; from ConfigMaps and Services to PVCs, everything related to the deployment of your application should be codified. In stateful applications, data is stored in PVs that are external to the cluster. Since you are separating data from configuration, restoring your applications to their previous state is as quick as reapplying their YAML files and reconnecting to your data.

Besides this, templating with Helm and continuous deployment with GitOps are optionally available to make this process even smoother. This automation ensures consistency across all your configurations, as changes will automatically be applied to environments for reduced manual intervention. A comprehensive cluster and application management approach indeed goes a long way toward simplifying disaster recovery, while also enhancing scalability, security, and operational efficiency.

In the following section, we will explore some of the cluster upgrade tasks and considerations.

Kubernetes Cluster Upgrades

Upgrading the Kubernetes cluster is one of the important tasks that keeps your environment secure, stable, and up-to-date with new features. Most of the managed Kubernetes distributions upgrade easily in cloud-based clusters since the underlying complexity is handled by the managed services. Examples of these include Amazon EKS, GKE, and Azure AKS. They have one-click upgrades that make it easy to upgrade to newer versions of Kubernetes with minimum or no downtime.

This will vary for on-premise or bespoke clusters; for example, kubeadm-built clusters have a documented (https://kubernetes.io/docs/tasks/administer-cluster/kubeadm/kubeadm-upgrade) upgrade path provided by Kubernetes that will walk you through steps to upgrade your control plane and nodes.

Whether you are working with cloud-based clusters or managing on-premises setups, following a structured upgrade process will be key. Here is a detailed overview of how to upgrade your Kubernetes cluster.

Pre-Upgrade Checklist

Before initiating the upgrade, it's essential to prepare your cluster. Here are some crucial steps:

- **Verify Compatibility**: Ensure that the new Kubernetes version is compatible with all your existing components and add-ons. Refer to the official Kubernetes documentation for compatibility matrices.
- **Back Up etcd**: etcd is the heart of your Kubernetes cluster. Always create a backup before proceeding with the upgrade to safeguard your cluster configuration.
- **Disable Swap:** Kubernetes requires swap to be disabled on all nodes. Ensure this setting is configured correctly to prevent potential issues.

Upgrade Process

The upgrade process typically involves several steps:

- **Drain Nodes:** Safely evict all pods from the nodes you plan to upgrade using kubectl drain <node-to-drain> --ignore-daemonsets. This ensures no new work is assigned to the node during the upgrade process.
- **Upgrade Control Plane:** Start by updating the control plane components, such as the API server, etcd, and controller-manager. Use your package manager's update and upgrade commands (e.g., apt-get or yum) to install the latest versions.
- **Upgrade kubeadm:** Update kubeadm to the desired version. This ensures compatibility with the new Kubernetes version.
- **Upgrade kubelet and kubectl:** After updating the control plane, upgrade kubelet and kubectl on each node. These components interact with the control plane and manage pods.
- **Uncordon Nodes:** Once a node is upgraded, re-enable it for scheduling pods using kubectl uncordon <node-name>.
- **Upgrade compute Nodes:** Perform a rolling upgrade of your worker nodes, following the same steps as for the control plane.

- **Upgrade CNI Plugin:** Ensure your **Container Network Interface (CNI)** plugin is compatible with the new Kubernetes version. Update it if necessary.

Post-Upgrade Tasks

The post-Upgrade tasks typically involve the following:

- **Verify Cluster Status:** Use `kubectl get nodes` to confirm that all nodes are in a Ready state.
- **Monitor etcd:** Keep an eye on etcd's health and performance during and after the upgrade.
- **Switch Package Repositories:** If you haven't already, update your package repositories to point to the new Kubernetes version's sources.

Rollback Plan

One important thing is that a rollback plan should be developed for those unexpected errors that can happen during the upgrade processes. It should include steps that are necessary for performing fallbacks to previous configurations and backup restorations. While inner changes to etcd's API and data structure make rollbacks hard, being prepared reduces the time and operational disruptions. Identifying what needs to be done and by whom within your team allows for a timely and coordinated response, even when the occurrence that requires such a plan to be implemented is infrequent.

Additional Tips

Some of the additional tips are as follows:

- **Test an Upgrade in a Staging Environment:** Before upgrading your production cluster, it is a good idea first to test the upgrade process on a staging or development environment.
- **Consider Using a Cluster Upgrade Tool:** Some tools automatically carry out some of the processes involved in upgrading; hence, there is less work for you to do manually with fewer chances of errors happening.
- **Monitor for Issues:** While the upgrade is in process and afterward, monitor your cluster for signs that something is not quite right.

It can be further supported by the inclusion of upgrade automation using Ansible, Terraform, AWS CloudFormation, and ARM templates, which will drive the upgrade process in place of node provision, deploy packages, and rolling updates.

In such a practical use case, one automates upgrading clusters in a multi-cloud environment. You can manage multi-cluster deployments here using tools such as ArgoCD or Fleet to make sure all clusters across different environments are upgraded consistently. The foregoing will be quite useful for an organization managing more than one cluster; hence, it reduces manual effort and maintains uniformity across the environments.

We will explore some of the well-known multi-cluster management tools in the next section.

Multi-Cluster Management

The exponential growth in organizations also raises the complexity of managing a number of Kubernetes clusters in diverse environments. It is here that multi-cluster management solutions help in offering a single control point that can deploy, monitor, and upgrade clusters. Many of these have features like automated cluster provisioning and rolling updates, which enable consistency and security across all managed clusters.

An example of this is that, in a multi-cloud environment, one may provision and manage a Kubernetes cluster using Terraform and ArgoCD on AWS, Azure, and Google Cloud. In such an environment, deployments and upgrades can be automated with minimal possibility for human error, while all clusters may have the same version of Kubernetes. It's especially useful in the case of a big organization with lots of teams or regions, where you really want the Kubernetes environment to be consistent and up-to-date for operational efficiency.

The following list contains some of the well-known Kubernetes multi-cluster management tools and services:

- **Rancher:** Rancher is an open-source platform designed to simplify Kubernetes management. It enables centralized management of clusters across different environments, whether on-premises or in the cloud. Rancher offers features such as multi-cluster application deployment, integrated monitoring, and RBAC for managing user permissions across clusters.

- **Lens:** Lens is a Kubernetes **integrated development environment** (IDE) that facilitates the management of multiple clusters from a single interface. It provides real-time insights, a built-in terminal, and resource management views, making it easier for developers and operators to visualize and control their Kubernetes environments.

- **Kops:** **Kubernetes Operations** (**Kops**) is a tool designed for managing the lifecycle of Kubernetes clusters, particularly on AWS. It automates the processes of creating, upgrading, and deleting clusters, and is well-regarded for its ability to streamline operations across various cloud platforms.

- **Red Hat Advanced Cluster Management for Kubernetes:** This tool provides a comprehensive solution for managing Kubernetes clusters across hybrid and multi-cloud environments. It includes features for policy-driven governance, application life cycle management, and cluster observability, ensuring that clusters are compliant and performing optimally.

- **Anthos (Google Cloud):** This is a multi-cloud and hybrid cloud management platform from Google Cloud that facilitates the management of Kubernetes clusters across different environments, whether they are on-premises or hosted on various cloud providers. Anthos provides centralized governance, security, and consistent application deployment across diverse infrastructure setups, ensuring a unified operational experience across all managed clusters.

- **Azure Arc:** This service extends Azure's management and governance capabilities to Kubernetes clusters running anywhere—on-premises, in other clouds, or at the edge. With Azure Arc, you can manage and secure Kubernetes clusters across multiple environments through a single interface, allowing for consistent policy enforcement, security management, and monitoring across your entire infrastructure.

In the following section, we will learn about the Kubernetes cluster hardening best practices.

Securing a Kubernetes Cluster — Best Practices

Securing a Kubernetes cluster is essential to prevent unauthorized access, data breaches, and disruptions. By implementing robust security measures, you can protect sensitive data and ensure smooth operation. This section outlines guidelines and best practices to help you secure your cluster against both accidental and malicious threats.

Certain concepts of security that will be discussed in this chapter have already been touched upon in *Chapter 18*, *Security in Kubernetes*. Here, we revisit those points to emphasize those as part of the Kubernetes best practices.

Controlling Access to the Kubernetes API

Since Kubernetes relies heavily on its API, controlling and limiting access is the first step in securing your cluster:

- **Use TLS for API Traffic:** Kubernetes encrypts API communication by default with TLS. Most installation methods handle the necessary certificates automatically. However, administrators should be aware of any unsecured local ports and secure them accordingly.

- **API Authentication:** Choose an authentication method that fits your needs. For smaller, single-user clusters, a simple certificate or static Bearer token might suffice. Larger clusters might require integration with existing authentication systems like OIDC or LDAP.

- **API Authorization:** After authentication, every API request must pass an authorization check. Kubernetes uses RBAC to match users or groups to a set of permissions defined in roles. These permissions are tied to specific actions on resources and can be scoped to namespaces or the entire cluster. For better security, use Node and RBAC authorizers together.

Controlling Access to the Kubelet

Kubelets, which manage nodes and containers, expose HTTPS endpoints that can grant significant control over the node. In production environments, ensure that Kubelet authentication and authorization are enabled.

To control access to the Kubelet in production, allow both the authentication and authorization of the Kubelet API to work effectively in limiting and ascribing permissions. By default, only requests performed through the Kubernetes API server are allowed; this blocks unauthorized direct access to the Kubelet. You can enhance this further by implementing RBAC policy settings for users and services, which define RBAC permissions with Kubelet, along with limiting the network exposure of the Kubelet endpoints by utilizing network policies or firewall rules.

Controlling Workload or User Capabilities at Runtime

Authorization in Kubernetes is high-level, but you can apply more granular policies to limit resource usage and control container privileges:

- **Limiting Resource Usage:** Use resource quotas and limit ranges to control the number of resources like CPU, memory, or disk space that a namespace can use. This prevents users from requesting unreasonably high or low resource values.

- **Controlling Container Privileges:** Pods can request access to run as specific users or with certain privileges. Most applications don't need root access, so it's recommended to configure your containers to run as non-root users.

- **Preventing Unwanted Kernel Modules:** To prevent attackers from exploiting vulnerabilities, block or uninstall unnecessary kernel modules from the node. You can also use a Linux Security Module like **SELinux** to prevent modules from loading for containers.

Restricting Network Access

Kubernetes allows you to control network access at various levels:

- **Network Policies:** Use network policies to restrict which pods in other namespaces can access resources in your namespace. You can also use quotas and limit ranges to control node port requests or load-balanced services.

- **Restricting Cloud Metadata API Access:** Cloud platforms often expose metadata services that can contain sensitive information. Use network policies to restrict access to these APIs and avoid using cloud metadata for secrets.

Protecting Cluster Components

To keep your cluster secure, it's important to protect critical components like etcd and ensure proper access control:

- **Restrict Access to etcd:** Gaining access to etcd can lead to full control of your cluster. Use strong credentials and consider isolating etcd servers behind a firewall. For example, for Kubernetes clusters in AWS, create a security group with restricted inbound rules that permit only Kubernetes control-plane IPs to reach the etcd on port 2379 in a private deployment. You can also configure etcd with --client-cert-auth and --trusted-ca-file flags, so only the control plane can connect over secured connections.

- **Enable Audit Logging:** Audit logging records API actions for later analysis. Enabling and securing these logs can help detect and respond to potential compromises. The Kubernetes cluster management team needs to define a custom audit policy in Kubernetes for the create, delete, and update events, and they can instruct logs securely stored in a secure logging tool like Elasticsearch. The following code snippet shows an example for the logging configuration in a kube-apiserver Pod manifest:

```
...
--audit-log-path=/var/log/audit.log
--audit-policy-file=/etc/kubernetes/audit-policy.yaml
...
```

- **Rotate Infrastructure Credentials Frequently:** Short-lived credentials reduce the risk of unauthorized access. Regularly rotate certificates, tokens, and other sensitive credentials to maintain security. For example, you can configure `cert-manager` (https://cert-manager.io/) to automate the renewal of TLS certificates and configure kubelet to periodically refresh its own certificate using the `RotateKubeletClientCertificate` and `RotateKubeletServerCertificate` flags.

- **Review Third-Party Integrations:** When adding third-party tools or integrations, review their permissions carefully. Restrict their access to specific namespaces where possible to minimize risk. For example, when installing tools such as Prometheus or Grafana, it is enough to allow read access by creating a read-only Role and binding the Role to the required namespaces with RoleBindings, thus limiting the amount of data exposure.

- **Encrypt Secrets at Rest:** Kubernetes supports encryption at rest for secrets stored in `etcd`. This ensures that even if someone gains access to the `etcd` data, they can't easily view the sensitive information. Configure `EncryptionConfig` in the Kubernetes apiserver configuration to use AES encryption for secrets stored in `etcd`, so that in the event of an `etcd` breach, the data is encrypted at an additional layer.

The following table summarizes some of the best practices for Kubernetes security hardening:

Section	Best Practices
Secure Cluster Setup	Enable RBAC and use dedicated service accounts. Keep Kubernetes components updated. Secure API server access with TLS.
Control Cluster Access	Use strong authentication methods. Enforce strict access controls and least privilege principles. Regularly audit and review access permissions.
Protect Network Communication	Encrypt internal communications. Implement network segmentation. Use secure network plugins and enforce network policies.
Secure Container Images	Use trusted container registries. Scan images for vulnerabilities. Enforce Pod Security Policies to restrict container privileges.
Monitor and Log Cluster Activity	Implement logging and monitoring solutions. Enable auditing. Regularly review logs for suspicious activities.

Regularly Update and Patch	Apply updates and patches promptly to address vulnerabilities.
	Follow a strict update management process.
Continuously Educate and Train	Educate your team on security best practices.
	Stay updated on the latest security developments.
	Promote a culture of security within your organization.

Table 21.1: Kubernetes cluster Security Best Practices

For more detailed guidance on Kubernetes security hardening, refer to official documentation and community resources. Additionally, consider reviewing comprehensive security hardening guidelines such as the Kubernetes Hardening Guidance, provided by the **Defense Information Systems Agency (DISA)** (https://media.defense.gov/2022/Aug/29/2003066362/-1/-1/0/CTR_KUBERNETES_HARDENING_GUIDANCE_1.2_20220829.PDF).

In the following section, we will learn some of the common Kubernetes troubleshooting methods.

Troubleshooting Kubernetes

Troubleshooting Kubernetes involves diagnosing and resolving issues that affect the functionality and stability of your cluster and applications. Common errors may include problems with Pod scheduling, container crashes, image pull issues, networking issues, or resource constraints. Identifying and addressing these errors efficiently is crucial for maintaining a healthy Kubernetes environment.

In the upcoming sections, we'll cover the essential skills you need to get started with Kubernetes troubleshooting.

Getting details about resources

When troubleshooting issues in Kubernetes, the `kubectl get` and `kubectl describe` commands are indispensable tools for diagnosing and understanding the state of resources within your cluster. You have already used these commands multiple times in the previous chapters; let us revisit the commands here again.

The `kubectl get` command provides a high-level overview of various resources in your cluster, such as pods, services, deployments, and nodes. For instance, if you suspect that a pod is not running as expected, you can use `kubectl get pods` to list all pods and their current statuses. This command will show you whether pods are running, pending, or encountering errors, helping you quickly identify potential issues.

On the other hand, `kubectl describe` dives deeper into the details of a specific resource. This command provides a comprehensive description of a resource, including its configuration, events, and recent changes. For example, if a Pod from the previous command is failing, you can use `kubectl describe pod todo-app` to get detailed information about why it might be failing.

This output includes the Pod's events, such as failed container startup attempts or issues with pulling images. It also displays detailed configuration data, such as resource limits and environment variables, which can help pinpoint misconfigurations or other issues.

To illustrate, suppose you're troubleshooting a deployment issue. Using `kubectl get deployments` can show you the deployment's status and number of replicas. If a deployment is stuck or not updating correctly, `kubectl describe deployment webapp` will provide detailed information about the deployment's rollout history, conditions, and errors encountered during updates.

In the next section, we will learn the important methods to find logs and events in Kubernetes to make our troubleshooting easy.

Kubernetes Logs and Events for troubleshooting

Kubernetes offers powerful tools like **Events** and **Audit Logs** to monitor and secure your cluster effectively. Events, which are cluster-wide resources of the **Event** kind, provide a real-time overview of key actions, such as pod scheduling, container restarts, and errors. These events help in diagnosing issues quickly and understanding the state of your cluster. You can view events using the `kubectl get events` command:

```
$ kubectl get events
```

This command outputs a timeline of events, helping you identify and troubleshoot problems. To focus on specific events, you can filter them by resource type, namespace, or time period. For example, to view events related to a specific pod, you can use the following:

```
$ kubectl get events --field-selector involvedObject.name=todo-pod
```

Audit Logs, represented by the Policy kind, are vital for ensuring compliance and security within your Kubernetes environment. These logs capture detailed records of API requests made to the Kubernetes API server, including the user, action performed, and outcome. This information is crucial for auditing activities like login attempts or privilege escalations. To enable audit logging, you need to configure the API server with an audit policy. Refer to the Auditing documentation (`https://kubernetes.io/docs/tasks/debug/debug-cluster/audit/`) to learn more.

When debugging Kubernetes applications, the `kubectl logs` command is an essential tool for retrieving and analyzing logs from specific containers within a pod. This helps in diagnosing and troubleshooting issues effectively.

To fetch logs from a pod, the basic command is as follows:

```
$ kubectl logs todo-app
```

This retrieves logs from the first container in the pod. If the pod contains multiple containers, specify the container name:

```
$ kubectl logs todo-app -c app-container
```

For real-time log streaming, akin to tail -f in Linux, use the -f flag:

```
$ kubectl logs -f todo-app
```

This is useful for monitoring live processes. If a pod has restarted, you can access logs from its previous instance using the following:

```
$ kubectl logs todo-app --previous
```

To filter logs based on labels, combine kubectl with tools like jq:

```
$ kubectl get pods -l todo -o json | jq -r '.items[] | .metadata.name' | xargs
-I {} kubectl logs {}
```

To effectively manage logs in Kubernetes, it's crucial to implement log rotation to prevent excessive disk usage, ensuring that old logs are archived or deleted as new ones are generated. Utilizing structured logging, such as JSON format, makes it easier to parse and analyze logs using tools like jq. Additionally, setting up a centralized logging system, like the **Elasticsearch, Fluentd, Kibana (EFK)** stack, allows you to aggregate and efficiently search logs across your entire Kubernetes cluster, providing a comprehensive view of your application's behavior.

Together, Kubernetes Events and Audit Logs provide comprehensive monitoring and security capabilities. Events offer insights into the state and behavior of your applications, while Audit Logs ensure that all actions within the cluster are tracked, helping you maintain a secure and compliant environment.

kubectl explain — the inline helper

The kubectl explain command is a powerful tool in Kubernetes that helps you understand the structure and fields of Kubernetes resources. Providing detailed information about a specific resource type allows you to explore the API schema directly from the command line. This is especially useful when writing or debugging YAML manifests, as it ensures that you're using the correct fields and structure.

For example, to learn about the Pod resource, you can use the following command:

```
$ kubectl explain pod
```

This command will display a high-level overview of the Pod resource, including a brief description. To dive deeper into specific fields, such as the spec field, you can extend the command like this:

```
$ kubectl explain pod.spec
```

This will provide a detailed explanation of the spec field, including its nested fields and the expected data types, helping you better understand how to configure your Kubernetes resources properly.

Interactive troubleshooting using kubectl exec

Using kubectl exec is a powerful way to troubleshoot and interact with your running containers in Kubernetes. This command allows you to execute commands directly inside a container, making it invaluable for debugging, inspecting the container's environment, and performing quick fixes. Whether you need to check logs, inspect configuration files, or even diagnose network issues, kubectl exec provides a direct way to interact with your applications in real time.

To use kubectl exec, you can start with a simple command execution inside the container (you may use kubectl apply -f trouble/blog-portal.yaml for testing):

```
$ kubectl get po -n trouble-ns
NAME                      READY   STATUS    RESTARTS   AGE
blog-675df44d5-gkrt2      1/1     Running   0          29m
```

For example, to list the environment variables of a container, you can use the following:

```
$ kubectl exec blog-675df44d5-gkrt2 -- env
```

If the pod has multiple containers, you can specify which one to interact with using the -c flag:

```
$ kubectl exec blog-675df44d5-gkrt2 -c blog -- env
```

One of the most common uses of kubectl exec is to open an interactive shell session within a container. This allows you to run diagnostic commands on the fly, such as inspecting log files or modifying configuration files. You can start an interactive shell (/bin/sh, /bin/bash, etc.), as demonstrated here:

```
$ kubectl exec -it blog-675df44d5-gkrt2 -n trouble-ns -- /bin/bash
root@blog-675df44d5-gkrt2:/app# whoami;hostname;uptime
root
blog-675df44d5-gkrt2
14:36:03 up 10:19,  0 user,  load average: 0.17, 0.07, 0.69
root@blog-675df44d5-gkrt2:/app#
```

Here, the following applies:

- -i: This is an interactive session.
- -t: This allocates pseudo-TTY.

This interactive session is particularly useful when you need to explore the container's environment or troubleshoot issues that require running multiple commands in sequence.

In addition to command execution, kubectl exec supports copying files to and from containers using kubectl cp. This can be particularly handy when you need to bring in a script or retrieve a log file for further analysis. For instance, here's how to copy a file from your local machine into a container:

```
$ kubectl cp troubles/test.txt blog-675df44d5-gkrt2:/app/test.txt -n trouble-ns
$ kubectl exec -it blog-675df44d5-gkrt2 -n trouble-ns -- ls -l /app
total 8
-rw-r--r-- 1 root root 902 Aug 20 16:52 app.py
-rw-r--r-- 1 1000 1000  20 Aug 31 14:42 test.txt
```

And to copy a file from a container to your local machine, you'd need the following:

```
$ kubectl cp blog-675df44d5-gkrt2:/app/app.py /tmp/app.py  -n trouble-ns
```

This capability simplifies the process of transferring files between your local environment and the containers running in your Kubernetes cluster, making troubleshooting and debugging more efficient.

In the next section, we will learn about ephemeral containers, which are very useful in Kubernetes troubleshooting tasks.

Ephemeral Containers in Kubernetes

Ephemeral containers are a special type of container in Kubernetes designed for temporary, on-the-fly tasks like debugging. Unlike regular containers, which are intended for long-term use within Pods, ephemeral containers are used for inspection and troubleshooting and are not automatically restarted or guaranteed to have specific resources.

These containers can be added to an existing Pod to help diagnose issues, making them especially useful when traditional methods like kubectl exec fall short. For example, if a Pod is running a distroless image with no debugging tools, an ephemeral container can be introduced to provide a shell and other utilities (e.g., nslookup, curl, mysql client, etc.) for inspection. Ephemeral containers are managed via a specific API handler and can't be added through kubectl edit or modified once set.

For example, in *Chapter 8*, *Exposing Your Pods with Services*, we used k8sutils (quay.io/iamgini/ k8sutils:debian12) as a separate Pod to test the services and other tasks. With ephemeral containers, we can use the same container image but insert the container inside the application Pod to troubleshoot.

Assume we have the Pod and Service called video-service running in the ingress-demo namespace (Refer to the ingress/video-portal.yaml file for deployment details). It is possible to start debugging utilizing the k8sutils container image as follows:

```
$ kubectl debug -it pod/video-7d945d8c9f-wkxc5 --image=quay.io/iamgini/
k8sutils:debian12 -c k8sutils -n ingress-demo

root@video-7d945d8c9f-wkxc5:/# nslookup video-service
Server:         10.96.0.10
Address:        10.96.0.10#53


Name:   video-service.ingress-demo.svc.cluster.local
Address: 10.109.3.177


root@video-7d945d8c9f-wkxc5:/# curl http://video-service:8080


    <!DOCTYPE html>
    <html>
    <head>
      <title>Welcome</title>
      <style>
        body {
```

```
        background-color: yellow;
        text-align: center;
...<removed for brevity>...
```

In summary, ephemeral containers offer a flexible way to investigate running Pods without altering the existing setup or relying on the base container's limitations.

In the following section, we will demonstrate some of the common Kubernetes troubleshooting tasks and methods.

Common troubleshooting tasks in Kubernetes

Troubleshooting Kubernetes can be complex and highly specific to your cluster setup and operations, as the list of potential issues can be extensive. Instead, let's focus on some of the most common Kubernetes problems and their troubleshooting methods to provide a practical starting point:

- **Pods are in Pending state:** The error message `Pending` indicates that the pod is waiting to be scheduled onto a node. This can be caused by insufficient resources or misconfigurations. To troubleshoot, use `kubectl describe pod <pod_name>` to check for events that describe why the pod is pending, such as resource constraints or node conditions. If the cluster doesn't have enough resources, the pod will remain in the pending state. You can adjust resource requests or add more nodes. (Try using `troubles/app-with-high-resource.yaml` to test this.)

- **CrashLoopBackOff or container errors:** The `CrashLoopBackOff` error occurs when a container repeatedly fails to start, possibly due to misconfigurations, missing files, or application errors. To troubleshoot, view the logs using `kubectl logs <pod_name>` or `kubectl describe pod <pod_name>` to identify the cause. Look for error messages or stack traces that can help diagnose the problem. If a container has an incorrect startup command, it will fail to start, leading to this error. Reviewing the container's exit code and logs will help fix any issues. (Apply `troubles/failing-pod.yaml` and test this scenario.)

- **Networking issues:** These types of errors suggest that network policies are blocking traffic to or from the pod. To troubleshoot, you can check the network policies affecting the pod using `kubectl describe pod <pod_name>`, and verify service endpoints with `kubectl get svc`. If network policies are too restrictive, necessary traffic might be blocked. For example, an empty ingress policy could prevent all traffic to a pod, and adjusting policies will allow the required services to communicate. (Use `troubles/networkpolicy.yaml` to test this scenario.)

- **Node not ready or unreachable:** The `NotReady` error indicates that a node is not in a ready state due to conditions like network issues. To troubleshoot, check the node status with `kubectl get nodes` and `kubectl describe node <node_name>`. This error may also be caused by node taints that prevent scheduling. If a node has the taint `NoSchedule`, it won't accept pods until the issue is resolved or the taint is removed.

- **Storage issues:** The PersistentVolumeClaim `Pending` error occurs when a **persistent volume claim (PVC)** is waiting for a matching **persistent volume (PV)** to be bound. To troubleshoot, check the status of PVs and PVCs with `kubectl get pv` and `kubectl get pvc`. For CSI, ensure the `storageClass` is configured properly and requested in the PVC definition accordingly. (Check `troubles/pvc.yaml` to explore this scenario.)

- **Service unavailability:** The `Service Unavailable` error means that a service is not accessible, potentially due to misconfigurations or networking issues. To troubleshoot, check the service details using `kubectl describe svc <service_name>`. Verify that the service is correctly configured and points to the appropriate pods by using appropriate labels. If the service is misconfigured, it may not route traffic to the intended endpoints, leading to unavailability. You can verify the Service endpoints (Pods) using the `kubectl describe svc <service_name>` command.

- **API server or control plane issues:** These errors typically point to connectivity problems with the API server, often due to issues within the control plane or network. Since `kubectl` commands won't work if the API server is down, you need to log in directly to the control plane server where the API server pods are running. Once logged in, you can check the status of the control plane components using commands like `crictl ps` (if you are using containerd) or `docker ps` (if you are using Docker) to ensure the API server Pod is up and running. Additionally, review logs and check the network connections to verify that all control plane components are functioning correctly.

- **Authentication and authorization problems:** The `Unauthorized` error indicates issues with user permissions or authentication. To troubleshoot, verify user permissions with `kubectl auth can-i <verb> <resource>`. For example, if a user lacks the required role or role binding, they will encounter authorization errors. Adjust roles and role bindings as needed to grant the necessary permissions.

- **Resource exhaustion:** The `ResourceQuota Exceeded` error occurs when a resource quota is exceeded, preventing the allocation of additional resources. To troubleshoot and monitor resource usage, use `kubectl get quota`, `kubectl top nodes`, and `kubectl top pods`. If a quota is too low, it may block new resource allocations. Adjusting resource quotas or reducing resource usage can alleviate this issue.

- **Ingress or load balancer issues:** The `IngressController Failed` error suggests that the ingress controller is not functioning correctly, impacting traffic routing. To troubleshoot, check the Ingress details using `kubectl describe ingress <ingress_name>`. Ensure that the ingress controller is properly installed and configured and that ingress rules correctly map to services. Misconfigurations in ingress rules can prevent proper traffic routing. Also, ensure the hostname DNS resolution is in place if you are using the optional host field in the Ingress configuration.

This was the last practical demonstration in this book, so let's now summarize what you have learned.

Summary

In this last chapter, we have explained advanced traffic routing approaches in Kubernetes using Ingress objects and Ingress Controllers. At the beginning, we did a brief recap of Kubernetes Service types. We refreshed our knowledge regarding `ClusterIP`, `NodePort`, and `LoadBalancer` Service objects. Based on that, we introduced Ingress objects and Ingress Controllers and explained how they fit into the landscape of traffic routing in Kubernetes. Now, you know that simple Services are commonly used when L4 load balancing is required, but if you have HTTP or HTTPS endpoints in your applications, it is better to use L7 load balancing offered by Ingress and Ingress Controllers. You learned how to deploy the nginx web server as an Ingress Controller and we tested this on example Deployments.

Lastly, we explained how you can approach Ingress and Ingress Controllers in cloud environments where you have native support for L7 load balancing outside of the Kubernetes cluster. As a demonstration, we deployed an AKS cluster with an **Application Gateway Ingress Controller** (AGIC) to handle Ingress objects.

We also saw how Kubernetes advances itself toward a platform where these cutting-edge technologies integrate well, such as Knative and KubeVirt, that extend Kubernetes' capabilities into areas including serverless cbvomputing, VM management, and machine learning. We saw the indispensable "day-2" operations that any Cluster Administrator performs, including Backup and Upgrades, foundational security best practices to fortify clusters, and some of the crucial troubleshooting techniques one could utilize to fix common issues that may come up within the cluster. These principles are the basic ones, based on which engineers are allowed to operate and secure the Kubernetes environments safely and effectively to keep the operations running non-stop for innovative solutions.

Congratulations! This has been a long journey into the exciting territory of Kubernetes and container orchestration. Good luck with your further Kubernetes journey and thanks for reading.

Further reading

- Ingress: `https://kubernetes.io/docs/concepts/services-networking/ingress/`
- Ingress Controllers: `https://kubernetes.io/docs/concepts/services-networking/ingress-controllers/`
- Ingress Installation Guide: `https://kubernetes.github.io/ingress-nginx/deploy`
- Set up Ingress on Minikube with the NGINX Ingress Controller: `https://kubernetes.io/docs/tasks/access-application-cluster/ingress-minikube/`
- Ephemeral Containers: `https://kubernetes.io/docs/concepts/workloads/pods/ephemeral-containers/`
- What is Application Gateway Ingress Controller: `https://learn.microsoft.com/en-us/azure/application-gateway/ingress-controller-overview`
- Operating etcd clusters for Kubernetes: `https://kubernetes.io/docs/tasks/administer-cluster/configure-upgrade-etcd/`
- Securing a Cluster: `https://kubernetes.io/docs/tasks/administer-cluster/securing-a-cluster/`
- Auditing: `https://kubernetes.io/docs/tasks/debug/debug-cluster/audit/`

For more information regarding autoscaling in Kubernetes, please refer to the following Packt books:

- *The Complete Kubernetes Guide*, by *Jonathan Baier, Gigi Sayfan, and Jesse White* (`https://www.packtpub.com/en-in/product/the-complete-kubernetes-guide-9781838647346`)
- *Getting Started with Kubernetes – Third Edition*, by *Jonathan Baier and Jesse White* (`https://www.packtpub.com/en-in/product/getting-started-with-kubernetes-9781788997263`)
- *Kubernetes for Developers*, by *Joseph Heck* (`https://www.packtpub.com/virtualization-and-cloud/kubernetes-developers`)

- *Hands-On Kubernetes on Windows*, by *Piotr Tylenda* (`https://www.packtpub.com/product/hands-on-kubernetes-on-windows/9781838821562`)

You can also refer to the following official documentation:

- Kubernetes documentation (`https://kubernetes.io/docs/home/`) is always the most up-to-date source of knowledge regarding Kubernetes in general.
- A list of many available Ingress Controllers can be found at `https://kubernetes.io/docs/concepts/services-networking/ingress-controllers/`.
- Similar to AKS, GKE offers a built-in, managed Ingress Controller called **GKE Ingress**. You can learn more in the official documentation at `https://cloud.google.com/kubernetes-engine/docs/concepts/ingress`. You can also check the Ingress features that are implemented in GKE at `https://cloud.google.com/kubernetes-engine/docs/how-to/ingress-features`.
- For Amazon EKS, there is **AWS Load Balancer Controller**. You can find more information in the official documentation at `https://docs.aws.amazon.com/eks/latest/userguide/alb-ingress.html`.

Join our community on Discord

Join our community's Discord space for discussions with the authors and other readers:

`https://packt.link/cloudanddevops`

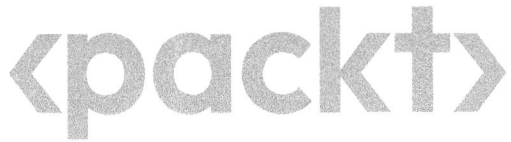

‹packt›

packt.com

Subscribe to our online digital library for full access to over 7,000 books and videos, as well as industry leading tools to help you plan your personal development and advance your career. For more information, please visit our website.

Why subscribe?

- Spend less time learning and more time coding with practical eBooks and Videos from over 4,000 industry professionals
- Improve your learning with Skill Plans built especially for you
- Get a free eBook or video every month
- Fully searchable for easy access to vital information
- Copy and paste, print, and bookmark content

At www.packt.com, you can also read a collection of free technical articles, sign up for a range of free newsletters, and receive exclusive discounts and offers on Packt books and eBooks.

Other Books You May Enjoy

If you enjoyed this book, you may be interested in these other books by Packt:

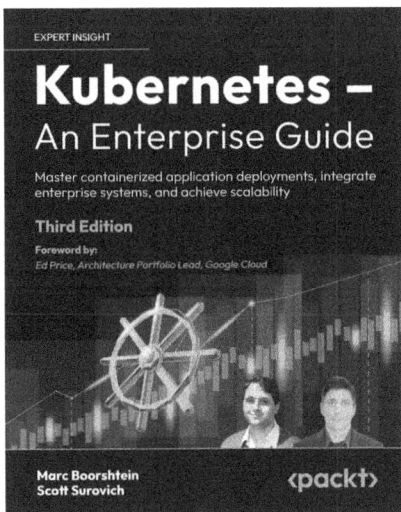

Kubernetes – An Enterprise Guide

Marc Boorshtein, Scott Surovich

ISBN: 9781835086957

- Manage Secrets with Vault and External Secret Operator
- Create multitenant clusters with vCluster for isolated environments
- Monitor clusters with Prometheus and visualize metrics using Grafana
- Aggregate and analyze logs centrally with OpenSearch for insights
- Build a developer platform integrating GitLab and ArgoCD for CI/CD
- Deploy applications in Istio service mesh and secure them with OPA and GateKeeper
- Secure your container runtime and halt hackers in their tracks with KubeArmor

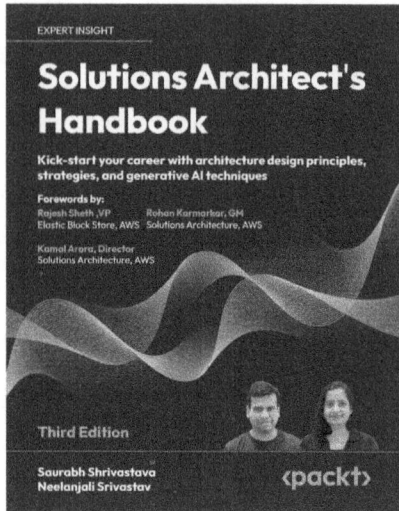

Solutions Architect's Handbook

Saurabh Shrivastava, Neelanjali Srivastav

ISBN: 9781835084236

- Explore various roles of a solutions architect in the enterprise
- Apply design principles for high-performance, cost-effective solutions
- Choose the best strategies to secure your architectures and boost availability
- Develop a DevOps and CloudOps mindset for collaboration, operational efficiency, and streamlined production
- Apply machine learning, data engineering, LLMs, and generative AI for improved security and performance
- Modernize legacy systems into cloud-native architectures with proven real-world strategies
- Master key solutions architect soft skills

Packt is searching for authors like you

If you're interested in becoming an author for Packt, please visit authors.packtpub.com and apply today. We have worked with thousands of developers and tech professionals, just like you, to help them share their insight with the global tech community. You can make a general application, apply for a specific hot topic that we are recruiting an author for, or submit your own idea.

Leave a Review!

Thank you for purchasing this book from Packt Publishing—we hope you enjoyed it! Your feedback is invaluable and helps us improve and grow. Please take a moment to leave an Amazon review; it will only take a minute, but it makes a big difference for readers like you.

https://packt.link/r/1835464718

Scan the QR code below to receive a free ebook of your choice.

https://packt.link/NzOWQ

Index

Download a free PDF copy of this book

Thanks for purchasing this book!

Do you like to read on the go but are unable to carry your print books everywhere?

Is your eBook purchase not compatible with the device of your choice?

Don't worry, now with every Packt book you get a DRM-free PDF version of that book at no cost.

Read anywhere, any place, on any device. Search, copy, and paste code from your favorite technical books directly into your application.

The perks don't stop there, you can get exclusive access to discounts, newsletters, and great free content in your inbox daily.

Follow these simple steps to get the benefits:

1. Scan the QR code or visit the link below:

https://packt.link/free-ebook/9781835464717

2. Submit your proof of purchase.
3. That's it! We'll send your free PDF and other benefits to your email directly.